Sequential Change Detection and Hypothesis Testing

MONOGRAPHS ON STATISTICS AND APPLIED PROBABILITY

Editors: F. Bunea, R. Henderson, N. Keiding, L. Levina, R. Smith, W. Wong

Recently Published Titles

Asymptotic Analysis of Mixed Effects Models
Theory, Applications, and Open Problems
Jiming Jiang 155

Missing and Modified Data in Nonparametric Estimation With R Examples
Sam Efromovich 156

Probabilistic Foundations of Statistical Network Analysis
Harry Crane 157

Multistate Models for the Analysis of Life History Data
Richard J. Cook and Jerald F. Lawless 158

Nonparametric Models for Longitudinal Data
with Implementation in R
Colin O. Wu and Xin Tian 159

Multivariate Kernel Smoothing and Its Applications
José E. Chacón and Tarn Duong 160

Sufficient Dimension Reduction
Methods and Applications with R
Bing Li 161

Large Covariance and Autocovariance Matrices
Arup Bose and Monika Bhattacharjee 162

The Statistical Analysis of Multivariate Failure Time Data: A Marginal Modeling Approach
Ross L. Prentice and Shanshan Zhao 163

Dynamic Treatment Regimes
Statistical Methods for Precision Medicine
Anastasios A. Tsiatis, Marie Davidian, Shannon T. Holloway, and Eric B. Laber 164

Sequential Change Detection and Hypothesis Testing
General Non-i.i.d. Stochastic Models and Asymptotically Optimal Rules
Alexander G. Tartakovsky 165

For more information about this series please visit: https://www.crcpress.com/Chapman--HallCRC-Monographs-on-Statistics--Applied-Probability/book-series/CHMONSTAAPP

Sequential Change Detection and Hypothesis Testing

General Non-i.i.d. Stochastic Models and Asymptotically Optimal Rules

Alexander G. Tartakovsky
Moscow, Russia and Los Angeles, USA

CRC Press is an imprint of the
Taylor & Francis Group, an **informa** business
A CHAPMAN & HALL BOOK

CRC Press
Taylor & Francis Group
6000 Broken Sound Parkway NW, Suite 300
Boca Raton, FL 33487-2742

First issued in paperback 2021

© 2020 by Taylor & Francis Group, LLC
CRC Press is an imprint of Taylor & Francis Group, an Informa business

No claim to original U.S. Government works

ISBN 13: 978-1-03-208435-0 (pbk)
ISBN 13: 978-1-4987-5758-4 (hbk)

This book contains information obtained from authentic and highly regarded sources. Reasonable efforts have been made to publish reliable data and information, but the author and publisher cannot assume responsibility for the validity of all materials or the consequences of their use. The authors and publishers have attempted to trace the copyright holders of all material reproduced in this publication and apologize to copyright holders if permission to publish in this form has not been obtained. If any copyright material has not been acknowledged please write and let us know so we may rectify in any future reprint.

Except as permitted under U.S. Copyright Law, no part of this book may be reprinted, reproduced, transmitted, or utilized in any form by any electronic, mechanical, or other means, now known or hereafter invented, including photocopying, microfilming, and recording, or in any information storage or retrieval system, without written permission from the publishers.

For permission to photocopy or use material electronically from this work, please access www.copyright.com (http://www.copyright.com/) or contact the Copyright Clearance Center, Inc. (CCC), 222 Rosewood Drive, Danvers, MA 01923, 978-750-8400. CCC is a not-for-profit organization that provides licenses and registration for a variety of users. For organizations that have been granted a photocopy license by the CCC, a separate system of payment has been arranged.

Trademark Notice: Product or corporate names may be trademarks or registered trademarks, and are used only for identification and explanation without intent to infringe.

Publisher's Note
The publisher has gone to great lengths to ensure the quality of this reprint but points out that some imperfections in the original copies may be apparent.

Visit the Taylor & Francis Web site at
http://www.taylorandfrancis.com

and the CRC Press Web site at
http://www.crcpress.com

IN MEMORY OF MY FATHER
GEORGIY P. TARTAKOVSKY
AS WELL AS
TO MY WIFE MARINA AND MY SON DANIEL

Contents

Preface

This monograph presents the theory of Sequential Analysis, a method of statistical inference proposed and developed by Abraham Wald in the 1940s. The book deals primarily with asymptotic behavior. Most of the book is devoted to sequential changepoint detection, or the problem of efficiently detecting abrupt changes in the distribution of a stochastic process when decisions are made sequentially (i.e., as observations arrive). Sequential changepoint detection has become especially important in recent years in a variety of challenging applications, including biomedical signal processing, quality control engineering, finance, link failure detection in communication networks, intrusion detection in computer networks, object detection in surveillance systems, computational molecular biology and bioinformatics, human activity recognition, rapid detection and tracking of malicious activity of terrorist groups, and many others. Changepoint detection problems originally arose from control charts, which were invented for monitoring quality of products in the late 1920s and early 1930s. The 1950s saw a considerable shift toward CUSUM (cumulative sum) tests, and in the 1960s Shiryaev and Shiryaev–Roberts detection procedures gained prominence. These procedures proved to be very efficient and even optimal for detecting abrupt changes in distributions in the case of independent and identically distributed (i.i.d.) observations with known pre-change and post-change distributions. Throughout the book, I will refer to this as the i.i.d. case. However, systematic investigation of more general non-i.i.d. models, where observations are dependent and non-identically distributed before and/or after the change, emerged much later (at the end of the 1990s and in the early 2000s).

This book is a continuation of Tartakovsky, Nikiforov, and Basseville's *Sequential Analysis: Hypothesis Testing and Changepoint Detection* [164], published in 2014. It is based on results from my papers and lectures over the last 5 years and contains both an expanded discussion of those results and new sections on quickest change detection for general stochastic models that include the non-i.i.d. case. In contrast to the previous book, which considered both discrete and continuous time models, this monograph addresses change detection in discrete time only. The purpose of the monograph is to describe a class of sequential hypothesis tests and changepoint detection rules for multiple data streams and to develop mathematical techniques for determining their properties. I focus on asymptotic optimality properties under very general conditions.

The previous book [164] is extended in two directions: (1) I consider a composite postchange hypothesis when the parameters of the post-change distribution are unknown, and (2) I describe a multi-stream scenario where it is unknown which streams are affected. While both scenarios have been addressed for i.i.d. data models, to my knowledge this is the first consideration of the non-i.i.d. case, which is important for many practical applications. Chapter 5 includes and builds on some recent results from George Moustakides and Moshe Pollak regarding the optimality of the Shewhart detection rule in the i.i.d. case.

The book is intended for researchers and advanced graduate students in applied probability and theoretical statistics. The general asymptotic theory of sequential change detection should be accessible to readers with a knowledge of probability and statistics, including a certain familiarity with complete and r-quick convergence, Markov random walks, renewal and nonlinear renewal theories, Markov renewal theory, and uniform ergodicity of Markov processes. The appendices provide basic prerequisites and results.

I wish to express my indebtedness to Professor Yury Sosulin who introduced me to the field of sequential analysis when I was a third-year undergraduate student, and I am especially indebted to my father, Professor George Tartakovsky, who further inspired and supported my graduate research when I was a Ph.D. student at the Moscow Institute of Physics and Technology in the late 1970s and early 1980s.

I am also grateful to the CRC Press editor Rob Calver for encouraging me to write this book and to Lara Spieker for expertly handling of the manuscript.

I would like to thank various U.S. and Russian agencies for supporting my work under multiple contracts.[1]

A special thanks to my wife, Marina, and my son, Daniel, for their help and inspiration.

Finally, I am thankful to two anonymous referees for thoughtfully reviewing the book and providing invaluable suggestions.

Alexander G. Tartakovsky
Los Angeles, California USA
and
Moscow, Russian Federation[2]
August, 2019

---

[1]In particular, my work was partially supported by the Russian Science Foundation under grant 18-19-00452, by the Russian Ministry of Education and Science 5-100 excellence project and the Arctic program at the Moscow Institute of Physics and Technology.

[2]I started to write the book when I was with the Department of Statistics at the University of Connecticut and completed after I became the head of the Space Informatics Laboratory at the Moscow Institute of Physics and Technology.

Notation and Symbols

Notation	Meaning
$X_t \xrightarrow[t\to\infty]{\mathsf{P}-\text{a.s.}} Y$	Almost sure convergence under P (or with probability 1).
ADD	Average delay to detection (detection delay).
$\mathbb{C}(\alpha,\beta)$	Class of sequential tests with probabilities of errors α and β.
$\mathbb{C}_\pi(\alpha)$	Class of Bayesian sequential change detection rules with prior distribution π and probability of false alarm α.
$\mathbb{C}_\gamma$	Class of sequential change detection rules with the average run length to false alarm γ.
$\mathbb{C}(\beta,m)$	Class of sequential change detection rules with the maximal probability of false alarm in the interval of length m equals β.
$[a,b]$	Closed interval.
$X_t \xrightarrow[t\to\infty]{\text{completely}} Y$	Complete convergence.
$(\Omega,\mathscr{F},\mathsf{P})$	Complete probability space.
CADD	Conditional average detection delay.
$\mathsf{E}[X \mid \mathscr{B}]$	Conditional expectation of the random variable X given sigma-algebra $\mathscr{B}$.
$X_t \xrightarrow[t\to\infty]{\text{law}} Y$	Convergence in distribution (or in law or weak).
$X_t \xrightarrow[t\to\infty]{\mathsf{P}} Y$	Convergence in probability.
$F(x)=\mathsf{P}(X\le x)$	Cumulative distribution function (cdf) of a random variable X.
$\det A$; $\lvert A\rvert$	Determinant of the matrix A.
δ	Decision rule, procedure, function.
$\lVert\mathbf{X}\rVert_2=\sqrt{\sum_{i=1}^n x_i^2}$	Euclidean norm.
E	Expectation.
ESS	Expected sample size (or average sample number).
$\{\mathscr{F}_t\}$	Filtration (a flow of sub-sigma-algebras $\mathscr{F}_t$).
$\dot{g}(x)$	First derivative of the function $x\mapsto g(x)$.
∇	Gradient (vector of first partial derivatives).

∇^2	Hessian (matrix of second partial derivatives).
$\mathbb{I}_n$	Identity matrix of size $n \times n$.
H_i	i^{th} hypothesis.
$\mathbb{1}_{\{A\}}$	Indicator of a set A.
A^{-1}	Inverse of the matrix A.
$\mathscr{K}$	Kullback–Leibler (KL) information (or distance or divergence).
$\mathbb{R}^\ell$	ℓ-dimensional Euclidean space.
$\Lambda = \frac{d\mathsf{P}}{d\mathsf{Q}}(\omega)$	Likelihood ratio (Radon–Nikodým derivative of measure P with respect to measure Q).
$\varkappa$	Limiting average overshoot.
$L(\theta, d)$	Loss function.
$X_t \xrightarrow[t\to\infty]{L^p} Y$	L^p-convergence (or in the pth mean).
$A = [a_{ij}]$	Matrix A of size $m \times n$ ($1 \le i \le m,\ 1 \le j \le n$).
$\mathbb{R}_+ = [0, \infty)$	Nonnegative real line.
$X_n, n \ge 1$	Observations in discrete time.
(a, b)	Open interval.
$\mathscr{P} = \{\mathsf{P}_\theta\}_{\theta\in\Theta}$	Parametric family of probability distributions.
$f_\theta(x), p_\theta(x)$	Parametrized probability density.
P	Probability measure.
$f(x), p(x)$	Probability density function (pdf).
θ	Parameter or vector of parameters.
ν	Point of change (or changepoint).
$\chi^2_m(p)$	p-quantile of the standard chi-squared distribution with m degrees of freedom.
$\mathbb{R} = (-\infty, \infty)$	Real line.
$X_t \xrightarrow[t\to\infty]{r-\text{quickly}} Y$	r-quick convergence.
$\ddot{g}(x)$	Second derivative of the function $x \mapsto g(x)$.
$\delta = (T, d)$	Sequential test (more generally rule).
$\mathbb{Z}_+ = \{0, 1, 2, \dots\}$	Set of nonnegative integers.
Ω	Set of elementary events ω.
$\{n : \dots\}$	Set of n such that ....

$\mathscr{F}$	Sigma algebra (field).
$\varphi(x)$	Standard normal density function.
$\Phi(x)$	Standard normal distribution function.
$\mathscr{N}(0,1)$	Standard normal random variable.
$(\Omega, \mathscr{F}, \mathscr{F}_t, \mathsf{P})$	Stochastic basis.
T	Stopping time.
SADD	Supremum average detection delay.
d	Terminal decision.
$\mathbf{X}_0^t = \{X_u, 0 \le u \le t\}$	Trajectory of a random process observed on the interval $[0,t]$.
$A^\top$	Transpose of the matrix A.
$\mathbf{X}_1^n = (X_1, X_2, \dots, X_n)$	Vector of observed n random variables.

Introduction

Statistical *Sequential Analysis* is a method of statistical inference whose main feature is that the number of observations (or more generally the number of stages) is not fixed in advance. Instead, the number of observations is random depending on the outcome of the experiment. In other words, at each stage, a statistician decides whether to continue observing or to stop and make a terminal decision. That decision could be to accept a hypothesis (when testing statistical hypotheses) or to form an estimate (when estimating a parameter). Thus, a sequential method is characterized by two components: (1) a stopping rule T, which at the nth stage depends on the previous n observations $(X_1, X_2, \dots, X_n)$ and decides whether to stop the observation process or to obtain an additional observation X_{n+1} for $n \ge 1$, that is, event $\{T = n\}$ depends on the sample $(X_1, \dots, X_n)$; and (2) a terminal decision rule $d(X_1, \dots, X_T)$ that specifies the action to be taken after the observation process has stopped. The pair $\delta = (T, d)$ is called the sequential decision rule (or procedure). The objective of the sequential methodology is to determine an optimal decision rule δ_{opt} that satisfies some optimization criterion.

An advantage of sequential methods over fixed sample size rules is that sequential methods often require, on average, substantially fewer observations. This is the case in hypothesis testing problems. However, there is a large class of surveillance problems, which can be reduced and effectively solved by the so-called *Quickest Changepoint Detection* methods, that are different. In these problems, a change in distribution occurs at an unknown point in time and has to be detected as quickly as possible for a given false alarm rate or risk. A change detection rule is a stopping time depending on the observations and the decision on no-change is equivalent to the decision on continuing observation. Usually, the observation process is not terminated after deciding that the change is in effect but rather renewed all over again, leading to a multicyclic detection procedure. This is practically always the case in surveillance applications and often in other applications.

In very general terms, there are two main practical motivations for sequential analysis. In some applications, sequential analysis is intrinsic: there can be no fixed sample size procedure. This is the case in industrial process control [19, 92, 174, 181, 182, 187], in the classic secretary problem [41], and in clinical trials when monitoring certain critical health parameters of a patient [183]. Most surveillance problems are also sequential in nature. Yet another intrinsically sequential problem is parameter estimation with the fixed given accuracy in the presence of nuisance parameters. In this case, one cannot construct a confidence region of a given size for a fixed sample size. However, sequential multistage estimation allows us to build a fixed precision estimate. (See, e.g., Stein [146, 147], Siegmund [140], Mukhopadhyay et al. [98]).

As mentioned above, in some statistical inference problems, sequential analysis is the most efficient solution in terms of the experiment's average sample size or cost. This is the case for the repeated significance test that maintains the flexibility of deciding sooner than the fixed sample size procedure at the price of some lower power [2, 190]. The sequential probability ratio test and the Kiefer–Weiss procedure also belong to the category of the most economic solutions, since they minimize the expected sample size (resp. the maximum expected sample size).

Perhaps the very first publication on sequential testing was the article by Dodge and Roming [29], who suggested a two-stage sampling scheme. This idea was generalized by Bartky [8] to multiple stages. But the real breakthrough was made by Abraham Wald in 1943 (classified report), who devised the sequential probability ratio test (SPRT) for testing two hypotheses. A detailed description of this test and its comparison with other sequential and non-sequential (fixed sample size) tests was given by Wald in his famous book *Sequential Analysis* [178] in 1947. A sequential test includes a stopping time and a terminal decision to achieve a tradeoff between the average sample size and the quality of the decision. Most efforts were devoted to testing two hypotheses, namely, to developing optimal strategies and obtaining lower bounds for the expected sample size necessary to decide between the two hypotheses with given error probabilities; see, e.g., Wald [177, 178], Wald and Wolfowitz [179, 180], Hoeffding [60, 61]. Also, these bounds were compared with the sample size of the best non-sequential, fixed sample size test. It has been established that the sequential test performs significantly better than the classical Neyman–Pearson test.

The problem of sequential testing of multiple hypotheses is substantially more difficult than that of testing two hypotheses. For multiple-decision problems, it is usually very difficult, if even possible, to obtain optimal solutions. The first results were established by Sobel and Wald [144], Armitage [1], and Paulson [105]. The lower bounds for the expected sample size were obtained by Simons [143]. A substantial part of the development of sequential multihypothesis testing has been directed towards the study of suboptimal procedures, basically multihypothesis modifications of a sequential probability ratio test, for independent and identically distributed (i.i.d.) observations. See, e.g., Armitage [1], Chernoff [25], Dragalin [31], Dragalin and Novikov [33], Kiefer and Sacks [69], Lorden [82, 85, 86], Pavlov [106, 107]. However, in many applications the i.i.d. assumption is too restrictive. Further advances in the asymptotic theory for the case of dependent and nonidentically distributed (non-i.i.d.) observations were made by Lai [73], Tartakovsky [149, 152, 154, 153], Golubev and Khas'minskii [51], Verdenskaya and Tartakovsky [176], Dragalin et al. [34], Tartakovsky et al. [164].

A separate branch of sequential analysis is quickest changepoint detection, which has many applications and is often identified with online surveillance. The goal is to detect a sudden change in distribution as soon as possible after it occurs, while controlling for false alarms. The sequential setting assumes the observations are made successively, and, as long as their behavior suggests that the process is in a normal state, the process is allowed to continue. However, if the state changes, the aim is to detect the change quickly to enable a timely and appropriate response. Historically, the subject of changepoint detection first emerged in the late 1920s, motivated by considerations of industrial quality control, due to the work of Walter Shewhart. Shewhart became the father of modern statistical quality control. Shewhart's work (in particular, Shewhart control charts) is highlighted in his books [127, 128], for which he gained recognition in the statistical community. However, efficient (optimal and quasioptimal) sequential detection procedures were developed much later in the 1950s and 1960s after the emergence of Wald's book [178]. It is also necessary to mention Page's (1954) seminal paper [104], where he proposed the now famous Cumulative Sum (CUSUM) detection procedure, followed by the series of works by Shiryaev (1961–1969) [130, 129, 131, 132, 133, 134, 135] and Lorden (1971) [84] where the first optimality results in Bayesian and non-Bayesian contexts were established. Further advances in the theory of change detection were made by Pollak [109] who established almost minimax optimality of the Shiryaev–Roberts rule and Moustakides [93] who established exact minimax optimality of the CUSUM rule in the class of rules with the given mean time to a false alarm in the i.i.d. case. In the late 90s and early 2000s, a generalization for non-i.i.d. models was presented in works by Lai [75], Fuh [44], Tartakovsky and Veeravalli [171], Baron and Tartakovsky [7], and more recently in Tartakovsky et al. [164], Tartakovsky [160], Fuh and Tartakovsky [47], Pergamenchtchikov and Tartakovsky [108].

The main subject of this book is sequential changepoint detection for very general non-i.i.d. stochastic models (Chapters 2–7). Chapter 1 considers sequential hypothesis testing in multiple data streams and non-i.i.d. models. While [Tartakovsky et al. [164]] covers general stochastic models, it does not consider the case of composite hypotheses in the non-i.i.d. scenario. The principal purpose of this book is to extend the asymptotic theory of changepoint detection to the case of a composite post-change hypothesis as well as for multi-stream data when the number of affected streams is unknown. This case is more relevant for practical applications, where the parameters of distributions are often unknown.

The book consists of seven core chapters (Chapters 1–7), one applied chapter (Chapter 8), and the Appendix.

Chapter 1 is concerned with the problem of sequential hypothesis testing in multiple data streams when observations are acquired one at a time in a number of data streams and the number and location of "patterns" of interest are either completely or partially unknown. We consider a general setup, in which the various streams may be coupled and correlated, and the observations in streams are non-i.i.d. We prove that the Generalized Sequential Likelihood Ratio Test and the Mixture Sequential Likelihood Ratio Test minimize asymptotically as the probabilities of errors go to zero moments of the sample size distribution under very general conditions.

In Chapter 2, we describe general changepoint models, changepoint mechanisms and several Bayesian and minimax criteria. Chapter 3 discusses the main conceptual ideas of the asymptotic Bayes change detection theory for a single data stream and establishes asymptotic optimality of mixture change detection rules for non-i.i.d. models; and Chapter 6 extends these results to multiple data streams when a number of affected streams is unknown. Chapter 7 addresses the semi-Bayesian joint change detection and identification (or isolation) problem when it is necessary not only to detect a change as quickly as possible but also to provide a detailed diagnosis of the occurred change – to determine which type of change is in effect. This problem is reduced to the multidecision change detection problem with multiple composite post-change hypotheses. A multihypothesis change detection–identification rule is proposed and shown to be asymptotically optimal for general non-i.i.d. models when the false alarm and misidentification rates are low. Asymptotic optimality in non-Bayesian settings, specifically pointwise and minimax, is considered in Chapter 4, where it is shown that the mixture-type Shiryaev–Roberts rule minimizes moments of the detection delay for all possible change points as well as in the worst-case scenario when the probability of a false alarm in a fixed time window becomes small. Chapter 5 is an exception, in the sense that, here, we consider the i.i.d. case. We establish that the Shewhart detection rule, which consists in comparing the running likelihood ratio to a threshold, is optimal with respect to Bayesian as well as maximin criteria, maximizing the instantaneous probability of detection. We also discuss possible extensions to the more practical case where one is interested in maximizing the detection probability in a window of any given size.

Chapter 8 discusses certain challenging applications to illustrate the general theory and emphasize some specific points. We show that the changepoint detection theory allows for the development of efficient algorithms for target track management in sonar systems, for detection of low-observable space objects, and for early detection of unauthorized break-ins in computer networks.

Appendices contain useful auxiliary results as well as necessary prerequisites and basic results related to Markov processes, convergence, nonlinear renewal theory, Markov random walks, Markov nonlinear renewal theory, etc. that are used in the book.

1

Sequential Hypothesis Testing in Multiple Data Streams

1.1 Introduction

Sequential hypothesis testing in multiple data streams (e.g., sensors, populations, multichannel systems) has important practical applications. To take just a few examples:

- *Public health*: Quickly detecting crucial to an epidemic present in only a fraction of hospitals and data sources [23, 145, 172].
- *Genomics*: Determining intervals of copy number variations, which are short and sparse, in multiple DNA sequences [141].
- *Environmental monitoring*: Rapidly discovering anomalies such as hazardous materials or intruders typically affecting only a small fraction of the many sensors covering a given area [42, 125].
- *Military defense*: Detecting an unknown number of objects in noisy observations obtained by radars, sonars, or optical sensors that are typically multichannel in range, velocity, and space [4, 162].
- *Cybersecurity*: Rapidly detecting and localizing malicious activity, such as distributed denial-of-service (DDoS) attacks, typically in multiple data streams [66, 65, 148, 159, 168, 169].

Motivated by these and many other applications, this chapter considers the sequential hypothesis testing problem where observations are acquired one at a time, in a number of data streams, and where the number and location of "patterns" of interest are either completely or partially unknown *a priori*. For simplicity, we will refer to patterns as *signals* with the understanding that there may be no physical signals in the noisy observations. Also, we will often call the corresponding hypothesis testing problem the *signal detection* problem. It is indeed convenient to use terminology from signal detection theory, referring to the null hypothesis as "absence of signals" in all streams and to the alternative hypothesis as "presence of signals" in some streams. It is worth noting that the hypothesis testing problem considered in this chapter is substantially different from the one considered in Chapter 6 in the context of quickest change detection in multiple streams since there is no change point and one has to stop observations not only when a decision on signal presence is made but also when the decision on signal absence is made.

Having said this, the goal is to quickly detect either the absence of all signals or the presence of an unknown subset of signals, while controlling the probabilities of false alarm (type-I error) and missed detection (type-II error). Two scenarios are of particular interest for applications. The first is when a single signal with an unknown location is distributed over a relatively small number of streams. For example, this may be the case when detecting an extended target with unknown location with a very high resolution sensor in a sequence of images. Following Siegmund [141], we call this the "structured" case, since there is a certain geometrical structure that can be approximately known.

A different, completely "unstructured" scenario is when an unknown number of signals affect the streams. For example, in many target detection applications, an unknown number of point targets present in different channels of a multichannel system, and it is unknown in which channels the signals are present [4, 162, 158].

We begin by considering a general multistream detection setup, in which the various streams may be coupled and correlated, and the observations in streams are not necessarily temporally i.i.d., but can be dependent and non-identically distributed. We focus on two multistream sequential tests, the Generalized Sequential Likelihood Ratio Test (GSLRT) and the Mixture Sequential Likelihood Ratio Test (MSLRT), which are based on the maximum and average likelihood ratio over all possible hypotheses regarding the number and location of signals, respectively. Our goal is to design these detection schemes in order to control the error probabilities below given target levels and also to establish their efficiency. In order to achieve these goals, we only require that the log-likelihood ratio statistics satisfy certain asymptotic stability properties. Specifically, we assume that the suitably normalized log-likelihood ratios between "signal present" and "no signal" hypotheses satisfy a Law of Large Numbers (LLN), converging under both regimes to positive and finite numbers, which we can think of as generalized Kullback–Leibler information numbers.

We establish asymptotic optimality properties of the two proposed sequential tests for any possible signal configuration. Specifically, if the LLR statistics have independent but not necessarily identically distributed increments, then the two tests minimize asymptotically all moments of the sample size distribution as the probabilities of errors vanish when the convergence in the LLN for the normalized log-likelihood ratio statistics is almost sure. In the general case when the log-likelihood ratio increments are not independent, we prove that the two proposed tests minimize asymptotically the first r moments of the sample size distribution when the convergence in the LLN is r-complete. While these results are established in a very general setup, in which the various streams may be coupled and correlated, the proposed procedures suffer from the curse of dimensionality when the number of possible signal configurations is very large. We show, however, that in the case of uncoupled and independent streams both proposed procedures are scalable with respect to the number of streams, even if there is complete uncertainty regarding location of the signals.

In the case of i.i.d. observations in streams, using nonlinear renewal theory, we obtain higher order asymptotic optimality properties of the sequential tests, higher order approximations for the expected sample size up to a vanishing term, and accurate asymptotic approximations for the error probabilities.

1.2 Sequential Multistream Hypothesis Testing Problem

Suppose that observations are sequentially acquired over time in N distinct streams (sources). The observations in the ith data stream correspond to a realization of a discrete-time stochastic process $X(i) = \{X_n(i)\}_{n \in \mathbb{N}}$, where $i \in \mathscr{N} := \{1,\dots,N\}$ and $\mathbb{N} = \{1,2,\dots\}$. Let P stand for the distribution of $\mathbf{X} = (X(1),\dots,X(N))$. Let H_0 be the null hypothesis according to which all N streams are not affected, i.e., there are no signals in all streams at all. For any given non-empty subset of components, $\mathscr{B} \subset \mathscr{N}$, let $\mathsf{H}_{\mathscr{B}}$ be the hypothesis according to which only the components $X(i)$ with i in $\mathscr{B}$ contain signals. Denote by P_0 and $\mathsf{P}_{\mathscr{B}}$ the distributions of $\mathbf{X}$ under hypotheses H_0 and $\mathsf{H}_{\mathscr{B}}$, respectively. Next, let $\mathscr{P}$ be a class of subsets of $\mathscr{N}$ that incorporates prior information that may be available regarding the subset of affected streams. Denote by $|\mathscr{B}|$ the size of a subset $\mathscr{B}$, i.e., the number of signals under $\mathsf{H}_{\mathscr{B}}$, and by $|\mathscr{P}|$ the size of class $\mathscr{P}$, i.e., the number of possible alternatives in $\mathscr{P}$. For example, if we know upper $\overline{K} \le N$ and lower $\underline{K} \ge 1$ bounds on the size of the affected subset, then

$$\mathscr{P} = \mathscr{P}_{\underline{K},\overline{K}} = \{\mathscr{B} \subset \mathscr{N} : \underline{K} \le |\mathscr{B}| \le \overline{K}\}. \tag{1.1}$$

When, in particular, we know that *exactly* K streams can be affected, we write $\mathscr{P} = \mathscr{P}_K$, whereas when we know that *at most* K streams can be affected, we write $\mathscr{P} = \overline{\mathscr{P}}_K$, where

$$\mathscr{P}_K \equiv \mathscr{P}_{K,K} = \{\mathscr{B} \subset \mathscr{N} : |\mathscr{B}| = K\}, \quad \overline{\mathscr{P}}_K \equiv \mathscr{P}_{1,K} = \{\mathscr{B} \subset \mathscr{N} : 1 \le |\mathscr{B}| \le K\}. \tag{1.2}$$

Note that $|\mathscr{P}|$ takes its maximum value when there is no prior information regarding the subset of affected streams, i.e., $\mathscr{P} = \overline{\mathscr{P}}_N$, in which case $|\mathscr{P}| = 2^N - 1$.

We are interested in testing H_0, the simple null hypothesis that there are no signals in all data streams, against the composite alternative, H_1, according to which the subset of streams with signals belongs to $\mathscr{P}$. In other words, the hypothesis testing problem is

$$\begin{aligned} \mathsf{H}_0 &: \quad \mathsf{P} = \mathsf{P}_0; \\ \mathsf{H}_1 = \bigcup_{\mathscr{B} \in \mathscr{P}} \mathsf{H}_{\mathscr{B}} &: \quad \mathsf{P} \in \{\mathsf{P}_{\mathscr{B}}\}_{\mathscr{B} \in \mathscr{P}}, \end{aligned} \tag{1.3}$$

and we want to distinguish between H_0 and H_1 as soon as possible as data from all streams are acquired, so we focus on sequential tests. To be more specific, let $\mathscr{F}_n = \sigma(\mathbf{X}_1, \dots, \mathbf{X}_n)$ denote the σ-algebra generated by the observations in all streams up to and including time n; $\mathbf{X}_n = (X_n(1), \dots, X_n(N))$. We say that a pair $\delta = (T, d)$ is a *sequential test* if T is an $\{\mathscr{F}_n\}$-stopping time and $d \in \{0, 1\}$ is a binary, $\mathscr{F}_T$-measurable random variable (terminal decision) such that $\{d = j\} = \{T < \infty, \mathsf{H}_j \text{ is selected}\}$, $j = 0, 1$. A sequential test should be designed in such a way that the type-I (false alarm) and type-II (missed detection) error probabilities are controlled, i.e., do not exceed given, user-specified levels. We denote by $\mathbb{C}_{\alpha,\beta}(\mathscr{P})$ the class of sequential tests with the probability of false alarm below $\alpha \in (0,1)$ and the probability of missed detection below $\beta \in (0,1)$, i.e.,

$$\mathbb{C}_{\alpha,\beta}(\mathscr{P}) = \left\{ \delta : \mathsf{P}_0(d = 1) \le \alpha \text{ and } \max_{\mathscr{B} \in \mathscr{P}} \mathsf{P}_{\mathscr{B}}(d = 0) \le \beta \right\}. \tag{1.4}$$

Given α and β we would like to utilize sequential tests that require (at least approximately) the smallest possible number of observations under any possible signal configuration, i.e., under H_0 and under $\mathsf{H}_{\mathscr{B}}$ for every $\mathscr{B} \in \mathscr{P}$. To be more precise, we would like to design a sequential test (T^*, d^*) in $\mathbb{C}_{\alpha,\beta}(\mathscr{P})$ that minimizes asymptotically as $\alpha, \beta \to 0$ to first order moments of the stopping time distribution under every possible scenario, that is,

$$\begin{aligned} \mathsf{E}_0[(T^*)^r] &\sim \inf_{\delta \in \mathbb{C}_{\alpha,\beta}(\mathscr{P})} \mathsf{E}_0[T^r], \\ \mathsf{E}_{\mathscr{B}}[(T^*)^r] &\sim \inf_{\delta \in \mathbb{C}_{\alpha,\beta}(\mathscr{P})} \mathsf{E}_{\mathscr{B}}[T^r] \quad \text{for all } \mathscr{B} \in \mathscr{P} \end{aligned} \tag{1.5}$$

for some $r \ge 1$, where E_0 and $\mathsf{E}_{\mathscr{B}}$ refer to expectation under P_0 and $\mathsf{P}_{\mathscr{B}}$, respectively.

Hereafter we use the notation $x_\alpha \sim y_\alpha$ as $\alpha \to 0$ when $\lim_{\alpha \to 0}(x_\alpha / y_\alpha) = 1$, $x_\alpha \le y_\alpha(1 + o(1))$ when $\limsup_{\alpha \to 0}(x_\alpha / y_\alpha) \le 1$, and $x_\alpha \ge y_\alpha(1 + o(1))$ when $\liminf_{\alpha \to 0}(x_\alpha / y_\alpha) \ge 1$.

1.3 Generalized Likelihood Ratio and Mixture Sequential Tests

Write $\mathsf{P}_0^{(n)} = \mathsf{P}_0|_{\mathscr{F}_n}$ and $\mathsf{P}_{\mathscr{B}}^{(n)} = \mathsf{P}_{\mathscr{B}}|_{\mathscr{F}_n}$ for restrictions of probability measures P_0 and $\mathsf{P}_{\mathscr{B}}$ to the σ-algebra $\mathscr{F}_n$ and let $p_0(\mathbf{X}^n)$ and $p_{\mathscr{B}}(\mathbf{X}^n)$ denote the corresponding probability densities of these measures with respect to some non-degenerate σ-finite measure μ_n, where $\mathbf{X}^n = (\mathbf{X}_1, \dots, \mathbf{X}_n)$ stands for the concatenation of the first n observations from all data streams. Then the hypothesis testing problem (1.3) can be equivalently re-written as

$$\begin{aligned} \mathsf{H}_0 : \quad & p(\mathbf{X}^n) = p_0(\mathbf{X}^n) = \prod_{t=1}^{n} p_0(\mathbf{X}_t|\mathbf{X}^{t-1}); \\ \mathsf{H}_1 = \bigcup_{\mathscr{B} \in \mathscr{P}} \mathsf{H}_{\mathscr{B}} : \quad & p_{\mathscr{B}}(\mathbf{X}^n) = \prod_{t=1}^{n} p_{\mathscr{B}}(\mathbf{X}_t|\mathbf{X}^{t-1}), \end{aligned} \tag{1.6}$$

where $p_0(\mathbf{X}_t|\mathbf{X}^{t-1})$ and $p_{\mathscr{B}}(\mathbf{X}_t|\mathbf{X}^{t-1})$ are conditional densities of $\mathbf{X}_t$ given the past $t-1$ observations $\mathbf{X}^{t-1}$, which certainly may depend on t, i.e., $p_0 = p_{0,t}$ and $p_{\mathscr{B}} = p_{\mathscr{B},t}$.

Let $\mathscr{P}$ be an arbitrarily class of subsets of $\mathscr{N}$. For any $\mathscr{B} \in \mathscr{P}$, let $\Lambda_{\mathscr{B}}(n)$ be the likelihood ratio of $\mathsf{H}_{\mathscr{B}}$ against H_0 given the observations from all streams up to time n, and let $\lambda_{\mathscr{B}}(n)$ be the corresponding log-likelihood ratio (LLR), i.e.,

$$\Lambda_{\mathscr{B}}(n) = \frac{\mathrm{dP}_{\mathscr{B}}^{(n)}}{\mathrm{dP}_0^{(n)}} = \prod_{t=1}^{n} \frac{p_{\mathscr{B}}(\mathbf{X}_t|\mathbf{X}^{t-1})}{p_0(\mathbf{X}_t|\mathbf{X}^{t-1})}, \quad \lambda_{\mathscr{B}}(n) = \log \Lambda_{\mathscr{B}}(n) = \sum_{t=1}^{n} \log \left[\frac{p_{\mathscr{B}}(\mathbf{X}_t|\mathbf{X}^{t-1})}{p_0(\mathbf{X}_t|\mathbf{X}^{t-1})} \right]. \tag{1.7}$$

Two natural statistics for testing H_0 against H_1 at time n are the maximum (generalized) log-likelihood ratio (GLLR) statistic

$$\widehat{\lambda}(n) = \max_{\mathscr{B} \in \mathscr{P}} \lambda_{\mathscr{B}}(n) \tag{1.8}$$

and the logarithm of the weighted (mixture) likelihood ratio statistic (MLLR)

$$\widetilde{\lambda}(\mathbf{p}; n) = \log \widetilde{\Lambda}(\mathbf{p}; n), \quad \widetilde{\Lambda}(\mathbf{p}; n) = \sum_{\mathscr{B} \in \mathscr{P}} p_{\mathscr{B}} \Lambda_{\mathscr{B}}(n), \tag{1.9}$$

where $\mathbf{p} = \{p_{\mathscr{B}}, \mathscr{B} \in \mathscr{P}\}$ is a probability mass function on $\mathscr{N}$ fully supported on $\mathscr{P}$, i.e.,

$$p_{\mathscr{B}} > 0 \quad \text{for all } \mathscr{B} \in \mathscr{P} \quad \text{and} \quad \sum_{\mathscr{B} \in \mathscr{P}} p_{\mathscr{B}} = 1. \tag{1.10}$$

These two statistics lead to two different sequential tests: the *Generalized Sequential Likelihood Ratio Test* (GSLRT)

$$\widehat{T} = \inf\left\{ n \geq 1 : \widehat{\lambda}(n) \notin (-a, b) \right\}, \quad \widehat{d} = \begin{cases} 1 & \text{when } \widehat{\lambda}(\widehat{T}) \geq b \\ 0 & \text{when } \widehat{\lambda}(\widehat{T}) \leq -a \end{cases}, \tag{1.11}$$

and the *Mixture Sequential Likelihood Ratio Test* (MSLRT)

$$\widetilde{T} = \inf\left\{ n \geq 1 : \widetilde{\lambda}(\mathbf{p}; n) \notin (-a, b) \right\}, \quad \widetilde{d} = \begin{cases} 1 & \text{when } \widetilde{\lambda}(\mathbf{p}; \widetilde{T}) \geq b \\ 0 & \text{when } \widetilde{\lambda}(\mathbf{p}; \widetilde{T}) \leq -a \end{cases}. \tag{1.12}$$

Here $a, b > 0$ are thresholds that should be selected appropriately for each scheme in order to guarantee the desired error probabilities, i.e., so that $\widehat{T}$ and $\widetilde{T}$ belong to class $\mathbb{C}_{\alpha,\beta}(\mathscr{P})$ for given α and β.

For the reason that will become clear later on (see Subsection 1.5.4) we shall consider a more general weighted version of the GSLRT based on the weighted GLLR statistic

$$\widehat{\lambda}(n; \mathbf{p}) = \max_{\mathscr{B} \in \mathscr{P}} \left(\lambda_{\mathscr{B}}(n) + \log p_{\mathscr{B}} \right). \tag{1.13}$$

Furthermore, let us generalize both tests (1.11) and (1.12) as follows:

$$\widehat{T} = \inf\{n \geq 1 : \widehat{\lambda}(n;\mathbf{p}_1) \geq b \text{ or } \widehat{\lambda}(n;\mathbf{p}_0) \leq -a\}, \quad \widehat{d} = \mathbb{1}_{\{\widehat{\lambda}(\widehat{T};\mathbf{p}_1) \geq b\}}, \tag{1.14}$$

$$\widetilde{T} = \inf\{n \geq 1 : \widetilde{\lambda}(n;\mathbf{p}_1) \geq b \text{ or } \widetilde{\lambda}(n;\mathbf{p}_0) \leq -a\}, \quad \widetilde{d} = \mathbb{1}_{\{\widetilde{\lambda}(\widetilde{T};\mathbf{p}_1) \geq b\}}, \tag{1.15}$$

where $\mathbf{p}_j = \{p_{j,\mathscr{B}}, \mathscr{B} \in \mathscr{P}\}$, $j = 0,1$ are arbitrarily, not necessarily identical weights and $\mathbb{1}_{\{\mathscr{A}\}}$ denotes the indicator of the event $\mathscr{A}$. Again, the reason for such generalization will become clear in Subsection (1.5.4) when considering the minimax problem in the i.i.d. case.

Note that the mixture likelihood ratio statistic $\widetilde{\Lambda}(\mathbf{p}_j)$ is the likelihood ratio (Radon-Nikodým derivative)

$$\widetilde{\Lambda}(\mathbf{p}_j;n) = \frac{\mathrm{d}\widetilde{\mathsf{P}}_{\mathbf{p}_j}^{(n)}}{\mathrm{d}\mathsf{P}_0^{(n)}}(\mathscr{F}_n), \tag{1.16}$$

where $\widetilde{\mathsf{P}}_{\mathbf{p}_j}$ is the mixture probability measure

$$\widetilde{\mathsf{P}}_{\mathbf{p}_j} = \sum_{\mathscr{B} \in \mathscr{P}} p_{j,\mathscr{B}} \mathsf{P}_{\mathscr{B}}. \tag{1.17}$$

Obviously, the GLLR and the MLLR statistics can be written as

$$\widehat{\lambda}(\mathbf{p};n) = \lambda_{\mathscr{B}}(n) + \widehat{\Xi}_{\mathscr{B}}(\mathbf{p};n) + \log p_{\mathscr{B}}, \tag{1.18}$$

$$\widetilde{\lambda}(\mathbf{p};n) = \lambda_{\mathscr{B}}(n) + \widetilde{\Xi}_{\mathscr{B}}(\mathbf{p};n) + \log p_{\mathscr{B}}, \tag{1.19}$$

where

$$\widehat{\Xi}_{\mathscr{B}}(\mathbf{p};n) = \log\left(\max\left\{1, \max_{\mathscr{A} \neq \mathscr{B}} \frac{p_{\mathscr{A}}}{p_{\mathscr{B}}} \frac{\Lambda_{\mathscr{A}}(n)}{\Lambda_{\mathscr{B}}(n)}\right\}\right), \tag{1.20}$$

$$\widetilde{\Xi}_{\mathscr{B}}(\mathbf{p};n) = \log\left(1 + \sum_{\mathscr{A} \neq \mathscr{B}} \frac{p_{\mathscr{A}}}{p_{\mathscr{B}}} \frac{\Lambda_{\mathscr{A}}(n)}{\Lambda_{\mathscr{B}}(n)}\right) \tag{1.21}$$

are non-negative processes. Note also that $\widehat{\Xi}_{\mathscr{B}}(\mathbf{p};n) \leq \widetilde{\Xi}_{\mathscr{B}}(\mathbf{p};n)$ and

$$\lambda_{\mathscr{B}}(n) + \log p_{\mathscr{B}} \leq \widehat{\lambda}(\mathbf{p};n) \leq \widetilde{\lambda}(\mathbf{p};n) \quad \text{for all } n \in \mathbb{N} \text{ and } \mathscr{B} \in \mathscr{P}. \tag{1.22}$$

The following elementary lemma states that the processes $\widehat{\Xi}_{\mathscr{B}}(\mathbf{p};n)$ and $\widetilde{\Xi}_{\mathscr{B}}(\mathbf{p};n)$ converge almost surely to 0 as $n \to \infty$.

Lemma 1.1.

$$\widehat{\Xi}_{\mathscr{B}}(\mathbf{p};n) \xrightarrow[n\to\infty]{\mathsf{P}_{\mathscr{B}}-a.s.} 0, \quad \widetilde{\Xi}_{\mathscr{B}}(\mathbf{p};n) \xrightarrow[n\to\infty]{\mathsf{P}_{\mathscr{B}}-a.s.} 0.$$

Proof. For every $\mathscr{A} \neq \mathscr{B}$, the ratio $\{\Lambda_{\mathscr{A}}(\mathbf{p};n)/\Lambda_{\mathscr{B}}(\mathbf{p};n)\}$ is a non-negative $\mathsf{P}_{\mathscr{B}}$-martingale with mean 1 that converges $\mathsf{P}_{\mathscr{B}}$-a.s. to 0 as $n \to \infty$. Consequently, $\exp\{\widetilde{\Xi}_{\mathscr{B}}(n)\}$ is a non-negative $\mathsf{P}_{\mathscr{B}}$-martingale with mean $1/p_{\mathscr{B}}$ that converges almost surely to 1 as $n \to \infty$, which implies that $\widetilde{\Xi}_{\mathscr{B}}(\mathbf{p};n) \to 0$ $\mathsf{P}_{\mathscr{B}}$-almost surely as $n \to \infty$. Since $0 \leq \widehat{\Xi}_{\mathscr{B}}(\mathbf{p};n) \leq \widetilde{\Xi}_{\mathscr{B}}(\mathbf{p};n)$, it follows that $\widehat{\Xi}_{\mathscr{B}}(\mathbf{p};n)$ converges $\mathsf{P}_{\mathscr{B}}$-a.s. to 0 as $n \to \infty$. □

1.4 Asymptotic Operating Characteristics and Near Optimality of GSLRT and MSLRT in the General Non-i.i.d. Case

1.4.1 Probabilities of Errors in the General Non-i.i.d. Case

In this subsection, we derive upper bounds and asymptotic approximations for the false alarm and missed detection probabilities of the two proposed sequential tests, which are used for the selection of thresholds a and b.

1.4.1.1 Upper Bounds on the Error Probabilities

The following lemma provides non-asymptotic upper bounds on the error probabilities, which does not require any distributional assumptions.

Lemma 1.2. *For any thresholds $a,b>0$,*

$$\mathsf{P}_0(\widehat{d}=1)\le e^{-b}, \quad \max_{\mathscr{B}\in\mathscr{P}} \mathsf{P}_{\mathscr{B}}(\widehat{d}=0)\le \frac{e^{-a}}{\min_{\mathscr{B}\in\mathscr{P}} p_{0,\mathscr{B}}} \tag{1.23}$$

and

$$\mathsf{P}_0(\widetilde{d}=1)\le e^{-b}, \quad \max_{\mathscr{B}\in\mathscr{P}} \mathsf{P}_{\mathscr{B}}(\widetilde{d}=0)\le \frac{e^{-a}}{\min_{\mathscr{B}\in\mathscr{P}} p_{0,\mathscr{B}}}. \tag{1.24}$$

Proof. We start with proving the upper bounds on the false alarm probabilities in (1.24) and (1.23). Let $\widetilde{\mathsf{E}}_{\mathbf{p}_i}$ be the expectation that corresponds to the mixture measure $\widetilde{\mathsf{P}}_{\mathbf{p}_i}$ defined in (1.17) for arbitrarily weights $\{p_{i,\mathscr{B}}, \mathscr{B}\in\mathscr{P}\}$ that satisfy (1.10). Since $\widetilde{\lambda}(\mathbf{p}_1;\widetilde{T})\ge b$ on $\{\widetilde{d}=1\}$, using Wald's likelihood ratio identity we obtain

$$\mathsf{P}_0(\widetilde{d}=1)=\widetilde{\mathsf{E}}_{\mathbf{p}_1}\left[\exp\{-\widetilde{\lambda}(\mathbf{p}_1;\widetilde{T})\};\widetilde{d}=1\right]\le e^{-b}\widetilde{\mathsf{P}}_{\mathbf{p}_1}(d=1)\le e^{-b}.$$

Now, since $\widehat{\lambda}(\mathbf{p}_1;\widehat{T})\ge b$ on $\{\widehat{d}=1\}$, it follows that on the event $\{\widehat{d}=1\}$

$$1\le e^{-b}\max_{\mathscr{B}\in\mathscr{P}} p_{1,\mathscr{B}}\Lambda_{\mathscr{B}}(\widehat{T})\le e^{-b}\sum_{\mathscr{B}\in\mathscr{P}} p_{1,\mathscr{B}}\Lambda_{\mathscr{B}}(\widehat{T}).$$

Therefore, for an arbitrarily $\mathscr{A}\in\mathscr{P}$ we obtain

$$\begin{aligned}\mathsf{P}_0(\widehat{d}=1)&=\mathsf{E}_{\mathscr{A}}\left[\frac{1}{\Lambda_{\mathscr{A}}(\widehat{T})};\widehat{d}=1\right]\le e^{-b}\mathsf{E}_{\mathscr{A}}\left[\sum_{\mathscr{B}\in\mathscr{P}}\frac{p_{1,\mathscr{B}}\Lambda_{\mathscr{B}}(\widehat{T})}{\Lambda_{\mathscr{A}}(\widehat{T})};\widehat{d}=1\right]\\&=e^{-b}\sum_{\mathscr{B}\in\mathscr{P}}\mathsf{E}_{\mathscr{A}}\left[\frac{p_{1,\mathscr{B}}\Lambda_{\mathscr{B}}(\widehat{T})}{\Lambda_{\mathscr{A}}(\widehat{T})};\widehat{d}=1\right]\\&=e^{-b}\sum_{\mathscr{B}\in\mathscr{P}} p_{1,\mathscr{B}}\mathsf{P}_{\mathscr{B}}(\widehat{d}=1)\le e^{-b},\end{aligned}$$

where the first and last equalities follow from Wald's likelihood ratio identity.

To prove the bounds on the missed detection probabilities in (1.23) and (1.24), it suffices to note that by (1.22) $\lambda_{\mathscr{B}}(\widetilde{T})+\log p_{0,\mathscr{B}}\le\widetilde{\lambda}(\mathbf{p}_0;\widetilde{T})\le -a$ on $\{\widetilde{d}=0\}$ and $\lambda_{\mathscr{B}}(\widehat{T})+\log p_{0,\mathscr{B}}\le\widehat{\lambda}(\mathbf{p}_0;\widehat{T})\le -a$ on $\{\widehat{d}=0\}$ for every $\mathscr{B}\in\mathscr{P}$ and to use Wald's likelihood ratio identity, which yields

$$\mathsf{P}_{\mathscr{B}}(\widetilde{d}=0)=\mathsf{E}_0\left[\exp\{\lambda_{\mathscr{B}}(\widetilde{T})\};\,\widetilde{d}=0\right]\le p_{0,\mathscr{B}}^{-1}e^{-a},$$

$$\mathsf{P}_{\mathscr{B}}(\widehat{d}=0)=\mathsf{E}_0\left[\exp\{\lambda_{\mathscr{B}}(\widehat{T})\};\widehat{d}=0\right]\le p_{0,\mathscr{B}}^{-1}e^{-a}.$$

Maximizing with respect to $\mathscr{B}\in\mathscr{P}$ completes the proof. □

1.4.1.2 Error Exponents

In the following theorem, we show that the error exponents in the inequalities of Lemma 1.2 are sharp under mild distributional conditions.

Theorem 1.1. *Suppose that for every $\mathscr{B} \in \mathscr{P}$ there are numbers $I_1^{\mathscr{B}}, I_0^{\mathscr{B}} > 0$ such that the normalized LLR $\lambda_{\mathscr{B}}(n)/n$ converges to $-I_0^{\mathscr{B}}$ P_0-completely and to $I_1^{\mathscr{B}}$ $\mathsf{P}_{\mathscr{B}}$-completely, i.e., for all $\varepsilon > 0$*

$$\sum_{n=1}^{\infty} \mathsf{P}_0\left(\left|\frac{\lambda_{\mathscr{B}}(n)}{n} + I_0^{\mathscr{B}}\right| > \varepsilon\right) < \infty, \quad \sum_{n=1}^{\infty} \mathsf{P}_{\mathscr{B}}\left(\left|\frac{\lambda_{\mathscr{B}}(n)}{n} - I_1^{\mathscr{B}}\right| > \varepsilon\right) < \infty. \tag{1.25}$$

Then, as $a, b \to \infty$ the following asymptotic approximations for the logarithms of the probabilities of errors of the GSLRT and the MSLRT hold:

$$\begin{aligned} &\log \mathsf{P}_0(\widehat{d} = 1) \sim -b \sim \log \mathsf{P}_0(\widetilde{d} = 1) \\ &\log\left[\max_{\mathscr{B}\in\mathscr{P}} \mathsf{P}_{\mathscr{B}}(\widehat{d} = 0)\right] \sim -a \sim \log\left[\max_{\mathscr{B}\in\mathscr{P}} \{\mathsf{P}_{\mathscr{B}}(\widetilde{d} = 0)\right]. \end{aligned} \tag{1.26}$$

Proof. We provide the proof only for the GSLRT, as the proof for the MSLRT is analogous. From Lemma 1.2 it follows that as $a, b \to \infty$

$$\log \mathsf{P}_0(\widehat{d} = 1) \le -b(1 + o(1)) \quad \text{and} \quad \log \max_{\mathscr{B}\in\mathscr{P}} \left[\mathsf{P}_{\mathscr{B}}(\widehat{d} = 0)\right] \le -a(1 + o(1)).$$

Therefore, it suffices to show that the reverse inequalities also hold under the complete convergence conditions (1.25), i.e., as $a, b \to \infty$

$$\log \mathsf{P}_0(\widehat{d} = 1) \ge -b(1 + o(1)), \quad \log\left[\max_{\mathscr{B}\in\mathscr{P}} \mathsf{P}_{\mathscr{B}}(\widehat{d} = 0)\right] \ge -a(1 + o(1)). \tag{1.27}$$

Recall that

$$\exp\{\widetilde{\lambda}(n; \mathbf{p})\} = \frac{d\widetilde{\mathsf{P}}_{\mathbf{p}}^{(n)}}{d\mathsf{P}_0^{(n)}}. \tag{1.28}$$

Using Wald's likelihood ratio identity and the obvious inequality $\widetilde{\lambda}(\mathbf{p}_1; n) \le \widehat{\lambda}(\mathbf{p}_1; n) + \log|\mathscr{P}|$ (for every n), we obtain

$$\mathsf{P}_0(\widehat{d} = 1) = \widetilde{\mathsf{E}}_{\mathbf{p}_1}\left[\exp\{-\widetilde{\lambda}(\mathbf{p}_1; \widehat{T})\}; \widehat{d} = 1\right] \ge |\mathscr{P}|\widetilde{\mathsf{E}}_{\mathbf{p}_1}\left[\exp\{-\widehat{\lambda}(\mathbf{p}_1; \widehat{T})\}; \widehat{d} = 1\right]. \tag{1.29}$$

Now, for any $\delta > 0$,

$$\begin{aligned} \widetilde{\mathsf{E}}_{\mathbf{p}_1}\left[\exp\{-\widehat{\lambda}(\mathbf{p}_1; \widehat{T})\}; \widehat{d} = 1\right] &\ge \widetilde{\mathsf{E}}_{\mathbf{p}_1}\left[\exp\{-\widehat{\lambda}(\mathbf{p}_1; \widehat{T})\}; \widehat{d} = 1, \widehat{\lambda}(\mathbf{p}_1; \widehat{T}) < b(1+\delta)\right] \\ &\ge e^{-b(1+\delta)}\, \widetilde{\mathsf{P}}_{\mathbf{p}_1}\left(\widehat{d} = 1, \widehat{\lambda}(\mathbf{p}_1; \widehat{T}) < b(1+\delta)\right). \end{aligned} \tag{1.30}$$

Next, we show that as $a, b \to \infty$

$$\widetilde{\mathsf{P}}_{\mathbf{p}_1}\left(\widehat{d} = 1, \widehat{\lambda}(\mathbf{p}_1; \widehat{T}) < b(1+\delta)\right) \to 1, \tag{1.31}$$

which along with (1.29) and (1.30) implies that for an arbitrarily $\delta > 0$

$$\liminf_{a,b\to\infty} \frac{1}{b} \log \mathsf{P}_0(\widehat{d} = 1) \ge -(1 + \delta),$$

and hence (1.27). In order to establish (1.31) it suffices to show that $\mathsf{P}_{\mathscr{B}}^{\mathbf{p}_1}(\widehat{d}=1,\widehat{\lambda}(\mathbf{p}_1;\widehat{T})<b(1+\delta))\to 1$ for every $\mathscr{B}\in\mathscr{P}$ as $a,b\to\infty$. Fix $\mathscr{B}\in\mathscr{P}$ and $\delta>0$ and observe that

$$\mathsf{P}_{\mathscr{B}}\left(\widehat{d}=1,\widehat{\lambda}(\mathbf{p}_1;\widehat{T})<b(1+\delta)\right)=1-\mathsf{P}_{\mathscr{B}}(\widehat{d}=0)\\ -\mathsf{P}_{\mathscr{B}}\left(\widehat{d}=1,\,\widehat{\lambda}(\mathbf{p}_1;\widehat{T})\geq b(1+\delta)\right).$$

By Lemma 1.2, $\mathsf{P}_{\mathscr{B}}(\widehat{d}=0)\to 0$ as $a\to\infty$ for every $\mathscr{B}\in\mathscr{P}$. Therefore, it suffices to show that the second probability on the right-hand side goes to 0 as well. By the definition of the stopping time $\widehat{T}$,

$$\mathsf{P}_{\mathscr{B}}\left(\widehat{d}=1,\,\widehat{\lambda}(\mathbf{p}_1;\widehat{T})\geq b(1+\delta)\right)=\mathsf{P}_{\mathscr{B}}\left(\widehat{\lambda}(\mathbf{p}_1;\widehat{T}-1)\in(-a,b),\,\widehat{\lambda}(\mathbf{p}_1;\widehat{T})\geq b(1+\delta)\right)\\ \leq\mathsf{P}_{\mathscr{B}}\left(\widehat{\lambda}(\mathbf{p}_1;\widehat{T}-1)<b,\,\widehat{\lambda}(\mathbf{p}_1;\widehat{T})\geq b(1+\delta)\right).$$

By Lemma B.2 and Remark B.2 in Appendix B, the upper bound goes to 0 as $b\to\infty$ when for every $\varepsilon>0$

$$\mathsf{P}_{\mathscr{B}}\left(\frac{\max_{1\leq s\leq n}\widehat{\lambda}(\mathbf{p}_1;s)}{n}-I_1^{\mathscr{B}}>\varepsilon\right)\to 0,\quad\sum_{n=1}^{\infty}\mathsf{P}_{\mathscr{B}}\left(\frac{1}{n}\widehat{\lambda}(\mathbf{p}_1;n)-I_1^{\mathscr{B}}<-\varepsilon\right)<\infty.\tag{1.32}$$

Since $\widehat{\lambda}(\mathbf{p}_1;n)\geq\lambda_{\mathscr{B}}(n)+\log p_{1,\mathscr{B}}$ for every $n\in\mathbb{N}$, we have

$$\sum_{n=1}^{\infty}\mathsf{P}_{\mathscr{B}}\left(\frac{\widehat{\lambda}(\mathbf{p}_1;n)}{n}-I_1^{\mathscr{B}}<-\varepsilon\right)\leq\sum_{n=1}^{\infty}\mathsf{P}_{\mathscr{B}}\left(\frac{\lambda_{\mathscr{B}}(n)+\log p_{1\mathscr{B}}}{n}-I_1^{\mathscr{B}}<-\varepsilon\right)<\infty,\tag{1.33}$$

where the finiteness of the upper bound follows from the second condition in (1.25). Moreover, by (1.19), $\widehat{\lambda}(\mathbf{p};n)=\lambda_{\mathscr{B}}(n)+\widehat{\Xi}_{\mathscr{B}}(\mathbf{p};n)+\log p_{\mathscr{B}}$. Hence, the first condition in (1.32) is satisfied whenever for any $\varepsilon>0$

$$\mathsf{P}_{\mathscr{B}}\left(\frac{\max_{1\leq s\leq n}\widehat{\Xi}_{\mathscr{B}}(\mathbf{p}_1;s)}{n}>\varepsilon\right)\to 0\tag{1.34}$$

and

$$\mathsf{P}_{\mathscr{B}}\left(\frac{\max_{1\leq s\leq n}\lambda_{\mathscr{B}}(s)}{n}-I_1^{\mathscr{B}}>\varepsilon\right)\to 0.\tag{1.35}$$

By Lemma 1.1, $\widehat{\Xi}_{\mathscr{B}}(\mathbf{p}_1;n)\}$ converges $\mathsf{P}_{\mathscr{B}}$-a.s. to 0 as $n\to\infty$, so by Lemma B.1 (1.34) holds. By Lemma B.1, (1.35) holds since $n^{-1}\lambda_{\mathscr{B}}(n)\to I_1^{\mathscr{B}}$ $\mathsf{P}_{\mathscr{B}}$-completely as $n\to\infty$ (the second condition in (1.25)). This completes the proof of the first inequality in (1.27).

We now turn to the proof of the second inequality in (1.27). Clearly, it suffices to show that

$$\log\mathsf{P}_{\mathscr{B}}(\widehat{d}=0)\geq -a(1+o(1))$$

for some $\mathscr{B}\in\mathscr{P}$ such that $I_0^{\mathscr{B}}=\min_{\mathscr{A}\in\mathscr{P}}I_0^{\mathscr{A}}$. Fix such a subset $\mathscr{B}$. Following a similar argument as before, for any $\delta>0$ we have

$$\begin{aligned}\mathsf{P}_{\mathscr{B}}(\widehat{d}=0)&=\mathsf{E}_0\left[\exp\{\lambda_{\mathscr{B}}(\widehat{T})\}\,;\widehat{d}=0\right]\\ &\geq\mathsf{E}_0\left[\exp\{\lambda_{\mathscr{B}}(\widehat{T})\}\,;\widehat{d}=0;-\lambda_{\mathscr{B}}(\widehat{T})<a(1+\delta)\right]\\ &\geq e^{-a(1+\delta)}\,\mathsf{P}_0\left(\widehat{d}=0,-\lambda_{\mathscr{B}}(\widehat{T})<a(1+\delta)\right)\\ &\geq e^{-a(1+\delta)}\left[1-\mathsf{P}_0(\widehat{d}=1)-\mathsf{P}_0\left(\widehat{d}=0,\,-\lambda_{\mathscr{B}}(\widehat{T})\geq a(1+\delta)\right)\right].\end{aligned}$$

Since δ is arbitrarily, it suffices to show that the two probabilities on the right-hand side go to 0. By Lemma 1.2, $\mathsf{P}_0(\widehat{d}=1) \to 0$ as $b \to \infty$. By the definition of the stopping time $\widehat{T}$, on the event $\{\widehat{d}=0\}$ there is a $\mathscr{B} \in \mathscr{P}$ such that $\lambda_{\mathscr{B}}(\widehat{T}-1) \in (-a,b)$, i.e.,

$$\left\{\widehat{d}=0,\ -\lambda_{\mathscr{B}}(\widehat{T}) \geq a(1+\delta)\right\} \subset \bigcup_{\mathscr{A} \in \mathscr{P}} \{-\lambda_{\mathscr{A}}(\widehat{T}-1) < a,\ -\lambda_{\mathscr{B}}(\widehat{T}) \geq a(1+\delta)\}.$$

Therefore, it suffices to show that for every $\mathscr{A} \in \mathscr{P}$

$$\mathsf{P}_0\left(-\lambda_{\mathscr{A}}(\widehat{T}-1) < a,\ -\lambda_{\mathscr{B}}(\widehat{T}) \geq a(1+\delta)\right) \to 0. \tag{1.36}$$

Since $I_0^{\mathscr{B}} \leq I_0^{\mathscr{A}}$ for every $\mathscr{A} \in \mathscr{P}$, from Lemma B.2 it follows that (1.36) holds whenever $n^{-1}\lambda_{\mathscr{B}}(n)$ converges P_0-completely to $-I_0^{\mathscr{B}}$ as $n \to \infty$ (the first condition in (1.25)), which completes the proof of the second inequality in (1.27). □

Remark 1.1. The proof shows that the assertions of Theorem 1.1 hold under the following set of one-sided conditions:

$$\mathsf{P}_0\left(\frac{1}{n}\max_{1\leq s\leq n}(-\lambda_{\mathscr{B}}(s)) - I_0^{\mathscr{B}} > \varepsilon\right) \xrightarrow[n\to\infty]{} 0, \quad \mathsf{P}_{\mathscr{B}}\left(\frac{1}{n}\max_{1\leq s\leq n}\lambda_{\mathscr{B}}(s) - I_1^{\mathscr{B}} > \varepsilon\right) \xrightarrow[n\to\infty]{} 0 \tag{1.37}$$

and

$$\sum_{n=1}^{\infty}\mathsf{P}_0\left(\frac{-\lambda_{\mathscr{B}}(n)}{n} - I_0^{\mathscr{B}} < -\varepsilon\right) < \infty, \quad \sum_{n=1}^{\infty}\mathsf{P}_{\mathscr{B}}\left(\frac{\lambda_{\mathscr{B}}(n)}{n} - I_1^{\mathscr{B}} < -\varepsilon\right) < \infty, \tag{1.38}$$

which are implied by the complete version of the SLLN (1.25).

Remark 1.2. The proof also reveals that the error exponent of the maximal missed detection probability is attained by those alternative hypotheses $\mathsf{H}_{\mathscr{B}}$ that are the closest to H_0 in the sense that $I_0^{\mathscr{B}} = I_0(\mathscr{P})$, where $I_0(\mathscr{P})$ is defined as follows

$$I_0(\mathscr{P}) = \min_{\mathscr{A}\in\mathscr{P}} I_0^{\mathscr{A}}. \tag{1.39}$$

This suggests that in practice it suffices to evaluate the missed detection probability, $\mathsf{P}_{\mathscr{B}}(\widehat{d}=0)$, only for subsets $\mathscr{B}$ such that $I_0^{\mathscr{B}} = I_0(\mathscr{P})$. The latter probabilities, as well as the false alarm probability, can be estimated reliably using an importance sampling method described in the next subsection.

1.4.1.3 Monte Carlo Importance Sampling

The non-asymptotic inequalities (1.23) and (1.24) produce threshold values for the GSLRT and the MSLRT that guarantee that their false alarm and missed detection probabilities do not exceed the prescribed error probabilities α and β, respectively. Specifically, they imply that $\widehat{\delta} \in \mathbb{C}_{\alpha,\beta}(\mathscr{P})$ and $\widetilde{\delta} \in \mathbb{C}_{\alpha,\beta}(\mathscr{P})$ when

$$b = |\log\alpha| \quad \text{and} \quad a = |\log\beta| + C, \tag{1.40}$$

where $C = \max_{\mathscr{B}\in\mathscr{P}}|\log p_{\mathscr{B}}|$. Note also that the inequalities (1.23)–(1.24) as well as the corresponding thresholds (1.40) are universal, in the sense that they do not depend on the distributions of the observations. However, this appealing property typically leads to conservative bounds, so that the resulting thresholds (1.40) do not make full use of the given error tolerance. In the absence of closed-form expressions, sharp bounds, or more accurate asymptotic approximations for the error probabilities, the most convenient and general option for obtaining accurate threshold values is to compute the error probabilities using Monte Carlo simulation. However, it is well known that the relative standard error of naive Monte Carlo simulation explodes as the probability of interest goes to 0, and that for very small probabilities it is preferable to rely on *importance sampling*. In the context of sequential testing, this approach was suggested initially by Siegmund [139] for the estimation of the error probabilities of Wald's SPRT for testing a simple null hypothesis versus a simple alternative hypothesis.

We now present the importance sampling estimators for the error probabilities in our multistream sequential testing problem, where the alternative hypothesis is composite. We focus on the GSLRT, since the approach for the MSLRT is identical. In order to compute the false alarm probability of the GSLRT, we change the probability measure from P_0 to $\widetilde{\mathsf{P}}_{\mathbf{p}_1}$ defined in (1.17), and from Wald's likelihood ratio identity we obtain

$$\mathsf{P}_0(\widehat{d}=1) = \widetilde{\mathsf{E}}_{\mathbf{p}_1}\left[\exp\{-\widetilde{\lambda}(\mathbf{p}_1,\widehat{T})\};\widehat{d}=1\right], \tag{1.41}$$

where, as before, $\widetilde{\mathsf{E}}_{\mathbf{p}_1}$ refers to expectation under $\widetilde{\mathsf{P}}_{\mathbf{p}_1}$ and $\widetilde{\lambda}(\mathbf{p};n)$ is defined in (1.9). This suggests estimating $\mathsf{P}_0(\widehat{d}=1)$ by averaging independent realizations of $\exp\{-\widetilde{\lambda}(\mathbf{p}_1;\widehat{T})\}\,\mathbb{1}_{\{\widehat{d}=1\}}$, simulated under $\widetilde{\mathsf{P}}_{\mathbf{p}_1}$. For the missed detection probability of the GSLRT under $\mathsf{H}_{\mathscr{B}}$, changing the measure $\mathsf{P}_{\mathscr{B}}$ to P_0 and using Wald's likelihood ratio identity, we obtain

$$\mathsf{P}_{\mathscr{B}}(\widehat{d}=0) = \mathsf{E}_0\left[\exp\{\lambda_{\mathscr{B}}(\widehat{T})\};\ \widehat{d}=0\right]. \tag{1.42}$$

This suggests that each missed detection probability $\mathsf{P}_{\mathscr{B}}(\widehat{d}=0)$ can be computed by averaging independent realizations of $\exp\{\lambda_{\mathscr{B}}(\widehat{T})\}\mathbb{1}_{\{\widehat{d}=0\}}$ simulated under P_0. Note that all these missed detection probabilities use the same realizations of $(\widehat{T},\widehat{d})$ under P_0.

We now establish asymptotic efficiency of the proposed importance sampling estimators under weak distributional conditions. In the special case of a simple null against a simple alternative, i.e., when class $\mathscr{P}$ is a singleton, $\mathscr{P}=\{\mathscr{B}\}$, and the LLR statistic $\lambda_{\mathscr{B}}$ is a random walk, from Siegmund [139] it follows that the relative error of the above importance sampling estimators remains bounded as the error probabilities go to 0. We now show that in the non-i.i.d. case under much more general complete convergence conditions (1.25) the proposed importance sampling estimators satisfy a weaker optimality property that we refer to as *logarithmic efficiency* [3, 126].

By the Cauchy–Schwartz inequality,

$$\begin{aligned}
\mathsf{P}_0(\widehat{d}=1) &= \widetilde{\mathsf{E}}_{\mathbf{p}_1}\left[\exp\{-\widetilde{\lambda}(\mathbf{p}_1;\widehat{T})\}\,\mathbb{1}_{\{\widehat{d}=1\}}\right] \le \sqrt{\widetilde{\mathsf{E}}_{\mathbf{p}_1}\left[\exp\{-2\widetilde{\lambda}(\mathbf{p}_1;\widehat{T})\}\,\mathbb{1}_{\{\widehat{d}=1\}}\right]} \\
&= \sqrt{\widetilde{\mathsf{Var}}\left[\exp\{-\widetilde{\lambda}(\mathbf{p}_1;\widehat{T})\}\,\mathbb{1}_{\{\widehat{d}=1\}}\right]} \\
\mathsf{P}_{\mathscr{B}}(\widehat{d}=0) &= \mathsf{E}_0\left[\exp\{\lambda_{\mathscr{B}}(\widehat{T})\}\,\mathbb{1}_{\{\widehat{d}=0\}}\right] \le \sqrt{\mathsf{E}_0\left[\exp\{2\lambda_{\mathscr{B}}(\widehat{T})\}\mathbb{1}_{\{\widehat{d}=0\}}\right]} \\
&= \sqrt{\mathsf{Var}_0\left[\exp\{\lambda_{\mathscr{B}}(\widehat{T})\}\,\mathbb{1}_{\{\widehat{d}=0\}}\right]},
\end{aligned}$$

where $\widetilde{\mathsf{Var}}$ and Var_0 denote variance under $\widetilde{\mathsf{P}}_{\mathbf{p}_1}$ and P_0, respectively. Using these inequalities, we obtain

$$\frac{\log \widetilde{\mathsf{Var}}\left[\exp\{-\widetilde{\lambda}(\mathbf{p}_1;\widehat{T})\}\,\mathbb{1}_{\{\widehat{d}=1\}}\right]}{\log \mathsf{P}_0(\widehat{d}=1)} \le 2, \tag{1.43}$$

$$\frac{\log \mathsf{Var}_0\left[\exp\{\lambda_{\mathscr{B}}(\widehat{T})\}\,\mathbb{1}_{\{\widehat{d}=0\}}\right]}{\log \mathsf{P}_{\mathscr{B}}(\widehat{d}=0)} \le 2. \tag{1.44}$$

Note that these (non-asymptotic) bounds are based exclusively on the unbiasedness of the importance sampling estimators. In order to make sure that their variances are controlled, we would like the bounds in (1.43)–(1.44) to be asymptotically sharp, i.e., as $a, b \to \infty$,

$$2\log \mathsf{P}_0(\widehat{d}=1) \sim \log \widetilde{\mathsf{Var}}\left[\exp\{-\widetilde{\lambda}(\mathbf{p}_1;\widehat{T})\}\,\mathbb{1}_{\{\widehat{d}=1\}}\right], \tag{1.45}$$

$$2\log \mathsf{P}_{\mathscr{B}}(\widehat{d}=0) \sim \log \mathsf{Var}_0\left[\exp\{\lambda_{\mathscr{B}}(\widehat{T})\}\,\mathbb{1}_{\{\widehat{d}=0\}}\right]. \tag{1.46}$$

In this case, we will say that the proposed importance sampling estimators are *asymptotically logarithmically efficient.*

In the next theorem we show that, under the complete convergence conditions (1.25) of Theorem 1.1, logarithmic asymptotics (1.45) and (1.46) hold.

Theorem 1.2. *Suppose that conditions* (1.37)–(1.38) *hold or, alternatively, that the complete convergence conditions* (1.25) *hold.*
(i) *The importance sampling estimator of the false alarm probability* $\mathsf{P}_0(\widehat{d}=1)$ *suggested by* (1.41) *is asymptotically logarithmically efficient, i.e.,* (1.45) *holds.*
(ii) *The importance sampling estimator of the missed detection probability* $\mathsf{P}_{\mathscr{B}}(\widehat{d}=0)$ *suggested by* (1.42) *is asymptotically logarithmically efficient, i.e.,* (1.46) *holds for* $\mathscr{B} \in \mathscr{P}$ *such that* $I_0^{\mathscr{B}} = I_0(\mathscr{P})$.

Proof. The proof of (i). It suffices to show that

$$\liminf_{a,b\to\infty} \frac{\log \widetilde{\mathsf{Var}}\left[\exp\{-\widetilde{\lambda}(\mathbf{p}_1;\widehat{T})\}\,\mathbb{1}_{\{\widehat{d}=1\}}\right]}{\log \mathsf{P}_0(\widehat{d}=1)} \geq 2. \tag{1.47}$$

Since $\widetilde{\lambda}(\mathbf{p}_1;\widehat{T}) \geq \widehat{\lambda}(\mathbf{p}_1;\widehat{T}) \geq b$ on $\{\widehat{d}=0\}$, we have

$$\begin{aligned}\widetilde{\mathsf{Var}}\left[\exp\{-\widetilde{\lambda}(\mathbf{p}_1;\widehat{T})\}\,\mathbb{1}_{\{\widehat{d}=1\}}\right] &= \widetilde{\mathsf{E}}_{\mathbf{p}_1}\left[\exp\{-2\widetilde{\lambda}(\mathbf{p}_1;\widehat{T})\};\widehat{d}=1\right] \\ &\leq \exp\{-2b\}\,\widetilde{\mathsf{P}}_{\mathbf{p}_1}(\widehat{d}=1) \leq \exp\{-2b\}.\end{aligned}$$

Consequently,

$$\limsup_{b\to\infty} \frac{1}{b}\log \widetilde{\mathsf{Var}}\left[\exp\{-\widetilde{\lambda}(\mathbf{p}_1;\widehat{T})\}\,\mathbb{1}_{\{\widehat{d}=1\}}\right] \leq -2. \tag{1.48}$$

Moreover, from Theorem 1.1 it follows that $\log \mathsf{P}_0(\widehat{d}=1) \sim -b$ as $a, b \to \infty$, which together with (1.48) implies (1.47).

The proof of (ii). It suffices to show that, for every $\mathscr{B} \in \mathscr{P}$ such that $I_0^{\mathscr{B}} = I_0(\mathscr{P})$,

$$\liminf_{a,b\to\infty} \frac{\mathsf{E}_0\left[\exp\{2\lambda_{\mathscr{B}}(\widehat{T})\};\widehat{d}=0\right]}{\log \mathsf{P}_{\mathscr{B}}\left(\widehat{d}=0\right)} \geq 2. \tag{1.49}$$

Since for any $a > 0$, $\lambda_{\mathscr{B}}(\widehat{T}) + \log p_{1,\mathscr{B}} \leq \widehat{\lambda}(\mathbf{p}_1;\widehat{T}) \leq -a$ on $\{\widehat{d}=0\}$, we obtain

$$\mathsf{E}_0\left[\exp\{2\lambda_{\mathscr{B}}(\widehat{T})\};\widehat{d}=0\right] \leq p_{1,\mathscr{B}}^{-2}\exp\{-2a\},$$

and consequently,

$$\limsup_{a\to\infty} \frac{1}{a}\log \mathsf{E}_0\left[\exp\{2\lambda_{\mathscr{B}}(\widehat{T})\};\widehat{d}=0\right] \leq -2. \tag{1.50}$$

Since $\mathscr{B} \in \mathscr{P}$ is such that $I_0^{\mathscr{B}} = I_0(\mathscr{P})$, by Theorem 1.1 and Remark 1.2, $\log \mathsf{P}_{\mathscr{B}}(\widehat{d}=0) \sim -a$ as $a \to \infty$, which together with (1.50) yields (1.49). □

We stress that asymptotic relation (1.46) holds for $\mathscr{B} \in \mathscr{P}$ such that $I_0^{\mathscr{B}} = I_0(\mathscr{P})$, i.e., for the alternative hypotheses of interest for the computation of the maximal missed detection probability (see Remark 1.2).

1.4.1.4 Asymptotic Optimality of GSLRT and MSLRT

In this section, we establish the first-order asymptotic optimality property of the GSLRT and the MSLRT with respect to positive moments of the stopping time distribution for general non-i.i.d. models as well as for independent observations. The proof of asymptotic optimality is based on the lower–upper bounding technique. To be more specific, we first obtain asymptotic lower bounds for moments of the stopping time of an arbitrarily hypothesis test $\delta = (T,d)$ from class $\mathbb{C}_{\alpha,\beta}(\mathscr{P})$, and then we show that under certain regularity conditions these bounds are attained (asymptotically) for the proposed sequential tests.

We thus begin by deriving asymptotic lower bounds for moments of the sample size, which are given in the following important theorem. This theorem plays a fundamental role in deriving asymptotic operating characteristics of the GSLRT and MSLRT and proving their asymptotic optimality properties. In what follows, we write $\alpha_{\max} = \max(\alpha,\beta)$. Recall the definition of $I_0(\mathscr{P})$ in (1.39)

Theorem 1.3. *Assume that $n^{-1}\lambda_{\mathscr{B}}(n) \to I_1^{\mathscr{B}}$ in $\mathsf{P}_{\mathscr{B}}$-probability and $n^{-1}\lambda_{\mathscr{B}}(n) \to -I_0^{\mathscr{B}}$ in P_0-probability as $n\to\infty$, where $I_1^{\mathscr{B}}$ and $I_0^{\mathscr{B}}$ are positive finite constants and that, for all $\varepsilon>0$ and for all $\mathscr{B}\in\mathscr{P}$,*

$$\begin{aligned} &\lim_{M\to\infty} \mathsf{P}_0\left(\frac{1}{M}\max_{1\le n\le M}\lambda_{\mathscr{B}}(n) \ge -(1+\varepsilon)I_0^{\mathscr{B}}\right) = 0, \\ &\lim_{M\to\infty} \mathsf{P}_{\mathscr{B}}\left(\frac{1}{M}\max_{1\le n\le M}\lambda_{\mathscr{B}}(n) \ge (1+\varepsilon)I_1^{\mathscr{B}}\right) = 0. \end{aligned} \tag{1.51}$$

Then, for all $r>0$,

$$\begin{aligned} &\liminf_{\alpha_{\max}\to 0} \frac{\inf_{\delta\in\mathbb{C}_{\alpha,\beta}(\mathscr{P})} \mathsf{E}_0[T^r]}{|\log\beta|^r} \ge \left(\frac{1}{I_0(\mathscr{P})}\right)^r, \\ &\liminf_{\alpha_{\max}\to 0} \frac{\inf_{\delta\in\mathbb{C}_{\alpha,\beta}(\mathscr{P})} \mathsf{E}_{\mathscr{B}}[T^r]}{|\log\alpha|^r} \ge \left(\frac{1}{I_1^{\mathscr{B}}}\right)^r \quad \text{for all } \mathscr{B}\in\mathscr{P}, \end{aligned} \tag{1.52}$$

where $\alpha_{\max} = \max(\alpha,\beta)$.

Proof. For any $\mathscr{B}\in\mathscr{P}$, let us denote by $\mathbb{C}_{\alpha,\beta}(\mathscr{B})$ the class of sequential tests $\mathbb{C}_{\alpha,\beta}(\mathscr{P})$, defined in (1.4), when $\mathscr{P}=\{\mathscr{B}\}$, i.e.,

$$\mathbb{C}_{\alpha,\beta}(\mathscr{B}) = \{\delta : \mathsf{P}_0(d=1)\le\alpha \text{ and } \mathsf{P}_{\mathscr{B}}(d=0)\le\beta\}.$$

By Lemma 3.4.1 in Tartakovsky *et al.* [164], the following asymptotic lower bounds hold under conditions (1.51):

$$\liminf_{\alpha_{\max}\to 0} \frac{\inf_{\delta\in\mathbb{C}_{\alpha,\beta}(\mathscr{B})} \mathsf{E}_{\mathscr{B}}[T^r]}{|\log\alpha|^r} \ge \left(\frac{1}{I_1^{\mathscr{B}}}\right)^r, \quad \liminf_{\alpha_{\max}\to 0} \frac{\inf_{\delta\in\mathbb{C}_{\alpha,\beta}(\mathscr{B})} \mathsf{E}_0[T^r]}{|\log\beta|^r} \ge \left(\frac{1}{I_0^{\mathscr{B}}}\right)^r. \tag{1.53}$$

Clearly, for any $\alpha,\beta\in(0,1)$, $\mathbb{C}_{\alpha,\beta}(\mathscr{P})\subset\mathbb{C}_{\alpha,\beta}(\mathscr{B})$, so that

$$\inf_{\delta\in\mathbb{C}_{\alpha,\beta}(\mathscr{P})} \mathsf{E}_0[T^r] \ge \inf_{\delta\in\mathbb{C}_{\alpha,\beta}(\mathscr{B})} \mathsf{E}_0[T^r], \quad \inf_{\delta\in\mathbb{C}_{\alpha,\beta}(\mathscr{P})} \mathsf{E}_{\mathscr{B}}[T^r] \ge \inf_{\delta\in\mathbb{C}_{\alpha,\beta}(\mathscr{B})} \mathsf{E}_{\mathscr{B}}[T^r]. \tag{1.54}$$

Inequalities (1.54) together with asymptotic inequalities (1.53) imply the lower bounds (1.52), and the proof is complete. □

Corollary 1.1. *Assume that the normalized LLR $\lambda_{\mathscr{B}}(n)/n$ converges as $n\to\infty$ to $-I_0^{\mathscr{B}}$ P_0-a.s. and to $I_1^{\mathscr{B}}$ $\mathsf{P}_{\mathscr{B}}$-a.s. Then asymptotic lower bounds (1.52) hold for all positive r.*

Proof. By Lemma B.1 in Appendix B, the almost sure convergence conditions postulated in the corollary imply the right-tail conditions (1.51), so that the assertion of the corollary follows. □

The following proposition establishes first-order asymptotic approximations for the moments of the stopping times of the GSLRT and the MSLRT for large threshold values. These asymptotic approximations may be useful, apart from proving asymptotic optimality in Theorem 1.4, for problems with different types of constraints, for example in Bayesian settings. We set $a_{\min} = \min(a,b)$.

Proposition 1.1. *Suppose that for every $\mathscr{B} \in \mathscr{P}$ there are positive and finite numbers $I_1^{\mathscr{B}}$ and $I_0^{\mathscr{B}}$ such that, for all $\mathscr{B} \in \mathscr{P}$,*

$$\mathsf{P}_0\left(\frac{1}{n}\lambda_{\mathscr{B}}(n) \xrightarrow[n\to\infty]{} -I_0^{\mathscr{B}}\right) = 1 \quad and \quad \mathsf{P}_{\mathscr{B}}\left(\frac{1}{n}\lambda_{\mathscr{B}}(n) \xrightarrow[n\to\infty]{} I_1^{\mathscr{B}}\right) = 1. \tag{1.55}$$

(i) *If the LLR $\{\lambda_{\mathscr{B}}(n)\}_{n\geq 1}$ has independent, but not necessarily identically distributed increments under both P_0 and $\mathsf{P}_{\mathscr{B}}$ for every $\mathscr{B} \in \mathscr{P}$, then the asymptotic approximations*

$$\begin{aligned} \mathsf{E}_0[\widehat{T}^r] &\sim \left(\frac{a}{I_0(\mathscr{P})}\right)^r \sim \mathsf{E}_0[\widetilde{T}^r], \\ \mathsf{E}_{\mathscr{B}}[\widehat{T}^r] &\sim \left(\frac{b}{I_1^{\mathscr{B}}}\right)^r \sim \mathsf{E}_{\mathscr{B}}[\widetilde{T}^r] \quad for\ all\ \mathscr{B} \in \mathscr{P} \end{aligned} \tag{1.56}$$

hold for all $r \geq 1$ as $a_{\min} \to \infty$.
(ii) *Let $r \geq 1$. If the LLR $\{\lambda_{\mathscr{B}}(n)\}_{n\geq 1}$ has arbitrarily (possibly dependent) increments and in addition to the a.s. convergence conditions (1.55) the following left-sided conditions*

$$\sum_{n=1}^{\infty} n^{r-1}\,\mathsf{P}_0\left(\frac{\lambda_{\mathscr{B}}(n)}{n} + I_0^{\mathscr{B}} < -\varepsilon\right) < \infty, \quad \sum_{n=1}^{\infty} n^{r-1}\,\mathsf{P}_{\mathscr{B}}\left(\frac{\lambda_{\mathscr{B}}(n)}{n} - I_1^{\mathscr{B}} < -\varepsilon\right) \tag{1.57}$$

are satisfied for all $0 < \varepsilon < \min\{I_0^{\mathscr{B}}, I_1^{\mathscr{B}}\}$, then the asymptotic approximations (1.56) hold for $1 \leq m \leq r$.

Proof. We provide a proof only for the GSLRT, as from representations (1.18) and (1.19) and Lemma 1.1 it is clear that the MSLRT stopping time has the same moments to first order. Since the almost sure convergence conditions (1.55) imply right-tail conditions (1.51) (see Lemma B.1) and since, by Lemma 1.2, $(\widehat{T}(a,b),\widehat{d}) \in \mathbb{C}_{\alpha,\beta}(\mathscr{P})$ for $\log\alpha = -a$ and $\log\beta = -b + \inf_{\mathscr{B}\in\mathscr{P}} \log p_{0,\mathscr{B}}$, applying Theorem 1.3 (or Corollary 1.1) we obtain the asymptotic lower bounds (for every $r \geq 1$):

$$\liminf_{a,b\to\infty} \frac{\mathsf{E}_0[\widehat{T}^r(a,b)]}{a^r} \geq \left(\frac{1}{I_0(\mathscr{P})}\right)^r, \tag{1.58}$$

$$\liminf_{a,b\to\infty} \frac{\mathsf{E}_{\mathscr{B}}[\widehat{T}^r(a,b)]}{b^r} \geq \left(\frac{1}{I_1^{\mathscr{B}}}\right)^r. \tag{1.59}$$

The second step is to prove that the asymptotic lower bounds (1.58) and (1.59) are sharp. Observe that

$$\widehat{T}(a,b) = \min\left\{\min_{\mathscr{B}\in\mathscr{P}} T_{\mathscr{B}}(b),\, T(a)\right\}, \tag{1.60}$$

where

$$T_{\mathscr{B}}(b) = \inf\left\{n \geq 1 : \lambda_{\mathscr{B}}(n) \geq b - \log\left(\inf_{\mathscr{B}\in\mathscr{P}} p_{0,\mathscr{B}}\right)\right\}, \quad T(a) = \inf\left\{n \geq 1 : \min_{\mathscr{B}\in\mathscr{P}}(-\lambda_{\mathscr{B}}(n)) \geq a\right\}, \tag{1.61}$$

so that $\widehat{T} \le T_{\mathscr{B}}(b)$ and $\widehat{T} \le T(a)$ and it suffices to show that, as $a_{\min} \to \infty$,

$$\mathsf{E}_{\mathscr{B}}[T^m_{\mathscr{B}}(b)] \le \left(\frac{b}{I_1^{\mathscr{B}}}\right)^m (1+o(1)), \quad \mathsf{E}_0[T^m(a)] \le \left(\frac{a}{I_0(\mathscr{P})}\right)^m (1+o(1)). \tag{1.62}$$

These asymptotic upper bounds follow from Theorem E.7(i) (Appendix E) for $1 \le m \le r$ when conditions (1.57) hold. Thus, all that remains to do is to show that (1.62) hold for every $m \ge 1$ when $\lambda_{\mathscr{B}}$ has independent increments. From Theorem E.7(ii) it follows that it suffices to show that there is a $\delta \in (0,1)$ such that

$$\sup_{n\in\mathbb{N}} \mathsf{E}_{\mathscr{B}}\left[\exp\{\delta(\Delta\lambda_{\mathscr{B}}(n))^-\}\right] < \infty, \quad \sup_{n\in\mathbb{N}} \mathsf{E}_0\left[\exp\{\delta(-\Delta\lambda_{\mathscr{B}}(n))^-\}\right] < \infty, \tag{1.63}$$

where $\Delta\lambda_{\mathscr{B}}(n) = \lambda_{\mathscr{B}}(n) - \lambda_{\mathscr{B}}(n-1)$, $n \in \mathbb{N}$. For any $\delta \in (0,1)$, by Jensen's inequality

$$\begin{aligned} \mathsf{E}_{\mathscr{B}}[\exp\{\delta\min\{\Delta\lambda_{\mathscr{B}}(n),0\}\}] &= \mathsf{E}_{\mathscr{B}}[\exp\{\delta\min\{\Delta\lambda_{\mathscr{B}}(n),0\}\}\,;\,\Delta\lambda_{\mathscr{B}}(n)\le 0] \\ &\quad + \mathsf{E}_{\mathscr{B}}[\exp\{\delta\min\{\Delta\lambda_{\mathscr{B}}(n),0\}\}\,;\,\Delta\lambda_{\mathscr{B}}(n)> 0] \\ &\le \mathsf{E}_{\mathscr{B}}[\exp\{-\delta\Delta\lambda_{\mathscr{B}}(n)\}] + 1 \\ &\le (\mathsf{E}_{\mathscr{B}}[\exp\{-\Delta\lambda_{\mathscr{B}}(n)\}])^{\delta} + 1 = 2, \end{aligned}$$

where the latter equality holds because the likelihood ratio $\exp\{-\Delta\lambda_{\mathscr{B}}(n)\}$ is a martingale with mean 1 under $\mathsf{P}_{\mathscr{B}}$. The second condition in (1.63) can be verified in a similar way. □

Remark 1.3. Note that both conditions (1.55) and (1.57) hold when the normalized LLR $\lambda_{\mathscr{B}}(n)/n$ converges to $-I_0^{\mathscr{B}}$ P_0-r-completely and to $I_1^{\mathscr{B}}$ $\mathsf{P}_1^{\mathscr{B}}$-$r$-completely, i.e., for every $\varepsilon > 0$

$$\sum_{n=1}^{\infty} n^{r-1}\mathsf{P}_0\left(\left|\frac{\lambda_{\mathscr{B}}(n)}{n} + I_0^{\mathscr{B}}\right| > \varepsilon\right) < \infty, \quad \sum_{n=1}^{\infty} n^{r-1}\mathsf{P}_{\mathscr{B}}\left(\left|\frac{\lambda_{\mathscr{B}}(n)}{n} - I_1^{\mathscr{B}}\right| > \varepsilon\right) < \infty. \tag{1.64}$$

We are now prepared to establish the asymptotic optimality property of the GSLRT $\widehat{\delta} = (\widehat{T}, \widehat{d})$ and the MSLRT $\widetilde{\delta} = (\widetilde{T}, \widetilde{d})$ with respect to positive moments of the stopping time distribution.

Theorem 1.4. *Consider an arbitrarily class of alternatives, $\mathscr{P}$, and suppose that thresholds a,b of the GSLRT and the MSLRT are selected in such a way that $\widehat{\delta}, \widetilde{\delta} \in \mathbb{C}_{\alpha,\beta}(\mathscr{P})$ and $b \sim |\log\alpha|$, $a \sim |\log\beta|$ as $\alpha_{\max} \to 0$, e.g., according to* (1.40).
(i) *Let the LLR $\{\lambda_{\mathscr{B}}(n)\}_{n\ge1}$ have independent but not necessarily identically distributed increments under P_0 and $\mathsf{P}_{\mathscr{B}}$ for every $\mathscr{B} \in \mathscr{P}$. If the a.s. convergence conditions* (1.55) *hold, then as $\alpha_{\max} \to 0$ for all $m \ge 1$*

$$\mathsf{E}_0[\widetilde{T}^m] \sim \mathsf{E}_0[\widehat{T}^m] \sim \left(\frac{|\log\beta|}{I_0(\mathscr{P})}\right)^m \sim \inf_{\delta\in\mathbb{C}_{\alpha,\beta}(\mathscr{P})} \mathsf{E}_0[T^m], \tag{1.65}$$

$$\mathsf{E}_{\mathscr{B}}[\widetilde{T}^m] \sim \mathsf{E}_{\mathscr{B}}[\widehat{T}^m] \sim \left(\frac{|\log\alpha|}{I_1^{\mathscr{B}}}\right)^m \sim \inf_{\delta\in\mathbb{C}_{\alpha,\beta}(\mathscr{P})} \mathsf{E}_{\mathscr{B}}[T^m] \quad \text{for all } \mathscr{B}\in\mathscr{P}. \tag{1.66}$$

(ii) *Let the LLR $\{\lambda_{\mathscr{B}}(n)\}_{n\ge1}$ have arbitrarily (possibly dependent) increments. If in addition to conditions* (1.55) *the left-tail conditions* (1.57) *hold for some $r \in \mathbb{N}$, then for all $1 \le m \le r$ the asymptotics* (1.65) *and* (1.66) *hold.*

Proof. When the thresholds in the proposed tests are selected in such a way that $b \sim |\log\alpha|$ and $a \sim |\log\beta|$, from Proposition 1.1 it follows that

$$\mathsf{E}_0[\widehat{T}^m] \sim \left(\frac{|\log\beta|}{I_0(\mathscr{P})}\right)^m, \quad \mathsf{E}_{\mathscr{B}}[\widehat{T}^m] \sim \left(\frac{|\log\alpha|}{I_1^{\mathscr{B}}}\right)^m \quad \text{as } \alpha_{\max} \to 0$$

for every $m \ge 1$ under conditions of (i) and for every $1 \le m \le r$ under conditions of (ii). These asymptotic approximations along with the lower bounds (1.52), which hold by Corollary 1.1, prove both assertions (i) and (ii). □

Remark 1.4. In the general non-i.i.d. case, the assertion of Theorem 1.4(ii) holds under the r-complete convergence conditions (1.64) since they guarantee both conditions (1.55) and (1.57).

Remark 1.5. The LLR process $\{\lambda_{\mathscr{B}}(n)\}_{n\ge1}$ may have independent increments even if the observations $\{X(n)\}$ are not independent over time. Indeed, certain models of dependent observations produce LLRs with independent increments. See, e.g., Example 1.1 in Section 1.4.3.

Remark 1.6. As expected, the tests are asymptotically first-order optimal for any weights $\mathbf{p}_1$ and $\mathbf{p}_0$. Thus, the unweighted GSLRT (1.11) and the MSLRT (1.12) (with an arbitrarily $\mathbf{p}$) are also optimal to first order. The choice of weights $\mathbf{p}_0$ and $\mathbf{p}_1$ matters when considering a minimax problem setting and higher-order optimality properties. See Subsection 1.5.4.3.

1.4.2 The Case of Independent Data Streams

In previous sections, we considered a very general multichannel sequential hypothesis testing problem where not only the observations in streams $X_n(i)$ may be non-i.i.d., but also the data between the various streams may be dependent. This generality comes at the price of computational complexity. Indeed, recall that the GSLRT and MSLRT statistics need to be computed at each time n, and this computation in general requires the computation of $|\mathscr{P}|$ LLR statistics. When $|\mathscr{P}|$ is very large, the two proposed schemes may not even be feasible. In this section, we consider a special case of independent streams, in which both proposed procedures become scalable with respect to the number of streams, even if there is complete uncertainty regarding the size of the affected subset.

Specifically, if the observations across streams are independent, i.e.,

$$\mathsf{P}_0 = \prod_{i=1}^{N} \mathsf{P}_0^i, \quad \mathsf{P}_{\mathscr{B}} = \left(\prod_{i \notin \mathscr{B}} \mathsf{P}_0^i\right) \times \left(\prod_{i \in \mathscr{B}} \mathsf{P}_1^i\right), \tag{1.67}$$

where P_0^i and P_1^i are distinct, mutually absolutely continuous probability measures when restricted to the σ-algebra $\mathscr{F}_i(n) = \sigma(X_t(i), 1 \le t \le n)$ of observations in the ith stream, then the hypothesis testing problem (1.6) is simplified as follows:

$$\begin{aligned} &\mathsf{H}_0: \quad p(\mathbf{X}^n) = \prod_{i=1}^{N}\prod_{t=1}^{n} p_{0,i}(\mathbf{X}_t(i)|\mathbf{X}^{t-1}(i)); \\ &\mathsf{H}_1 = \bigcup_{\mathscr{B}\in\mathscr{P}} \mathsf{H}_{\mathscr{B}}, \quad \mathsf{H}_{\mathscr{B}}: p(\mathbf{X}^n) = \begin{cases} \prod_{i\notin\mathscr{B}}\prod_{t=1}^{n} p_{0,i}(\mathbf{X}_t(i)|\mathbf{X}^{t-1}(i)) & \text{if } i \notin \mathscr{B} \\ \prod_{i\in\mathscr{B}}\prod_{t=1}^{n} p_{1,i}(\mathbf{X}_t(i)|\mathbf{X}^{t-1}(i)) & \text{if } i \in \mathscr{B} \end{cases}, \end{aligned} \tag{1.68}$$

where $p_{0,i}(\mathbf{X}_t(i)|\mathbf{X}^{t-1}(i))$ and $p_{1,i}(\mathbf{X}_t(i)|\mathbf{X}^{t-1}(i))$ are conditional densities of $\mathbf{X}_t(i)$ given the past $t-1$ observations $\mathbf{X}^{t-1}(i)$ from the ith stream when there is no signal and there is a signal in this stream, respectively.

If we denote by $\Lambda_i(n)$ the Radon–Nikodým derivative (likelihood ratio) of P_1^i with respect to P_0^i given $\mathscr{F}_i(n)$ and by $\lambda_i(n)$ the corresponding LLR, i.e.,

$$\Lambda_i(n) = \frac{d\mathsf{P}_1^i}{d\mathsf{P}_0^i}(\mathscr{F}_i(n)) \quad \text{and} \quad \lambda_i(n) = \log \Lambda_i(n), \tag{1.69}$$

then from (1.68) it follows that

$$\Lambda_{\mathscr{B}}(n) = \prod_{i\in\mathscr{B}} \Lambda_i(n) \quad \text{and} \quad \lambda_{\mathscr{B}}(n) = \sum_{i\in\mathscr{B}} \lambda_i(n), \tag{1.70}$$

where

$$\Lambda_i(n) = \prod_{t=1}^{n} \frac{p_{1,i}(\mathbf{X}_t(i)|\mathbf{X}^{t-1}(i))}{p_{0,i}(\mathbf{X}_t(i)|\mathbf{X}^{t-1}(i))}, \quad \lambda_i(n) = \sum_{t=1}^{n} \log \left[\frac{p_{1,i}(\mathbf{X}_t(i)|\mathbf{X}^{t-1}(i))}{p_{0,i}(\mathbf{X}_t(i)|\mathbf{X}^{t-1}(i))} \right]. \tag{1.71}$$

1.4.2.1 Asymptotic Optimality in the Case of Independent Streams

Before discussing the implementation of the proposed schemes in this case, we first formulate sufficient conditions for their asymptotic optimality. Due to (1.70) it is clear that the SLLN (1.55) are implied by the corresponding SLLN for the marginal LLRs in the streams. That is, (1.55) holds for every $\mathscr{B} \in \mathscr{P}$ when there are positive numbers I_0^i and I_1^i, $i = 1, \dots, N$, such that

$$\mathsf{P}_1^i \left(\frac{1}{n} \lambda_i(n) \xrightarrow[n\to\infty]{} I_1^i \right) = 1 \text{ and } \mathsf{P}_0^i \left(\frac{1}{n} \lambda_i(n) \xrightarrow[n\to\infty]{} -I_0^i \right) = 1 \text{ for all } i \in \mathscr{N}, \tag{1.72}$$

in which case

$$I_0^{\mathscr{B}} = \sum_{i \in \mathscr{B}} I_0^i \quad \text{and} \quad I_1^{\mathscr{B}} = \sum_{i \in \mathscr{B}} I_1^i. \tag{1.73}$$

Also, it is easy to show that the r-complete convergence conditions (1.64) are implied by the corresponding conditions for the marginal LLRs in the streams, i.e.,

$$\frac{1}{n} \lambda_i(n) \xrightarrow[n\to\infty]{\mathsf{P}_0^i - r\text{-completely}} -I_0^i \text{ and } \frac{1}{n} \lambda_i(n) \xrightarrow[n\to\infty]{\mathsf{P}_1^i - r\text{-completely}} I_1^i, \quad i = 1, \dots, N. \tag{1.74}$$

Thus, applying Theorem 1.4, we obtain the following asymptotic optimality result for the proposed schemes in the case of independent streams.

Corollary 1.2. *Consider an arbitrarily class of alternatives, $\mathscr{P}$, and suppose that the thresholds a, b of the GSLRT and the MSLRT are selected in such a way that $\widehat{\delta}, \widetilde{\delta} \in \mathbb{C}_{\alpha,\beta}(\mathscr{P})$ and $b \sim |\log \alpha|$, $a \sim |\log \beta|$ as $\alpha_{\max} \to 0$, e.g., according to (1.40). Assume that the data are independent across streams, i.e., X_i and X_j are mutually independent for all $i \neq j$, $i, j = 1, \dots, N$.*
(i) *Let $\{\lambda_i(n)\}_{n \geq 1}$ have independent, but not necessarily identically distributed increments under P_0^i and P_1^i for every $i \in \mathscr{N}$. If the a.s. convergence conditions (1.72) hold, then as $\alpha_{\max} \to 0$ for all $m \geq 1$*

$$\mathsf{E}_0[\widetilde{T}^m] \sim \mathsf{E}_0[\widehat{T}^m] \sim \left(\frac{|\log \beta|}{I_0(\mathscr{P})} \right)^m \sim \inf_{\delta \in \mathbb{C}_{\alpha,\beta}(\mathscr{P})} \mathsf{E}_0[T^m], \tag{1.75}$$

$$\mathsf{E}_{\mathscr{B}}[\widetilde{T}^m] \sim \mathsf{E}_{\mathscr{B}}[\widehat{T}^m] \sim \left(\frac{|\log \alpha|}{I_1^{\mathscr{B}}} \right)^m \sim \inf_{\delta \in \mathbb{C}_{\alpha,\beta}(\mathscr{P})} \mathsf{E}_{\mathscr{B}}[T^m] \quad \text{for all } \mathscr{B} \in \mathscr{P}, \tag{1.76}$$

where $I_0^{\mathscr{B}}$ and $I_1^{\mathscr{B}}$ are given by (1.73) and $I_0(\mathscr{P}) = \inf_{\mathscr{B} \in \mathscr{P}} I_0^{\mathscr{B}}$.
(ii) *Let $\{\lambda_i(n)\}_{n \geq 1}$ have arbitrarily (possibly dependent) increments. If the r-complete convergence conditions (1.74) hold for some $r \geq 1$, then asymptotics (1.75) and (1.76) are satisfied for all $1 \leq m \leq r$.*

1.4.2.2 Scalability in the Case of Independent Streams

The implementation of both the GSLRT and the MSLRT is greatly simplified in the case of independent streams, and the two procedures become feasible even with a large number of streams and under complete uncertainty with respect to the affected subset. Indeed, recalling (1.70), the weighted GLLR statistic takes the form

$$\widehat{\lambda}(\mathbf{p}; n) = \max_{\mathscr{B} \in \mathscr{P}} \sum_{i \in \mathscr{B}} [\lambda_i(n) + \log p_i],$$

and its computation becomes very easy for class $\mathscr{P}_{\underline{K},\overline{K}}$, i.e., when there are at least $\underline{K}$ and at most $\overline{K}$ signals in the N streams, as long as there are positive weights $\{w_i\}_{1\le i\le N}$ such that the prior distribution has the form

$$p_{\mathscr{B}} = \prod_{i\in\mathscr{B}} p_i, \quad p_i = \frac{w_i}{\sum_{\mathscr{B}\in\mathscr{P}} \prod_{j\in\mathscr{B}} w_j}. \tag{1.77}$$

In order to see this, let $\lambda_i^*(p_i;n) = \lambda_i(n) + \log p_i$ and let us use the following notation for the order statistics: $\lambda_{(1)}^*(\mathbf{p};n) \ge \ldots \ge \lambda_{(N)}^*(\mathbf{p};n)$, i.e., $\lambda_{(1)}^*(\mathbf{p};n) = \max_{1\le i\le N} \lambda_i^*(p_i;n)$ is the top local statistic and $\lambda_{(N)}(\mathbf{p};n) = \min_{1\le i\le N} \lambda_i^*(p_i;n)$ is the smallest local statistic at time n. When the size of the affected subset is known in advance, i.e., $\underline{K} = \overline{K} = m$, then

$$\widehat{\lambda}(\mathbf{p};n) = \sum_{i=1}^{m} \lambda_{(i)}^*(\mathbf{p};n). \tag{1.78}$$

In the more general case that $\underline{K} < \overline{K}$, we have

$$\widehat{\lambda}(\mathbf{p};n) = \sum_{i=1}^{\underline{K}} \lambda_{(i)}^*(\mathbf{p};n) + \sum_{i=\underline{K}+1}^{\overline{K}} \max\left\{\lambda_{(i)}^*(\mathbf{p};n), 0\right\},$$

and the GSLRT takes the following form:

$$\begin{aligned} \widehat{T} &= \inf\left\{n \ge 1 : \sum_{i=1}^{\overline{K}} \max\left\{\lambda_{(i)}^*(\mathbf{p}_1;n), 0\right\} \ge b \text{ or } \sum_{i=1}^{\underline{K}} \lambda_{(i)}^*(\mathbf{p}_0;n) \le -a\right\}, \\ \widehat{d} &= \begin{cases} 1 & \text{when } \sum_{i=1}^{\overline{K}} \max\left\{\lambda_{(i)}^*(\mathbf{p}_1;\widehat{T}), 0\right\} \ge b \\ 0 & \text{when } \sum_{i=1}^{\underline{K}} \lambda_{(i)}^*(\mathbf{p}_0;\widehat{T}) \le -a \end{cases}. \end{aligned} \tag{1.79}$$

Similarly, in the case of independent streams, the MSLRT is computationally feasible even for large N when the weights are selected according to (1.77). Indeed, when there is an upper and a lower bound on the size of the affected subset, i.e., $\mathscr{P} = \mathscr{P}_{\underline{K},\overline{K}}$ for some $1 \le \underline{K} \le \overline{K} \le N$, the mixture likelihood ratio statistic takes the form

$$\widetilde{\Lambda}(\mathbf{p};n) = \sum_{m=\underline{K}}^{\overline{K}} \sum_{\mathscr{B}\in\mathscr{P}_m} \prod_{i\in\mathscr{B}} p_i \Lambda_i(n), \tag{1.80}$$

and its computational complexity is polynomial in the number of streams, N. Furthermore, note that in the special case of complete uncertainty ($\underline{K} = 1, \overline{K} = N$), the MSLRT requires only $O(N)$ operations. Indeed, setting $p_i = p = w/C(\overline{\mathscr{P}}_N)$ in (1.77), the mixture likelihood ratio in (1.80) admits the following representation for class $\mathscr{P} = \overline{\mathscr{P}}_N$:

$$\widetilde{\Lambda}(n) = C(\overline{\mathscr{P}}_N)\left[(1-\pi)^{-N} \overline{\Lambda}(n) - 1\right], \tag{1.81}$$

where $C(\overline{\mathscr{P}}_N) = \sum_{\mathscr{B}\in\overline{\mathscr{P}}_N} w^{|\mathscr{B}|}$, $\pi = w/(1+w)$ and

$$\overline{\Lambda}(n) = \prod_{i=1}^{N} \left[1 - \pi + \pi\,\Lambda_i(n)\right]. \tag{1.82}$$

Remark 1.7. The statistic $\overline{\Lambda}(n)$ has an appealing statistical interpretation – it is the likelihood ratio that corresponds to the case where each stream belongs to the affected subset with probability $\pi \in (0,1)$. The decision statistic (1.81) exploits information from the observations via the LR statistic (1.82) as well as incorporates prior information by an appropriate selection of π. Thus, if we know the exact size of the affected subset, say $\mathscr{P} = \mathscr{P}_m$, we may set $\pi = m/N$; if we know that at most m streams may be affected, i.e, $\mathscr{P} = \overline{\mathscr{P}}_m$, then we may set $\pi = m/(2N)$.

1.4.3 Examples

In this section, we consider four particular examples to which the previous results apply. The first three examples deal with independent streams in the context of Section 1.4.2. The last example assumes dependent streams.

Example 1.1 (*A linear Gaussian state-space model*)**.** Assume that the data in the N streams $X(1),\dots,X(N)$ are independent, and the ℓ-dimensional observed vector in the ith stream at time n, $X_n(i) = (X_n^1(i),\dots,X_n^\ell(i))^\top$, is governed by the following hidden Markov model under P_j^i ($j=0,1$):

$$X_n(i) = H_i\,\theta_n(i) + V_n(i) + j\,b_x(i),$$
$$\theta_n(i) = F_i\,\theta_{n-1}(i) + W_{n-1}(i) + j\,b_\theta(i), \quad \theta_0(i) = 0,$$

where $\theta_n(i) = (\theta_n^1(i),\dots,\theta_n^m(i))^\top$ is an unobserved m-dimensional Markov vector and where $W_n(i)$ and $V_n(i)$ are zero-mean Gaussian i.i.d. vectors having covariance matrices $K_W(i)$ and $K_V(i)$, respectively; $b_\theta(i) = (b_\theta^1(i),\dots,b_\theta^m(i))^\top$ and $b_x(i) = (b_x^1(i),\dots,b_x^\ell(i))^\top$ are the mean values; F_i is the $(m\times m)$ state transition matrix; H_i is the $(\ell\times m)$ matrix, and $j=0$ if the mean values in the ith stream are not affected and $j=1$ otherwise. The components $X_n(i)$ and $\theta_n(i)$ are assumed to be independent for different streams i.

It can be shown that under the null hypothesis H_0^i the observed sequence $\{X_n(i)\}$ has an equivalent representation

$$X_n(i) = H_i\widehat{\theta}_n(i) + \xi_n(i), \quad n\in\mathbb{N}$$

with respect to the "innovation" sequence $\xi_n(i) = X_n(i) - H_i\,\widehat{\theta}_n(i)$, where $\xi_n(i)\sim\mathcal{N}(0,\Sigma_n(i))$, $n=1,2,\dots$ are independent Gaussian vectors and $\widehat{\theta}_n(i) = \mathsf{E}_0^i[\theta_n(i)\,|\,X_1(i),\dots,X_{n-1}(i)]$ is the optimal one-step ahead predictor in the mean-square sense, i.e., the estimate of $\theta_n(i)$ based on observing $X_1(i),\dots,X_{n-1}(i)$, which can be obtained by the Kalman filter (cf., e.g., [5]). Under H_1^i the observed sequence $\{X_n(i)\}$ admits the following representation

$$X_n(i) = \Upsilon_n(i) + H_i\,\widehat{\theta}_n(i) + \xi_n(i), \quad n\in\mathbb{N},$$

where $\Upsilon_n(i)$ can be computed using relations given in [11, pp. 282-283]. Consequently, the local LLR $\lambda_i(n)$ can be written as

$$\lambda_i(n) = \sum_{t=1}^{n}\Upsilon_t^\top(i)\Sigma_t^{-1}(i)\,\xi_t(i) - \frac{1}{2}\sum_{t=1}^{n}\Upsilon_t^\top(i)\Sigma_t^{-1}(i)\Upsilon_t(i),$$

where $\Sigma_n(i)$, $n\in\mathbb{N}$ are given by Kalman's equations (see [11, Eq. (3.2.20)]). Thus, each local LLR $\lambda_i(n)$ has independent Gaussian increments, and it can be shown that the normalized LLR $n^{-1}\lambda_i(n)$ converges almost surely as $n\to\infty$ to I_i under P_1^i and $-I_i$ under P_0^i, where

$$I_i = \frac{1}{2}\lim_{n\to\infty}\frac{1}{n}\sum_{t=1}^{n}\Upsilon_t^\top(i)\Sigma_t^{-1}(i)\Upsilon_t(i).$$

By Corollary 1.2(i), the GSLRT and the MSLRT are asymptotically optimal with respect to all moments of the sample size.

Example 1.2 (*An autoregression model with unknown correlation coefficient*)**.** Assume that the data in the N streams $X(1),\dots,X(N)$ are independent, and each $X(i)$ is a Markov Gaussian (AR(1)) process of the form

$$X_n(i) = \rho^i X_{n-1}(i) + \xi_n(i), \quad n\in\mathbb{N}, \quad X_0(i) = x, \tag{1.83}$$

where $\{\xi_n(i)\}_{n\in\mathbb{N}}$, $i=1,\dots,N$, are mutually independent sequences of i.i.d. normal random variables with zero mean and unit variance and x is either a fixed number or a random variable. Suppose that $\rho^i = \rho^i_j$ under H^i_j, $j=0,1$, where ρ^i_j are known constants. Then, the transition densities are $f^i_j(X_n(i)|X_{n-1}(i)) = \varphi(X_n(i) - \rho^i_j X_{n-1}(i))$, $j=0,1$, where φ is the density of the standard normal distribution, and the LLR in the ith stream can be written as

$$\lambda_i(n) = \sum_{t=1}^{n} \psi_i\left(X_t(i), X_{t-1}(i)\right), \tag{1.84}$$

where

$$\psi_i(y,x) = \log\left[\frac{\varphi(y-\rho^i_1 x)}{\varphi(y-\rho^i_0 x)}\right] = (\rho^i_1 - \rho^i_0)x\left[y - \frac{\rho^i_1+\rho^i_0}{2}x\right]. \tag{1.85}$$

In order to show that $\{n^{-1}\lambda_i(n)\}$ converges asymptotically as $n\to\infty$, further assume that $|\rho^i_j|<1$, $1\le i\le N$, $j=0,1$, so that $X(i)$ is stable. Let π^i_j be the stationary distribution of $X(i)$ under H^i_j, which coincides with the distribution of

$$w^i_j = \sum_{n=1}^{\infty} (\rho^i_j)^{n-1}\,\xi_n(i), \quad j=0,1. \tag{1.86}$$

Since $\mathsf{E}^i_j|\lambda_i(1)|^r<\infty$ for any $r\ge 1$, by Lemma D.1 (Appendix D), the normalized LLR process $\{n^{-1}\lambda_i(n)\}$ converges under P^i_1 as $n\to\infty$ r-completely for every $r\ge 1$ to

$$I^i_1 = \int_{-\infty}^{\infty}\left(\int_{-\infty}^{\infty} \psi_i(y,x)\,\varphi(y-\rho^i_1 x)\mathrm{d}y\right)\mathrm{dQ}_i(x),$$

where Q_i is the stationary distribution of the process $\{X_n(i)\}$ when the signal is present in the ith stream. In the Gaussian case considered, Q_i is $\mathcal{N}(0,(1-(\rho^i_1)^2)^{-1})$, so I^i_1 can be calculated explicitly as

$$I^i_1 = \frac{(\rho^i_1-\rho^i_0)^2}{2[1-(\rho^i_1)^2]}.$$

By symmetry, under P^i_0 the normalized LLR $\{n^{-1}\lambda_i(n)\}$ converges r-completely for all $r\ge 1$ to $-I^i_0$ with $I^i_0 = (\rho^i_1-\rho^i_0)^2/2[1-(\rho^i_0)^2]$.

By Corollary 1.2(ii), both the GSLRT and the MSLRT are asymptotically optimal, minimizing all moments of the stopping time distribution as the error probabilities go to 0.

Example 1.3 (*Multistream invariant sequential t-tests*). Suppose that the data in the N streams $X(1),\dots,X(N)$ are independent, and each $X(i)$ is given by

$$X_n(i) = \begin{cases} \mu_i + \xi_n(i) & \text{under } \mathsf{H}^i_1,\ 1\le i\le N \\ \xi_n(i) & \text{under } \mathsf{H}^i_0 \end{cases}, \quad n\in\mathbb{N}, \tag{1.87}$$

where $\mu_i>0$ and $\xi_n(i)\sim\mathcal{N}(0,\sigma_i^2)$, $n\in\mathbb{N}$ are zero-mean normal i.i.d. (mutually independent) sequences (noises) with unknown variances σ_i^2. In other words, under the local null hypothesis in the ith stream, H^i_0, there is no signal in the ith stream, i.e., $\mu_i=0$, and under the local alternative hypothesis in the ith stream, there is a signal $\mu_i>0$ in the ith stream. Therefore, the hypotheses H^i_0, H^i_1 are not simple and our results cannot be directly applied. Nevertheless, if we assume that the value of the "signal-to-noise" ratio $Q_i=\mu_i/\sigma_i$ is known, we can transform this into a testing problem of simple hypotheses in the streams by using the principle of invariance, since the problem is invariant under the group of scale changes. Indeed, the maximal invariant statistic in the ith stream is

$$\mathbf{Y}_n(i) = \left(1, \frac{X_2(i)}{X_1(i)}, \dots, \frac{X_n(i)}{X_1(i)}\right),$$

and it can be shown [164, Sec 3.6.2] that the invariant LLR, which is built based on the maximal invariant $\mathbf{Y}_n(i)$, is given by $\lambda_i(n) = \log[J_n(Q_i T_n(i))/J_n(0)]$, where

$$T_n(i) = \frac{n^{-1}\sum_{t=1}^n X_t(i)}{\sqrt{n^{-1}\sum_{t=1}^n (X_t(i))^2}} \tag{1.88}$$

and

$$J_n(z) = \int_0^\infty \frac{1}{u} \exp\left\{\left[-\frac{1}{2}u^2 + zu + \log u\right] n\right\} du.$$

Note that $T_n(i)$ is the Student t-statistic, which is the basis for the Student's t-test in the fixed sample size setting. For this reason, we refer to the sequential tests (1.11) and (1.12) that are based on the invariant LLRs as t-tests, specifically as the t-GSLRT and the t-MSLRT, respectively. Although the invariant LLR $\lambda_i(n)$ is difficult to calculate explicitly, it can be approximated by $\psi_i(T_n(i))\, n$, using a uniform version of the Laplace asymptotic integration technique, where

$$\psi_i(x) = \frac{1}{4}x\left(x + \sqrt{4+x^2}\right) + \log\left(x + \sqrt{4+x^2}\right) - \log 2 - \frac{1}{2}Q_i^2, \quad x \in \mathbb{R}.$$

Indeed, as shown in [164, Sec 3.6.2], there is a finite positive constant C such that for all $n \ge 1$

$$\left|n^{-1}\lambda_i(n) - \psi_i(T_n(i))\right| \le C/n. \tag{1.89}$$

It follows from (1.89) that if under P_j^i the t-statistic $T_n(i)$ converges r-completely to a constant V_j^i as $n \to \infty$, then the normalized LLR $n^{-1}\lambda_i(n)$ converges in a similar sense to $\psi_i(V_j^i)$, $j = 0, 1$. Therefore, it suffices to study the limiting behavior of $T_n(i)$. Since $\mathsf{E}_j^i[|X_1(i)|^r] < \infty$, $j = 0, 1$ (for all $r \ge 1$), we obtain that for every $r \ge 1$

$$T_n(i) \xrightarrow[n\to\infty]{\mathsf{P}_1^i - r-\text{completely}} \frac{\mathsf{E}_1^i[X_1(i)]}{\sqrt{\mathsf{E}_1^i[(X_1(i))^2]}} = \frac{Q_i}{\sqrt{1+Q_i^2}}, \qquad T_n(i) \xrightarrow[n\to\infty]{\mathsf{P}_0^i - r-\text{completely}} 0,$$

which implies that the r-complete convergence conditions (1.64) for the normalized LLR $\{n^{-1}\lambda_i(n)\}$ hold for all $r \ge 1$ with

$$I_1^i = \psi_i\left(\frac{Q_i}{\sqrt{1+Q_i^2}}\right) \quad \text{and} \quad I_0^i = \frac{1}{2}Q_i^2.$$

It is easy to verify that $I_1^i > 0$ and $I_0^i > 0$.

By Corollary 1.2(ii), the invariant t-GSLRT and t-MSLRT asymptotically minimize all moments of the stopping time distribution.

Example 1.4 (*Detection of deterministic signals in correlated Gaussian streams*)**.** We now give an example in which the data in streams are dependent. Assume that $\mathbf{X}_n = (X_n(1), \dots, X_n(N))$, $n \in \mathbb{N}$, are independent Gaussian random vectors with (invertible) covariance matrix Σ_n and mean vector $(\theta_1 \mu_1(n), \dots, \theta_N \mu_N(n))$, where $\mu_1(n), \dots, \mu_N(n)$ can be interpreted as deterministic signals observed in Gaussian noise. Then $\lambda_{\mathscr{B}}(n) = \sum_{t=1}^n \Delta\lambda_{\mathscr{B}}(t)$, where

$$\Delta\lambda_{\mathscr{B}}(t) = [\Sigma_t^{-1} \cdot \mu_{\mathscr{B}}(t)] \cdot \mathbf{X}_t - \frac{1}{2}[\Sigma_t^{-1} \cdot \mu_{\mathscr{B}}(t)] \cdot \mu_{\mathscr{B}}(t),$$

and $\mu_{\mathscr{B}}(t)$ is a N-dimensional vector whose ith component is 0 if $i \notin \mathscr{B}$ and $\mu_i(t)$ if $i \in \mathscr{B}$. For every $n \in \mathbb{N}$,

$$\mathsf{E}_{\mathscr{B}}[\lambda_{\mathscr{B}}(n)] = -\mathsf{E}_0[\lambda_{\mathscr{B}}(n)] = \frac{1}{2}\sum_{t=1}^n [\Sigma_t^{-1} \cdot \mu_{\mathscr{B}}(t)] \cdot \mu_{\mathscr{B}}(t).$$

It is easily shown that if for every $\mathscr{B} \in \mathscr{P}$ there is a positive number $I_{\mathscr{B}} > 0$ such that

$$\frac{1}{2}\sum_{t=1}^{n}[\Sigma_t^{-1} \cdot \mu_{\mathscr{B}}(t)] \cdot \mu_{\mathscr{B}}(t) \to I_{\mathscr{B}} \quad \text{as } n \to \infty,$$

then $n^{-1}\lambda_{\mathscr{B}}(n) \to I_{\mathscr{B}}$ $\mathsf{P}_{\mathscr{B}}$-almost surely and $n^{-1}\lambda_{\mathscr{B}}(n) \to -I_{\mathscr{B}}$ P_0-almost surely.

Hence, by Theorem 1.4(i), the proposed sequential tests minimize asymptotically all moments of the sample size.

Let us now specialize the above setup in the case of two streams with the cross stream correlation coefficient ρ_n. Additional notation:

$$\overline{X}_n(i) = X_n(i)/\sigma_n(i), \quad \theta_n(i) = \mu_i(n)/\sigma_n(i), \quad \sigma_n(i) = \sqrt{\mathsf{Var}[X_n(i)]}.$$

We have

$$\Delta\lambda_{\mathscr{B}}(n) = \begin{cases} [\theta_n(1) - \rho_n\,\theta_n(2)]\overline{X}_n(1) + [\theta_n(2) - \rho_n\,\theta_n(1)]\overline{X}_n(2) & \\ \qquad -\frac{1}{2}[(\theta_n(1))^2 + (\theta_n(2))^2 - 2\rho_n\theta_n(1)\,\theta_n(2)] & \text{if } \mathscr{B} = \{1,2\} \\ -\rho_n\theta_n(2)\overline{X}_n(1) + \theta_n(2)\overline{X}_n(2) - \frac{1}{2}(\theta_n(2))^2 & \text{if } \mathscr{B} = \{2\} \\ -\rho_n\,\theta_n(1)\overline{X}_n(2) + \theta_n(1)\overline{X}_n(1) - \frac{1}{2}(\theta_n(1))^2 & \text{if } \mathscr{B} = \{1\} \end{cases},$$

and consequently,

$$\mathsf{E}_{\mathscr{B}}[\Delta\lambda_{\mathscr{B}}(n)] = -\mathsf{E}_0[\Delta\lambda_{\mathscr{B}}(n)] = \begin{cases} \frac{1}{2}\left[(\theta_n(1))^2 + (\theta_n(2))^2 - 2\rho_n\theta_n(1)\theta_n(2)\right] & \text{if } \mathscr{B} = \{1,2\} \\ \frac{1}{2}(\theta_n(2))^2 & \text{if } \mathscr{B} = \{2\} \\ \frac{1}{2}(\theta_n(1))^2 & \text{if } \mathscr{B} = \{1\} \end{cases}.$$

We conclude that if

$$\lim_{n\to\infty}\frac{1}{n}\sum_{t=1}^{n}(\theta_t(i))^2 = Q_i, \quad i = 1,2$$

and

$$\lim_{n\to\infty}\frac{1}{n}\sum_{t=1}^{n}\rho_t\,\theta_t(1)\theta_t(2) = Q_{12},$$

then $n^{-1}\lambda_{\mathscr{B}}(n) \to I_{\mathscr{B}}$ $\mathsf{P}_1^{\mathscr{B}}$-almost surely and $n^{-1}\lambda_{\mathscr{B}}(n) \to -I_{\mathscr{B}}$ $\mathsf{P}_0^{\mathscr{B}}$-almost surely as $n \to \infty$, where

$$I_{\mathscr{B}} = \begin{cases} \frac{1}{2}[Q_1 + Q_2 - 2Q_{12}] & \text{if } \mathscr{B} = \{1,2\} \\ \frac{1}{2}Q_i & \text{if } \mathscr{B} = \{i\},\ i = 1,2 \end{cases}.$$

1.4.4 Monte Carlo Simulations

We now present the results of a simulation study whose goal is to compare the performance of the GSLRT and the MSLRT under different signal configurations and to quantify the effect of prior information on the detection performance. Simulations are performed for the autoregressive model of Example 1.2 with independent streams. That is, the process in each stream is given by (1.83) and the corresponding marginal LLR by (1.84). In simulations, we set $\rho_0^i = 0$ and $\rho_1^i = \rho = 0.5$, in which case the corresponding Kullback–Leibler numbers take the form $I_1^i \equiv I_1 = (1/2)\rho^2/(1-\rho^2)$ and $I_0^i \equiv I_0 = (1/2)\rho^2$. We compare the GSLRT and the MSLRT for two different scenarios regarding the available prior information – when the size of the affected subset is known, i.e., $\mathscr{P} = \mathscr{P}_{|\mathscr{B}|}$ where $|\mathscr{B}|$ is the cardinality of the true affected subset $\mathscr{B}$, and when there is complete uncertainty regarding the affected subset, i.e., $\mathscr{P} = \overline{\mathscr{P}}_N$.

Given the class $\mathscr{P}$ that reflects the available information, the next step is to select the thresholds b and a for the two tests. These will be determined by α and β, the desired type-I and type-II error probabilities. To this end, (1.40) can be used to determine the relationship of the two thresholds given the desired ratio α/β. For simplicity, we set $\alpha = \beta$. Therefore, (1.40) suggests selecting the thresholds as $a = b + \log|\mathscr{P}|$, since for the design of both tests we use uniform weights, i.e., $p_{0,\mathscr{B}} = p_{1,\mathscr{B}} = 1/|\mathscr{P}|$ for every $\mathscr{B} \in \mathscr{P}$. For the computation of the operating characteristics of the GSLRT (resp. MSLRT), we compute the expected sample sizes under H_0 and $\mathsf{H}_{\mathscr{B}}$ for every $\mathscr{B} \in \mathscr{P}$ using plain Monte Carlo simulation, as well as the error probabilities, using the importance sampling algorithms of Subsection 1.4.1.3, for different values of a (resp. b).

From Lemma 1.2 it follows that the error probabilities of the GSLRT are upper bounded by $e^{-b} = |\mathscr{P}|e^{-a}$. In Tables 1.1 and 1.2 we present the operating characteristics of the GSLRT when $b = 8$ and $a = 8 + \log|\mathscr{P}|$, where $\mathscr{P}$ is the class of possibly affected subsets, in which case error probabilities of both tests are bounded by $\exp(-8) = 3.35 \cdot 10^{-4}$. From these tables we can see that the upper bound of Lemma 1.2 for the false alarm probability of the GSLRT is much more conservative than the corresponding upper bound for the false alarm probability of the MSLRT, which is expected.

TABLE 1.1
Error probabilities and the expected sample size under the null hypothesis of the GSLRT and the MSLRT with thresholds $b = 8$ and $a = b + \log|\mathscr{P}|$.

	$\mathsf{P}_0(d=1)$		$\mathsf{E}_0[T]$		$\mathsf{P}_1(d=0)$	
$\mathscr{P}$	GSLRT	MSLRT	GSLRT	MSLRT	GSLRT	MSLRT
$\overline{\mathscr{P}}_N$	$2.39 \cdot 10^{-5}$	$9.3 \cdot 10^{-5}$	135.8	140.0	$3.7 \cdot 10^{-5}$	$2.54 \cdot 10^{-5}$
$\mathscr{P}_1$	$2.00 \cdot 10^{-4}$	$2.00 \cdot 10^{-4}$	134.0	113.0	$2.15 \cdot 10^{-5}$	$2.42 \cdot 10^{-4}$
$\mathscr{P}_3$	$1.17 \cdot 10^{-4}$	$1.71 \cdot 10^{-4}$	52.4	38.9	$6.08 \cdot 10^{-6}$	$2.3 \cdot 10^{-4}$
$\mathscr{P}_6$	$4.34 \cdot 10^{-5}$	$1.45 \cdot 10^{-4}$	23.3	17.00	$5.1 \cdot 10^{-6}$	$1.8 \cdot 10^{-4}$
$\mathscr{P}_9$	$5.46 \cdot 10^{-5}$	$1.2 \cdot 10^{-4}$	10.75	9.17	$3.21 \cdot 10^{-5}$	$1.42 \cdot 10^{-4}$

TABLE 1.2
The expected sample size of tests for various scenarios under the alternative hypothesis. Same setup as in Table 1.1.

	$\mathsf{E}_{\mathscr{B}}[T]$			
$\vert\mathscr{B}\vert$	GSLRT		MSLRT	
	$\overline{\mathscr{P}}_N$	$\mathscr{P}_{\vert\mathscr{B}\vert}$	$\overline{\mathscr{P}}_N$	$\mathscr{P}_{\vert\mathscr{B}\vert}$
1	96.8 (0.6)	69.6 (0.5)	94.8 (0.6)	71.2 (0.5)
3	32.4 (0.2)	29.1 (0.2)	29.7 (0.2)	28.0 (0.2)
6	16.57 (0.09)	15.39 (0.09)	14.26 (0.08)	13.97 (0.09)
9	11.57 (0.06)	9.28 (0.09)	9.37 (0.05)	8.67 (0.05)

We illustrate our findings graphically in Figures 1.1, 1.2, 1.3. In all these graphs, the dashed lines correspond to the case that the size of the affected subset is known, i.e., $\mathscr{P} = \mathscr{P}_{|\mathscr{B}|}$, whereas the solid lines correspond to the case of no prior information, i.e., $\mathscr{P} = \overline{\mathscr{P}}_N$. Moreover, in all these graphs, the dark lines correspond to MSLRT, whereas the grey lines correspond to GSLRT.

In Figure 1.1, for both the GSLRT and the MSLRT we plot the expected sample size of the test against the logarithm of the type-I error probability for different cases regarding the size of the affected subset. Specifically, if $\mathscr{B}$ is the affected subset, we plot $\mathsf{E}_{\mathscr{B}}[\widehat{T}]$ against $|\log_{10} \mathsf{P}_0(\widehat{d} = 1)|$ and $\mathsf{E}_{\mathscr{B}}[\widetilde{T}]$ against $|\log_{10} \mathsf{P}_0(\widetilde{d} = 1)|$ for the following cases: $|\mathscr{B}| = 1, 3, 6, 9$. We observe that when we know the size of the affected subset, then the performance of the two tests is essentially the same.

However, when we have no prior information regarding the signal, the GSLRT performs slightly better (resp. worse) than the MSLRT when the signal is present in a small (resp. large) number of streams, for large and moderate error probabilities. Note, however, that when the number of affected streams is large, the signal-to-noise ratio is high, which means that the "absolute" loss of the GSLRT in these cases is small. Nevertheless, for small error probabilities, we see that the performance of the two tests is essentially the same.

In Figure 1.2 we plot the *normalized* expected sample size of each test under the alternative hypothesis against the logarithm of the type-I error probability for different cases regarding the size of the affected subset. Specifically, we plot $|\mathscr{B}| I_1 \mathsf{E}_{\mathscr{B}}[\widehat{T}]/|\log \mathsf{P}_0(\widehat{d}=1)|$ against $|\log_{10} \mathsf{P}_0(\widehat{d}=1)|$ and $|\mathscr{B}| I_1 \mathsf{E}_{\mathscr{B}}[\widetilde{T}]/|\log \mathsf{P}_0(\widetilde{d}=1)|$ against $|\log_{10} \mathsf{P}_0(\widetilde{d}=1)|$ when $|\mathscr{B}| = 1,3,6,9$. Again, the dashed lines correspond to the versions of the two schemes when the size of the affected subset is known ($\mathscr{P}_{|\mathscr{B}|}$). The solid lines correspond to the versions of the two schemes with no prior information ($\widetilde{\mathscr{P}}_N$). The dark lines correspond to the MSLRT, whereas the gray lines correspond to the GSLRT. Our asymptotic theory suggests that the curves in Figure 1.2 converge to 1 as the error probabilities go to 0. Our numerical results suggest that this convergence is relatively slow in most cases. Note, however, that we do not normalize the expected sample sizes by the optimal performance, but with an asymptotic lower bound on it, which explains to some extent why these normalized expected sample sizes are far from 1 even for very small error probabilities.

Similarly, in Figure 1.3 we plot the expected sample size, as well as its *normalized* version, against the logarithm of the maximal missed detection (type-II error) probability for the GSLRT and the MSLRT with respect to class $\mathscr{P}$. Specifically, on the left-hand side we plot $\mathsf{E}_0[\widehat{T}]$ against $|\log_{10} \mathsf{P}_1(\widehat{d}=0)|$ and $\mathsf{E}_0[\widetilde{T}]$ against $|\log_{10} \mathsf{P}_1(\widetilde{d}=0)|$, whereas on the right-hand side we plot $I_0(\mathscr{P})\, \mathsf{E}_0[\widehat{T}]/|\log \mathsf{P}_1(\widehat{d}=0)|$ against $|\log_{10} \mathsf{P}_1(\widehat{d}=0)|$ and $I_0(\mathscr{P})\, \mathsf{E}_0[\widetilde{T}]/|\log \mathsf{P}_1(\widetilde{d}=0)|$ against $|\log_{10} \mathsf{P}_1(\widetilde{d}=0)|$. Our asymptotic theory suggests that the curves on the right-hand side of Figure 1.3 should converge to 1 as the error probabilities go to 0. Again, our numerical results show that this convergence is relatively slow, which again can be explained by the fact that we do not normalize the expected sample sizes by the optimal performance, but with an asymptotic lower bound to it.

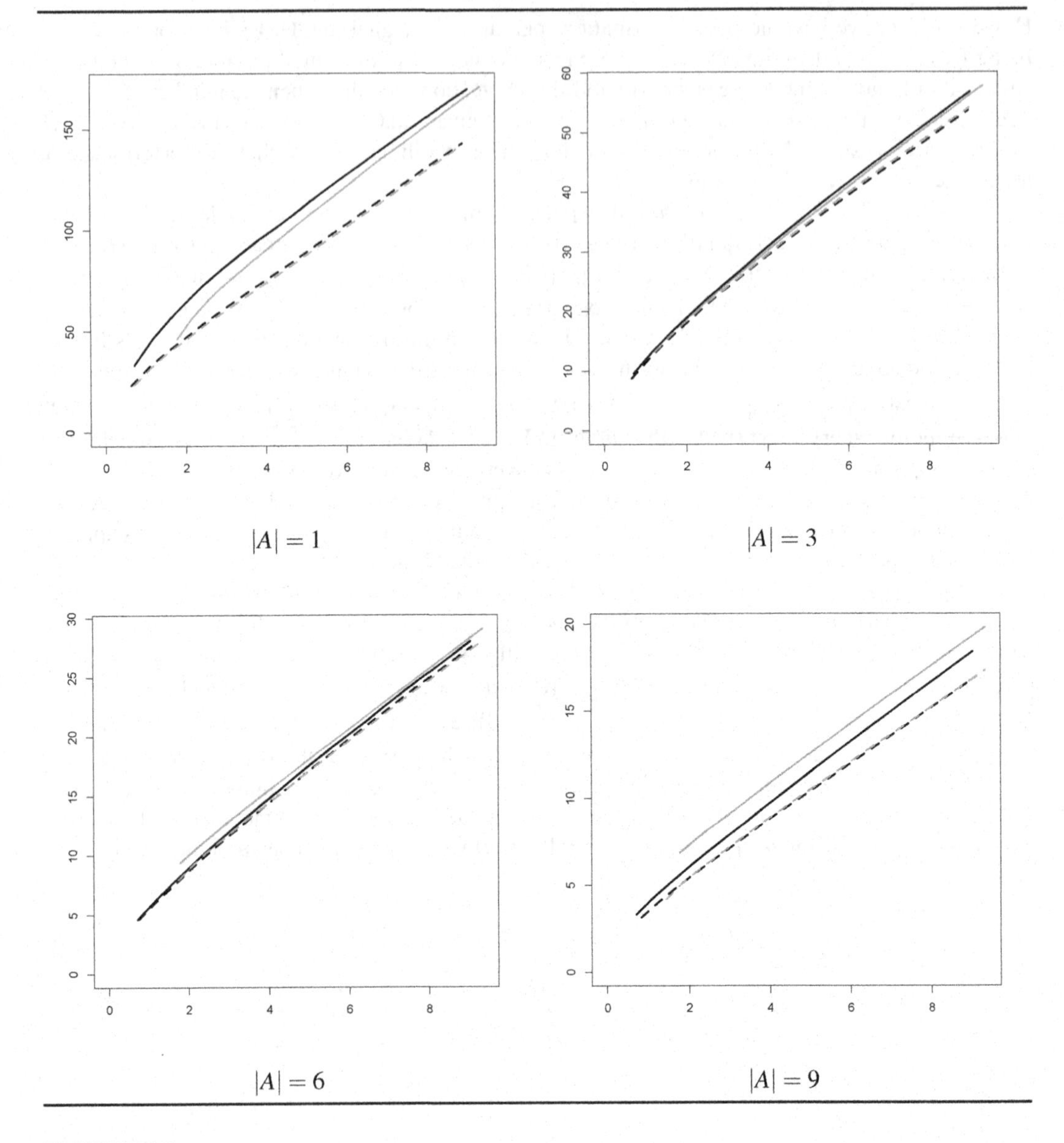

FIGURE 1.1
Expected sample size against the false alarm (type-I) error probability in log-scale for GSLRT and MSLRT with respect to class $\mathscr{P}$. That is, if δ is the sequential test of interest and $\mathscr{B}$ the affected subset, we plot $\mathsf{E}_{\mathscr{B}}[T]$ against $|\log \mathsf{P}_0(d=1)|$ for the following cases: $|\mathscr{B}| = 1, 3, 6, 9$. Light lines refer to to the GSLRT and dark lines to MSLRT. Solid lines refer to the case of no prior information ($\mathscr{P} = \overline{\mathscr{P}}_N$), whereas dashed lines refer to the case that the size of the affected subset is known in advance ($\mathscr{P} = \mathscr{P}_{|\mathscr{B}|}$).

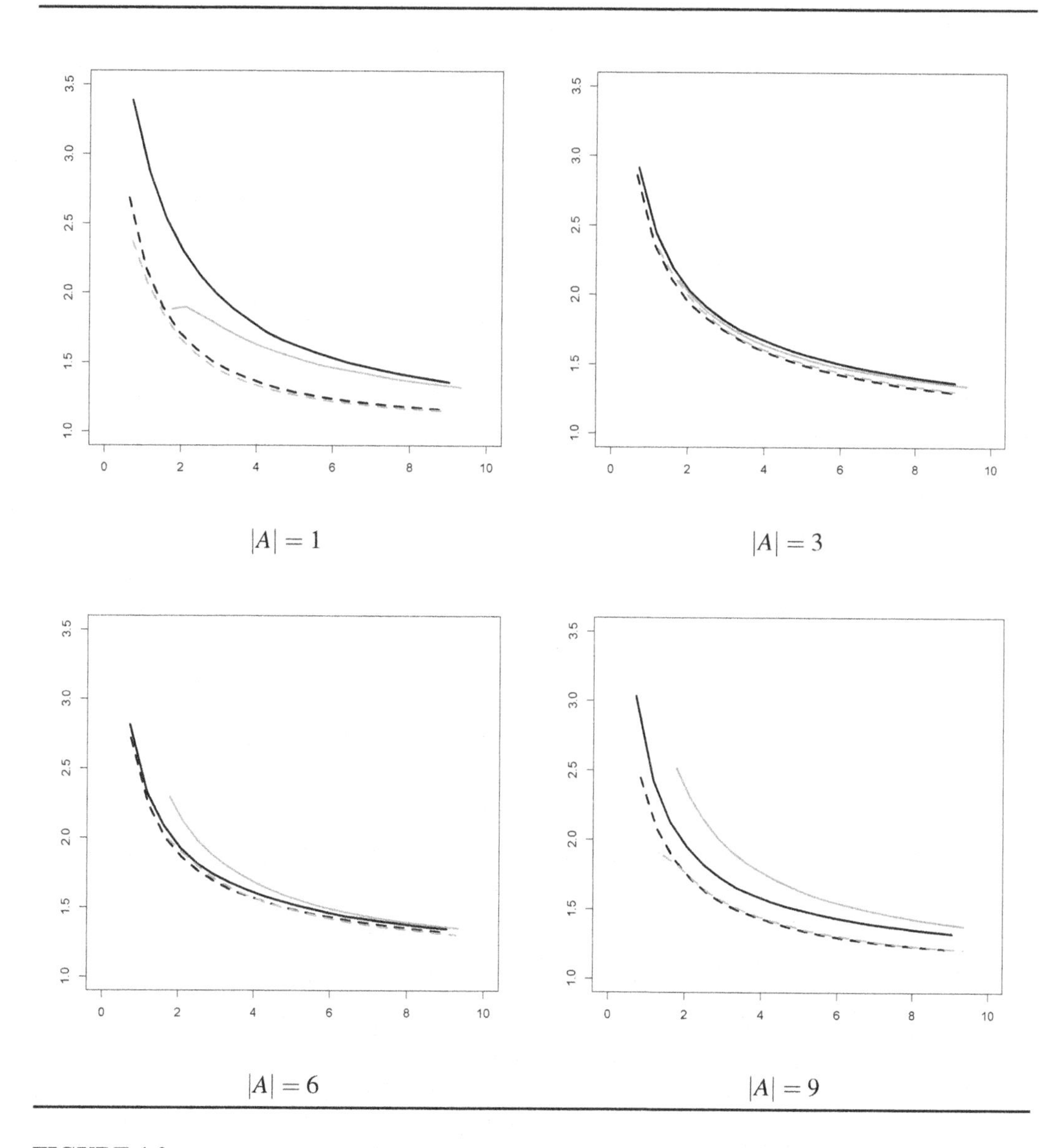

FIGURE 1.2

Normalized expected sample size under $\mathsf{H}_{\mathscr{B}}$ against the false alarm probability in log-scale for the GSLRT and the MSLRT. That is, denoting by $\delta = (T,d)$ the sequential test of interest and by $\mathscr{B}$ the affected subset, we plot $|\mathscr{B}| I_1 \mathsf{E}_{\mathscr{B}}[T] / |\log \mathsf{P}_0(d=1)|$ against $|\log_1 0\mathsf{P}_0(d=1)|$ for the following cases: $|\mathscr{B}| = 1,3,6,9$. Light lines refer to to the GSLRT and dark lines to MSLRT. For both tests, solid lines refer to the case of no prior information ($\overline{\mathscr{P}}_N$), whereas dashed lines refer to the case that the size of the affected subset is known in advance ($\mathscr{P}_{|\mathscr{B}|}$).

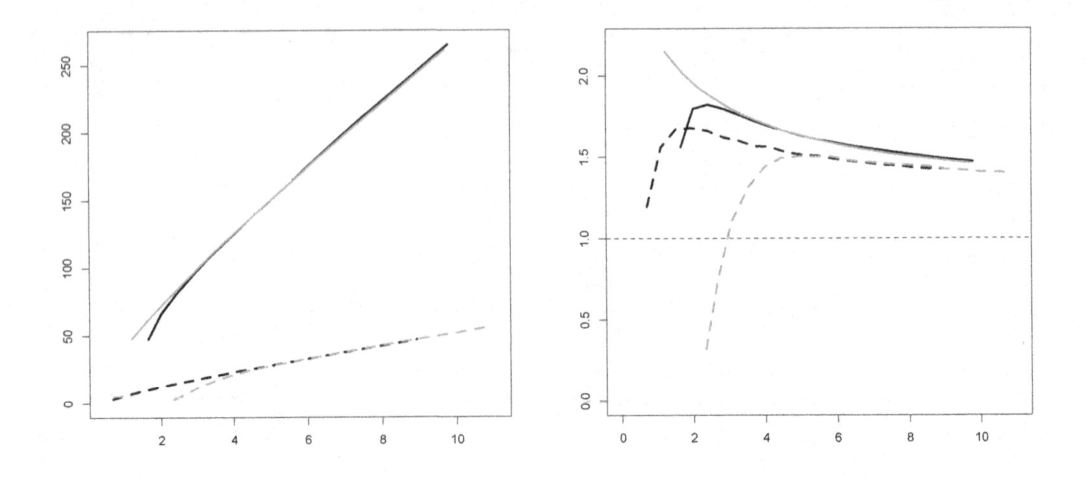

FIGURE 1.3
On the left (right)-hand side, we plot the (normalized) expected sample size under H_0 against the maximal missed detection probability in log-scale for the GSLRT and the MSLRT with respect to class $\mathscr{P}$. That is, denoting δ the sequential test of interest, on the left-hand side we plot $\mathsf{E}_0[T]$ against $|\log_{10} \mathsf{P}_1(d=0)|$ and on the right-hand side we plot $I_0(\mathscr{P})\, \mathsf{E}_0[T]/|\log \mathsf{P}_1(d=0)|$ against $|\log_{10} \mathsf{P}_1(d=0)|$. Solid lines refer to the case of no prior information ($\mathscr{P} = \overline{\mathscr{P}}_N$), where $I_0(\mathscr{P}) = I_0$, whereas dashed lines refer to the case that the size of the affected subset is known in advance ($\mathscr{P} = \mathscr{P}_{|\mathscr{B}|}$), in particular $|\mathscr{B}| = 5$, where $I_0(\mathscr{P}) = |\mathscr{B}| I_0$.

1.5 Higher Order Approximations for the Expected Sample Size and Optimality in the i.i.d. Case

The formulas for the expected sample size and the probabilities of errors can be substantially refined when the observations are i.i.d., or more generally, when the LLR processes can be well approximated by random walks. The goal of this section is to improve first-order approximations (1.90) and the approximations for the probabilities of errors for an arbitrarily choice of weights $\mathbf{p}_0$ and $\mathbf{p}_1$ as well as to show that a particular choice of weights leads to a third order optimality property with respect to the expected sample size. It is worth mentioning again that first-order approximations and optimality do not depend on the choice of weights.

1.5.1 Preliminaries

We now proceed with the hypothesis testing problem (1.68) where not only data across streams are independent but also the data in streams $X_n(i)$, $n = 1, 2, \dots$ $(i = 1, \dots, N)$ are i.i.d. with densities $f_0^i(x)$ and $f_1^i(x)$ when there is no signal and there is a signal in the ith stream. In other words, we go on to tackle the hypothesis testing problem $\mathsf{H}_0 = \cap_{i=1}^N \mathsf{H}_0^i$ versus $\mathsf{H}_1 = \underset{\mathscr{B} \in \mathscr{P}}{\cup} \mathsf{H}^{\mathscr{B}}$, where $\mathsf{H}_0^i : f_i = f_0^i$

and

$$\mathsf{H}^{\mathscr{B}}: f_i = \begin{cases} f_0^i & \text{if } i \notin \mathscr{B} \\ f_1^i, & \text{if } i \in \mathscr{B} \end{cases}.$$

In this case, the LLRs are

$$\lambda_i(n) = \sum_{t=1}^{n} \Delta\lambda_i(t), \quad \Delta\lambda_i(n) = \log \frac{f_1^i(X_n(i))}{f_0^i(X_n(i))}, \quad i = 1,\dots,N,$$

and the numbers $I_0^i = -\mathsf{E}_0^i[\lambda_i(1)]$ and $I_1^i = \mathsf{E}_1^i[\lambda_i(1)]$ are the Kullback–Leibler (KL) information numbers. By Corollary 1.2(i), the GSLRT and the MSLRT minimize asymptotically all moments of the stopping time when the KL numbers are positive and finite and first-order asymptotic approximations (1.75) and (1.76) hold with $I_0^{\mathscr{B}} = \sum_{i\in\mathscr{B}} I_0^i$, $I_1^{\mathscr{B}} = \sum_{i\in\mathscr{B}} I_1^i$, and $I_0(\mathscr{P}) = \min_{\mathscr{B}\in\mathscr{P}} I_0^{\mathscr{B}}$. Also, using Proposition 1.1(i), we obtain the following first-order approximations for the expected sample sizes of the tests for large threshold values:

$$\begin{aligned} \mathsf{E}_0[\widehat{T}] &\sim \frac{a}{I_0(\mathscr{P})} \sim \mathsf{E}_0[\widetilde{T}], \\ \mathsf{E}_{\mathscr{B}}[\widehat{T}] &\sim \frac{b}{I_1^{\mathscr{B}}} \sim \mathsf{E}_{\mathscr{B}}[\widetilde{T}] \quad \text{for all } \mathscr{B} \in \mathscr{P} \end{aligned} \tag{1.90}$$

as $a_{\min} \to \infty$.

Introduce the $\mathscr{L}$-number

$$\mathscr{L}_{\mathscr{B}} = \exp\left\{ -\sum_{n=1}^{\infty} \frac{1}{n} \left[\mathsf{P}_0(\lambda_{\mathscr{B}}(n) > 0) + \mathsf{P}_{\mathscr{B}}(\lambda_{\mathscr{B}}(n) \le 0) \right] \right\}. \tag{1.91}$$

Throughout the rest of this section without further special mentioning we assume that the LLR $\lambda_{\mathscr{B}}(1)$ is non-arithmetic under P_0 and $\mathsf{P}_{\mathscr{B}}$ (see Definition E.1), which is the case if all local LLRs $\lambda_i(1)$, $i = 1,\dots,N$, are non-arithmetic under P_0 and P_1^i. Then, if we define the first hitting times

$$\tau_c^{\mathscr{B}} = \inf\{n \ge 1 : \lambda_{\mathscr{B}}(n) \ge c\}, \quad \sigma_c^{\mathscr{B}} = \inf\{n \ge 1 : \lambda_{\mathscr{B}}(n) \le -c\}, \quad c > 0,$$

the overshoots $\kappa_{\mathscr{B}}^1(c) = \lambda_{\mathscr{B}}(\tau_c^{\mathscr{B}}) - c$ and $\kappa_{\mathscr{B}}^0(c) = |\lambda_{\mathscr{B}}(\sigma_c^{\mathscr{B}}) + c|$ have well-defined asymptotic distributions under $\mathsf{P}_{\mathscr{B}}$ and P_0, respectively,

$$\mathscr{H}_{\mathscr{B}}(x) = \lim_{c\to\infty} \mathsf{P}^{\mathscr{B}}(\kappa_{\mathscr{B}}^1(c) \le x), \quad \mathscr{H}_0^{\mathscr{B}}(x) = \lim_{c\to\infty} \mathsf{P}_0(\kappa_{\mathscr{B}}^0(c) \le x), \quad x > 0,$$

and, consequently, we can define the following Laplace transforms

$$\gamma_{\mathscr{B}} = \int_0^{\infty} e^{-x}\, \mathscr{H}_{\mathscr{B}}(\mathrm{d}x), \quad \gamma_0^{\mathscr{B}} = \int_0^{\infty} e^{-x}\, \mathscr{H}_0^{\mathscr{B}}(\mathrm{d}x),$$

which connect the KL-numbers with the $\mathscr{L}$-number as follows:

$$\mathscr{L}_{\mathscr{B}} = \gamma_{\mathscr{B}} I_{\mathscr{B}} = \gamma_0^{\mathscr{B}} I_0^{\mathscr{B}} \tag{1.92}$$

(see Theorem E.4). This identity turns out to be very useful for computing the constants $\gamma_{\mathscr{B}}$ and $\gamma_0^{\mathscr{B}}$.

Define also mean values of the limiting distributions of the overshoots

$$\varkappa_{\mathscr{B}} = \int_0^{\infty} x\, \mathscr{H}_{\mathscr{B}}(\mathrm{d}x), \quad \varkappa_0^{\mathscr{B}} = \int_0^{\infty} x\, \mathscr{H}_0^{\mathscr{B}}(\mathrm{d}x). \tag{1.93}$$

Next, introduce the one-sided stopping times

$$\widehat{T}_1(\mathbf{p}_1; b) = \inf\left\{n : \widehat{\lambda}(\mathbf{p}_1; n) \ge b\right\}, \quad \widehat{T}_0(\mathbf{p}_0; a) = \inf\left\{n : \widehat{\lambda}(\mathbf{p}_0; n) \le -a\right\}, \tag{1.94}$$

$$\widetilde{T}_1(\mathbf{p}_1;b) = \inf\left\{n : \widetilde{\lambda}(\mathbf{p}_1;n) \ge b\right\}, \quad \widetilde{T}_0(\mathbf{p}_0;a) = \inf\left\{n : \widetilde{\lambda}(\mathbf{p}_0;n) \le -a\right\}, \tag{1.95}$$

and the corresponding overshoots

$$\hat{\kappa}_b^1 = \widehat{\lambda}(\mathbf{p}_1;\widehat{T}_1(\mathbf{p}_1;b)) - b, \quad \hat{\kappa}_a^0 = |\widehat{\lambda}(\mathbf{p}_0;\widehat{T}_0(\mathbf{p}_0;b)) + a|,$$
$$\tilde{\kappa}_b^1 = \widetilde{\lambda}(\mathbf{p}_1;\widetilde{T}_1(\mathbf{p}_1;b)) - b, \quad \tilde{\kappa}_a^0 = |\widetilde{\lambda}(\mathbf{p}_0;\widetilde{T}_0(\mathbf{p}_0;b)) + a|.$$

Obviously, the GSLRT and the MSLRT can be written via these one-sided Markov times as

$$\widehat{T} = \min\left\{\widehat{T}_1(\mathbf{p}_1;b), \widehat{T}_0(\mathbf{p}_0;a)\right\}, \quad \widehat{d} = i \text{ if } \widehat{T} = \widehat{T}_i, \quad i = 0,1;$$
$$\widetilde{T} = \min\left\{\widetilde{T}_1\mathbf{p}_1;b), \widetilde{T}_0(\mathbf{p}_0;a)\right\}, \quad \widetilde{d} = i \text{ if } \widetilde{T} = \widetilde{T}_i, \quad i = 0,1$$

and the overshoots in the GSLRT and the MSLRT as

$$\hat{\chi}(a,b) = \hat{\kappa}_b^1 \mathbb{1}_{\{\widehat{d}=1\}} + \hat{\kappa}_a^0 \mathbb{1}_{\{\widehat{d}=0\}},$$
$$\tilde{\chi}(a,b) = \tilde{\kappa}_b^1 \mathbb{1}_{\{\widetilde{d}=1\}} + \tilde{\kappa}_a^0 \mathbb{1}_{\{\widetilde{d}=0\}}.$$

For the null hypothesis H_0, the asymptotic analysis and operating characteristics change dramatically depending on whether the KL distances $I_0^i = \mathsf{E}_0[-\lambda_i(1)]$ from f_0^i to f_1^i are different for different streams i (asymmetric case) or the same (symmetric case). Thus, these cases should be considered separately. In the asymmetric case, the number $i^* = \arg\min_{1\le i\le N} I_0^i$ for which I_0^i attains its minimum is unique. To simplify notation and without loss of generality assume that streams (densities $f_1^1,\dots,f_1^i$) are ordered with respect to the KL distances to the null hypothesis, i.e., $I_0^1 \le I_0^2 \le \cdots \le I_0^N$. Write ℓ for a number of densities with minimal KL distance from f_0^i, i.e.,

$$I_0 \equiv \min_{1\le i\le N} I_0^i = I_0^1 = \cdots = I_0^\ell < I_0^{\ell+1} \le \cdots \le I_0^N. \tag{1.96}$$

Note that in the aforementioned asymmetric situation $\ell = 1$ and by our assumption a unique index $i^* = 1$, while in the completely symmetric situation when I_0^i is the same for every $1 \le i \le N$ the value of $\ell = N$. The latter case occurs, for example, when $f_1^i = f_1$ and $f_0^i = f_0$.

The following results will be used for the asymptotic analysis of tests in the i.i.d. case. Recall that we always assume that the LLRs are non-arithmetic. The precise definition of the slowly changing sequence is given in Section F.1 (see (F.5)–(F.6)).

Lemma 1.3. **(i)** *For any $\mathscr{B} \in \mathscr{P}$,*

$$\mathsf{E}_{\mathscr{B}}[e^{-\hat{\chi}(a,b)}\, \mathbb{1}_{\{\widehat{d}=1\}}] \to \gamma_{\mathscr{B}}, \quad \mathsf{E}_{\mathscr{B}}[e^{-\tilde{\chi}(a,b)}\, \mathbb{1}_{\{\widetilde{d}=1\}}] \to \gamma_{\mathscr{B}} \quad \text{as } a_{\max} \to \infty. \tag{1.97}$$

In the asymmetric case where $\ell = 1$ in (1.96),

$$\mathsf{E}_0[e^{-\hat{\chi}(a,b)}\, \mathbb{1}_{\{\widehat{d}=0\}}] \to \gamma_0^1, \quad \mathsf{E}_0[e^{-\tilde{\chi}(a,b)}\, \mathbb{1}_{\{\widetilde{d}=0\}}] \to \gamma_0^1 \quad \text{as } a_{\max} \to \infty. \tag{1.98}$$

The values of $\gamma_{\mathscr{B}}$ and γ_0^1 can be computed as in (F.8).
(ii) *If $\mathsf{E}_i[\lambda_i^2(1)] < \infty$ for $1 \le i \le N$, then as $a_{\min} \to \infty$*

$$\mathsf{E}_{\mathscr{B}}[\hat{\chi}(a,b)] \to \mathsf{E}_{\mathscr{B}}[\hat{\kappa}_b^1] \to \varkappa_{\mathscr{B}}, \quad \mathsf{E}_{\mathscr{B}}[\tilde{\chi}(a,b)] \to \mathsf{E}_{\mathscr{B}}[\tilde{\kappa}_a^0] \to \varkappa_{\mathscr{B}}. \tag{1.99}$$

If $\mathsf{E}_0[\lambda_i^2(1)] < \infty$ for $1 \le i \le N$ and if $\ell = 1$ in (1.96), *then as $a_{\min} \to \infty$*

$$\mathsf{E}_0[\hat{\chi}(a,b)] \to \mathsf{E}_0[\hat{\kappa}_b^1] \to \varkappa_0^1, \quad \mathsf{E}_0[\tilde{\chi}(a,b)] \to \mathsf{E}_0[\tilde{\kappa}_a^0] \to \varkappa_0^1. \tag{1.100}$$

The values of $\varkappa_{\mathscr{B}}$ and $\varkappa_0^1$ can be computed as in (F.9).

Proof. The proof of (i). We prove only the first assertions in (1.97)–(1.100) for the GSLRT, since the second ones for the MSLRT can be proven analogously.

By (1.18), $\widehat{\lambda}(\mathbf{p}_1;n) = \lambda_{\mathscr{B}}(n) + \widehat{\Xi}_{\mathscr{B}}(\mathbf{p}_1;n) + \log p_{1,\mathscr{B}}$, so that the Markov time $\widehat{T}_1(\mathbf{p}_1,b)$ can be written as

$$\widehat{T}_1(\mathbf{p}_1;b) = \inf\left\{n \geq 1 : \lambda_{\mathscr{B}}(n) + \widehat{\Xi}_{\mathscr{B}}(\mathbf{p}_1;n) + \log p_{1,\mathscr{B}} \geq b\right\}. \tag{1.101}$$

From Lemma 1.1 it follows that $\widehat{\Xi}_{\mathscr{B}}(\mathbf{p}_1;n) \to 0$ almost surely under $\mathsf{P}_{\mathscr{B}}$. Hence, the sequence $\{\widehat{\Xi}_{\mathscr{B}}(\mathbf{p}_1;n)\}_{n\in\mathbb{N}}$ is slowly changing and $\widehat{\lambda}(\mathbf{p}_1;n)$ is a perturbed random walk. Since $\widehat{T} = \widehat{T}_1(\mathbf{p}_1;b)$ and $\hat{\chi}(a,b) = \hat{\kappa}_b^1$ on $\{\widehat{d} = 1\}$, from the First Nonlinear Renewal Theorem for perturbed random walks (Theorem F.1) it follows that the overshoot $\hat{\chi}(a,b)$ converges (as $a_{\min} \to \infty$) weakly to the overshoot $\kappa_{\mathscr{B}}^1(b)$ under $\mathsf{P}_{\mathscr{B}}$ on $\{\widehat{d} = 1\}$. Therefore, the Bounded Convergence Theorem yields $\mathsf{E}_{\mathscr{B}}[e^{-\hat{\chi}}\,\mathbb{1}_{\{\widehat{d}=1\}}] \to \gamma_{\mathscr{B}}$.

Next, $\widehat{T} = \widehat{T}_0(\mathbf{p}_0;a)$ and $\hat{\chi}(a,b) = \hat{\kappa}_0(a)$ on the event $\{\widehat{d} = 0\}$ and the Markov time $\widehat{T}_0(\mathbf{p}_0;b)$ can be represented as

$$\widehat{T}_0(\mathbf{p}_0;a) = \inf\left\{n \geq 1 : -\lambda_1(n) - \widehat{\Xi}_1(\mathbf{p}_0;n) - \log p_{0,1} \geq a\right\}, \tag{1.102}$$

where

$$\widehat{\Xi}_1(\mathbf{p}_0;n) = \max\left\{0, \max_{\mathscr{B}\neq\{1\}}\left[\log(p_{0,\mathscr{B}}/p_{0,1}) + \sum_{i\in\mathscr{B}\neq\{1\}} \lambda_i(n) - \lambda_1(n)\right]\right\}, \quad n \geq 1.$$

Since, by the assumption that $\ell = 1$ in (1.96), $I_0^1 - I_0^i < 0$ for all $i \neq 1$,

$$\mathsf{E}_0\left[\sum_{i\in\mathscr{B}\neq\{1\}} \lambda_i(n) - \lambda_1(n)\right] = \left[-\sum_{i\in\mathscr{B}\neq\{1\}} I_0^i + I_0^1\right] n \to -\infty \quad \text{as } n \to \infty.$$

Hence,

$$\widehat{\Xi}_1(\mathbf{p}_0;n) \xrightarrow[n\to\infty]{\mathsf{P}_0-\text{a.s.}} 0, \tag{1.103}$$

so that $\widehat{\Xi}_1(\mathbf{p}_0;n)$, $n \geq 1$ are slowly changing when $\ell = 1$. In just the same way as above, using Theorem F.1, we obtain that $\mathsf{E}_0[e^{-\hat{\chi}(a,b)}\,\mathbb{1}_{\{\widehat{d}=0\}}] \to \gamma_0^1$. This completes the proof of (i).

The proof of (ii). Again, using Theorem F.1, the First Nonlinear Renewal Theorem for perturbed random walks, we obtain that under the second moment condition $\mathsf{E}_{\mathscr{B}}[\lambda_{\mathscr{B}}^2(1)] < \infty$, which holds by assumption $\mathsf{E}_i[\lambda_i^2(1)] < \infty$ for $i \in \mathscr{N}$, the limiting average overshoot $\lim_{b\to\infty} \mathsf{E}_{\mathscr{B}}[\hat{\kappa}_b^1]$ is equal to the limiting average overshoot $\varkappa_{\mathscr{B}} = \lim_{c\to\infty} \mathsf{E}_{\mathscr{B}}[\kappa_{\mathscr{B}}^1(c)]$. Since $\hat{\chi}(a,b) = \hat{\kappa}_b^1\,\mathbb{1}_{\{\widehat{d}=1\}}$ and $\mathsf{P}_{\mathscr{B}}(\widehat{d} = 1) \to 1$ as $a_{\max} \to 0$, the convergence in (1.99) follows.

The convergence in (1.100) is proved in the same way using the fact that under condition $\ell = 1$ the sequence $\{\widehat{\Xi}_1(\mathbf{p}_1;n)\}$ is slowly changing, so that by Theorem F.1, $\lim_{a\to\infty} \mathsf{E}_0[\hat{\kappa}_a^0] = \lim_{c\to\infty} \mathsf{E}_0[\kappa_1^0(c)] = \varkappa_0^1$ and that $\hat{\chi}(a,b) = \hat{\kappa}_a^0\,\mathbb{1}_{\{\widehat{d}=1\}}$, where $\mathsf{P}_0(\widehat{d} = 0) \to 1$ as $a_{\min} \to 0$. □

1.5.2 Asymptotic Approximations for the Probabilities of Errors

The following theorem provides asymptotic approximations for the probabilities of false alarm and improved upper bounds on the probabilities of missed detection in the asymmetric case when $\ell = 1$ in (1.96).

Theorem 1.5. *As $a_{\min} \to \infty$,*

$$\mathsf{P}_0(\widetilde{d} = 1) = \left(\sum_{\mathscr{B}\in\mathscr{P}} p_{1,\mathscr{B}}\,\gamma_{\mathscr{B}}\right) e^{-b}(1 + o(1)), \tag{1.104}$$

$$\mathsf{P}_0(\widehat{d}=1) \le \Big(\sum_{\mathscr{B}\in\mathscr{P}} p_{1,\mathscr{B}}\gamma_{\mathscr{B}}\Big) e^{-b}(1+o(1)). \tag{1.105}$$

In the asymmetric case when $\ell = 1$ in (1.96), for every $\mathscr{B}\in\mathscr{P}$,

$$\mathsf{P}_{\mathscr{B}}(\widetilde{d}=0) \le \frac{\gamma_0^1}{p_{0,\mathscr{B}}} e^{-a}(1+o(1)), \quad \mathsf{P}_{\mathscr{B}}(\widehat{d}=0) \le \frac{\gamma_0^1}{p_{0,\mathscr{B}}} e^{-a}(1+o(1)). \tag{1.106}$$

Therefore, if the thresholds are selected as

$$a_\beta(\mathbf{p}_0) = \log\left(\frac{\gamma_0^1}{\beta \min_{\mathscr{B}\in\mathscr{P}} p_{0,\mathscr{B}}}\right), \quad b_\alpha(\mathbf{p}_1) = \log\left(\frac{\sum_{\mathscr{B}\in\mathscr{P}} p_{1,\mathscr{B}}\gamma_{\mathscr{B}}}{\alpha}\right), \tag{1.107}$$

then as $\max(\alpha,\beta)\to 0$

$$\mathsf{P}_0(\widetilde{d}=1) = \alpha(1+o(1)), \quad \mathsf{P}_0(\widehat{d}=1) \le \alpha(1+o(1)),$$

and if additionally $\ell = 1$, then

$$\max_{\mathscr{B}\in\mathscr{P}} \mathsf{P}_{\mathscr{B}}(\widetilde{d}=0) \le \beta(1+o(1)), \quad \max_{\mathscr{B}\in\mathscr{P}} \mathsf{P}_{\mathscr{B}}(\widehat{d}=0) \le \beta(1+o(1)).$$

Proof. Changing the measure and using Wald's likelihood ratio identity and the fact that $\widetilde{\lambda}(\widetilde{T}) = b + \tilde{\chi}(a,b)$ on $\{\widetilde{d}=1\}$, we obtain

$$\begin{aligned}\mathsf{P}_0(\widetilde{d}=1) &= \mathsf{E}_0\left[\mathbb{1}_{\{\widetilde{d}=1\}}\right] = \widetilde{\mathsf{E}}^{\mathbf{p}_1}\left[e^{-\widetilde{\lambda}(\mathbf{p}_1;\widetilde{T})}\mathbb{1}_{\{\widetilde{d}=1\}}\right] \\ &= \sum_{\mathscr{B}\in\mathscr{P}} p_{1,\mathscr{B}}\mathsf{E}_{\mathscr{B}}\left[e^{-\widetilde{\lambda}(\mathbf{p}_1;\widetilde{T})}\mathbb{1}_{\{\widetilde{d}=1\}}\right] = e^{-b}\sum_{\mathscr{B}\in\mathscr{P}} p_{1,\mathscr{B}}\mathsf{E}_{\mathscr{B}}\left[e^{-\tilde{\chi}(a,b)}\mathbb{1}_{\{\widetilde{d}=1\}}\right].\end{aligned}$$

Since by Lemma 1.3(i),

$$\mathsf{E}_{\mathscr{B}}\left[e^{-\tilde{\chi}(a,b)}\mathbb{1}_{\{\widetilde{d}=1\}}\right] = \gamma_{\mathscr{B}}(1+o(1)) \quad \text{as } a_{\min}\to\infty,$$

(1.104) follows.

Since $\widehat{\lambda}(\mathbf{p}_1;\widehat{T}) \le \widetilde{\lambda}(\mathbf{p}_1;\widehat{T})$ and $\widehat{\lambda}(\mathbf{p}_1;\widehat{T}) = b + \hat{\chi}(a,b)$ on $\{\widehat{d}=1\}$ and since by Lemma 1.3(i),

$$\mathsf{E}_{\mathscr{B}}\left[e^{-\hat{\chi}(a,b)}\mathbb{1}_{\{\widehat{d}=1\}}\right] = \gamma_{\mathscr{B}}(1+o(1)) \quad \text{as } a_{\max}\to 0,$$

we have

$$\begin{aligned}\mathsf{P}_0(\widehat{d}=1) &= \widetilde{\mathsf{E}}^{\mathbf{p}_1}\left[e^{-\widetilde{\lambda}(\mathbf{p}_1;\widehat{T})}\mathbb{1}_{\{\widehat{d}=1\}}\right] \le \widetilde{\mathsf{E}}^{\mathbf{p}_1}\left[e^{-\widehat{\lambda}(\mathbf{p}_1;\widehat{T})}\mathbb{1}_{\{\widehat{d}=1\}}\right] = \sum_{\mathscr{B}\in\mathscr{P}} p_{1,\mathscr{B}}\mathsf{E}_{\mathscr{B}}\left[e^{-\widehat{\lambda}(\mathbf{p}_1;\widehat{T})}\mathbb{1}_{\{\widehat{d}=1\}}\right] \\ &= e^{-b}\sum_{\mathscr{B}\in\mathscr{P}} p_{1,\mathscr{B}}\mathsf{E}_{\mathscr{B}}\left[e^{-\hat{\chi}(a,b)}\mathbb{1}_{\{\widehat{d}=1\}}\right] = e^{-b}\sum_{\mathscr{B}\in\mathscr{P}} p_{1,\mathscr{B}}\gamma_{\mathscr{B}}(1+o(1)),\end{aligned}$$

which proves (1.105).

To prove the first asymptotic bound in (1.106), note that, by (1.19), $\lambda_{\mathscr{B}}(n) = \widetilde{\lambda}(\mathbf{p}_0;n) - \widetilde{\Xi}_{\mathscr{B}}(\mathbf{p}_0;n) - \log p_{0,\mathscr{B}}$, where $\widetilde{\Xi}_{\mathscr{B}}(\mathbf{p}_0;n) \ge 0$, and that $\widetilde{\lambda}(\mathbf{p}_0;\widetilde{T}) = -a - \tilde{\chi}(a,b)$ on $\{\widetilde{d}=0\}$. Hence, $\lambda_{\mathscr{B}}(\widetilde{T}) \le -a - \log p_{0,\mathscr{B}} - \tilde{\chi}(a,b)$ on $\{\widetilde{d}=0\}$ for every $\mathscr{B}\in\mathscr{P}$ and using Wald's likelihood ratio identity, we obtain

$$\mathsf{P}_{\mathscr{B}}(\widetilde{d}=0) = \mathsf{E}_0\left[e^{\lambda_{\mathscr{B}}(\widetilde{T})};\, \widetilde{d}=0\right] \le p_{0,\mathscr{B}}^{-1} e^{-a}\mathsf{E}_0\left[e^{-\tilde{\chi}(a,b)};\, \widetilde{d}=0\right].$$

Since by Lemma 1.3(ii) when $\ell = 1$,

$$\mathsf{E}_0\left[e^{-\tilde{\chi}(a,b)};\widetilde{d}=0\right] = \gamma_0^1(1+o(1)) \quad \text{as } a_{\min}\to\infty,$$

this yields the first inequality in (1.106). The second one for the GSLRT is proved absolutely analogously. □

1.5.3 Third-Order Asymptotic Approximations for the ESS

Consider the hypothesis $\mathsf{H}_{\mathscr{B}}$ and the GSLRT. Observe that the stopping time $\widehat{T}_1(\mathbf{p}_1, b)$ has the form (1.101), where $\widehat{\Xi}_{\mathscr{B}}(\mathbf{p}_1; n)$, $n \geq 1$ are slowly changing under $\mathsf{P}_{\mathscr{B}}$ and converge to 0 as $n \to \infty$ and that, by Lemma 1.3(ii), the average overshoots $\mathsf{E}_{\mathscr{B}}[\hat{\chi}(a,b)]$ and $\mathsf{E}_{\mathscr{B}}[\hat{\kappa}_b^1]$ converge to the same limiting value $\varkappa_{\mathscr{B}}$ (see (1.99)) under the second moment condition for the LLRs. So we may apply the Second Nonlinear Renewal Theorem for perturbed random walks (Theorem F.2) to obtain

$$\mathsf{E}_{\mathscr{B}}[\widehat{T}_1(\mathbf{p}_1, b)] = \frac{1}{I_{\mathscr{B}}}(b + \varkappa_{\mathscr{B}} - \log p_{1,\mathscr{B}}) + o(1) \quad \text{as } b \to \infty \tag{1.108}$$

as long as conditions of this theorem hold. Hence, if we now establish that $\mathsf{E}_{\mathscr{B}}[\widehat{T})] = \mathsf{E}_{\mathscr{B}}[\widehat{T}_1(\mathbf{p}_1, b)] + o(1)$ as $a_{\min} \to \infty$, then the right-hand side is the third-order asymptotic approximation for the ESS $\mathsf{E}_{\mathscr{B}}[\widehat{T}]$. The same intuitive argument applies to the MSLRT.

Consider now the null hypothesis H_0. As mentioned above, the asymmetric case where the condition (1.96) holds with $\ell = 1$, i.e., when the minimal KL distance I_0^1 is unique, and the general case of $\ell > 1$ are dramatically different. If $\ell = 1$, then we may use the same argument as above for the hypothesis $\mathsf{H}_{\mathscr{B}}$. Indeed, in this case, the stopping time $\widehat{T}_0(\mathbf{p}_0, a)$ has the form (1.102), where $\widehat{\Xi}_1(\mathbf{p}_0; n)$, $n \geq 1$ are slowly changing under P_0 and, by Lemma 1.3(ii), the average overshoots $\mathsf{E}_0[\hat{\chi}(a,b)]$ and $\mathsf{E}_0[\hat{\kappa}_a^0]$ converge to $\varkappa_0^1$, so that we may apply the Second Nonlinear Renewal Theorem for perturbed random walks to obtain

$$\mathsf{E}_0[\widehat{T}_0(\mathbf{p}_1, a)] = \frac{1}{I_0}\left(a + \varkappa_0^1 + \log p_{0,1}\right) + o(1) \quad \text{as } a \to \infty, \tag{1.109}$$

where $I_0 = \min_{1 \leq i \leq N} I_0^i (= I_0^1)$. If we now establish that $\mathsf{E}_0[\widehat{T}] = \mathsf{E}_0[\widehat{T}_0(\mathbf{p}_1, a)] + o(1)$ as $a_{\min} \to \infty$, then the right-hand side is the third-order asymptotic approximation for the ESS $\mathsf{E}_0[\widehat{T}]$. However, if $\ell > 1$, then $\widehat{\Xi}_1(\mathbf{p}_0; n)$, $n \geq 1$ are not slowly changing and a different argument is needed. As we will see later on, the General Second Nonlinear Renewal Theorem (Theorem F.4) can be used in this case.

1.5.3.1 Asymptotic Approximations for the ESS Under Hypothesis $\mathsf{H}_{\mathscr{B}}$ and Under Hypothesis H_0 in the Asymmetric Case

We begin with establishing third-order asymptotic approximations for the ESS under the hypothesis $\mathsf{H}_{\mathscr{B}}$ and under the null hypothesis H_0 in the asymmetric case $\ell = 1$. The following lemma and theorem make the above heuristic argument precise.

Lemma 1.4. *Assume that the KL information numbers I_1^i and I_0^i are positive and finite for all $i = 1, \dots, N$ and that the thresholds a and b approach infinity in such a way that $b^2 e^{-a} \to 0$ and $a^2 e^{-b} \to 0$ as $a_{\min} = \min(a,b) \to \infty$. Then,*

$$\mathsf{E}_{\mathscr{B}}[\widehat{T}] = \mathsf{E}_{\mathscr{B}}[\widehat{T}_1(\mathbf{p}_1, b)] + o(1) \quad \text{as } a_{\min} \to \infty \quad \text{for all } \mathscr{B} \in \mathscr{P}, \tag{1.110}$$

$$\mathsf{E}_0[\widehat{T}] = \mathsf{E}_0[\widehat{T}_0(\mathbf{p}_0, a)] + o(1) \quad \text{as } a_{\min} \to \infty. \tag{1.111}$$

Proof. Since $\widehat{T}_1(\mathbf{p}_1, b) - \widehat{T} = 0$ on $\{\widehat{d} = 1\}$ and $\widehat{T}_1(\mathbf{p}_1, b) - \widehat{T} = \widehat{T}_1(\mathbf{p}_1, b) - \widehat{T}_0(\mathbf{p}_0, a)$ on $\{\widehat{d} = 0\}$, we have

$$\widehat{T}_1(\mathbf{p}_1, b) - \widehat{T} = (\widehat{T}_1(\mathbf{p}_1, b) - \widehat{T}_0(\mathbf{p}_1, a))\mathbb{1}_{\{\widehat{d}=0\}} \leq \widehat{T}_1(\mathbf{p}_1, b)\mathbb{1}_{\{\widehat{d}=0\}},$$

and applying the Cauchy–Schwartz inequality yields

$$\mathsf{E}_{\mathscr{B}}[\widehat{T}_1 - \widehat{T}] \leq \mathsf{E}_{\mathscr{B}}[\widehat{T}_1 \mathbb{1}_{\{\widehat{d}=0\}}] \leq \sqrt{\mathsf{E}_{\mathscr{B}}[\widehat{T}_1^2]\mathsf{P}_{\mathscr{B}}(\widehat{d} = 0)}. \tag{1.112}$$

Obviously, for any $\mathscr{B}$,

$$\widehat{T}_1 \le T_{\mathscr{B}}(b) = \inf\{n \ge 1 : \lambda_{\mathscr{B}}(n) \ge b - \log p_{1,\mathscr{B}}\}.$$

But by (1.62),

$$\mathsf{E}_{\mathscr{B}}[T_{\mathscr{B}}^m(b)] \le \left(\frac{b}{I_1^{\mathscr{B}}}\right)^m (1+o(1)), \quad b \to \infty.$$

In the i.i.d. case, this inequality holds for any $m > 0$ whenever the KL numbers are positive and finite. Setting $m = 2$ yields

$$\mathsf{E}_{\mathscr{B}}[\widehat{T}_1^2] \le \mathsf{E}_{\mathscr{B}}[T_{\mathscr{B}}^2] \le \left(\frac{b}{I_{\mathscr{B}}}\right)^2 (1+o(1)), \quad b \to \infty.$$

Since by Lemma 1.2, $\mathsf{P}_{\mathscr{B}}(\widehat{d}=0) \le e^{-a}/p_{0,\mathscr{B}}$ and by the assumption of the theorem $b^2 e^{-a} \to 0$, we have

$$\mathsf{E}_{\mathscr{B}}[\widehat{T}_1^2 \mathsf{P}_{\mathscr{B}}(\widehat{d}=0)] \le O(b^2 e^{-a}) \to 0 \quad \text{as } a_{\min} \to 0.$$

The relation (1.110) follows.

To establish (1.111), note first that, as in (1.112),

$$\mathsf{E}_0[\widehat{T}_0(\mathbf{p}_0,a) - \widehat{T}] \le \mathsf{E}_0[\widehat{T}_0(\mathbf{p}_0,a)\mathbb{1}_{\{\widehat{d}=1\}}] \le \sqrt{\mathsf{E}_0[\widehat{T}_0^2(\mathbf{p}_0,a)]\mathsf{P}_0(\widehat{d}=1)},$$

where $\mathsf{P}_0(\widehat{d}=1) \le e^{-b}$ by Lemma 1.2. Obviously,

$$\widehat{T}_0(\mathbf{p}_0,a) \le T(a) = \inf\left\{n \ge 1 : \min_{\mathscr{B}\in\mathscr{P}}(-\lambda_{\mathscr{B}}(n)) \ge a\right\},$$

so that by (1.62),

$$\mathsf{E}_0[\widehat{T}_0^2(\mathbf{p}_0,a)] \le \left(\frac{a}{\min_{\mathscr{B}\in\mathscr{P}} I_0^{\mathscr{B}}}\right)^2 (1+o(1)) \quad \text{as } a \to \infty.$$

Therefore,

$$\mathsf{E}_0[\widehat{T}_0(\mathbf{p}_0,a) - \widehat{T}] \le O(a e^{-b/2}) \quad \text{as } a_{\min} \to 0$$

By the assumption $a^2 e^{-b} \to 0$, so $O(a e^{-b/2}) \to 0$ as $a_{\min} \to 0$ and we obtain that $\mathsf{E}_0[\widehat{T}_0(\mathbf{p}_0,a) - \widehat{T}] \to 0$, i.e., (1.111). □

It is worth noting that the asymptotic relationship (1.111) holds regardless of the value of ℓ, i.e., not only in the asymmetric situation when $\ell = 1$ but in the general case of $1 \le \ell \le N$ as well. For the sake of simplicity of presentation, we exclude class $\mathscr{P}_{\underline{K},\overline{K}}$ with minimal possible number of signals $\underline{K} > 1$ from the consideration. For this class, the results have to be modified.

Theorem 1.6. *Let $\mathsf{E}_i[\lambda_i^2(1)] < \infty$ and $\mathsf{E}_0[\lambda_i^2(1)] < \infty$ for all $i \in \mathscr{N}$. Let thresholds a and b go to infinity in such a way that $b^2 e^{-a} \to 0$ and $a^2 e^{-b} \to 0$ as $a_{\min} = \min(a,b) \to \infty$.*

(i) *Then, as $a_{\min} \to \infty$, the following third-order asymptotic approximations for the ESS under the hypothesis $\mathsf{H}_{\mathscr{B}}$ hold for any $\mathscr{B} \in \mathscr{P}$:*

$$\mathsf{E}_{\mathscr{B}}[\widehat{T}] = \frac{1}{I_{\mathscr{B}}}(b + \varkappa_{\mathscr{B}} - \log p_{1,\mathscr{B}}) + o(1), \quad \mathsf{E}_{\mathscr{B}}[\widetilde{T}] = \frac{1}{I_{\mathscr{B}}}(b + \varkappa_{\mathscr{B}} - \log p_{1,\mathscr{B}}) + o(1). \quad (1.113)$$

(ii) *In the asymmetric situation when $\ell = 1$ in (1.96), as $a_{\min} \to \infty$, the following third-order asymptotic approximations for the ESS under the hypothesis H_0 hold:*

$$\mathsf{E}_0[\widehat{T}] = \frac{1}{I_0}(a + \varkappa_0^1 + \log p_{0,1}) + o(1), \quad \mathsf{E}_0[\widetilde{T}] = \frac{1}{I_0}(a + \varkappa_0^1 + \log p_{0,1}) + o(1). \quad (1.114)$$

Proof. We provide the proof only for the ESS of the GSLRT since the proof for the MSLRT is essentially the same.

The proof of (i). Due to Lemma 1.4 to prove the asymptotic expansion (1.113) it suffices to establish the approximation (1.108) for $\mathsf{E}_{\mathscr{B}}[\widehat{T}_1(\mathbf{p}_1,b)]$. To this end, we use the Second Nonlinear Renewal Theorem for perturbed random walks, i.e., Theorem F.2. Recall that the stopping time $\widehat{T}_1(\mathbf{p}_1,b)$ can be represented as in (1.101). In order to apply this theorem, we have to verify conditions (F.11)–(F.17). Since in our case, $\ell_n = \log p_{1,\mathscr{B}}$ and $\eta_n = \widehat{\Xi}_{\mathscr{B}}(\mathbf{p}_1;n) \geq 0$ the following conditions should be checked:

$$\max_{0\leq i\leq n} \widehat{\Xi}_{\mathscr{B}}(\mathbf{p}_1;n+i),\ n\geq 1 \quad \text{are uniformly integrable under } \mathsf{P}_{\mathscr{B}}, \tag{1.115}$$

$$\sum_{n=1}^{\infty} \mathsf{P}_{\mathscr{B}}\left(\widehat{\Xi}_{\mathscr{B}}(\mathbf{p}_1;n) \leq -\varepsilon n\right) < \infty \quad \text{for some } 0<\varepsilon<I_{\mathscr{B}}, \tag{1.116}$$

$$\widehat{\Xi}_{\mathscr{B}}(\mathbf{p}_1;n) \xrightarrow[n\to\infty]{\text{law}} \eta, \tag{1.117}$$

$$\lim_{b\to\infty} b\,\mathsf{P}_{\mathscr{B}}\left(\widehat{T}_1(\mathbf{p}_1;b) \leq (1-\varepsilon)b/I_{\mathscr{B}}\right) = 0 \quad \text{for some } 0<\varepsilon<1. \tag{1.118}$$

Since, for every $n\geq 0$, $\widehat{\Xi}_{\mathscr{B}}(\mathbf{p}_1;n) \leq \widetilde{\Xi}_{\mathscr{B}}(\mathbf{p}_1;n)$, we have

$$\mathsf{P}_{\mathscr{B}}\left(\max_{0\leq t\leq n} \widehat{\Xi}_{\mathscr{B}}(\mathbf{p}_1;t) > x\right) \leq \mathsf{P}_{\mathscr{B}}\left(\max_{0\leq t\leq n} \exp\left\{\widetilde{\Xi}_{\mathscr{B}}(\mathbf{p}_1;t)\right\} > e^x\right),$$

where $\exp\{\widetilde{\Xi}_{\mathscr{B}}(\mathbf{p}_1;t)\}$, $t\geq 1$ is a non-negative $\mathsf{P}_{\mathscr{B}}$-martingale with mean $1/p_{1,\mathscr{B}}$. Using Doob's submartingale inequality, yields

$$\mathsf{P}_{\mathscr{B}}\left(\max_{0\leq t\leq n} \widehat{\Xi}_{\mathscr{B}}(\mathbf{p}_1;t) > x\right) \leq p_{1,\mathscr{B}}^{-1} e^{-x}, \tag{1.119}$$

which implies that $\mathsf{E}_{\mathscr{B}}[\sup_n \widehat{\Xi}_{\mathscr{B}}(\mathbf{p}_1;n)] < \infty$ and hence condition (1.115). Condition (1.116) holds trivially since $\widehat{\Xi}_{\mathscr{B}}(\mathbf{p}_1;n) \geq 0$. By Lemma 1.1, condition (1.117) holds with $\eta \equiv 0$.

It remains to check condition (1.118), which immediately follows from Lemma F.2 since (1.119) implies that for any $\varepsilon > 0$

$$n\,\mathsf{P}_{\mathscr{B}}\left(\max_{1\leq t\leq n} \widehat{\Xi}_{\mathscr{B}}(\mathbf{p}_1;t) + \log p_{1,\mathscr{B}} \geq \varepsilon n\right) = O(n e^{-\varepsilon n}) \to 0 \quad \text{as } n\to\infty.$$

Therefore, the asymptotic approximation (1.108) follows from Theorem F.2, and the proof of (i) is complete.

The proof of (ii). Since by Lemma 1.4, $\mathsf{E}_0[\widehat{T})] = \mathsf{E}_0[\widehat{T}_0(\mathbf{p}_0,a)] + o(1)$, it suffices to establish the approximation (1.109) for $\mathsf{E}_0[\widehat{T}_0(\mathbf{p}_0,a)]$. To this end, we again use Theorem F.2, the Second Nonlinear Renewal Theorem for perturbed random walks. Since $\ell_n = \log p_{0,1}$ and $\eta_n = -\widehat{\Xi}_1(\mathbf{p}_0;n) \leq 0$, to apply Theorem F.2 we have to verify the following conditions:

$$\max_{0\leq i\leq n} [\widehat{\Xi}_1(\mathbf{p}_0;n+i)],\ n\geq 1 \quad \text{are uniformly integrable under } \mathsf{P}_0, \tag{1.120}$$

$$\sum_{n=1}^{\infty} \mathsf{P}_0\left\{\widehat{\Xi}_1(\mathbf{p}_0;n) \geq \varepsilon n\right\} < \infty \quad \text{for some } 0<\varepsilon<I_0, \tag{1.121}$$

$$\widehat{\Xi}_1(\mathbf{p}_0;n) \xrightarrow[n\to\infty]{\mathsf{P}_0-\text{law}} \eta, \quad \mathsf{E}_0[\eta] < \infty, \tag{1.122}$$

$$\lim_{a\to\infty} a\,\mathsf{P}_0\left\{\widehat{T}_0(\mathbf{p}_0;a) \leq (1-\varepsilon)a/I_0\right\} = 0 \quad \text{for some } 0<\varepsilon<1. \tag{1.123}$$

Additional notation: $v_{\mathscr{B}} = \log(p_{0,\mathscr{B}}/p_{0,1})$,

$$\mu_{\mathscr{B}} = I_0 - \sum_{i \in \mathscr{B} \neq \{1\}} I_0^i, \quad W_{\mathscr{B}}(n) = \sum_{i \in \mathscr{B} \neq \{1\}} \lambda_i(n) - \lambda_1(n), \quad \widetilde{W}_{\mathscr{B}}(n) = W_{\mathscr{B}}(n) - \mu_{\mathscr{B}} n.$$

Note that $\mu_{\mathscr{B}} < 0$ by assumption that $\ell = 1$ in (1.96).

To verify the uniform integrability condition (1.120) we observe that

$$\begin{aligned} \mathsf{E}_0\left[\sup_{n \geq 1} \widehat{\Xi}_1(\mathbf{p}_0; n)\right] &= \int_0^\infty \mathsf{P}_0\left\{\max_{\mathscr{B} \neq \{1\}} \left[v_{\mathscr{B}} + \sup_{n \geq 1} W_{\mathscr{B}}(n)\right] > y\right\} \mathrm{d}y \\ &\leq \sum_{\mathscr{B} \neq \{1\}} \int_0^\infty \mathsf{P}_0\left\{v_{\mathscr{B}} + \sup_{n \geq 1} W_{\mathscr{B}}(n) > y\right\} \mathrm{d}y. \end{aligned}$$

By Theorem 10.7 (page 100) in Gut [55],

$$\mathsf{E}_0\left[\max\left(0, \sup_{n \geq 1} W_{\mathscr{B}}(n)\right)\right] = \int_0^\infty \mathsf{P}_0\left\{\sup_{n \geq 1} W_{\mathscr{B}}(n) > y\right\} \mathrm{d}y < \infty,$$

which implies that $\mathsf{E}_0\left[\sup_{n \geq 1} \widehat{\Xi}_1(\mathbf{p}_0; n)\right] < \infty$ and hence uniform integrability (1.120).

Next, we check condition (1.121). Note that $\{\widetilde{W}_{\mathscr{B}}(n)\}$ is a zero-mean random walk with finite variance, so that by the Baum–Katz rate of convergence in the law of large numbers (cf. implications (B.10) in Appendix B),

$$\sum_{n=1}^\infty \mathsf{P}_0\left\{\frac{1}{n}|\widetilde{W}_{\mathscr{B}}(n)| > \varepsilon\right\} < \infty \quad \text{for all } \varepsilon > 0. \tag{1.124}$$

We now show that this result can be used to deduce condition (1.121). Since $\mu_{\mathscr{B}} < 0$, it follows that

$$\begin{aligned} \widehat{\Xi}_1(\mathbf{p}_0; n) &= \max\left\{0, \max_{\mathscr{B} \neq \{1\}} \left[\widetilde{W}_{\mathscr{B}}(n) + v_{\mathscr{B}} + \mu_{\mathscr{B}} n\right]\right\} \leq \max\left\{0, \max_{\mathscr{B} \neq \{1\}} \left[\widetilde{W}_{\mathscr{B}}(n) + v_{\mathscr{B}}\right]\right\} \\ &\leq \max\left\{0, \max_{\mathscr{B} \neq \{1\}} \widetilde{W}_{\mathscr{B}}(n) + \max_{\mathscr{B} \neq \{1\}} v_{\mathscr{B}}\right\} \leq \max_{\mathscr{B} \neq \{1\}} |\widetilde{W}_{\mathscr{B}}(n)| + \max_{\mathscr{B} \neq \{1\}} |v_{\mathscr{B}}| \end{aligned}$$

and, consequently,

$$\begin{aligned} \mathsf{P}_0\left\{\widehat{\Xi}_1(\mathbf{p}_0; n) \geq \varepsilon n\right\} &\leq \mathsf{P}_0\left\{\max_{\mathscr{B} \neq \{1\}} |\widetilde{W}_{\mathscr{B}}(n)| \geq \varepsilon n - \bar{v}\right\} \leq \mathsf{P}_0\left\{\frac{1}{n} \max_{\mathscr{B} \neq \{1\}} |\widetilde{W}_{\mathscr{B}}(n)| \geq \varepsilon^*\right\} \\ &\leq \sum_{\mathscr{B} \neq \{1\}} \mathsf{P}_0\left\{\frac{1}{n}|\widetilde{W}_{\mathscr{B}}(n)| > \varepsilon^*\right\} \quad \text{for all } n \geq 1 + \lfloor \bar{v}/\varepsilon \rfloor, \end{aligned}$$

where $\bar{v} = \max_{\mathscr{B} \neq \{1\}} |v_{\mathscr{B}}|$ and $\varepsilon^* = \varepsilon[1 - \bar{v}/(\varepsilon + \bar{v})] > 0$. This inequality along with inequality (1.124) proves that

$$\sum_{n=1}^\infty \mathsf{P}_0\left\{\widehat{\Xi}_1(\mathbf{p}_0; n) \geq \varepsilon n\right\} < \infty \quad \text{for all } \varepsilon > 0,$$

i.e., condition (1.121).

It follows from (1.103) that condition (1.122) holds with $\eta \equiv 0$.

Let $N_a = (1 - \varepsilon)a/I_0$. To check condition (1.123) note that since $\widehat{\Xi}_1(\mathbf{p}_0; n) \geq 0$, for any $a > |\log p_{0,\mathscr{B}}|$ we have

$$\mathsf{P}_0\left(\widehat{T}_0(\mathbf{p}_0; a) \leq N_a\right) = \mathsf{P}_0\left(\max_{1 \leq n \leq N_a} [-\lambda_1(n) - \widehat{\Xi}_1(\mathbf{p}_0; n)] - \log p_{0,1} \geq a\right)$$

$$\le \mathsf{P}_0\left(\max_{1\le n\le N_a}[-\lambda_1(n)]\ge a\right)\le \mathsf{P}_0\left(\max_{1\le n\le N_a}[-\lambda_1(n)]\ge (1-\varepsilon^2)a\right)$$
$$=\mathsf{P}_0\left(\max_{1\le n\le N_a}[-\lambda_1(n)]\ge (1+\varepsilon)I_0N_a\right)\le \mathsf{P}_0\left(\max_{1\le n\le N_a}[-\lambda_1(n)-I_0n]\ge \varepsilon I_0N_a\right).$$

By Lemma F.1, for every $\varepsilon\in(0,1)$,

$$a\mathsf{P}_0\left(\max_{1\le n\le N_a}[-\lambda_1(n)-I_0n]\ge \varepsilon I_0N_a\right)\to 0\quad \text{as } a\to\infty,$$

so that condition (1.123) is satisfied.

Theorem F.2 implies asymptotic approximation (1.109), and the proof of (ii) is complete. □

The following corollary, which will be useful in the sequel, follows directly from Theorems 1.5 and 1.6.

Corollary 1.3. *Let $\mathsf{E}_i|\lambda_i(1)|^2<\infty$ and $\mathsf{E}_0|\lambda_i(1)|^2<\infty$, $i=1,\dots,N$. Let α and β approach 0 so that $\beta/|\log\alpha|^2\to 0$ and $\alpha/|\log\beta|^2\to 0$. Assume that thresholds a and b are so selected that $\mathsf{P}_0(\tilde{d}=1)\sim \mathsf{P}_0(\hat{d}=1)\sim\alpha$ and $\max_{\mathscr{B}\in\mathscr{P}}\mathsf{P}_{\mathscr{B}}(\tilde{d}=0)\sim\max_{\mathscr{B}\in\mathscr{P}}\mathsf{P}_{\mathscr{B}}(\hat{d}=0)\sim\beta$ as $\alpha_{\max}\to 0$.*

(i) *Then*

$$\mathsf{E}_{\mathscr{B}}[\tilde{T}]=\frac{1}{I_{\mathscr{B}}}\left[|\log\alpha|+\log\left(\sum_{\mathscr{A}\in\mathscr{P}}p_{1,\mathscr{A}}\gamma_{\mathscr{A}}\right)+\varkappa_{\mathscr{B}}-\log p_{1,\mathscr{B}}\right]+o(1), \tag{1.125}$$

$$\mathsf{E}_{\mathscr{B}}[\hat{T}]\le\frac{1}{I_{\mathscr{B}}}\left[|\log\alpha|+\log\left(\sum_{\mathscr{A}\in\mathscr{P}}p_{1,\mathscr{A}}\gamma_{\mathscr{A}}\right)+\varkappa_{\mathscr{B}}-\log p_{1,\mathscr{B}}\right]+o(1). \tag{1.126}$$

(ii) *If in addition $\ell=1$ in (1.96), then*

$$\mathsf{E}_0[\tilde{T}]\le\frac{1}{I_0}\left[|\log\beta|+\log\left(\frac{\gamma_0^1}{\min_{\mathscr{B}\in\mathscr{P}}p_{0,\mathscr{B}}}\right)+\varkappa_0^1+\log p_{0,1}\right]+o(1), \tag{1.127}$$

$$\mathsf{E}_0[\hat{T}]\le\frac{1}{I_0}\left[|\log\beta|+\log\left(\frac{\gamma_0^1}{\min_{\mathscr{B}\in\mathscr{P}}p_{0,\mathscr{B}}}\right)+\varkappa_0^1+\log p_{0,1}\right]+o(1). \tag{1.128}$$

Proof. The proof of (i). From Theorem 1.5 (see (1.104)–(1.105)) and the assumption that $\mathsf{P}_0(\tilde{d}=1)\sim\mathsf{P}_0(\hat{d}=1)\sim\alpha$ it follows that, as $\alpha_{\max}\to 0$,

$$b=\log\left(\frac{\sum_{\mathscr{B}\in\mathscr{P}}p_{1,\mathscr{B}}\gamma_{\mathscr{B}}}{\alpha}\right)+o(1)$$

for the MSLRT and

$$b\le\log\left(\frac{\sum_{\mathscr{B}\in\mathscr{P}}p_{1,\mathscr{B}}\gamma_{\mathscr{B}}}{\alpha}\right)+o(1)$$

for the GSLRT. These relationships and Theorem 1.6(i) imply (1.125)–(1.126).

The proof of (ii). From Theorem 1.5 (see (1.106)) and the assumption that $\max_{\mathscr{B}\in\mathscr{P}}\mathsf{P}_{\mathscr{B}}(\tilde{d}=0)\sim\max_{\mathscr{B}\in\mathscr{P}}\mathsf{P}_{\mathscr{B}}(\hat{d}=0)\sim\beta$ it follows that, as $\alpha_{\max}\to 0$,

$$a\le\log\left(\frac{\gamma_0^1}{\beta\min_{\mathscr{B}\in\mathscr{P}}p_{0,\mathscr{B}}}\right)+o(1),$$

which along with Theorem 1.6(ii) proves (1.127)–(1.128). □

1.5.3.2 Asymptotic Approximations for the ESS Under Hypothesis H_0 in the General Case

Consider now the general case (1.96) where $1 < \ell \le N$. In this case, the sequence $\{\widehat{\Xi}_1(\mathbf{p}_0;n)\}$ in (1.102) is not slowly changing and Theorem F.2 cannot be applied. However, the GLLR and the MLLR statistics can still be represented in a form that allows us to use the Second General Nonlinear Renewal Theorem (Theorem F.4) in order to derive a higher order approximation for the ESS. Indeed, for $i \in \mathscr{N}$, let $W_i(n) \equiv -\lambda_i(n) = \sum_{t=1}^n \Delta W_i(t)$, where $\Delta W_i(t) = \log[f_0^i(X_t(i))/f_1^i(X_t(i))]$. Also, set

$$D = \sum_{i=1}^N I_0^i - I_0 \quad \text{and} \quad \widetilde{W}_{\mathscr{B}}(n) = \sum_{i \in \mathscr{N} \setminus \mathscr{B}} W_i(n) - Dn.$$

Note that

$$\widetilde{W}_0(n) := \sum_{i \in \mathscr{N} \setminus \{0\}} W_i(n) - Dn = \sum_{i=1}^N W_i(n) - Dn.$$

Obviously,

$$\sum_{i \in \mathscr{N} \setminus \mathscr{B}} W_i(n) - \sum_{i=1}^N W_i(n) = -\sum_{i \in \mathscr{B}} W_i(n) = \lambda_{\mathscr{B}}(n),$$

and therefore,

$$\begin{aligned}\left\{\max_{\mathscr{B}}[\lambda_{\mathscr{B}}(n) + \log p_{0,\mathscr{B}}] \le -a\right\} &= \left\{\sum_{i=1}^N W_i(n) - \max_{\mathscr{B}}\left[\sum_{i \in \mathscr{N} \setminus \mathscr{B}} W_i(n) + \log p_{0,\mathscr{B}}\right] \ge a\right\} \\ &= \left\{\widetilde{W}_0(n) - \max_{\mathscr{B}}\left[\widetilde{W}_{\mathscr{B}}(n) + \log p_{0,\mathscr{B}}\right] \ge a\right\}.\end{aligned}$$

Thus, using this notation, the Markov time $\widehat{T}_0(\mathbf{p}_0, a)$ can be written as

$$\widehat{T}_0(\mathbf{p}_0, a) = \inf\left\{n \ge 1 : \widetilde{W}_0(n) - \max_{\mathscr{B}}\left[\widetilde{W}_{\mathscr{B}}(n) + \log p_{0,\mathscr{B}}\right] \ge a\right\}, \tag{1.129}$$

where $\{\widetilde{W}_0(n)\}$ is a random walk under P_0 with increments having the mean value $\mathsf{E}_0[\widetilde{W}_0(1)] = I_0$ and $\{\widetilde{W}_{\mathscr{B}}(n)\}$ are random walks with expectations

$$\mathsf{E}_0[\widetilde{W}_{\mathscr{B}}(1)] = I_0 - \sum_{i \in \mathscr{B}} I_0^i.$$

Evidently, $\mathsf{E}_0[\widetilde{W}_{\mathscr{B}}(1)] = 0$ for $\mathscr{B} = \{i\}, i = 1, \dots, \ell$ and $\mathsf{E}_0[\widetilde{W}_{\mathscr{B}}(1)] < 0$ otherwise.

Define an ℓ-dimensional vector $\mathbf{Y} = (Y_1, Y_2, \dots, Y_\ell)$ with components

$$Y_i = \widetilde{W}_i(1) = \sum_{j \in \mathscr{N} \setminus \{i\}} W_j(1) - D, \quad i = 1, \dots, \ell.$$

Note that $\mathbf{Y}$ is zero-mean. Let $\mathbf{V} = \mathsf{Cov}(\mathbf{Y})$ denote its covariance matrix with respect to P_0 and let

$$\phi_{\mathbf{V}}(\mathbf{y}) = \left[(2\pi)^\ell |\mathbf{V}|\right]^{-1/2} \exp\left\{-\tfrac{1}{2}\mathbf{y}\mathbf{V}^{-1}\mathbf{y}^\top\right\}$$

be the density of a multivariate normal distribution function with covariance matrix $\mathbf{V}$.

The asymptotic expansions in the general case are derived using normal approximations [14]. Specifically, introduce the variables

$$h_\ell = \int_{\mathbb{R}^\ell} \max_{1 \le i \le \ell}\{y_i\} \phi_{\mathbf{V}}(\mathbf{y})\, \mathrm{d}\mathbf{y} \tag{1.130}$$

and

$$C_\ell = \int_{\mathbb{R}^\ell} \max_{1\le i\le \ell}\{y_i\} \left(\mathscr{P}(\mathbf{y}) + \mathbf{w}\mathbf{V}^{-1}\mathbf{y}^\top\right) \phi_{\mathbf{V}}(\mathbf{y})\, d\mathbf{y}, \tag{1.131}$$

where

$$\mathbf{w} = (w_1, \dots, w_\ell); \quad w_i = \log p_{0,i}, \tag{1.132}$$

and where $\mathscr{P}(\mathbf{y})$ is a polynomial in $\mathbf{y} \in \mathbb{R}^\ell$ of the third degree whose coefficients involve $\mathbf{V}$ and the P_0-cumulants of $\mathbf{Y}$ up to order 3 and is given explicitly by formula (7.19) in Bhattacharya and Rao [14]. Note that the constant h_ℓ is the expected value of the maximum of ℓ zero-mean normal random variables with density $\phi_{\mathbf{V}}(\mathbf{y})$.

Adding the term $h_\ell\sqrt{n}$ on both sides in (1.129), we obtain the following representation for the Markov time $\widehat{T}_0(\mathbf{p}_0, a)$:

$$\widehat{T}_0(\mathbf{p}_0, a) = \inf\left\{n \ge 1 : \widetilde{W}_0(n) + \xi_n \ge b_n(a)\right\}, \tag{1.133}$$

where $b_n(a) = a + h_\ell\sqrt{n}$ and

$$\xi_n = -\max_{\mathscr{B}}\left[\widetilde{W}_{\mathscr{B}}(n) + \log p_{0,\mathscr{B}}\right] + h_\ell\sqrt{n}. \tag{1.134}$$

Denote by χ_a the excess of the process $\widetilde{W}_0(n) + \xi_n$ over the boundary $b_n(a)$ at time $n = \widehat{T}_0$, i.e.,

$$\chi_a = \widetilde{W}_0(\widehat{T}_0) + \xi_{\widehat{T}_0} - b_{\widehat{T}_0}(a) \quad \text{on } \{\widehat{T}_0 < \infty\}.$$

We will need the following additional notation:

$$N_a := \sup\{n \ge 1 : b_n(a) \ge I_0 n\} = \frac{a}{I_0} + \frac{h_\ell}{I_0}\sqrt{\frac{a}{I_0} + \frac{h_\ell^2}{4I_0^2}} + \frac{h_\ell^2}{2I_0^2}; \tag{1.135}$$

$$\dot{b}_t(a) := \frac{\partial b_t(a)}{\partial t} = \frac{h_\ell}{2\sqrt{t}}; \quad \ddot{b}_t(a) := \frac{\partial^2 b_t(a)}{\partial t^2} = -\frac{h_\ell}{4t^{3/2}};$$

$$d_a := \dot{b}_t(a)|_{t=N_a} = \frac{h_\ell}{2\sqrt{N_a}}; \quad d_{\sup} := \sup_{t \ge N_a, a \ge 0} \dot{b}_t(a) = \frac{h_\ell}{2\sqrt{N_0}} = \frac{\sqrt{I_0}}{2};$$

$$G_{\widehat{T}_0} = b_{\widehat{T}_0}(a) - I_0 N_a - d_a(\widehat{T}_0 - N_a).$$

Also, define the stopping time

$$\tau_a = \inf\left\{n \ge 1 : \widetilde{W}_0(n) - d_a n > (I_0 - d_a)N_a\right\} \tag{1.136}$$

and the associated overshoot

$$\kappa_a = (\widetilde{W}_0(\tau_a) - d_a\tau_a) - (I_0 - d_a)N_a \quad \text{on } \{\tau_a < \infty\}.$$

Let $\varkappa = \lim_{a\to\infty} \mathsf{E}_0[\kappa_a]$ be the limiting average overshoot. Since $\widetilde{W}_0(n) - d_a n$ is the random walk with drift $I_0 - d_a$ and $\lim_{a\to\infty} d_a = 0$, by renewal theory, under the second moment condition $\mathsf{E}_0[W_i^2(1)] < \infty$

$$\varkappa = \frac{\mathsf{E}_0[\widetilde{W}_0^2(\tau_0)]}{2\mathsf{E}_0[\widetilde{W}_0(\tau_0)]} = \frac{\mathsf{E}_0[\widetilde{W}_0^2(1)]}{2I_0} - \sum_{n=1}^{\infty} \frac{1}{n}\mathsf{E}_0[(\widetilde{W}_0(n))^-] \tag{1.137}$$

(cf. Theorems E.1 and E.3 in Appendix E).

Using these definitions, we obtain

$$\widetilde{W}_0(\widehat{T}_0) - \widetilde{W}_0(\tau_a) - d_a(\widehat{T}_0 - \tau_a) = \chi_a - \kappa_a - \xi_{\widehat{T}_0} + G_{\widehat{T}_0}.$$

Taking expectations on both sides yields

$$(I_0 - d_a)\mathsf{E}_0[\widehat{T}_0] - (I_0 - d_a)\mathsf{E}_0[\tau_a] = \mathsf{E}_0[\chi_a - \kappa_a] - \mathsf{E}_0[\xi_{\widehat{T}_0}] + \mathsf{E}_0[G_{\widehat{T}_0}],$$

which implies

$$\mathsf{E}_0[\widehat{T}_0] = \mathsf{E}_0[\tau_a] + \frac{1}{I_0 - d_a}\left\{\mathsf{E}_0[\chi_a - \kappa_a] - \mathsf{E}_0[\xi_{\widehat{T}_0}] + \mathsf{E}_0[G_{\widehat{T}_0}]\right\}. \tag{1.138}$$

Since $d_a \to 0$, by renewal theory,

$$\mathsf{E}_0[\tau_a] = N_a + I_0^{-1}\varkappa + o(1) \quad \text{as } a \to \infty \tag{1.139}$$

(cf. Theorem E.6 in Appendix E).

Using (1.138) and (1.139), we obtain that for large a,

$$\mathsf{E}_0[\widehat{T}_0] \approx \frac{1}{I_0}\left\{a + h_\ell\sqrt{\frac{a}{I_0} + \frac{h_\ell^2}{4I_0}} + \frac{h_\ell^2}{2I_0} + \varkappa + \lim_{a\to\infty}\mathsf{E}_0[\chi_a - \kappa_a] - \lim_{a\to\infty}\mathsf{E}_0[\xi_{\widehat{T}_0}] + \lim_{a\to\infty}\mathsf{E}_0[G_{\widehat{T}_0}]\right\}. \tag{1.140}$$

We iterate that, under P_0, the process $\{\widetilde{W}_0(n)\}$ is a random walk with mean $I_0 n$, the processes $\{\widetilde{W}_i(n)\}$ are zero-mean random walks for $i = 1,\dots,\ell$ and $\{\widetilde{W}_{\mathscr{B}}(n)\}$ are random walks with negative means for $\mathscr{B} \neq \{i\}$, $i = 1,\dots,\ell$. Below, we argue that $\{\xi_n\}_{n\geq 1}$ converges to a random variable η with finite expectation. In fact, maximization in (1.134) can be replaced by maximization over closest values $\mathscr{B} = \{i\}$, $i = 1,\dots,\ell$ for which random walks $\widetilde{W}_i(n)$ have zero mean values and exclude the rest subsets for which random walks have negative mean values. This seems intuitively obvious. Under certain conditions, $\mathsf{E}_0[\xi_{\widehat{T}_0}] = -C_\ell + o(1)$, $\mathsf{E}_0[\chi_a - \kappa_a] = o(1)$, and $\mathsf{E}_0[G_{\widehat{T}_0}] = o(1)$ as $a \to \infty$, where C_ℓ is defined above in (1.131). Therefore, using (1.140) along with the fact that $\mathsf{E}_0[\widehat{T}] = \mathsf{E}_0[\widehat{T}_0] - o(1)$, we expect that, as $a_{\min} \to \infty$,

$$\mathsf{E}_0[\widehat{T}] = \frac{1}{I_0}\left(a + h_\ell\sqrt{\frac{a}{I_0} + \frac{h_\ell^2}{4I_0}} + \frac{h_\ell^2}{2I_0} + \varkappa + C_\ell\right) + o(1). \tag{1.141}$$

A similar argument shows that under certain conditions formulated below in Theorem 1.7 the following expansion (as $a_{\min} \to \infty$) for the ESS of the MSLRT holds:

$$\mathsf{E}_0[\widetilde{T}] = \frac{1}{I_0}\left(a + h_\ell\sqrt{\frac{a}{I_0} + \frac{h_\ell^2}{4I_0}} + \frac{h_\ell^2}{2I_0} + \varkappa + C_\ell - R_\ell\right) + o(1), \tag{1.142}$$

where the constant $0 \leq R_\ell \leq \log \ell$.

While the previous argument formally fits well into the Second General Renewal Theorem (see Theorem F.4 in Appendix F), this theorem cannot be applied in full since the following two conditions do not hold in our case:

$$\left\{\max_{1\leq m\leq n}|\xi_{n+m}|, n \geq 1\right\} \quad \text{is uniformly integrable;}$$

$$\lim_{n\to\infty}\mathsf{P}_0\left(\max_{1\leq m\leq\sqrt{n}}|\xi_{n+m} - \xi_n| \geq \varepsilon\right) = 0 \quad \text{for every } \varepsilon > 0.$$

A rigorous proof of (1.141) and (1.142), which uses a modified version of the Second General Renewal Theorem (see Theorem F.5) is given below. Recall that we always assume that the LLRs $\lambda_i(1)$, $i = 1,\dots,N$, are non-arithmetic.

Theorem 1.7. *Let $\ell > 1$ in (1.96). Let thresholds a and b go to infinity so that $b^2 e^{-a} \to 0$ and $a^2 e^{-b} \to 0$ as $a_{\min} \to \infty$. Assume that the covariance matrix $\mathbf{V}$ of the vector $\mathbf{Y}$ is positive definite, that $\mathsf{E}_0[||\mathbf{Y}||^3] < \infty$ and that the Cramér condition on the joint characteristic function of the vector $\mathbf{Y}$*

$$\limsup_{||\mathbf{t}|| \to \infty} \mathsf{E}_0 \left[\exp\{i \cdot (\mathbf{t}, \mathbf{Y})\}\right] < 1 \tag{1.143}$$

is satisfied. Then the third-order asymptotic approximations (1.141) and (1.142) for the ESS of the GSLRT and the MSLRT hold.

Proof. Consider the GSLRT. Since by Lemma 1.4, $\mathsf{E}_0[\widehat{T}] = \mathsf{E}_0[\widehat{T}_0] + o(1)$ whenever $b^2 e^{-a} \to 0$ and $a^2 e^{-b} \to 0$ as $a_{\min} = \min(a,b) \to \infty$, it suffices to prove that the expansion (1.141) holds for the expected value $\mathsf{E}_0[\widehat{T}_0(\mathbf{p}_0, a)]$. This can be done using the modified Second General Nonlinear Renewal Theorem (see Theorem F.5). Since the stopping time $\widehat{T}_0$ can be written as in (1.133), by Theorem F.5 (assuming that all required conditions hold),

$$\mathsf{E}_0[\widehat{T}_0] = N_a + \frac{1}{I_0}(\varkappa_0^1 - \mathsf{E}_0[\eta]) + o(1) \quad \text{as } a \to \infty, \tag{1.144}$$

where N_a is defined in (1.135), $\varkappa$ is given by (1.137), and η is the limiting value (as $n \to \infty$) of η_n defined below in (1.152). When deriving (1.144) from (F.45) we used the fact that $\partial b_t(a)/\partial t = \frac{1}{2} h_\ell t^{-1/2}$ and $|N_a \partial^2 b_t(a)/\partial t^2| = N_a h_\ell / 8t^{3/2}$, so that conditions (F.38)–(F.39) hold and condition (F.40) holds with $d = 0$.

Since $\ell_n = 0$ in (F.33), in order to deduce asymptotic approximation (1.141) from (1.138)–(1.139), applying Theorem F.5, we have to verify the following conditions:

$$\xi_n = \eta_n \quad \text{for } n \ge L, \quad \mathsf{E}_0[L] < \infty; \tag{1.145}$$

$$\lim_{n \to \infty} n \,\mathsf{P}_0 \left(\max_{0 \le m \le n} \eta_{n+m} \ge \varepsilon n \right) = 0 \quad \text{for all } \varepsilon > 0; \tag{1.146}$$

$$\sum_{n=1}^{\infty} \mathsf{P}_0 \left(\eta_n \le -\varepsilon n\right) < \infty \quad \text{for some } 0 < \varepsilon < I_0; \tag{1.147}$$

$$\lim_{n \to \infty} \mathsf{E}_0[\eta_n] = \mathsf{E}_0[\eta] < \infty; \tag{1.148}$$

$$\lim_{a \to \infty} \mathsf{E}_0[\chi_a - \kappa_a] = 0; \tag{1.149}$$

$$\lim_{a \to \infty} \mathsf{E}_0[G_{\widehat{T}_0}] = 0; \tag{1.150}$$

$$\lim_{a \to \infty} a \,\mathsf{P}_0 \left\{ \widehat{T}_0(\mathbf{p}_0; a) \le \varepsilon a / I_0 \right\} = 0 \quad \text{for some } 0 < \varepsilon < 1. \tag{1.151}$$

In order to verify condition (1.145), we first show that maximization over $\mathscr{B}$ in (1.134) can be replaced by maximization over $i \in \{1, \dots, \ell\}$. To be more specific, we will show that $\xi_n = \eta_n$ on $\{n > L\}$, where

$$\eta_n = -\max_{1 \le i \le \ell} \left[\widetilde{W}_i(n) + w_i \right] + h_\ell \sqrt{n} \tag{1.152}$$

and

$$L = \sup\left\{ n \ge 1 : \max_{\mathscr{B} \in \mathscr{B}_\ell} \left[w_{\mathscr{B}} + \widetilde{W}_{\mathscr{B}}(n) \right] > \max_{1 \le j \le \ell} \left[w_j + \widetilde{W}_j(n) \right] \right\}.$$

Here $w_j = \log p_{0,j}$, $w_{\mathscr{B}} = \log p_{0,\mathscr{B}}$, and $\mathscr{B}_\ell = \{\ell+1, \dots, N\}$. Therefore, to establish condition (1.145) it suffices to show that the expectation of the random variable L is finite, $\mathsf{E}_0[L] < \infty$. Additional notation:

$$\mu_{\mathscr{B}} = I_0 - \sum_{i \in \mathscr{B}} I_0^i, \quad \mu = -\max_{\mathscr{B} \in \mathscr{B}_\ell} \mu_{\mathscr{B}}, \quad W_{\mathscr{B}}^*(n) = \widetilde{W}_{\mathscr{B}}(n) - \mu_{\mathscr{B}} n, \quad V_{\mathscr{B}}(n) = W_{\mathscr{B}}^*(n) - W_1(n).$$

Note that $\mu_{\mathscr{B}} < 0$, $\mu > 0$, and $\{W^*_{\mathscr{B}}(n)\}$ and $\{V_{\mathscr{B}}(n)\}$ are zero-mean random walks with finite variances.

Indeed, due to our basic assumption (1.96), we have

$$
\begin{aligned}
\mathsf{P}_0(L \ge t) &= \mathsf{P}_0\left\{\max_{\mathscr{B}\in\mathscr{B}_\ell}\left[w_{\mathscr{B}}+\widetilde{W}_{\mathscr{B}}(t)\right] > \max_{1\le j\le \ell}\left[w_j+\widetilde{W}_j(t)\right] \text{ for some } t\ge n\right\} \\
&\le \mathsf{P}_0\left\{\max_{\mathscr{B}\in\mathscr{B}_\ell}\left[w_{\mathscr{B}}+\widetilde{W}_{\mathscr{B}}(t)\right] > w_1+\widetilde{W}_1(t) \text{ for some } t\ge n\right\} \\
&\le \mathsf{P}_0\left\{\max_{\mathscr{B}\in\mathscr{B}_\ell}\left[\widetilde{W}_{\mathscr{B}}(t)-\widetilde{W}_1(t)\right] > w_1-\max_{\mathscr{B}\in\mathscr{B}_\ell} w_{\mathscr{B}} \text{ for some } t\ge n\right\} \\
&= \mathsf{P}_0\left\{\max_{\mathscr{B}\in\mathscr{B}_\ell}\left[W^*_{\mathscr{B}}(t)-\widetilde{W}_1(t)+\mu_{\mathscr{B}}t\right] > w_1-\max_{\mathscr{B}\in\mathscr{B}_\ell} w_{\mathscr{B}} \text{ for some } t\ge n\right\} \\
&\le \mathsf{P}_0\left\{\max_{\mathscr{B}\in\mathscr{B}_\ell} V_{\mathscr{B}}(t) > -\max_{\mathscr{B}\in\mathscr{B}_\ell}\mu_{\mathscr{B}}t+w_1 \text{ for some } t\ge n\right\} \\
&\le \sum_{\mathscr{B}\in\mathscr{B}_\ell}\mathsf{P}_0\left\{\frac{V_{\mathscr{B}}(t)}{t} > \mu-\frac{|w_1|}{t} \text{ for some } t\ge n\right\} \\
&\le \sum_{\mathscr{B}\in\mathscr{B}_\ell}\mathsf{P}_0\left\{\sup_{t\ge n}\left[\frac{|V_{\mathscr{B}}(t)|}{t}\right] > \mu-\frac{|w_1|}{n}\right\}.
\end{aligned}
$$

Since $\mu - |w_1|/n > 0$ for $n \ge n_0 = \lfloor |w_1|/\mu \rfloor + 1$, denoting $\mu - |w_1|/n_0 = \varepsilon$, we obtain

$$
\mathsf{P}_0(L\ge n) \le \sum_{\mathscr{B}\in\mathscr{B}_\ell}\mathsf{P}_0\left\{\sup_{t\ge n}\left[\frac{|V_{\mathscr{B}}(t)|}{t}\right] > \varepsilon\right\} \quad \text{for } n\ge n_0,
$$

where $\varepsilon > 0$. Thus,

$$
\mathsf{E}_0[L] = \sum_{n=1}^{\infty}\mathsf{P}_0(L\ge n) \le M + \sum_{\mathscr{B}\in\mathscr{B}_\ell}\sum_{n=n_0}^{\infty}\mathsf{P}_0\left\{\sup_{t\ge n}\left[\frac{|V_{\mathscr{B}}(t)|}{t}\right] > \varepsilon\right\},
$$

where

$$
M = \sum_{\mathscr{B}\in\mathscr{B}_\ell}\sum_{n=1}^{n_0-1}\mathsf{P}_0\left\{\sup_{t\ge n}\left[\frac{|V_{\mathscr{B}}(t)|}{t}\right] > \mu-\frac{|w_1|}{n}\right\} < \infty.
$$

As mentioned above, $\{V_{\mathscr{B}}(n)\}$ is a zero-mean random walk with finite variance. Hence, by the rate of convergence in the SLLN (see Lemma B.4 in Appendix B.2)

$$
\sum_{n=1}^{\infty}\mathsf{P}_0\left\{\sup_{t\ge n}\left[\frac{|V_{\mathscr{B}}(t)|}{t}\right] > \varepsilon\right\} < \infty \quad \text{for all } \varepsilon > 0,
$$

which along with the previous inequality proves that $\mathsf{E}_0[L] < \infty$.

Thus, for $n > L$, we can write ξ_n of (1.134) as

$$
\xi_n = \eta_n = -\sqrt{n}(\zeta_n - h_\ell) \tag{1.153}
$$

where

$$
\zeta_n = \frac{1}{\sqrt{n}}\max_{1\le i\le \ell}\left[\widetilde{W}_i(n)+w_i\right]. \tag{1.154}
$$

To check condition (1.146), let $M_\varepsilon = \varepsilon n - h_\ell\sqrt{2n}$. We have

$$
\mathsf{P}_0\left\{\max_{1\le j\le n}\eta_{n+j} \ge \varepsilon n\right\} \le \mathsf{P}_0\left\{\max_{1\le j\le 2n}\eta_j \ge \varepsilon n\right\}
$$

$$\le \mathsf{P}_0\left\{\max_{1\le j\le 2n}\min_{1\le i\le \ell}\left[-\widetilde{W}_i(j)+w_i\right]\ge M_\varepsilon\right\}$$
$$\le \mathsf{P}_0\left\{\max_{1\le j\le 2n}\left[-\widetilde{W}_1(j)\right]\ge M_\varepsilon - w_1\right\},$$

where by Lemma F.1,

$$n\,\mathsf{P}_0\left\{\max_{1\le j\le 2n}\left[-\widetilde{W}_1(j)\right]\ge M_\varepsilon - w_1\right\}\to 0 \quad \text{as } n\to\infty \quad \text{for all } \varepsilon>0,$$

which yields (1.146).

Next, since $\{W_i(n)\}$, $i=1,\dots,\ell$, are random walks with mean zero and finite variances, by the rate of convergence in the law of large numbers (see Lemma B.4)

$$\sum_{n=1}^{\infty}\mathsf{P}_0\left\{|\widetilde{W}_i(n)|\ge \varepsilon n\right\}<\infty \quad \text{for all } \varepsilon>0,$$

so denoting $M_\varepsilon=\varepsilon n+h_\ell\sqrt{n}$, we obtain

$$\sum_{n=1}^{\infty}\mathsf{P}_0(\eta_n\le -\varepsilon n)\le \sum_{n=1}^{\infty}\mathsf{P}_0\left\{\max_{1\le i\le \ell}[\widetilde{W}_i(n)-w_i]\ge M_\varepsilon\right\}$$
$$\le \sum_{i=1}^{\ell}\sum_{n=1}^{\infty}\mathsf{P}_0\left\{\widetilde{W}_i(n)-w_i\ge M_\varepsilon\right\}\le \sum_{i=1}^{\ell}\sum_{n=1}^{\infty}\mathsf{P}_0\left\{\widetilde{W}_i(n)-w_i\ge \varepsilon n\right\}<\infty.$$

Hence, condition (1.147) holds.

Next, we prove convergence (1.148). For the expectation of the random variable ζ_n defined in (1.154), under the assumption of our theorem, using Theorem 20.1 of Bhattacharya and Rao [14] with $f(\mathbf{y})=\max_{1\le i\le \ell}\{y_i\}$ and $s=3$, we obtain

$$\begin{aligned}\mathsf{E}_0[\zeta_n]=\mathsf{E}&\left[\max_{1\le i\le \ell}\left(\varsigma_i+\frac{w_i}{\sqrt{n}}\right)\right]\\&+\frac{1}{\sqrt{n}}\int_{\mathbb{R}^\ell}\max_{1\le i\le \ell}\left\{y_i+\frac{w_i}{\sqrt{n}}\right\}\mathscr{P}(\mathbf{y})\phi_{\mathbf{V}}(\mathbf{y})\,\mathrm{d}\mathbf{y}+o(n^{-1/2}),\end{aligned}\tag{1.155}$$

where $\varsigma=(\varsigma_1,\dots,\varsigma_\ell)\sim\mathscr{N}_\ell(0,\mathbf{V})$ is a random variable having the multivariate normal distribution with mean zero and covariance $\mathbf{V}$. Transformation of variables and first-order Taylor expansion for $\phi_{\mathbf{V}}(\mathbf{y})$ in the first integral yield (as $n\to\infty$)

$$\begin{aligned}\mathsf{E}\left[\max_{1\le i\le \ell}\left(\varsigma_i+\frac{w_i}{\sqrt{n}}\right)\right]&=\int_{\mathbb{R}^\ell}\max_{1\le i\le \ell}\{y_i\}\phi_{\mathbf{V}}(\mathbf{y}-\mathbf{w}/\sqrt{n})\,\mathrm{d}\mathbf{y}\\&=\int_{\mathbb{R}^\ell}\max_{1\le i\le \ell}\{y_i\}\phi_{\mathbf{V}}(\mathbf{y})\,\mathrm{d}\mathbf{y}+\frac{1}{\sqrt{n}}\int_{\mathbb{R}^\ell}\max_{1\le i\le \ell}\{y_i\}\mathbf{w}\mathbf{V}^{-1}\mathbf{y}^\top\phi_{\mathbf{V}}(\mathbf{y})\,\mathrm{d}\mathbf{y}+o(n^{-1/2}).\end{aligned}\tag{1.156}$$

By

$$J_2=\frac{1}{\sqrt{n}}\int_{\mathbb{R}^\ell}\max_{1\le i\le \ell}\left\{y_i+\frac{w_i}{\sqrt{n}}\right\}\mathscr{P}(\mathbf{y})\phi_{\mathbf{V}}(\mathbf{y})\,\mathrm{d}\mathbf{y}$$

denote the second integral in (1.155). Obviously, we have the following inequalities

$$\frac{1}{\sqrt{n}}\int_{\mathbb{R}^\ell}\max_{1\le i\le \ell}\{y_i\}\mathscr{P}(\mathbf{y})\phi_{\mathbf{V}}(\mathbf{y})\,\mathrm{d}\mathbf{y}+\frac{1}{n}\min_{1\le i\le \ell}\{w_i\}\int_{\mathbb{R}^\ell}\mathscr{P}(\mathbf{y})\phi_{\mathbf{V}}(\mathbf{y})\,\mathrm{d}\mathbf{y}\le J_2$$
$$\le \frac{1}{\sqrt{n}}\int_{\mathbb{R}^\ell}\max_{1\le i\le \ell}\{y_i\}\mathscr{P}(\mathbf{y})\phi_{\mathbf{V}}(\mathbf{y})\,\mathrm{d}\mathbf{y}+\frac{1}{n}\max_{1\le i\le \ell}\{w_i\}\int_{\mathbb{R}^\ell}\mathscr{P}(\mathbf{y})\phi_{\mathbf{V}}(\mathbf{y})\,\mathrm{d}\mathbf{y},$$

which show that

$$J_2 = \frac{1}{\sqrt{n}} \int_{\mathbb{R}^\ell} \max_{1 \le i \le \ell} \{y_i\} \mathscr{P}(\mathbf{y}) \phi_{\mathbf{V}}(\mathbf{x}) \, d\mathbf{y} + o(n^{-1/2}) \quad \text{as } n \to \infty. \tag{1.157}$$

Consequently, relations (1.155)–(1.157) yield

$$\mathsf{E}_0[\zeta_n] = h_\ell + \frac{1}{\sqrt{n}} \int_{\mathbb{R}^\ell} \max_{1 \le i \le \ell} \{y_i\} \left[\mathbf{w} \mathbf{V}^{-1} \mathbf{y}^\top + \mathscr{P}(\mathbf{y})\right] \phi_{\mathbf{V}}(\mathbf{y}) \, d\mathbf{y} + o(n^{-1/2}) \quad \text{as } n \to \infty,$$

which along with (1.153) shows that

$$\lim_{n \to \infty} \mathsf{E}_0[\xi_n] = \lim_{n \to \infty} \mathsf{E}_0[\eta_n] = \mathsf{E}_0[\eta] = -C_\ell, \tag{1.158}$$

where C_ℓ is defined in (1.131).

Note that uniform integrability of the overshoot κ_a follows from renewal theory, so that to verify condition (1.149) we have to establish uniform integrability of the overshoot χ_a. Observe that $\chi_a = \min_{\mathscr{B}} \left[W_{\mathscr{B}}(\widehat{T}_0) - \log p_{0,\mathscr{B}}\right] - a$ and define the stopping times

$$T_{\mathscr{B}}(a) = \inf\{n \ge 1 : W_{\mathscr{B}}(n) - \log p_{0,\mathscr{B}} \ge a\}, \quad \mathscr{B} \in \mathscr{P}.$$

Obviously,

$$\chi_a \le \sum_{\mathscr{B} \in \mathscr{P}} \left[W_{\mathscr{B}}(T_{\mathscr{B}}(a)) - \log p_{0,\mathscr{B}} - a\right],$$

and, by Theorem 1 of Lorden [83],

$$\sup_{a \ge 0} \mathsf{E}_0 \left[W_{\mathscr{B}}(T_{\mathscr{B}}(a)) - \log p_{0,\mathscr{B}} - a\right] \le \frac{\mathsf{E}_0[\Delta W_{\mathscr{B}}(1)^+]^2}{\mathsf{E}_0[\Delta W_{\mathscr{B}}(1)]},$$

so that

$$\sup_{a \ge 0} \mathsf{E}_0[\chi_a] \le \sum_{\mathscr{B} \in \mathscr{P}} \sup_{a \ge 0} \mathsf{E}_0 \left[W_{\mathscr{B}}(T_{\mathscr{B}}(a)) - \log p_{0,\mathscr{B}} - a\right] \le \sum_{\mathscr{B} \in \mathscr{P}} \frac{\mathsf{E}_0|\Delta W_{\mathscr{B}}(1)|^2}{\mathsf{E}_0[\Delta W_{\mathscr{B}}(1)]},$$

which is finite by the second moment assumption $\mathsf{E}_0|\Delta W_i(1)|^2 < \infty$. Thus, $\{\chi_a\}_{a>0}$ is uniformly integrable, which implies (1.149).

To verify condition (1.151) recall that the stopping time $\widehat{T}_0(\mathbf{p}_0; a)$ can be written as

$$\widehat{T}_0(\mathbf{p}_0; a) = \inf\left\{n \ge 1 : \widetilde{W}_0(n) + g(n) \ge a\right\},$$

where

$$g(n) := -\max_{\mathscr{B}} \left[\widetilde{W}_{\mathscr{B}}(n) + w_{\mathscr{B}}\right] = \min_{\mathscr{B}} \left[-\widetilde{W}_{\mathscr{B}}(n) - w_{\mathscr{B}}\right]$$

(see (1.129)). Recall also that $\{\widetilde{W}_0(n)\}$ is the random walk with increments having mean I_0 and finite variance and $\{\widetilde{W}_{\mathscr{B}}(n)\}$ are zero-mean random walks for $\mathscr{B} = \{1\}, \dots, \{\ell\}$ also with finite variance. The idea is to use Lemma F.2 with $m = 2$ and these random walk properties for establishing the desired result. In order to use Lemma F.2 it suffices to show that the probability $\mathsf{P}_0\{\max_{1 \le t \le n} g(t) \ge \varepsilon n\} = o(1/n)$ as $n \to \infty$. Since

$$\max_{1 \le t \le n} g(t) \le \max_{1 \le t \le n} \left[-\widetilde{W}_1(t)\right] - w_1$$

this probability can be upper-bounded as follows:

$$\mathsf{P}_0\left\{\max_{1 \le t \le n} g(t) \ge \varepsilon n\right\} \le \mathsf{P}_0\left\{\max_{1 \le t \le n} \left[-\widetilde{W}_1(t)\right] \ge \varepsilon n - |w_1|\right\}$$

$$= \mathsf{P}_0\left\{\frac{1}{n}\max_{1\le t\le n}\left[-\widetilde{W}_1(t)\right] \ge \varepsilon - \frac{|w_1|}{n}\right\}$$

$$\le \mathsf{P}_0\left\{\frac{1}{n}\max_{1\le t\le n}\left[-\widetilde{W}_1(t)\right] \ge \varepsilon/2\right\} \quad \text{for any } n \ge 1+\frac{2|w_1|}{\varepsilon}.$$

It follows from Lemma F.1 that for any $\varepsilon > 0$

$$n\mathsf{P}_0\left\{\frac{1}{n}\max_{1\le t\le n}\left[-\widetilde{W}_1(t)\right] \ge \varepsilon/2\right\} \to 0 \quad \text{as } n\to\infty,$$

so that condition (1.151) is satisfied.

Finally, substituting the limiting value for $\mathsf{E}_0[\eta_n]$ given in (1.158) in (1.144), the required asymptotic approximation (1.141) follows.

Consider the MSLRT. Note that similarly to (1.133) the Markov time $\widetilde{T}_0(\mathbf{p}_0,a)$ can be written as

$$\widetilde{T}_0(\mathbf{p}_0,a) = \inf\left\{n\ge 1 : \widetilde{W}_0(n) + \tilde{\xi}_n \ge a + h_\ell\sqrt{n}\right\}, \tag{1.159}$$

where

$$\tilde{\xi}_n = -\log\left(\sum_{\mathscr{B}} p_{0,\mathscr{B}} \exp\left\{\widetilde{W}_{\mathscr{B}}(n)\right\}\right) + h_\ell\sqrt{n}.$$

Next, in just the same way as in (1.153), the sequence $\{\tilde{\xi}_n\}$ can be replaced by

$$\tilde{\eta}_n = -\log\left(\sum_{i=1}^{\ell} p_{0,i} \exp\left\{\widetilde{W}_i(n)\right\}\right) + h_\ell\sqrt{n}.$$

Since $0 \le \eta_n - \tilde{\eta}_n \le \log \ell$ and $\lim_{n\to\infty} \mathsf{E}_0[\eta_n] = -C_\ell$ it follows that

$$\lim_{n\to\infty} \mathsf{E}_0[\tilde{\eta}_n] = R_\ell - C_\ell, \quad \text{where } 0 \le R_\ell \le \log\ell.$$

Since by Lemma 1.4, $\mathsf{E}_0[\widetilde{T}] = \mathsf{E}_0[\widetilde{T}_0] + o(1)$, approximation (1.142) follows from (1.141). □

Remark 1.8. Consider the asymmetric case where $\ell = 1$. Then, obviously, $h_1 = 0$ and $R_1 = 0$. Also, as shown in (7.21) of [14], $\mathscr{P}(y) = \mathsf{E}_0[Y_1]^3\,(y^3 - 3y)/6$ and since for the standard normal random variable $\mathsf{E}[X]^4 = 3\mathsf{E}[X^2]$, we see from (1.131) that $C_1 = \log p_{0,1}$. Thus, the resulting expressions for the expected sample sizes (1.141) and (1.142) are consistent with expressions (1.114) obtained in Theorem 1.6(ii) under weaker second moment conditions $\mathsf{E}_0|\lambda_i(1)|^2 < \infty$, $i = 1,\dots,N$.

Remark 1.9. It can be shown that under second moment conditions $\mathsf{E}_0|\lambda_i(1)|^2 < \infty$, $i = 1,\dots,N$, the following second-order asymptotic expansions hold as $a_{\max} \to \infty$:

$$\mathsf{E}_0[\widehat{T}] = \frac{1}{I_0}\left(a + h_\ell\sqrt{\frac{a}{I_0}}\right) + O(1), \quad \mathsf{E}_0[\widetilde{T}] = \frac{1}{I_0}\left(a + h_\ell\sqrt{\frac{a}{I_0}}\right) + O(1) \tag{1.160}$$

(cf. Theorem 2(c) in [40]).

In order to use asymptotic approximations (1.141) and (1.142), one has to be able to compute the constants h_ℓ and C_ℓ. Computing the constant h_ℓ, which is the expected value of the maximum of ℓ zero-mean normal random variables with density $\phi_{\mathbf{V}}(\mathbf{y})$, is straightforward, since the integral in (1.130) involves only the covariance matrix of the vector $\mathbf{Y}$. Computing the constant C_ℓ is, in general, quite difficult due to the fact that the polynomial $\mathscr{P}(\mathbf{y})$ is a complicated function of the cumulants of $\mathbf{Y}$. Note, however, that so far we did not assume any specific structure of the covariance matrix $\mathbf{V}$, i.e., the results of the theorem hold for an arbitrarily positive definite covariance matrix.

Let $\mathbb{I}$ denote the identity matrix. Simplification is possible in our multistream detection problem in the symmetric situation when $w_i = w = \log p_{0,1}$ for $i = 1, \ldots, \ell$ and $\{Y_i, i = 1, \ldots, \ell\}$ are identically distributed (but not necessarily independent). This is the case, e.g., when densities $f_0^i(X_n(i)) = f_0(X_n(i))$ and $f_1^i(X_n(i)) = f_1(X_n(i))$ do not depend on i for $i = 1, \ldots, \ell$. More specifically, suppose that the covariance matrix of the vector $\mathbf{Y} = (Y_1, \ldots, Y_\ell)$ is of the form

$$\mathbf{V} = \mathrm{v}^2\mathbb{I} + \varepsilon, \quad \varepsilon > -\mathrm{v}^2/\ell. \tag{1.161}$$

We now show that, in this case, as $a \to \infty$

$$\begin{aligned}
\mathsf{E}_0[\widehat{T}] &= \frac{1}{I_0}\left(a + w + \mathrm{v}\tilde{h}_\ell\sqrt{\frac{a+w}{I_0} + \frac{\mathrm{v}^2\tilde{h}_\ell^2}{4I_0}} + \frac{\mathrm{v}^2\tilde{h}_\ell^2}{2I_0} + \varkappa_0^1 + \frac{\mathsf{E}_0[\widetilde{Y}_1^3]}{6\mathrm{v}^2}\widetilde{C}_\ell\right) + o(1), \\
\mathsf{E}_0[\widetilde{T}] &= \frac{1}{I_0}\left(a + w + \mathrm{v}\tilde{h}_\ell\sqrt{\frac{a+w}{I_0} + \frac{\mathrm{v}^2\tilde{h}_\ell^2}{4I_0}} + \frac{\mathrm{v}^2\tilde{h}_\ell^2}{2I_0} + \varkappa_0^1 + \frac{\mathsf{E}_0[\widetilde{Y}_1^3]}{6\mathrm{v}^2}\widetilde{C}_\ell - R_\ell\right) + o(1).
\end{aligned} \tag{1.162}$$

Introduce the following notation:

$$\mathrm{v}^2 = \mathsf{Var}_0[W_1(1)], \quad \tilde{h}_\ell = \ell\int_{-\infty}^{\infty} y\varphi(y)\Phi(y)^{\ell-1}\,\mathrm{d}y,$$

$$\widetilde{C}_\ell = \ell\int_{-\infty}^{\infty} y\varphi(y)\Phi(y)^{\ell-2}\Big[(\ell-1)\varphi(y)(1-y^2) + (y^3-3y)\Phi(y)\Big]\,\mathrm{d}y,$$

where $\varphi(y) = (2\pi)^{-1/2}\exp(-y^2/2)$ and $\Phi(y) = \int_{-\infty}^{y}\varphi(x)\mathrm{d}x$ are respectively standard normal density and distribution functions and where $\mathsf{Var}_0[\cdot]$ is the variance relative to the density f_0. Note that $\tilde{h}_\ell$ is the expected value of the ℓth order statistic from the standard normal distribution. We assume that $\mathsf{Var}_0[W_1(1)] = \cdots = \mathsf{Var}_0[W_\ell(1)]$.

Observe that when $w_i = w$ for $i = 1, \ldots, \ell$, the value of w can be simply subtracted from the slowly changing sequence η_n and added to the threshold a, so that replacing a by $a + w$ in (1.141) and (1.142), we obtain the equivalent approximations with a new constant C_ℓ given by (1.131) with $\mathbf{w} = 0$, i.e.,

$$C_\ell = \int_{\mathbb{R}^\ell} \max_{1\le i\le \ell}\{y_i\}\mathscr{P}(\mathbf{y})\phi_{\mathbf{V}}(\mathbf{y})\,\mathrm{d}\mathbf{y}.$$

For $\mathbf{V}$ to be positive definite, we must require that $\varepsilon > -\mathrm{v}^2/\ell$. Evidently, we can write $Y_i = \widetilde{Y}_i + \delta\overline{Y}$, where $\overline{Y} = \frac{1}{\ell}\sum_{i=1}^{\ell}\widetilde{Y}_i$, $\delta = \sqrt{1+\varepsilon\ell/\mathrm{v}^2} - 1$ and

$$\widetilde{Y}_i = Y_i - \frac{\delta}{1+\delta}\overline{Y} \tag{1.163}$$

is zero-mean with covariance $\mathbf{V} = \mathrm{v}^2\mathbb{I}$. Thus, for $\varepsilon \neq 0$, the slowly changing term ζ_n in equation (1.154) simply gets modified, relative to the case $\varepsilon = 0$, by the addition of the term $\delta\overline{Y}$. It is easy to see that the addition of $\delta\overline{Y}$ does not affect the conditions (1.145)–(1.148). Moreover, since $\mathsf{E}_0[\overline{Y}] = 0$, the limiting expectation $\lim_{n\to\infty}\mathsf{E}_0[\xi_n] = \mathsf{E}[\eta]$ is also unaffected. It is not difficult to show that for any (allowable) $\varepsilon > -\mathrm{v}^2/\ell$ the constant C_ℓ is given by

$$C_\ell = \frac{\mathsf{E}_0[\widetilde{Y}_1^3]}{6\mathrm{v}^2}\widetilde{C}_\ell. \tag{1.164}$$

Also, for any $\varepsilon > -\mathrm{v}^2/\ell$,

$$h_\ell = \mathrm{v}\tilde{h}_\ell \tag{1.165}$$

since if $(X_1, \ldots, X_\ell) \sim \mathscr{N}(0, \mathbf{V})$ with $\mathbf{V} = \mathrm{v}^2\mathbb{I} + \varepsilon$, then

$$\mathsf{E}[\max\{X_1, \ldots, X_\ell\}] = \mathsf{E}[\max\{\widetilde{X}_1 + \delta\overline{X}, \ldots, \widetilde{X}_\ell + \delta\overline{X}\}] = \mathsf{E}[\max\{\widetilde{X}_1, \ldots, \widetilde{X}_r\}] + \delta\mathsf{E}[\overline{X}],$$

where $(\widetilde{X}_1,\dots,\widetilde{X}_\ell) \sim \mathscr{N}(0, \mathrm{v}^2\mathbb{I})$ and $\overline{X} = \frac{1}{\ell}\sum_{i=1}^{\ell}\widetilde{X}_i$ is zero mean. This completes the proof of the asymptotic approximations (1.162) whenever the weights $w_i = w = \log p_{0,1}$ are the same for $i = 1,\dots,\ell$ and $\mathbf{Y}$ has the covariance matrix defined in (1.161).

The values of universal constants $\tilde{h}_\ell$ and $\widetilde{C}_\ell$ for $\ell = 2,3,\dots,1000$ are given in Tables 1.3 and 1.4.

TABLE 1.3
Expected values of standard normal order statistics.

ℓ	$\tilde{h}_\ell$	ℓ	$\tilde{h}_\ell$	ℓ	$\tilde{h}_\ell$
2	0.56418 95835	14	1.70338 1555	80	2.42677 4421
3	0.84628 43753	15	1.73591 3445	90	2.46970 0479
4	1.02937 5373	16	1.76599 1393	100	2.50759 3639
5	1.16296 4473	17	1.79394 1981	200	2.74604 2451
6	1.26720 6361	18	1.82003 1880	300	2.87776 6853
7	1.35217 8376	19	1.84448 1512	400	2.96817 8187
8	1.42360 0306	20	1.86747 5060	500	3.03669 9351
9	1.48501 3162	30	2.04276 0846	600	3.09170 2266
10	1.53875 2731	40	2.16077 7180	700	3.13754 7901
11	1.58643 6352	50	2.24907 3631	800	3.17679 1412
12	1.62922 7640	60	2.31927 8210	900	3.21105 5997
13	1.66799 0177	70	2.37735 9241	1000	3.24143 5777

TABLE 1.4
Values of the absolute constant $\widetilde{C}_\ell$ in the case $\mathbf{V} = \mathbb{I}$.

ℓ	$\widetilde{C}_\ell$	ℓ	$\widetilde{C}_\ell$	ℓ	$\widetilde{C}_\ell$
2	0.0	14	2.20924	80	5.08274
3	0.27566	15	2.31444	90	5.28802
4	0.55133	16	2.41374	100	5.47243
5	0.80002	17	2.50776	200	6.70147
6	1.02174	18	2.59705	300	7.43096
7	1.22030	19	2.68205	400	7.95237
8	1.39953	20	2.76316	500	8.35874
9	1.56262	30	3.41871	600	8.69193
10	1.71210	40	3.89695	700	8.97438
11	1.85003	50	4.27404	800	9.21958
12	1.97802	60	4.58561	900	9.43625
13	2.09740	70	4.85120	1000	9.63036

Consider now a fully symmetric case where densities $f_0^i(X_n(i)) = f_0(X_n(i))$ and $f_1^i(X_n(i)) = f_1(X_n(i))$ do not depend on i for all $i = 1,\dots,N$. Then the KL distances $I_0^i = I_0$ are the same for all $i = 1,\dots,N$, where

$$I_0 = \int \log\left[\frac{f_0(x)}{f_1(x)}\right] f_0(x)\,\mathrm{d}x. \tag{1.166}$$

Further, assume that the prior probability of the null hypothesis (absence of any signal) $\Pr(\mathsf{H}_0) = p_0$, $0 < p_0 < 1$, and $p_{1,\mathscr{B}} = p_{0,\mathscr{B}} = p_{\mathscr{B}}$, where $p_{\mathscr{B}}$ is the prior distribution of location of signals. Let the latter distribution be uniform, i.e.,

$$p_{\mathscr{B}} = \frac{(1-p_0)\omega^{|\mathscr{B}|}}{\sum_{\mathscr{B}\in\mathscr{P}}\omega^{|\mathscr{B}|}},$$

where ω is some positive number.

By the complete symmetry of the problem, $\ell = N$, and it is easy to show that the covariance matrix $\mathbf{V}$ has the form (1.161) where $\varepsilon = (N-2)v^2$. Furthermore,

$$w_i = w = \log p_i = \log\left[\frac{(1-p_0)\omega}{\sum_{\mathscr{B}\in\mathscr{P}}\omega^{|\mathscr{B}|}}\right], \quad i = 1,\dots,N.$$

Therefore, the expected sample sizes of the GSLRT and MSLRT can be computed using approximations (1.162), where w is given by the above formula, $C_\ell = C_N$ is given by (1.164) with

$$\widetilde{Y}_1 = Y_1 - \frac{\sqrt{1+N(N-2)}-1}{\sqrt{1+N(N-2)}}\left(\frac{1}{N}\sum_{i=1}^{N}\widetilde{Y}_i\right)$$

(see (1.163)) and where $0 \le R_N \le \log N$.

Note also that, by Theorem 1.6(i), the expected sample sizes $\mathsf{E}_{\mathscr{B}}[\widehat{T}]$ and $\mathsf{E}_{\mathscr{B}}[\widetilde{T}]$ under $\mathsf{P}_{\mathscr{B}}$ are computed using approximations (1.113), where

$$\log p_{1,\mathscr{B}} = \log\left[\frac{(1-p_0)\omega^{|\mathscr{B}|}}{\sum_{\mathscr{B}\in\mathscr{P}}\omega^{|\mathscr{B}|}}\right]$$

and $I_{\mathscr{B}} = I_1|\mathscr{B}|$,

$$I_1 = \int \log\left[\frac{f_1(x)}{f_0(x)}\right] f_1(x)\,\mathrm{d}x.$$

Example 1.5 (*Testing for a multistream exponential model*). Consider a fully symmetric exponential model with densities $f_0^i(X_n(i)) = f_0(X_n(i))$ and $f_1^i(X_n(i)) = f_1(X_n(i))$ $(i = 1,\dots,N)$ of the form

$$f_0(x) = \exp(-x)\,\mathbb{1}_{\{[0,\infty)\}}(x), \quad f_1(x) = \frac{1}{1+q}\exp\left(-\frac{x}{1+q}\right)\mathbb{1}_{\{[0,\infty)\}}(x). \tag{1.167}$$

As already discussed in Example 6.3, this model arises in many applications. Aside from testing for Poisson processes based on observing intervals between events, it is useful in radar applications when detecting targets in multichannel systems (range and doppler streams). In particular, suppose that one wants to detect a fluctuating signal in additive white Gaussian noise from data at the output of a pre-processing scheme which consists of a match filter and square-law detector [4]. For the Swerling II model, under the assumption that the signal has slow Gaussian fluctuations within pulses and fast fluctuations between pulses, the pre-processed observations are exponentially distributed and independent. After appropriate normalization, we obtain the above model, where q is the signal-to-noise ratio.

In the rest of this example, we will be interested in the case where only a single signal can be present in one of N streams, i.e., we consider class $\mathscr{P} = \overline{\mathscr{P}}_1$. It is easy to see that the KL numbers I_0 and I_1 are given by:

$$I_0^i = I_0 = \log(1+q) - q/(1+q), \quad I_1^i = I_1 = q - \log(1+q), \quad i = 1,\dots,N \tag{1.168}$$

and that

$$\Delta\lambda_i(n) = \frac{q}{1+q}X_n(i) - \log(1+q).$$

Recall that

$$\varkappa_i = \int_0^\infty y\,\mathrm{d}\mathscr{H}_i(y), \quad \gamma_i = \int_0^\infty e^{-y}\,\mathrm{d}\mathscr{H}_i(y), \quad \varkappa_0^i = \int_0^\infty y\,\mathrm{d}\mathscr{H}_0^i(y), \quad \gamma_0^i = \int_0^\infty e^{-y}\,\mathrm{d}\mathscr{H}_0^i(y),$$

where

$$\mathscr{H}_i(y) = \lim_{c\to\infty} \mathsf{P}_i(\kappa_i^1(c) \le y), \quad \mathscr{H}_0^i(y) = \lim_{c\to\infty} \mathsf{P}_0(\kappa_i^0(c) \le y), \quad y > 0,$$

are limiting distribution of the overshoots $\kappa_i(c) = \lambda_i(\tau_c^i) - c$ and $\kappa_i^0(c) = -\lambda_i(\sigma_c^i) - c = W_i(\sigma_c^i) - c$ in the one-sided sequential tests $\tau_c^i = \inf\{n : \lambda_i(n) \ge c\}$ and $\sigma_c^i = \inf\{n : W_i(n) \ge c\}$, respectively. Due to the symmetry of the problem in further computations it suffices to focus on $i = 1$.

Since under P_1 the distribution of $\Delta\lambda_1(1)$ has an exponential right tail,

$$\mathsf{P}_1\{\Delta\lambda_1(1) > z\} = \frac{1}{(1+q)^{1/q}} \exp\left\{-\frac{1}{q}z\right\} \mathbb{1}_{\{[-\log(1+q),\infty)\}}(z), \tag{1.169}$$

the distribution of the overshoot $\kappa_1(c)$ is exponential for all $c > 0$:

$$\mathsf{P}_1\{\kappa_1(c) > y\} = \exp\left\{-\frac{1}{q}y\right\} \mathbb{1}_{\{[0,\infty)\}}(y) = 1 - \mathscr{H}_1(y)$$

(cf. Example E.2). Thus,

$$\varkappa_1 = q, \quad \gamma_1 = \int_0^\infty e^{-y}\, \mathrm{d}\mathscr{H}_1(y) = 1/(1+q).$$

To compute the constant γ_0^1 it suffices to use the identity (1.92), which yields $\gamma_0^1 = I_1\gamma_1/I_0$.

To compute $\varkappa_0^1$, we will exploit the following representation for the density of the overshoot $\kappa_1^0(c)$ under H_0:

$$p_0(y) = \frac{1}{I_0}\mathsf{P}_0\left\{\inf_{n\ge 1} W_1(n) > y\right\} = \frac{1}{I_0}\mathsf{P}_0\{W_1(n) > y, n \ge 1\}.$$

For $y \le \log(1+q)$, introduce the stopping time

$$\tau_-(y) = \inf\{n \ge 1 : W_1(n) \le y\},$$

and let $\chi_-(y) = y - W_1(\tau_-)$ denote the overshoot at stopping. Noting that

$$\exp\{W_1(n)\} = \frac{\mathrm{d}\mathsf{P}_0^n}{\mathrm{d}\mathsf{P}_1^n}$$

and using Wald's likelihood ratio identity yields

$$\begin{aligned}\mathsf{P}_0\{W_1(n) > y, n \ge 1\} &= \mathsf{P}_0\{\tau_-(y) = \infty\} = 1 - \mathsf{P}_0\{\tau_-(y) < \infty\} \\ &= 1 - \mathsf{E}_1\left[\exp\{W_1(\tau_-)\}\mathbb{1}_{\{\tau_-<\infty\}}\right] = 1 - e^y \mathsf{E}_1\left[e^{-\chi_-(y)}\right].\end{aligned}$$

It follows from (1.169) that for any $y \le \log(1+q)$

$$\mathsf{P}_1\{\chi_-(y) \ge z\} = \exp\left\{-\frac{z}{q}\right\} \mathbb{1}_{\{[0,\infty)\}}(z), \quad \mathsf{E}_1\left[e^{-\chi_-(y)}\right] = (1+q)^{-1},$$

and hence,

$$p_0(y) = \frac{1}{I_0}\left(1 - \frac{e^y}{1+q}\right) \mathbb{1}_{\{[0,\log(1+q)]\}}(y).$$

Using this last expression, we obtain

$$\varkappa_0^1 = \frac{1}{2I_0}[\log(1+q)]^2 - 1.$$

The constant $v^2 = q^2/(1+q)^2$. It is easy to show that

$$Y_i(n) = \frac{q}{1+q} \sum_{j \in \mathcal{N} \setminus \{i\}} (1 - X_n(j)).$$

Now, using the definition of $\tilde{Y}_1$ given in (1.163) and performing straightforward calculation yield

$$\mathsf{E}_0\left[\tilde{Y}_1^3\right] = \frac{2q^3}{(1+q)^3}\left[1 + \frac{4}{N^2} - \frac{6}{N}\right].$$

Using these constants, we can compute the expected sample sizes for both tests for any given priors and threshold values. Specifically, by (1.113), the higher order (HO) approximations for the ESS of both tests when the signal is located in the ith stream are

$$\mathsf{E}_i[\widehat{T}] \approx \mathsf{E}_i[\widetilde{T}] \approx \max\left\{1, \frac{1}{q - \log(1+q)}(b - \log p_{1,i} + q)\right\}, \quad i = 1, \ldots, N.$$

Also, assume uniform prior $p_i = (1 - p_0)/N$, $i = 1, \ldots, N$ ($p_{1,i} = p_{0,i}$), where $p_0 = \Pr(H_0)$ is the probability of signal absence in all streams. Then, using (1.162), we obtain that the HO approximation for the ESS of the MSLRT under H_0 is

$$\mathsf{E}_0[\widetilde{T}] \approx \frac{1}{I_0(q)}\left\{a - \log[(N/1 - p_0)] + \frac{q\tilde{h}_N}{1+q}\sqrt{\frac{a - \log[(N/1 - p_0)]}{I_0(q)} + \frac{q^2\tilde{h}_K^2}{4(1+q)^2 I_0(q)}}\right.$$
$$\left. + \frac{q^2\tilde{h}_\ell^2}{2(1+q)^2 I_0(q)} + \frac{[\log(1+q)]^2}{2I_0(q)} + \frac{q}{3(1+q)}\left(1 + \frac{4}{N^2} - \frac{6}{N}\right) - R_N\right\},$$

where $I_0(q) = \log(1+q) - q/(1+q)$, and the ESS of the GSLRT is given by the same approximate formula with $R_N = 0$.

Recall that the first-order (FO) approximations are

$$\mathsf{E}_i[\widehat{T}] \approx \mathsf{E}_i[\widetilde{T}] \approx \max\left\{1, \frac{b}{q - \log(1+q)}\right\}, \quad i = 1, \ldots, N;$$

$$\mathsf{E}_0[\widehat{T}] \approx \mathsf{E}_0[\widetilde{T}] \approx \max\left\{1, \frac{a}{\log(1+q) - q/(1+q)}\right\}.$$

The results of computations as well as MC simulations are given in Table 1.5 for the class of alternatives $\overline{\mathscr{P}}_1$, i.e., when only a single signal (with unknown location) can be present in the multichannel system, assuming uniform prior $p_i = (1 - p_0)/N$, $i = 1, \ldots, N$. In the table, MCESS correspond to MC simulations and FOESS and HOESS correspond to first-order and higher-order approximations, respectively. The values of $\hat{\alpha}$ and $\hat{\beta}$ correspond to the probabilities of errors computed by MC simulations. The values of $\varepsilon_{\rm fo}$ and $\varepsilon_{\rm ho}$ stand for the relative errors of the first-order and higher-order approximations (compared to MC estimates). The analysis of simulation results in the table shows that while the first order approximations for the expected sample size are fairly inaccurate in most cases, the derived higher-order approximations (up to a vanishing term) are accurate not only for large but also for moderate sample sizes. This is especially true for the null hypothesis. But even for the hypothesis H_1, the HO approximations are substantially more accurate compared to the FO ones.

For the MSLRT, the error probability constraints are sharply met by setting the thresholds as specified in Theorem 1.5; see (1.107). This is true not only for the false alarm probability (as may be expected from the asymptotic equality (1.104)) but also for the missed detection probability, which satisfies only the asymptotic inequality (1.106). For the GSLRT, however, it is much more difficult to meet the error probabilities constraints, as expected from Theorem 1.5. However, if the thresholds for the GSLRT are chosen (by trial and error) to meet the constraints tightly, then the expected sample sizes are practically the same for both tests.

TABLE 1.5
Results of computations for $N = 4$, $p_0 = 0.5$, and $q = 0.5$ for class of alternatives $\overline{\mathscr{P}}_1$.

Results for MSLRT									
	Error Probab. & Thresholds			Expected Sample Size					
	α,β	a,b	$\hat{\alpha},\hat{\beta}$	MCESS	FOESS	$\varepsilon_{\text{fo}}\%$	HOESS	$\varepsilon_{\text{ho}}\ \%$
H_0	0.01	3.78	9.1×10^{-3}	77.48	52.36	32.41	75.81	2.15
H_1	0.001	4.42	9.0×10^{-4}	66.78	46.78	29.93	66.74	0.06
H_0	0.001	6.08	7.8×10^{-4}	118.92	84.28	29.13	118.10	0.68
H_1	0.0001	6.72	9.4×10^{-5}	92.01	71.14	22.68	92.02	1.00
Results for GSLRT									
	α,β	a,b	$\hat{\alpha},\hat{\beta}$	MCESS	FOESS	$\varepsilon_{\text{fo}}\%$	HOESS	$\varepsilon_{\text{ho}}\ \%$
H_0	0.01	4.47	7.0×10^{-3}	81.64	61.97	24.09	88.89	8.88
H_1	0.001	4.42	1.1×10^{-3}	65.83	46.78	28.93	66.74	1.38
H_0	0.001	6.77	5.7×10^{-4}	124.41	93.89	24.53	130.35	4.76
H_1	0.0001	6.72	1.0×10^{-4}	91.28	71.14	22.06	91.10	0.21

Example 1.6 (*Signal detection in Gaussian noise*). Consider the additive model where $X_n(i) = S_i + V_n(i)$ if a signal S_i is present in the ith stream and $X_n(i) = V_n(i)$ otherwise. We will suppose that the signal S_i is constant and sensor noise $V_n(i)$ in the ith stream is the sequence of i.i.d. Gaussian random variables with mean zero and variance σ_i^2 ($V_n(i)$ and $V_n(j)$ are independent for $i \neq j$). In radar applications, S_i usually represents the result of preprocessing (attenuation and matched filtering) of modulated pulses in the ith stream.

In this case, the LLR in the ith stream is of the form

$$\lambda_i(n) = \frac{S_i}{\sigma_i^2} \sum_{t=1}^{n} X_t(i) - \frac{S_i^2}{2\sigma_i^2} n$$

and has i.i.d. Gaussian increments. The KL numbers are given by

$$I_1^i = I_0^i = \frac{S_i^2}{2\sigma_i^2} = q_i, \quad i = 1, \dots, N,$$

where q_i can be interpreted as signal-to-noise ratios (SNR) in streams. Thus, the approximations (1.104)–(1.106) and (1.113) can be used to evaluate the probabilities of errors and the expected sample sizes $\mathsf{E}_{\mathscr{B}}[\widehat{T}]$ and $\mathsf{E}_{\mathscr{B}}[\widetilde{T}]$, where $I_{\mathscr{B}} = \sum_{i \in \mathscr{B}} q_i$, $I_0 = \min_{1 \le i \le N} q_i$ and the values of $\gamma_{\mathscr{B}}$, γ_0^1, $\varkappa_{\mathscr{B}}$, and $\varkappa_0^1$ are calculated using renewal-theoretic techniques described in Section E.1 (Appendix E). Specifically, define $\gamma(y)$ and $\varkappa(y)$ as follows:

$$\gamma(y) = \frac{1}{y} \exp\left\{ -2 \sum_{n=1}^{\infty} \frac{1}{n} \Phi\left(-\sqrt{\frac{y}{2} n} \right) \right\}, \tag{1.170}$$

$$\varkappa(y) = 1 + \frac{y}{2} - \sqrt{2y} \sum_{n=1}^{\infty} \left[\frac{1}{\sqrt{n}} \varphi\left(\sqrt{\frac{y}{2} n} \right) - \sqrt{\frac{y}{2} n}\, \Phi\left(-\sqrt{\frac{y}{2} n} \right) \right]. \tag{1.171}$$

Then

$$\gamma_{\mathscr{B}} = \gamma(I_{\mathscr{B}}), \quad \gamma_0^1 = \gamma(I_0), \tag{1.172}$$

$$\varkappa_{\mathscr{B}} = \varkappa(I_{\mathscr{B}}), \quad \varkappa_0^1 = \varkappa(I_0). \tag{1.173}$$

In Table 1.6, we present the values of $\gamma(y)$ and $\varkappa(y)$ for $y = 0.05(0.05)1.0$. For relatively small y, one can also use the following simple Siegmund's corrected Brownian motion approximations [140]:

$$\gamma(y) \approx \gamma_{\text{app}}(y) = \exp\{-\alpha\sqrt{2y}\}, \quad \varkappa(y) \approx \varkappa_{\text{app}}(y) = y/4 + \alpha\sqrt{2y}, \tag{1.174}$$

where

$$\alpha = -\frac{1}{\pi}\int_0^\infty \frac{1}{t^2}\log\left\{\frac{2\left(1-e^{-t^2/2}\right)}{t^2}\right\} dt \approx 0.582597.$$

The remaining term in the approximations (1.174) is of the order $o(y)$ for small y, and therefore, their accuracy grows when SNR decreases. The values of $\gamma_{\text{app}}(y)$ and $\varkappa_{\text{app}}(y)$ are also given in Table 1.6. It is seen that the approximations (1.174) are quite accurate and can be used when y is not very large.

TABLE 1.6
The values of $\gamma(y)$ and $\varkappa(y)$

y	γ	γ_{app}	$\varkappa$	$\varkappa_{\text{app}}$
0.05	0.83183	0.83174	0.19706	0.19673
0.10	0.77087	0.77063	0.28647	0.28555
0.15	0.72721	0.72680	0.35830	0.35660
0.20	0.69240	0.69179	0.42109	0.41847
0.25	0.66316	0.66235	0.47812	0.47446
0.30	0.63784	0.63681	0.53109	0.52628
0.35	0.61544	0.61420	0.58100	0.57494
0.40	0.59534	0.59387	0.62849	0.62109
0.45	0.57709	0.57539	0.67403	0.66520
0.50	0.56037	0.55845	0.71794	0.70760
0.55	0.54495	0.54279	0.76046	0.74853
0.60	0.53063	0.52824	0.80179	0.78820
0.65	0.51728	0.51465	0.84208	0.82676
0.70	0.50477	0.50191	0.88145	0.86434
0.75	0.49301	0.48991	0.92000	0.90103
0.80	0.48191	0.47858	0.95783	0.93693
0.85	0.47142	0.46785	0.99499	0.97211
0.90	0.46146	0.45766	1.03156	1.00664
0.95	0.45200	0.44796	1.06757	1.04055
1.00	0.44298	0.43871	1.10309	1.07392

In the symmetric case of $\ell = N$ when SNRs are the same for all streams (i.e., $q_i = q$ for $i = 1,\dots,N$) and $w_i = w$ the approximation (1.162) can be used for evaluating the ESS under hypothesis H_0. Evidently, the vectors $\mathbf{Y}$ and $\widetilde{\mathbf{Y}}$ are both Gaussian and zero mean. Hence $\mathsf{E}_0[Y_1]^3 = 0$. The constant $\mathrm{v}^2 = 2q$. Therefore, by (1.162), the ESS of the GSLRT under H_0 can be approximated as

$$\mathsf{E}_0[\widehat{T}] = \frac{1}{q}\left(a + w + \tilde{h}_N\sqrt{2\,(a + w + q\tilde{h}_N^2)} + \tilde{h}_\ell^2 + \varkappa_0^1\right) + o(1), \tag{1.175}$$

where $\varkappa_0^1$ is given by (1.173), and the ESS of the MSLRT by the same expression subtracting R_N, $0 \le R_N \le \log N$.

Below we present the results of MC simulations in the fully symmetric case for the GSLRT for class $\mathscr{P} = \overline{\mathscr{P}}_1$, i.e., in the scenario where only a single signal can appear in one of N streams (it is unknown in which one), assuming also that the weights are the same, $p_{1,i} = p_{0,i} = p_i = 1/N$,

$i = 1,\dots,N$. In this case, $w = -\log N$ in (1.175), so

$$\mathsf{E}_0[\widehat{T}] \approx \max\left\{1, \frac{1}{q}\left(a - \log N + \tilde{h}_N\sqrt{2\left(a - \log N + q\tilde{h}_N^2\right)} + \tilde{h}_\ell^2 + \varkappa(q)\right)\right\}. \tag{1.176}$$

Note that due to the symmetry, $\varkappa_0^1 = \varkappa_1 = \varkappa(q)$ and $\gamma_1 = \gamma_0^1 = \gamma(q)$, where $\varkappa(q)$ and $\gamma(q)$ are given by (1.170) and (1.171) and tabulated in Table 1.6. By (1.113), the higher order (HO) approximation for the ESS under P_i (signal is located in the ith stream) is

$$\mathsf{E}_i[\widehat{T}] \approx \max\left\{1, \frac{1}{q}\left(b - \log N + \varkappa(q)\right)\right\}, \quad i = 1,\dots,N. \tag{1.177}$$

For comparison, we also used the first order (FO) approximations for the expected sample sizes

$$\mathsf{E}_i[\widehat{T}] \approx \max\{1, b/q\}, \quad \mathsf{E}_0[\widehat{T}] \approx \max\{1, a/q\}. \tag{1.178}$$

In Table 1.7, we present the MC estimates of expected sample sizes $\mathsf{E}_0[\widehat{T}]$ and $\mathsf{E}_1[\widehat{T}]$ along with the theoretical approximate values computed according to (1.176)–(1.178) for the 10-stream system ($N = 10$). The abbreviations MCESS, FOESS, and HOESS are used for the expected sample sizes obtained by the MC experiment, by FO approximations (1.178) and by HO approximations (1.176)–(1.177), respectively. We also show the MC estimates for the probabilities of errors $\mathsf{PFA} = \mathsf{P}_0(\widehat{d} = 1)$ and $\mathsf{PMS} = \mathsf{P}_1(\widehat{d} = 0)(= \max_{1\le i\le N}\mathsf{P}_i(\widehat{d} = 0))$. The results in the table correspond to the thresholds

$$a = \log[N\gamma(q)/\beta], \quad b = \log[\gamma(q)/\alpha]$$

(see (1.107)) that account for overshoots and correspond to asymptotic upper bounds in (1.105) and (1.106). It is seen that these formulas allow one to obtain very accurate approximations for the true probability of false alarm as long as the ESS is not too small, but not for the probability of missed detection. As mentioned in previous sections, this can be expected. FOESS (1.178) has a satisfactory accuracy for the hypothesis H_1 but is poor for the hypothesis H_0. This could be expected, since the FO approximation for $\mathsf{E}_0[\widehat{T}]$ neglects the second term that increases at the rate of the square root of the threshold. HOESS gives fairly accurate estimates in all cases where SNR q is not very large, i.e., ESS is not very small.

TABLE 1.7
Results for 10-stream system with $b = \log(\gamma/\alpha)$ and $a = \log(10\gamma/\beta)$ for $\alpha = 10^{-3}$ and $\beta = 10^{-1}$ and various values of the SNR q.

	H_1					H_0				
q	PMS	MCESS	FOESS	HOESS	$\mathscr{E}_1$	PFA	MCESS	FOESS	HOESS	$\mathscr{E}_0$
0.2	0.020	176.8	180.5	181.5	1.41	$9.1\cdot10^{-4}$	148.4	42.37	154.0	1.68
0.4	0.020	86.75	89.50	89.37	1.44	$4.6\cdot10^{-4}$	72.34	20.42	75.06	1.73
1	0.021	33.20	35.20	34.11	1.51	$2.3\cdot10^{-4}$	27.08	7.57	27.92	1.85
2	0.021	15.24	17.26	15.70	1.64	$2.1\cdot10^{-4}$	12.12	3.45	12.33	2.06
4	0.020	6.31	8.40	6.50	1.98	$9.6\cdot10^{-5}$	4.74	1.49	4.61	2.64
8	0.012	2.22	4.04	1.91	2.81	$8.0\cdot10^{-5}$	1.74	1.00	1.00	3.59

In addition, in the columns $\mathscr{E}_1$ and $\mathscr{E}_0$, we show the results of the comparison with the multichannel fixed sample size (FSS) detection algorithm, which is based on the GLR statistic:

$$d_n = \begin{cases} 1 & \text{if } \max_{1\le i\le N}\lambda_i(n) \ge c \\ 0 & \text{if } \max_{1\le i\le N}\lambda_i(n) < c, \end{cases}$$

where c is a threshold. The efficiency of the sequential detection algorithm compared to the FSS algorithm is defined as

$$\mathscr{E}_i(\mathsf{PFA},\mathsf{PMS},q) = \frac{n_0(\mathsf{PFA},\mathsf{PMS},q)}{\mathsf{E}_i[\widehat{T}(\mathsf{PFA},\mathsf{PMS},q)]}, \quad i = 0,1,$$

where $n_0(\mathsf{PFA},\mathsf{PMS},q)$ is the sample size which is required to guarantee the probabilities of errors $\mathsf{PFA} = \mathsf{P}_0(d_{n_0} = 1)$ and $\mathsf{PMS} = \mathsf{P}_1(d_{n_0} = 0)$ in the FSS detection procedure. We now show that

$$\begin{aligned} n_0 &= \frac{1}{2q}\left\{\Phi^{-1}\left[(1-\mathsf{PFA})^{1/N}\right] + \Phi^{-1}\left[1 - \frac{\mathsf{PMS}}{(1-\mathsf{PMS})^{1-1/N}}\right]\right\}^2, \\ c &= \left\{\Phi^{-1}\left[(1-\mathsf{PFA})^{1/N}\right] + \Phi^{-1}\left[1 - \frac{\mathsf{PMS}}{(1-\mathsf{PMS})^{1-1/N}}\right]\right\}^2, \end{aligned} \tag{1.179}$$

where $\Phi^{-1}(\alpha) = \{x : \Phi(x) = \alpha\}$ is the α-quantile of the standard normal distribution. Indeed, it is easily seen that

$$\mathsf{P}_0(d_n = 1) = \mathsf{P}_0\left\{\max_{1\le i\le N} \lambda_i(n) \ge c\right\} = 1 - \prod_{i=1}^N \mathsf{P}_0\{\lambda_i(n) < c\} = 1 - \left[\Phi\left(\frac{c-qn}{\sqrt{2qn}}\right)\right]^N,$$

$$\begin{aligned} \mathsf{P}_1(d_n = 0) &= \mathsf{P}_1\left\{\max_{1\le i\le N} \lambda_i(n) < c\right\} = \mathsf{P}_1\{\lambda_1(n) < c\}\prod_{i=2}^N \mathsf{P}_0\{\lambda_i(n) < c\} \\ &= \Phi\left(\frac{c+qn}{\sqrt{2qn}}\right)\left[\Phi\left(\frac{c-qn}{\sqrt{2qn}}\right)\right]^{N-1}, \end{aligned}$$

and we obtain the following two equations for the threshold c and sample size n:

$$\Phi\left(\frac{c-qn}{\sqrt{2qn}}\right) = (1-\mathsf{PFA})^{1/N}, \quad (1-\mathsf{PFA})^{1-1/N}\Phi\left(\frac{c+qn}{\sqrt{2qn}}\right) = \mathsf{PMS}.$$

Solving these equations yields (1.179).

The data in Table 1.7 allow us to conclude that the GSLRT requires much smaller ESS compared to the sample size of the GLR FSS test. Typically, the sample size is $1.7-3.6$ times smaller when there is no signal and $1.4-2.8$ times smaller when there is a signal.

1.5.4 High-Order Asymptotic Optimality

1.5.4.1 Uniform Asymptotic Optimality

We begin by proving that both tests minimize the ESS to second order (i.e., within an $O(1)$ term) under $\mathsf{P}_{\mathscr{B}}$ for all $\mathscr{B} \in \mathscr{P}$ and also under P_0 in the asymmetric case where $\ell = 1$. We formulate the results only for the MSLRT since for the GSLRT the results are the same.

Theorem 1.8. *Let* $\mathsf{E}_i|\lambda_i(1)|^2 < \infty$ *and* $\mathsf{E}_0|\lambda_i(1)|^2 < \infty$, $i = 1,\dots,N$. *Let* α *and* β *approach* 0 *so that* $\beta/|\log\alpha|^2 \to 0$ *and* $\alpha/|\log\beta|^2 \to 0$. *Assume that thresholds* a *and* b *are so selected that* $\tilde{\delta} = (\tilde{T},\tilde{d}) \in \mathbb{C}_{\alpha,\beta}(\mathscr{P})$ *and, as* $\alpha_{\max} \to 0$,

$$\begin{aligned} |\log\alpha| + o(1) &\le |\log \mathsf{P}_0(\tilde{d} = 1)| \le |\log\alpha| + |\log c| + o(1), \\ |\log\beta| + o(1) &\le |\max_{\mathscr{B}\in\mathscr{P}} \log \mathsf{P}_{\mathscr{B}}(\tilde{d} = 0)| \le |\log\beta| + |\log c| + o(1), \end{aligned} \tag{1.180}$$

where $c \in (0,1]$.

(i) *Then*

$$\mathsf{E}_{\mathscr{B}}[\widetilde{T}] = \inf_{\delta\in\mathbb{C}_{\alpha,\beta}(\mathscr{P})} \mathsf{E}_{\mathscr{B}}[T] + O(1) \quad \text{for all } \mathscr{B}\in\mathscr{P} \quad \text{as } \alpha_{\max}\to 0, \tag{1.181}$$

and if in addition $\ell = 1$ *in* (1.96), *then*

$$\mathsf{E}_0[\widetilde{T}] = \inf_{\delta\in\mathbb{C}_{\alpha,\beta}(\mathscr{P})} \mathsf{E}_0[T] + O(1) \quad \text{as } \alpha_{\max}\to 0. \tag{1.182}$$

(ii) *If* $\ell > 1$ *in* (1.96), *then*

$$\mathsf{E}_0[\widetilde{T}] = \inf_{\delta\in\mathbb{C}_{\alpha,\beta}(\mathscr{P})} \mathsf{E}_0[T](1+o(1)) \quad \text{as } \alpha_{\max}\to 0. \tag{1.183}$$

Proof. Proof of part (i). It follows from the well-known Wald's lower bounds for the ESS (cf., e.g., Theorem 3.2.2 (page 149) in [164]) that, as $\alpha_{\max}\to 0$,

$$\inf_{\delta\in\mathbb{C}_{\alpha,\beta}(\mathscr{P})} \mathsf{E}_{\mathscr{B}}[T] \ge \frac{|\log\alpha|}{I_{\mathscr{B}}} + o(1), \tag{1.184}$$

$$\inf_{\delta\in\mathbb{C}_{\alpha,\beta}(\mathscr{P})} \mathsf{E}_0[T] \ge \frac{|\log\beta|}{I_0(\mathscr{P})} + o(1), \tag{1.185}$$

where $I_0(\mathscr{P}) = \min_{\mathscr{B}\in\mathscr{P}} I_0^{\mathscr{B}}$. But it follows from Theorem 1.6(i) that

$$\mathsf{E}_{\mathscr{B}}[\widetilde{T}] = \frac{b}{I_{\mathscr{B}}} + O(1) = \frac{|\log\alpha|}{I_{\mathscr{B}}} + O(1)$$

when thresholds are selected so that inequalities (1.180) are satisfied. This proves (1.181).

Now, from Theorem 1.6(ii) it follows that when $\ell = 1$

$$\mathsf{E}_0[\widetilde{T}] = \frac{a}{I_0} + O(1) = \frac{|\log\beta|}{I_0} + O(1)$$

when thresholds are selected so that inequalities (1.180) are satisfied. Since $I_0 = \min_{\mathscr{B}\in\mathscr{P}} I_0^{\mathscr{B}}$ when $\ell = 1$, comparing with the lower bound (1.185) yields (1.182).

Proof of part (ii). The assertion (ii) follows directly from Corollary 1.2. □

While we have no proof, we conjecture that both tests minimize the ESS $\mathsf{E}_0[T]$ to second order even for $\ell > 1$ under conditions of Theorem 1.7.

1.5.4.2 Bayesian-type Asymptotic Optimality

Denote by $\widetilde{\delta}^*(\mathbf{p}) = (\widetilde{T}^*(\mathbf{p}), \widetilde{d}^*(\mathbf{p}))$ and $\widehat{\delta}^*(\mathbf{p}) = (\widehat{T}^*(\mathbf{p}), \widehat{d}^*(\mathbf{p}))$ the MSLRT and the GSLRT with weights

$$p_{1,\mathscr{B}} = \frac{p_{\mathscr{B}}}{\mathscr{L}_{\mathscr{B}}\overline{p}_1} \quad \text{and} \quad p_{0,\mathscr{B}} = \frac{p_{\mathscr{B}}\mathscr{L}_{\mathscr{B}}}{\overline{p}_0}, \quad \mathscr{B}\in\mathscr{P}, \tag{1.186}$$

where $\overline{p}_1 = \sum_{\mathscr{B}\in\mathscr{P}}(p_{\mathscr{B}}/\mathscr{L}_{\mathscr{B}})$, $\overline{p}_0 = \sum_{\mathscr{B}\in\mathscr{P}}(p_{\mathscr{B}}\mathscr{L}_{\mathscr{B}})$, $\mathbf{p} = \{p_{\mathscr{B}}, \mathscr{B}\in\mathscr{P}\}$, $p_{\mathscr{B}} > 0$ for every $\mathscr{B}\in\mathscr{P}$, and $\sum_{\mathscr{B}\in\mathscr{P}} p_{\mathscr{B}} = 1$ and where the $\mathscr{L}$-number $\mathscr{L}_{\mathscr{B}}$, which takes into account the overshoot, is defined in (1.91).

From Corollary 1.3(i) and the fact that $\mathscr{L}_{\mathscr{B}} = \gamma_{\mathscr{B}} I_{\mathscr{B}}$ (cf. identity (1.92)), it follows that if thresholds a and b are selected so that $\mathsf{P}_0(\widetilde{d}^*(\mathbf{p}) = 1) \sim \alpha$ and $\mathsf{P}_0(\widehat{d}^*(\mathbf{p}) = 1) \sim \alpha$, then

$$\mathsf{E}_{\mathscr{B}}[\widetilde{T}^*(\mathbf{p})] = \frac{1}{I_{\mathscr{B}}}\left[|\log\alpha| + \varkappa_{\mathscr{B}} + \log\gamma_{\mathscr{B}} + C_{\mathscr{B}}(\mathbf{p})\right] + o(1), \tag{1.187}$$

$$\mathsf{E}_{\mathscr{B}}[\widehat{T}^*(\mathbf{p})] \le \frac{1}{I_{\mathscr{B}}}[|\log\alpha| + \varkappa_{\mathscr{B}} + \log\gamma_{\mathscr{B}} + C_{\mathscr{B}}(\mathbf{p})] + o(1), \tag{1.188}$$

where

$$C_{\mathscr{B}}(\mathbf{p}) = \log\left(\sum_{\mathscr{A}\in\mathscr{P}} \frac{p_{\mathscr{A}}}{I_{\mathscr{A}}}\right) - \log\frac{p_{\mathscr{B}}}{I_{\mathscr{B}}}, \quad \mathscr{B}\in\mathscr{P}. \tag{1.189}$$

The next theorem states that $\widetilde{\delta}^*(\mathbf{p})$ and $\widehat{\delta}^*(\mathbf{p})$ are third-order asymptotically optimal, minimizing the average ESS $\mathsf{E}^{\mathbf{p}}[T]$ within an $o(1)$ term, where $\mathsf{E}^{\mathbf{p}}$ is expectation with respect to the probability measure $\mathsf{P}^{\mathbf{p}} = \sum_{\mathscr{B}\in\mathscr{P}} p_{\mathscr{B}} \mathsf{P}_{\mathscr{B}}$.

Theorem 1.9. *Let* $\mathsf{E}_i|\lambda_i(1)|^2 < \infty$ *and* $\mathsf{E}_0|\lambda_i(1)|^2 < \infty$, $i = 1,\dots,N$. *Let* α *and* β *approach* 0 *so that* $|\log\beta|/|\log\alpha| \to 1$. *Then, as* $\alpha_{\max} \to 0$,

$$\inf_{\delta\in\mathbb{C}_{\alpha,\beta}(\mathscr{P})} \mathsf{E}^{\mathbf{p}}[T] = \sum_{\mathscr{B}\in\mathscr{P}} \frac{p_{\mathscr{B}}}{I_{\mathscr{B}}}[|\log\alpha| + \varkappa_{\mathscr{B}} + \log\gamma_{\mathscr{B}} + C_{\mathscr{B}}(\mathbf{p})] + o(1). \tag{1.190}$$

If thresholds a and b are so selected that conditions (1.180) *hold, then, as* $\alpha_{\max} \to 0$,

$$\inf_{\delta\in\mathbb{C}_{\alpha,\beta}(\mathscr{P})} \mathsf{E}^{\mathbf{p}}[T] = \mathsf{E}^{\mathbf{p}}[\widetilde{T}^*(\mathbf{p})] + o(1), \qquad \inf_{\delta\in\mathbb{C}_{\alpha,\beta}(\mathscr{P})} \mathsf{E}^{\mathbf{p}}[T] = \mathsf{E}^{\mathbf{p}}[\widehat{T}^*(\mathbf{p})] + o(1). \tag{1.191}$$

If thresholds a and b are selected so that $\widetilde{\delta}^*(\mathbf{p})$ *and* $\widehat{\delta}^*(\mathbf{p})$ *belong to* $\mathbb{C}_{\alpha,\beta}(\mathscr{P})$ *and* $\mathsf{P}_0(\widetilde{d}^*(\mathbf{p}) = 1) \sim \alpha$ *and* $\mathsf{P}_0(\widehat{d}^*(\mathbf{p}) = 1) \sim \alpha$, *then asymptotic relationships* (1.191) *are satisfied, and hence, both sequential tests are asymptotically optimal to third order in class* $\mathbb{C}_{\alpha,\beta}(\mathscr{P})$.

Proof. To prove this theorem, consider the purely Bayesian sequential testing problem with $|\mathscr{P}|+1$ states, $\mathsf{H}_0 : f^i = f_0^i$ for all $i = 1,\dots,N$ and $\mathsf{H}_1^{\mathscr{B}} : f^{\mathscr{B}} = f_1^{\mathscr{B}}$, $\mathscr{B}\in\mathscr{P}$, and two terminal decisions $d = 0$ (accept H_0) and $d = 1$ (accept $\mathsf{H}_1 = \cup_{\mathscr{B}\in\mathscr{P}} H_1^{\mathscr{B}}$). Specifically, let c denote the sampling cost per observation and let the loss $L(d,\mathsf{H})$ associated with making a decision d when the hypothesis H is correct be

$$L(d=i,\mathsf{H}=\mathsf{H}_j) = \begin{cases} L_1 & \text{if } i=0, j=1 \\ L_0 & \text{if } i=1, j=0 \\ 0 & \text{otherwise} \end{cases}.$$

Let $\pi = \Pr(\mathsf{H}_0)$ be the prior probability of the hypothesis H_0 and $p_{\mathscr{B}} = \Pr(\mathsf{H}_1^{\mathscr{B}}|\mathsf{H}_1)$ be the prior probability of $f^{\mathscr{B}} = f_1^{\mathscr{B}}$ given that H_1 is correct. Define the probability measure $\mathbf{P} = \pi\,\mathsf{P}_0 + (1-\pi)\,\mathsf{P}^{\mathbf{p}}$ and let $\mathbf{E}$ denote the corresponding expectation.

The average risk of a sequential test $\delta = (T,d)$ is $\mathscr{R}(\delta) = \mathscr{R}_s(\delta) + c\,\mathbf{E}[T]$, where $c\,\mathbf{E}[T]$ is the average cost of sampling and $\mathscr{R}_s(\delta)$ is the average risk due to a wrong decision upon stopping,

$$c\,\mathbf{E}[T] = c\,\{\pi\,\mathsf{E}_0[T] + (1-\pi)\,\mathsf{E}^{\mathbf{p}}[T]\},$$
$$\mathscr{R}_s(\delta) = \mathbf{E}[L(d,H)] = \pi L_0\,\mathsf{P}_0(d=1) + (1-\pi)L_1\,\mathsf{P}^{\mathbf{p}}(d=0).$$

Denote by $\mathscr{R}^* = \inf_\delta \mathscr{R}(\delta)$ the minimal Bayes risk that corresponds to an unknown optimal Bayes sequential test. Let thresholds a and b be chosen as

$$a_c = \log\left(\frac{1-\pi}{\pi}\frac{L_1}{c}\right) \quad \text{and} \quad b_c = \log\left(\frac{\pi}{1-\pi}\frac{L_0}{c}\right) \tag{1.192}$$

and let $\widetilde{\delta}_c^*(\mathbf{p})$ and $\widehat{\delta}_c^*(\mathbf{p})$ denote the sequential tests $\widetilde{\delta}^*(\mathbf{p})$ and $\widehat{\delta}^*(\mathbf{p})$ whose thresholds are given by a_c and b_c. It follows from Lorden [86] that $\widetilde{\delta}_c^*(\mathbf{p})$ and $\widehat{\delta}_c^*(\mathbf{p})$ are *almost Bayes* for a small cost c under the second moment conditions $\mathsf{E}_i|\lambda_i(1)|^2 < \infty$:

$$\mathscr{R}(\widetilde{\delta}_c^*(\mathbf{p})) - \mathscr{R}^* = o(c) \quad \text{and} \quad \mathscr{R}(\widehat{\delta}_c^*(\mathbf{p})) - \mathscr{R}^* = o(c) \quad \text{as } c\to 0. \tag{1.193}$$

In the rest of the proof, we focus only on the MSLRT. For the GSLRT, the proof is essentially the same.

Using Corollary 1.3, we immediately obtain that if thresholds a, b are selected so that $\mathsf{P}_0(\widetilde{d}^* = 1) \sim \alpha$ and $\mathsf{P}_0(\widehat{d}^* = 1) \sim \alpha$, then as $\alpha_{\max} \to 0$

$$\mathbf{E}[\widetilde{T}^*(\mathbf{p})] = \sum_{\mathscr{B} \in \mathscr{P}} \frac{p_{\mathscr{B}}}{I_{\mathscr{B}}} [|\log \alpha| + \varkappa_{\mathscr{B}} + \log \gamma_{\mathscr{B}} + C_{\mathscr{B}}(\mathbf{p})] + o(1)$$

(which is the right-hand side in (1.190)). Hence, if for $\inf_{\delta \in \mathbb{C}_{\alpha,\beta}(\mathscr{P})} \mathsf{E}^{\mathbf{p}}[T]$ approximation (1.190) holds and additionally the MSLRT belongs to $\mathbb{C}_{\alpha,\beta}(\mathscr{P})$, then infimum $\inf_{\delta \in \mathbb{C}_{\alpha,\beta}(\mathscr{P})} \mathsf{E}^{\mathbf{p}}[T]$ is attained by the MSLRT to within a negligible term $o(1)$. Thus, it suffices to establish (1.190).

Consider the class of sequential tests

$$\mathbb{C}^{\mathbf{p}}_{\alpha,\beta}(\mathscr{P}) = \{\delta : \mathsf{P}_0(d = 1) \le \alpha \quad \text{and} \quad \mathsf{P}^{\mathbf{p}}(d = 0) \le \beta\}.$$

Since $\mathbb{C}_{\alpha,\beta}(\mathscr{P}) \subset \mathbb{C}^{\mathbf{p}}_{\alpha,\beta}(\mathscr{P})$, we have

$$\inf_{\delta \in \mathbb{C}_{\alpha,\beta}(\mathscr{P})} \mathsf{E}^{\mathbf{p}}[T] \ge \inf_{\delta \in \mathbb{C}^{\mathbf{p}}_{\alpha,\beta}(\mathscr{P})} \mathsf{E}^{\mathbf{p}}[T].$$

Thus, it suffices to show that

$$\inf_{\delta \in \mathbb{C}^{\mathbf{p}}_{\alpha,\beta}(\mathscr{P})} \mathsf{E}^{\mathbf{p}}[T] = \sum_{\mathscr{B} \in \mathscr{P}} \frac{p_{\mathscr{B}}}{I_{\mathscr{B}}} \Big[|\log \alpha| + \varkappa_{\mathscr{B}} + \log \gamma_{\mathscr{B}} + C_{\mathscr{B}}(\mathbf{p})\Big] + o(1). \tag{1.194}$$

Consider now the sequential test $\widetilde{\delta}^*_c(\mathbf{p})$ with thresholds a_c and b_c selected so that $\mathsf{P}_0(\widetilde{d}^*_c = 1) = \alpha$ and $\mathsf{P}^{\mathbf{p}}(\widetilde{d}^*_c = 0) = \beta$. It follows from Corollary 1.3(i) that $\mathsf{E}^{\mathbf{p}}[\widetilde{T}^*_c]$ is equal to the right-hand side in (1.194) as $c \to 0$, so it suffices to show that

$$\inf_{\delta \in \mathbb{C}^{\mathbf{p}}_{\alpha,\beta}(\mathscr{P})} \mathsf{E}^{\mathbf{p}}[T] = \mathsf{E}^{\mathbf{p}}[\widetilde{T}^*_c] + o(1) \quad \text{as } c \to 0. \tag{1.195}$$

More specifically, if δ is an arbitrarily sequential test in $\mathbb{C}^{\mathbf{p}}_{\alpha,\beta}(\mathscr{P})$, we need to show that, for a sufficiently small c, $|\mathsf{E}^{\mathbf{p}}[T] - \mathsf{E}^{\mathbf{p}}[\widetilde{T}^*_c]|$ is bounded above by an arbitrarily small, but fixed number.

Observe that

$$\mathscr{R}_s(\delta) = \pi L_0 \, \mathsf{P}_0(d = 1) + (1 - \pi) w_1 \, \mathsf{P}^{\mathbf{p}}(d = 0) \le \pi L_0 \, \alpha + (1 - \pi) w_1 \, \beta = \mathscr{R}_s(\widetilde{d}^*_c), \tag{1.196}$$

where the inequality is due to the fact that $\delta \in \mathbb{C}^{\mathbf{p}}_{\alpha,\beta}(\mathscr{P})$ and the second equality follows from the assumption that $\mathsf{P}_0(\widetilde{d}^*_c = 1) = \alpha$ and $\mathsf{P}^{\mathbf{p}}(\widetilde{d}^*_c = 0) = \beta$.

Using the upper bounds for the probabilities of errors (1.24) in Lemma 1.2 and the definition of a_c and b_c in (1.192), we obtain

$$\begin{aligned}
\mathscr{R}_s(\widetilde{\delta}^*_c) &= \pi L_0 \, \mathsf{P}_0(\widetilde{d}^*_c = 1) + (1 - \pi) L_1 \, \mathsf{P}^{\mathbf{p}}(\widetilde{d}^*_c = 0) \\
&\le \pi L_0 e^{-b_c} + (1 - \pi) L_1 \sum_{\mathscr{B} \in \mathscr{P}} \frac{p_{\mathscr{B}}}{p_{0,\mathscr{B}}} e^{-a_c} \\
&\le (1 - \pi) c + \sum_{\mathscr{B} \in \mathscr{P}} \frac{p_{\mathscr{B}}}{p_{0,\mathscr{B}}} \pi c \le (Q - 1) c,
\end{aligned} \tag{1.197}$$

where $Q > 1$ is some constant that does not depend on c or π.

For an arbitrarily fixed $\varepsilon > 0$, introduce the sequential test $\delta_{\varepsilon c} = (T_{\varepsilon c}, d_{\varepsilon c})$ as

$$T_{\varepsilon c} = \min\{\widetilde{T}^*_{\varepsilon c}, T\}, \quad d_{\varepsilon c} = d \mathbb{1}_{\{T \le \widetilde{T}^*_{\varepsilon c}\}} + \widetilde{d}^*_{\varepsilon c} \mathbb{1}_{\{T > \widetilde{T}^*_{\varepsilon c}\}}.$$

Obviously,

$$\mathscr{R}_s(\delta_{\varepsilon c}) \le \mathscr{R}_s(\delta) + \mathscr{R}_s(\widetilde{\delta}^*_{\varepsilon c}) \le \mathscr{R}_s(\widetilde{\delta}^*_c) + \mathscr{R}_s(\widetilde{\delta}^*_{\varepsilon c}) \le \mathscr{R}_s(\widetilde{\delta}^*_c) + (Q-1)c\varepsilon, \tag{1.198}$$

where the second inequality is due to (1.196) and the third is due to (1.197).

Since $\widetilde{\delta}^*_c$ is almost Bayes (see (1.193)), for a sufficiently small c,

$$\mathscr{R}_s(\widetilde{\delta}^*_c) + c\mathbf{E}[\widetilde{T}^*_c] \le \mathscr{R}_s(\delta_{\varepsilon c}) + c\mathbf{E}[T_{\varepsilon c}] + c\varepsilon. \tag{1.199}$$

From (1.198) it follows that $c\mathbf{E}[\widetilde{T}^*_c] \le c\mathbf{E}[T_{\varepsilon c}] + Qc\varepsilon$ and since $T_{\varepsilon c} = \min\{\widetilde{T}^*_{\varepsilon c}, T\}$, we obtain

$$\pi \mathsf{E}_0[\widetilde{T}^*_c] + (1-\pi)\mathsf{E}^{\mathbf{p}}[\widetilde{T}^*_c] \le \pi \mathsf{E}_0[T_{\varepsilon c}] + (1-\pi)\,\mathsf{E}^{\mathbf{p}}[T_{\varepsilon c}] + Q\varepsilon \le \pi \mathsf{E}_0[\widetilde{T}^*_{\varepsilon c}] + (1-\pi)\,\mathsf{E}^{\mathbf{p}}[T] + Q\varepsilon. \tag{1.200}$$

Rearranging terms, we obtain from (1.200) that

$$\mathsf{E}^{\mathbf{p}}[\widetilde{T}^*_c] - \mathsf{E}^{\mathbf{p}}[T] \le \frac{\pi}{1-\pi}\left(\mathsf{E}_0[\widetilde{T}^*_{\varepsilon c}] - \mathsf{E}_0[\widetilde{T}^*_c]\right) + \frac{Q\varepsilon}{1-\pi}. \tag{1.201}$$

Since the last inequality holds for any $\pi \in (0,1)$, setting $\pi = \varepsilon/(1+\varepsilon)$ yields $b_c = \log(\varepsilon L_0/c)$ and $a_c = \log[L_1/(\varepsilon c)]$, whereas (1.201) becomes

$$\mathsf{E}^{\mathbf{p}}[\widetilde{T}^*_c] - \mathsf{E}^{\mathbf{p}}[T] \le \varepsilon\,(\mathsf{E}_0[\widetilde{T}^*_{\varepsilon c}] - \mathsf{E}_0[\widetilde{T}^*_c]) + Q\varepsilon(1+\varepsilon). \tag{1.202}$$

Since the right-hand side in (1.202) does not depend on T, we also have

$$\mathsf{E}^{\mathbf{p}}[\widetilde{T}^*_c] - \inf_{\delta \in \mathbb{C}^{\mathbf{p}}_{\alpha,\beta}(\mathscr{P})} \mathsf{E}^{\mathbf{p}}[T] \le \varepsilon\,(\mathsf{E}_0[\widetilde{T}^*_{\varepsilon c}] - \mathsf{E}_0[\widetilde{T}^*_c]) + Q\varepsilon(1+\varepsilon).$$

But from asymptotic expansions (1.114) and (1.160) it follows that, as $c \to 0$,

$$I_0\,(\mathsf{E}_0[\widetilde{T}^*_{\varepsilon c}] - \mathsf{E}_0[\widetilde{T}^*_c]) = O(a_{\varepsilon c} - a_c)$$

and, from (1.192), we have $a_{\varepsilon c} - a_c = |\log\varepsilon| + O(1)$ as $c \to 0$. Since ε can be arbitrarily small, the asymptotic relationship (1.195) follows and the proof is complete. □

It is worth stressing that including the $\mathscr{L}$-numbers in the weights (1.186) is important. These numbers are necessary corrections for overshoots that make the resulting tests nearly optimal up to third order. Any other choice leads to second-order optimality at best.

Remark 1.10. A similar argument applies to show that if $\mathsf{P}_0(\widetilde{d}^*(\mathbf{p}) = 1) = \alpha$ and $\mathsf{P}^{\mathbf{p}}(\widetilde{d}^*(\mathbf{p}) = 0) = \beta$, then

$$\inf_{\delta \in \mathbb{C}_{\alpha,\beta}(\mathscr{P})} \mathsf{E}_0[T] \ge \inf_{\delta \in \mathbb{C}^{\mathbf{p}}_{\alpha,\beta}(\mathscr{P})} \mathsf{E}_0[T] = \mathsf{E}_0[\widetilde{T}^*(\mathbf{p})] + o(1),$$

and similarly for the GSLRT. Both tests remain third-order optimal in class $\delta \in \mathbb{C}^{\mathbf{p}}_{\alpha,\beta}(\mathscr{P})$, but not in class $\mathbb{C}_{\alpha,\beta}(\mathscr{P})$, since the right-hand side in this asymptotic lower bound is generally not attained by $\widetilde{\delta}^*(\mathbf{p})$ or $\widehat{\delta}^*(\mathbf{p})$ when their thresholds are selected so that $\widetilde{\delta}^*(\mathbf{p}), \widehat{\delta}^*(\mathbf{p}) \in \mathbb{C}_{\alpha,\beta}(\mathscr{P})$.

Remark 1.11. While we have no rigorous proof, we believe that the assertions of Theorem 1.9 (as well as of Theorem 1.10 below) hold true in the more general case where α and β approach zero in such a way that the ratio $\log\alpha/\log\beta$ is bounded away from zero and infinity, which allows one to cover the asymptotically asymmetric case where $\log\alpha/\log\beta \to d$ with $0 < d < \infty$ as well.

1.5.4.3 Asymptotic Minimax Properties with Respect to Kullback–Leibler Information

For any stopping time T and any $\mathscr{B} \in \mathscr{P}$, let $\mathscr{I}_{\mathscr{B}}(T) = I_{\mathscr{B}} \mathsf{E}_{\mathscr{B}}[T] = \mathsf{E}_{\mathscr{B}}[\lambda_{\mathscr{B}}(T)]$ denote the expected KL information (divergence between $\mathsf{P}_{\mathscr{B}}$ and P_0) that has been accumulated up to time T.

Let $\hat{\mathbf{p}} = \{\hat{p}_{\mathscr{B}}, \mathscr{B} \in \mathscr{P}\}$ denote the prior distribution in (1.186) given by

$$\hat{p}_{\mathscr{B}} = \frac{\mathscr{L}_{\mathscr{B}} e^{\varkappa_{\mathscr{B}}}}{\sum_{\mathscr{A} \in \mathscr{P}} \mathscr{L}_{\mathscr{A}} e^{\varkappa_{\mathscr{A}}}}, \quad \mathscr{B} \in \mathscr{P}. \tag{1.203}$$

Using Theorem 1.6(i), we obtain

$$\mathscr{I}_{\mathscr{B}}(\widetilde{T}^*(\hat{\mathbf{p}})) = b + \log\left(\sum_{\mathscr{A} \in \mathscr{P}} e^{\mathscr{L}_{\mathscr{A}} \varkappa_{\mathscr{A}}}\right) + o(1), \tag{1.204}$$

$$\mathscr{I}_{\mathscr{B}}(\widehat{T}^*(\hat{\mathbf{p}})) = b + \log\left(\sum_{\mathscr{A} \in \mathscr{P}} e^{\mathscr{L}_{\mathscr{A}} \varkappa_{\mathscr{A}}}\right) + o(1), \tag{1.205}$$

where only negligible terms $o(1)$ may depend on $\mathscr{B}$. In other words, the weights (1.186) with $\hat{\mathbf{p}}$ (almost) equalize the KL information, accumulated by both the MSLRT and the GSLRT until stopping, in the sense that $\mathscr{I}_{\mathscr{B}}(\widetilde{T}^*(\hat{\mathbf{p}}))$ and $\mathscr{I}_{\mathscr{B}}(\widehat{T}^*(\hat{\mathbf{p}}))$ are independent of $\mathscr{B}$ up to an $o(1)$ term. Now, if b is selected so that $\mathsf{P}(\widetilde{d}^*(\hat{\mathbf{p}}) = 1) \sim \alpha$ and $\mathsf{P}(\widehat{d}^*(\hat{\mathbf{p}}) = 1) \sim \alpha$, then (1.125)–(1.126) imply that for all $\mathscr{B} \in \mathscr{P}$,

$$\mathscr{I}_{\mathscr{B}}(\widetilde{T}^*(\hat{\mathbf{p}})) = |\log \alpha| + \log\left(\sum_{\mathscr{A} \in \mathscr{P}} \gamma_{\mathscr{A}} e^{\varkappa_{\mathscr{A}}}\right) + o(1), \tag{1.206}$$

$$\mathscr{I}_{\mathscr{B}}(\widehat{T}^*(\hat{\mathbf{p}})) \le |\log \alpha| + \log\left(\sum_{\mathscr{A} \in \mathscr{P}} \gamma_{\mathscr{A}} e^{\varkappa_{\mathscr{A}}}\right) + o(1), \tag{1.207}$$

so that for the maximal expected KL information $\widehat{\mathscr{I}}(T) = \max_{\mathscr{B} \in \mathscr{P}} \mathscr{I}_{\mathscr{B}}(T)$ of the MSLRT and the GSLRT we have the asymptotic (as $\alpha_{\max} \to 0$) approximations

$$\widehat{\mathscr{I}}(\widetilde{T}^*(\hat{\mathbf{p}})) = |\log \alpha| + \log\left(\sum_{\mathscr{B} \in \mathscr{P}} \gamma_{\mathscr{B}} e^{\varkappa_{\mathscr{B}}}\right) + o(1), \tag{1.208}$$

$$\widehat{\mathscr{I}}(\widehat{T}^*(\hat{\mathbf{p}})) \le |\log \alpha| + \log\left(\sum_{\mathscr{B} \in \mathscr{P}} \gamma_{\mathscr{B}} e^{\varkappa_{\mathscr{B}}}\right) + o(1). \tag{1.209}$$

Since by the general decision theory, the minimax test should be an equalizer, we may expect that both tests $\widetilde{\delta}^*(\hat{\mathbf{p}}) = (\widetilde{T}^*(\hat{\mathbf{p}}), \widetilde{d}^*(\hat{\mathbf{p}}))$ and $\widehat{\delta}^*(\hat{\mathbf{p}}) = (\widehat{T}^*(\hat{\mathbf{p}}), \widehat{d}^*(\hat{\mathbf{p}}))$ are almost minimax in the KL sense. The following theorem justifies this conjecture.

Theorem 1.10. *Let* $\mathsf{E}_i|\lambda_i(1)|^2 < \infty$ *and* $\mathsf{E}_0|\lambda_i(1)|^2 < \infty$, $i = 1, \dots, N$. *Let* α *and* β *approach* 0 *so that* $|\log \beta| / |\log \alpha| \to 1$. *Then, as* $\alpha_{\max} \to 0$,

$$\inf_{\delta \in \mathbb{C}_{\alpha,\beta}(\mathscr{P})} \widehat{\mathscr{I}}(T) = |\log \alpha| + \log\left(\sum_{\mathscr{B} \in \mathscr{P}} \gamma_{\mathscr{B}} e^{\varkappa_{\mathscr{B}}}\right) + o(1). \tag{1.210}$$

If thresholds a and b are so selected that conditions (1.180) *hold, then, as* $\alpha_{\max} \to 0$,

$$\inf_{\delta \in \mathbb{C}_{\alpha,\beta}(\mathscr{P})} \widehat{\mathscr{I}}(T) = \widehat{\mathscr{I}}(\widetilde{T}^*(\hat{\mathbf{p}})) + o(1), \qquad \inf_{\delta \in \mathbb{C}_{\alpha,\beta}(\mathscr{P})} \widehat{\mathscr{I}}(T) = \widehat{\mathscr{I}}(\widehat{T}^*(\hat{\mathbf{p}})) + o(1). \tag{1.211}$$

If thresholds a and b are selected so that $\widetilde{\delta}^*(\hat{\mathbf{p}})$ *and* $\widehat{\delta}^*(\hat{\mathbf{p}})$ *belong to* $\mathbb{C}_{\alpha,\beta}(\mathscr{P})$ *and* $\mathsf{P}_0(\widetilde{d}^*(\mathbf{p}) = 1) \sim \alpha$ *and* $\mathsf{P}_0(\widehat{d}^*(\mathbf{p}) = 1) \sim \alpha$, *then asymptotic relationships* (1.211) *are satisfied, and hence, both sequential tests are asymptotically minimax to third order in class* $\mathbb{C}_{\alpha,\beta}(\mathscr{P})$.

Proof. Consider the MSLRT $\widetilde{\delta}^*(\hat{\mathbf{p}})$. By (1.208), the maximal KL information $\mathscr{I}_{\mathscr{B}}(\widetilde{T}^*(\hat{\mathbf{p}}))$ of the test $\widetilde{\delta}^*(\hat{\mathbf{p}})$ is equal to the right-hand side of equality (1.210) whenever thresholds a and b are so selected that conditions (1.180) hold, as well as if $\widetilde{\delta}^*(\hat{\mathbf{p}}) \in \mathbb{C}_{\alpha,\beta}(\mathscr{P})$ and $\mathsf{P}_0(\widetilde{d}^*(\mathbf{p}) = 1) \sim \alpha$. Therefore, to prove (1.210) it suffices to show that the following lower bound holds:

$$\inf_{\delta \in \mathbb{C}_{\alpha,\beta}(\mathscr{P})} \widehat{\mathscr{I}}[T] \geq |\log \alpha| + \log\left(\sum_{\mathscr{B} \in \mathscr{P}} \gamma_{\mathscr{B}} e^{\varkappa_{\mathscr{B}}}\right) + o(1). \tag{1.212}$$

Obviously,

$$\left(\sum_{\mathscr{B} \in \mathscr{P}} \frac{\hat{p}_{\mathscr{B}}}{I_{\mathscr{B}}}\right) \inf_{\delta \in \mathbb{C}_{\alpha,\beta}(\mathscr{P})} \widehat{\mathscr{I}}(T) \geq \inf_{\delta \in \mathbb{C}_{\alpha,\beta}(\mathscr{P})} \sum_{\mathscr{B} \in \mathscr{P}} \frac{\hat{p}_{\mathscr{B}}}{I_{\mathscr{B}}} \mathscr{I}_{\mathscr{B}}(T) = \inf_{\delta \in \mathbb{C}_{\alpha,\beta}(\mathscr{P})} \sum_{\mathscr{B} \in \mathscr{P}} \hat{p}_{\mathscr{B}} \mathsf{E}_{\mathscr{B}}[T].$$

If a and b are selected so that $\widetilde{\delta}^*(\hat{\mathbf{p}}) \in \mathbb{C}_{\alpha,\beta}(\mathscr{P})$ and $\mathsf{P}_0(\widetilde{d}^*(\hat{\mathbf{p}}) = 1) \sim \alpha$, then it follows from Theorem 1.9 that

$$\inf_{\delta \in \mathbb{C}_{\alpha,\beta}(\mathscr{P})} \sum_{\mathscr{B} \in \mathscr{P}} \hat{p}_{\mathscr{B}} \mathsf{E}_{\mathscr{B}}[T] = \sum_{\mathscr{B} \in \mathscr{P}} \hat{p}_{\mathscr{B}} \mathsf{E}_{\mathscr{B}}[\widetilde{T}^*(\hat{\mathbf{p}})] + o(1)$$

and from (1.206) that

$$\sum_{\mathscr{B} \in \mathscr{P}} \hat{p}_{\mathscr{B}} \mathsf{E}_{\mathscr{B}}[\widetilde{T}^*(\hat{\mathbf{p}})] = \left(\sum_{\mathscr{B} \in \mathscr{P}} \frac{\hat{p}_{\mathscr{B}}}{I_{\mathscr{B}}}\right) \left[|\log \alpha| + \log\left(\sum_{\mathscr{B} \in \mathscr{P}} \gamma_{\mathscr{B}} e^{\varkappa_{\mathscr{B}}}\right) + o(1)\right].$$

which yields (1.212), and therefore, (1.210).

Asymptotic equality (1.211) follows from (1.208) and (1.210).

For the GSLRT, the argument is the same. □

1.5.4.4 Further Optimization and MC Simulations

Typically the prior distribution $\{p_{\mathscr{B}}\}$ needed for defining the weights (1.186), and therefore, for the design of the tests is not known in practice. One way is to select it as in (1.82), which leads to nearly minimax tests with respect to the KL information. However, for practical reasons, it is useful to consider possible alternative ways of specification of the prior distribution $\mathbf{p}$ that determines the weights $\mathbf{p}_0$ and $\mathbf{p}_1$ of the MSLRT and the GSLRT that allow for certain optimization of performance in terms of the expected sample sizes.

We propose to quantify the resulting performance loss for the tests under $\mathsf{P}_{\mathscr{B}}$ by the measure

$$\Delta_{\mathscr{B}}(\mathbf{p}) = \frac{\mathsf{E}_{\mathscr{B}}[\widetilde{T}^*(\mathbf{p})] - \mathsf{E}_{\mathscr{B}}[T_{\mathscr{B}}]}{\mathsf{E}_{\mathscr{B}}[T_{\mathscr{B}}]}, \quad \mathscr{B} \in \mathscr{P},$$

where

$$T_{\mathscr{B}} = \inf\{n \geq 1 : \lambda_{\mathscr{B}}(n) \notin (-a_{\alpha,\beta}, b_{\alpha,\beta})\}$$

is the stopping time of the SPRT $(T_{\mathscr{B}}, d_{\mathscr{B}})$ for testing f_0 against $f_{\mathscr{B}}$ with type-I and type-II error probabilities α and β, respectively. Since the SPRT is strictly optimal when the pattern $\mathscr{B}$ is known, the measure $\Delta_{\mathscr{B}}(\mathbf{p})$ represents the *additional* expected sample size due to the uncertainty in the alternative hypothesis divided by the smallest possible expected sample size that is required for testing f_0 against $f_{\mathscr{B}}$.

Using the following well-known asymptotic (as $a_{\alpha,\beta}, b_{\alpha,\beta} \to \infty$) approximations for the probabilities of errors and ESS of the SPRT

$$\mathsf{P}_0(d_{\mathscr{B}} = 1) = \gamma_{\mathscr{B}} e^{-b}(1 + o(1), \quad \mathsf{P}_{\mathscr{B}}(d_{\mathscr{B}} = 0) = \gamma_0 e^{-a}(1 + o(1), \quad \mathsf{E}_{\mathscr{B}}[T_{\mathscr{B}}] = \frac{1}{I_{\mathscr{B}}}(b + \varkappa_{\mathscr{B}}) + o(1)$$

(cf. Theorem 3.1.2 (page 129) and Theorem 3.1.4 (page 131) in [164]), we obtain that under the second moment conditions $\mathsf{E}_i|\lambda_i(1)|^2 < \infty$ and $\mathsf{E}_0|\lambda_i(1)|^2 < \infty$, $i = 1, \dots, N$, assumed in Theorem 1.9, the following asymptotic expansion for the ESS of the SPRT holds:

$$\mathsf{E}_{\mathscr{B}}[T_{\mathscr{B}}] = \frac{1}{I_{\mathscr{B}}}\left(|\log\alpha| + \varkappa_{\mathscr{B}} + \log\gamma_{\mathscr{B}}\right) + o(1), \quad \alpha_{\max} \to 0. \tag{1.213}$$

So, under the conditions of Theorem 1.9, from (1.187) and (1.213) it follows that

$$\Delta_{\mathscr{B}}(\mathbf{p}) \approx \frac{C_{\mathscr{B}}(\mathbf{p})}{|\log\alpha| + \varkappa_{\mathscr{B}} + \log\gamma_{\mathscr{B}}} = \frac{\log\left[\sum_{\mathscr{A}\in\mathscr{P}}(p_{\mathscr{A}}/I_{\mathscr{A}})\right] + \log I_{\mathscr{B}} - \log p_{\mathscr{B}}}{|\log\alpha| + \varkappa_{\mathscr{B}} + \log\gamma_{\mathscr{B}}}. \tag{1.214}$$

From this expression, we can see that the magnitude of $\Delta_{\mathscr{B}}(\mathbf{p})$ is mainly determined by $|\mathscr{P}|$, the cardinality of class $\mathscr{P}$, and the probability of type-I error, α. In particular, for every $\mathscr{B}$ and $\mathbf{p}$, $\Delta_{\mathscr{B}}(\mathbf{p})$ is "small" when $|\log\alpha|$ is much larger than $\log|\mathscr{P}|$, which implies that the choice of $\mathbf{p}$ may make a difference only when $|\log\alpha|$ is not much larger than $\log|\mathscr{P}|$. This is typically the case in our problem.

Below we compare several priors with respect to this criterion, in particular $\hat{\mathbf{p}}$, defined in (1.82), as well as $\mathbf{p}^I$, $\mathbf{p}^{\mathscr{L}}$, and $\mathbf{p}^u$ defined as

$$p^I_{\mathscr{B}} = \frac{I_{\mathscr{B}}}{\sum_{\mathscr{A}\in\mathscr{P}} I_{\mathscr{A}}}, \quad p^{\mathscr{L}}_{\mathscr{B}} = \frac{\mathscr{L}_{\mathscr{B}}}{\sum_{\mathscr{A}\in\mathscr{P}} \mathscr{L}_{\mathscr{A}}}, \quad p^u_{\mathscr{B}} = \frac{1}{|\mathscr{P}|}.$$

Note that $\mathbf{p}^{\mathscr{L}}$, $\mathbf{p}^I$, and $\hat{\mathbf{p}}$ are ranked in the sense that $\mathscr{L}_{\mathscr{B}} \le I_{\mathscr{B}} \le e^{\varkappa_{\mathscr{B}}}\mathscr{L}_{\mathscr{B}}$, since $\mathscr{L}_{\mathscr{B}} = \gamma_{\mathscr{B}} I_{\mathscr{B}}$ and $\gamma_{\mathscr{B}} \le 1 \le e^{\varkappa_{\mathscr{B}}}\gamma_{\mathscr{B}}$. Thus, $\mathbf{p}^{\mathscr{L}}$ (resp. $\hat{\mathbf{p}}$) assigns relatively less (resp. more) weight than $\mathbf{p}^I$ to a hypothesis as its "signal-to-noise ratio" increases. Note also that $\mathbf{p}^{\mathscr{L}}$ and $\hat{\mathbf{p}}$ reduce to $\mathbf{p}^I$ when there is no overshoot effect, in which case $\varkappa_{\mathscr{B}} = 0$ and $\gamma_{\mathscr{B}} = 1$, whereas all these priors reduce to $\mathbf{p}^u$ in the symmetric case where the KL numbers $I_i \equiv I$, $i \in \mathscr{N}$, do not depend on i.

In order to make some concrete comparisons, we focus on Example 1.5 with the exponential model (1.167). As shown in Example 1.5, the values of I_1^i, $\varkappa_i$, and γ_i are given by

$$I_1^i = q_i - \log(1+q_i), \quad \varkappa_i = q_i, \quad \gamma_i = 1/(1+q_i).$$

In the computations, we assume that $N = 2$, $q_1 = 4$ and we let q_2 vary. Thus, the signal-to-noise ratio in the first (resp. second) stream is stronger (resp. weaker) than that in the second (resp. first) stream when $q_2 < 4$ (resp. $q_2 > 4$). In Figure 1.4, we plot $\Delta_1(\mathbf{p})$ and $\Delta_2(\mathbf{p})$, the inflicted performance loss when signal is present in the first and second stream, respectively, as a function of q_2. For computations, we used approximation (1.214), in which we set $\alpha = 10^{-4}$. The results presented in Figure 1.4 show that $\mathbf{p} = \hat{\mathbf{p}}$ (resp. $\mathbf{p} = \mathbf{p}^u$) leads to a better performance when signal is present in the stream with stronger (resp. weaker) signal-to-noise ratio. However, the inflicted performance loss when the signal is present in the other stream can be very high. On the other hand, $\mathbf{p} = \mathbf{p}^I$ or $\mathbf{p} = \mathbf{p}^{\mathscr{L}}$ lead to a more robust behavior, since the resulting performance loss is relatively low and stable for various signal strengths.

Recall that this exponential example has practical meaning in radar applications, as discussed in Example 1.5.

We now present a simulation study to verify the accuracy of the asymptotic approximations (1.187) and (1.188) for the expected sample sizes and compare the MSLRT with the GSLRT for realistic probabilities of errors. We considered the multistream setup with $N = 3$ streams and the exponential model (1.167). The parameter values are selected according to Table 1.8. Since our main emphasis is on the fast signal detection, we set the missed detection probability $\beta = 10^{-2}$ and considered different values of the probability of false alarm α. We chose the thresholds a and b according to (1.107) and selected the weights according to (1.186) with prior $\mathbf{p} = \mathbf{p}^I$, i.e., in this case

$$p_{1,i} = \frac{I_i}{\mathscr{L}_k \overline{p}_1} \quad \text{and} \quad p_{0,i} = \frac{I_i \mathscr{L}_i}{\overline{p}_0}, \quad i = 1, 2, 3,$$

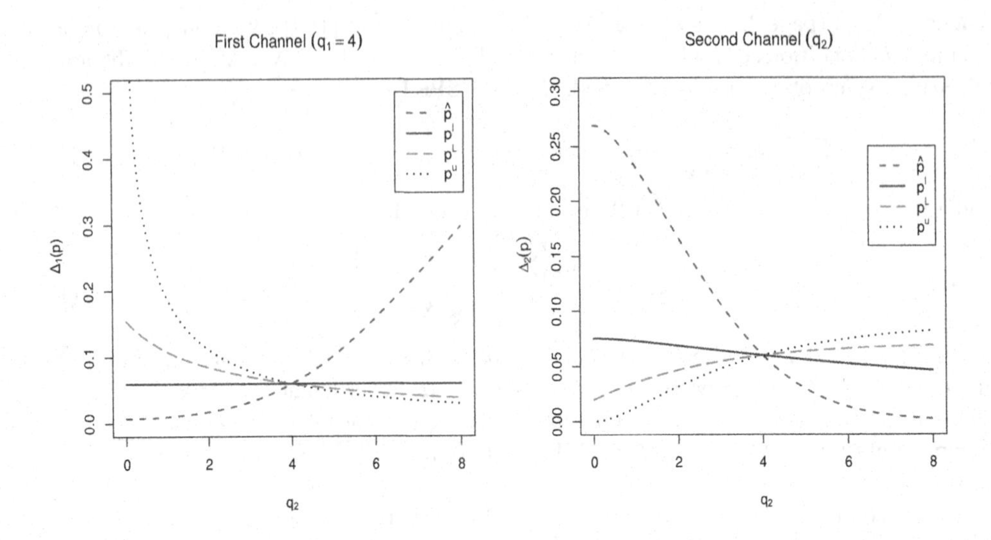

FIGURE 1.4
Performance loss for different prior distributions.

TABLE 1.8
Parameter values in a multichannel problem with exponential data.

q_i	I_i	$\varkappa_i$	γ_i	$p_{1,i} \cdot \overline{p}_1$	$p_{0,i} \cdot \overline{p}_0$
0.5	0.095	0.5	0.67	0.308	0.013
1	0.584	1	0.4	0.837	0.078
2	0.901	2	0.33	1.380	0.138

In the first three columns of Table 1.9, we compare the false alarm probabilities of the GSLRT and the MSLRT, estimated by Monte Carlo, with the given value α. More specifically, these error probabilities were computed using the asymptotic formulas (1.104), (1.105), i.e.,

$$\mathsf{P}_0(\widetilde{d}^* = 1) \approx \mathsf{P}_0(\widehat{d}^* = 1) \approx \left(\sum_{i=1}^{N} p_{1,\mathscr{B}} \gamma_i\right) e^{-b},$$

as well as MC simulations, using *importance sampling*. The results indicate that selecting threshold b according to (1.107) leads to false alarm probabilities very close to α for both tests, even for not too small α.

TABLE 1.9
Probabilities of false alarm and the expected sample sizes under P_i, $i = 1,2,3$ for different values of the target probability α when $\beta = 10^{-2}$.

α	$\frac{\mathsf{P}_0(\widetilde{d}^*=1)}{\alpha}$	$\frac{\mathsf{P}_0(\widehat{d}^*=1))}{\alpha}$	$\mathsf{E}_1[\widetilde{T}^*]$	$\mathsf{E}_1[\widehat{T}^*]$	$\mathsf{E}_2[\widetilde{T}^*]$	$\mathsf{E}_2[\widehat{T}^*]$	$\mathsf{E}_3[\widetilde{T}^*]$	$\mathsf{E}_3[\widehat{T}^*]$
10^{-2}	1.051	0.994	59.9	59.4	17.8	19.4	6.2	7.3
10^{-3}	1.033	0.995	84.1	84.1	25.7	27.1	9.0	9.9
10^{-4}	1.025	0.996	108.5	108.3	33.7	34.6	11.7	12.4
10^{-5}	1.017	0.996	132.5	132.3	41.4	42.0	14.3	15.0

In Table 1.9, we also present the MC estimates of the ESS under P_i, $i = 1,2,3$. In Figure 1.5, we plot these values against the corresponding (simulated) type-I error probabilities. In these graphs, we also superimpose asymptotic approximation (1.125) (dashed lines), as well as the asymptotic performance of the SPRT (solid lines), given by (1.213). Triangles correspond to the GSLRT and circles to the MSLRT. From these results, we can see that the approximation (1.125) is very accurate for both tests. Also, the two tests have almost the same performance. In particular, their performance is practically identical when the signal is present in the stream with the smallest signal strength. In the other two cases, the MSLRT performs slightly better.

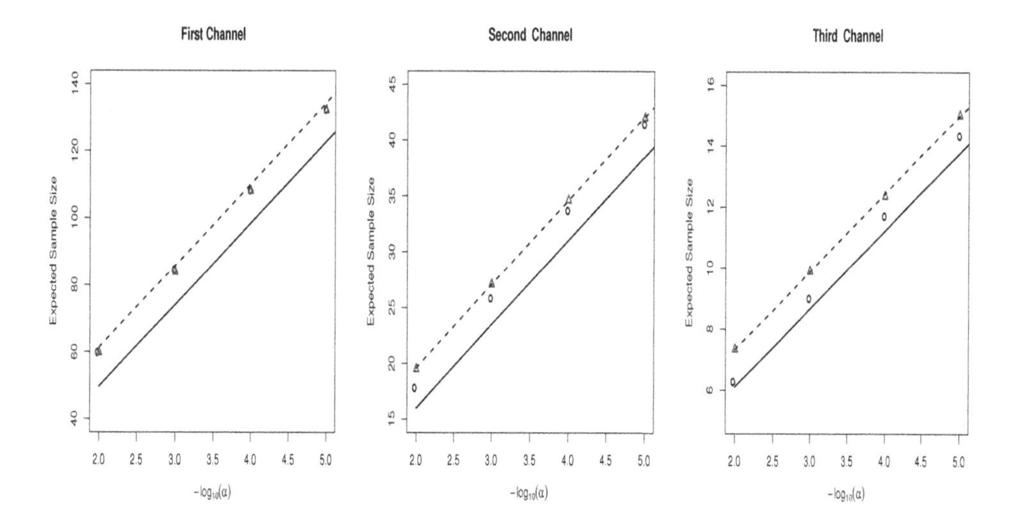

FIGURE 1.5
Simulated ESS of MSLRT (circles) and GSLRT (triangles) under P_i against type-I error probability in logarithmic scale, $i = 1,2,3$. The dashed line represents asymptotic approximation (1.125) and the solid line refers to (1.213), the asymptotic performance of the corresponding SPRT.

2

Sequential Detection of Changes: Changepoint Models, Performance Metrics and Optimality Criteria

2.1 Introduction

Sequential changepoint detection (or quickest disorder detection) is an important branch of Sequential Analysis. In the sequential setting, one assumes that the observations are made successively, one at a time, and as long as their behavior suggests that the process of interest is in a normal state, the process is allowed to continue; if the state is believed to have become anomalous, the goal is to detect the change in distribution or anomaly as rapidly as possible. Quickest change detection problems have an enormous number of important applications, e.g., object detection in noise and clutter, industrial quality control, environment surveillance, failure detection, navigation, seismology, computer network security, genomics (see, e.g., [4, 11, 65, 66, 162, 148, 159, 168, 169, 141]). Several challenging application areas are discussed in the book by Tartakovsky, Nikiforov, and Basseville [164] as well as in Chapter 8.

In the beginning of the 1960s, Shiryaev [130, 131] developed a Bayesian sequential changepoint detection theory in the i.i.d. case assuming that the observations are independent and identically distributed (i.i.d.) according to a distribution G pre-change and another distribution F post-change and that the prior distribution of the change point is geometric. In particular, Shiryaev [131] proved that the detection procedure that is based on thresholding the posterior probability of the change being active before the current time is strictly optimal, minimizing the average delay to detection in the class of procedures with a given probability of false alarm. Tartakovsky and Veeravalli [171] and Tartakovsky [160] generalized Shiryaev's theory for the non-i.i.d. case that covers very general discrete-time non-i.i.d. stochastic models and a wide class of prior distributions that include distributions with both exponential tails and heavy tails. In particular, it was proved that the Shiryaev detection procedure is asymptotically optimal – it minimizes the average delay to detection as well as higher moments of the detection delay as the probability of a false alarm is small. Baron and Tartakovsky [7] developed an asymptotic Bayesian theory for general continuous-time stochastic processes.

The key assumption in general asymptotic theories developed in [7, 160, 171] is a certain stability property of the log-likelihood ratio process between the "change" and "no-change" hypotheses, which was expressed in the form of the strong law of large numbers with a positive and finite number and its strengthened r-quick or r-complete versions.

In this chapter, we describe very general stochastic models for the observations, general change point mechanisms, and formulate optimization problems that will be addressed in subsequent chapters for finding optimal and asymptotically optimal change detection rules.

2.2 Changepoint Models

To formulate the general quickest changepoint detection problem, we begin with introducing a changepoint model, which consists of two components – the structure of the monitored process (or observations) and the model for the changepoint.

2.2.1 Models for Observed Processes

2.2.1.1 A Single Stream Scenario

Let $X_1, X_2, \dots$ denote the series of observations related to a stochastic process $\{X_n\}_{n>-\infty}$, and let ν be the changepoint at which the process changes statistical properties. In this book, we assume that $X_{\nu+1}$ is the *first* post-change observation, i.e., ν is a serial number of the *last* pre-change observation. We assume that ν may take negative values $-1,-2,\dots$, which means that the change has already occurred before the observations become available. Let P_k and E_k denote the probability measure and expectation when $\nu = k$, and P_∞ and E_∞ will stand for the case where $\nu = \infty$, i.e., when there is no change. We will assume that the values of the process $X_{-\infty}, \dots, X_{-1}, X_0$ are not observed but may affect the properties of the observations $\{X_n\}_{n\ge 1}$ through the initial condition X_0. For example, if there is a system that works for a long time and produces a stationary process, then the initial value X_0 has a stationary distribution of this process, which we will attach to the observed sample $\mathbf{X}_1^n = (X_1, \dots, X_n)$ forming an extended sample $\mathbf{X}_0^n = (X_0, X_1, \dots, X_n)$. For $0 \le t \le n$, write $\mathbf{X}_t^n = (X_t, \dots, X_n)$.

Let $p_\nu(\mathbf{X}_0^n) = p(X_0, X_1, \dots, X_n | \nu)$ be the joint probability density of $X_0, X_1, \dots, X_n$ for a fixed changepoint ν. Let $\{g_n(X_n|\mathbf{X}_0^{n-1})\}_{n\ge 1}$ and $\{f_n(X_n|\mathbf{X}_0^{n-1})\}_{n\ge 1}$ be two sequences of conditional densities of X_n given $\mathbf{X}_0^{n-1}$. In what follows we usually omit the subscript n in densities that shows that they may depend on time, as it is the case where observations are independent but non-identically distributed. In the most general non-i.i.d. case, we have

$$
\begin{aligned}
p_\nu(\mathbf{X}_0^n) &= p_\infty(\mathbf{X}_0^n) = \prod_{t=0}^{n} g(X_t|\mathbf{X}_0^{t-1}) \quad \text{for } \nu \ge n, \\
p_\nu(\mathbf{X}_0^n) &= p_\infty(\mathbf{X}_0^\nu) \times p_0(\mathbf{X}_{\nu+1}^n|\mathbf{X}_0^\nu) = \prod_{t=0}^{\nu} g(X_t|\mathbf{X}_0^{t-1}) \times \prod_{t=\nu+1}^{n} f(X_t|\mathbf{X}_0^{t-1}) \quad \text{for } \nu < n
\end{aligned}
\tag{2.1}
$$

where $g(X_0|\mathbf{X}_0^{-1}) = g(X_0)$, $f(X_0|\mathbf{X}_0^{-1}) = f(X_0)$. We will refer to the densities $g(X_n|\mathbf{X}_0^{n-1})$ and $f(X_n|\mathbf{X}_0^{n-1})$, $n \ge 1$ as the *pre-change* and *post-change* conditional densities, respectively. In other words, under the measure P_∞ the conditional density of X_n given $\mathbf{X}_0^{n-1}$ is $g(X_n|\mathbf{X}_0^{n-1})$ for every $n \ge 1$ and under the measure P_ν for any $0 \le \nu < \infty$ the conditional density of X_n is $g(X_n|\mathbf{X}_0^{n-1})$ if $n \le \nu$ and is $f(X_n|\mathbf{X}_0^{n-1})$ if $n > \nu$. Therefore, if the change occurs at time ν, then the conditional density of the $(\nu+1)$th observation changes from $g(X_{\nu+1}|\mathbf{X}_0^\nu)$ to $f(X_{\nu+1}|\mathbf{X}_0^\nu)$. Graphically this changepoint scenario is show in Figure 2.1.

Note that in general the post-change densities may and often do depend on the changepoint ν, namely $f(X_n|\mathbf{X}^{n-1}) = f^{(\nu)}(X_n|\mathbf{X}^{n-1})$, $n > \nu$. This is typically the case for state–space models and hidden Markov models due to the propagation of the change and the fact that the pre-change model affects the post-change model. In the sequel, for brevity we omit the superscript ν.

So far we have considered the case where both the pre-change and post-change conditional densities $g(X_n|\mathbf{X}_0^{n-1})$ and $f(X_n|\mathbf{X}_0^{n-1})$ are completely specified and only the change point ν is unknown, in which case the hypotheses $\mathsf{H}_\infty : \nu = \infty$ and $\mathsf{H}_k : \nu = k$ (for a fixed k) are simple. A more practical case typical for most applications is where pre-change densities $g(X_n|\mathbf{X}_0^{n-1})$ are known but there is a parametric uncertainty in the post-change scenario, i.e., post-change densities $f_\theta(X_n|\mathbf{X}_0^{n-1})$ are

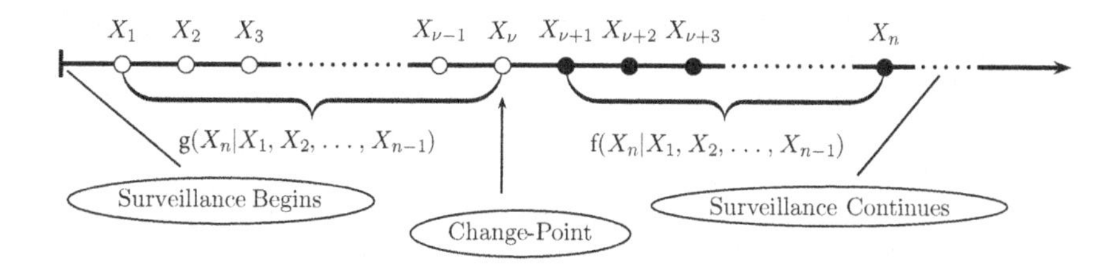

FIGURE 2.1
Changepoint scenario.

specified up to an unknown parameter $\theta \in \Theta$. While the simple setting yields a benchmark for the best one can hope, it is important to consider a more realistic scenario where the post-change parameter θ (possibly multidimensional) is unknown, since the putative value of θ is rarely representative. While in most applications the change occurs from a known value θ_0 to some unknown value $\theta \neq \theta_0$, this is not necessarily the case. So we will focus on a more general case where the pre-change densities $g(X_n|\mathbf{X}_0^{n-1})$ are known but may not belong to the same parametric class $\{f_\theta(X_n|\mathbf{X}_0^{n-1})\}_{\theta \in \Theta}$ as the post-change densities. In this more general and realistic situation the pre-change hypothesis H_∞ remains simple but the post-change hypothesis $\mathsf{H}_{k,\vartheta} : \nu = k, \theta = \vartheta$ ($\vartheta \in \Theta$) becomes composite even for the fixed k. Note that in general the post-change parameter value $\theta = \theta(n, \nu)$ may depend on time and the change point. Therefore, taking into account (2.1) and using the convention that $\prod_{t=m}^n Z_t = 1$ for $m > n$, for any $0 \le \nu \le \infty$, the most general parametric changepoint model that allows the observations to be arbitrarily dependent and nonidentically distributed can be written as

$$p_{\nu,\theta}(\mathbf{X}_0^n) = \prod_{t=0}^{\nu} g(X_t|\mathbf{X}_0^{t-1}) \times \prod_{t=\nu+1}^{n} f_\theta(X_t|\mathbf{X}_0^{t-1}), \tag{2.2}$$

where $f_\theta(X_t|\mathbf{X}_0^{t-1})$ may depend on ν.

A particular case, which we will refer to as the *i.i.d. case* (with some abuse of terminology) is where the observations $\{X_n\}_{n \ge 1}$ are all independent, $X_1, \dots, X_\nu$ are i.i.d. with a common density $g(x)$ and $X_{\nu+1}, X_{\nu+2}, \dots$ are i.i.d. with another common density $f_\theta(x)$. In this case, there is no need to add the unobserved initial condition X_0 to the observed sample $\mathbf{X}_1^n$, so that the model (2.2) reduces to

$$p_{\nu,\theta}(\mathbf{X}_1^n) = \prod_{t=1}^{\nu} g(X_t) \times \prod_{t=\nu+1}^{n} f_\theta(X_t). \tag{2.3}$$

The model (2.2) is indeed very general and includes practically all possible scenarios. In the early stages, several publications (e.g., [6]) attempted to generalize the i.i.d. model (2.3) to the scenario where there are two independent parallel processes that switch at an unknown time, so a change occurs from one non-i.i.d. model to another non-i.i.d. model, which are mutually independent. Obviously, this is a particular case of the model (2.2) where the post-change conditional densities $f_\theta(X_t|\mathbf{X}_0^{t-1})$, $t \ge \nu + 1$ have to be replaced with $f_\theta(X_t|\mathbf{X}_{\nu+1}^{t-1})$:

$$p_{\nu,\theta}(\mathbf{X}_0^n) = p_\infty(\mathbf{X}_0^\nu) \times p_{0,\theta}(\mathbf{X}_{\nu+1}^n) = \prod_{t=0}^{\nu} g(X_t|\mathbf{X}_0^{t-1}) \times \prod_{t=\nu+1}^{n} f_\theta(X_t|\mathbf{X}_{\nu+1}^{t-1}). \tag{2.4}$$

2.2.1.2 A Multistream Scenario

In certain applications, it is of importance to generalize the single-stream changepoint model to the case of multiple data streams. To be more specific, suppose there are N data streams $\{X_n(i)\}_{n \ge 1}$, $i = 1, \dots, N$, observed sequentially in time subject to a change at an unknown time ν, so that

$X_1(i),\dots,X_\nu(i)$ are generated by one stochastic model and $X_{\nu+1}(i),X_{\nu+2}(i),\dots$ by another model when the change occurs in the ith stream. The change happens at a subset of streams $\mathscr{B} \in \{1,\dots,N\}$ of a size (cardinality) $1 \le |\mathscr{B}| \le K \le N$, where K is an assumed maximal number of streams that can be affected. Let $\mathsf{H}_{\nu,\mathscr{B},\theta}$ stand for the hypothesis that the change occurs at time ν in the subset of streams $\mathscr{B}$ with the post-change parameter θ.

Assume first that the observations are independent across data streams, but have a fairly general stochastic structure within streams. So if we let $\mathbf{X}_1^n(i) = (X_1(i),\dots,X_n(i))$ denote the sample of size n in the ith stream and if $\{f_{\theta_i}(X_n(i)|\mathbf{X}_0^{n-1}(i))\}_{n\ge1}$, $\theta_i \in \Theta_i$ is a parametric family of conditional densities of $X_n(i)$ given $\mathbf{X}_0^{n-1}(i)$, then when $\nu = \infty$ (there is no change) the parameter θ_i is equal to the known value $\theta_{0,i}$, i.e., $f_{\theta_i}(X_n(i)|\mathbf{X}_0^{n-1}(i)) = f_{\theta_{i,0}}(X_n(i)|\mathbf{X}_0^{n-1}(i))$ for all $n \ge 1$ and when $\nu = k < \infty$, then $\theta_i = \theta_{i,1} \ne \theta_{i,0}$, i.e., $f_{\theta_i}(X_n(i)|\mathbf{X}_0^{n-1}(i)) = f_{\theta_{0,i}}(X_n(i)|\mathbf{X}_0^{n-1}(i))$ for $n \le k$ and $f_{\theta_i}(X_n(i)|\mathbf{X}_0^{n-1}(i)) = f_{\theta_{i,1}}(X_n(i)|\mathbf{X}_0^{n-1}(i))$ for $n > k$. Not only the point of change ν, but also the subset $\mathscr{B}$, its size $|\mathscr{B}|$, and the post-change parameters $\theta_{i,1}$ are unknown. More generally, we may assume that under the hypothesis H_∞ when there is no change the pre-change densities $f_{\theta_{i,0},n}(X_n(i)|\mathbf{X}_0^{n-1}(i)) = g_i(X_n(i)|\mathbf{X}_0^{n-1}(i))$. Thus, in the multistream scenario the single stream model (2.2) generalizes as

$$
\begin{aligned}
p(\mathbf{X}_0^n|H_{\nu,\mathscr{B},\boldsymbol{\theta}_\mathscr{B}}) = &\prod_{t=0}^{\nu}\prod_{i=1}^{N} g_i(X_t(i)|\mathbf{X}_0^{t-1}(i)) \times \\
&\prod_{t=\nu+1}^{n}\prod_{i\in\mathscr{B}} f_{i,\theta_i}(X_t(i)|\mathbf{X}_0^{t-1}(i)) \prod_{i\notin\mathscr{B}} g_i(X_t(i)|\mathbf{X}_0^{t-1}(i)),
\end{aligned}
\tag{2.5}
$$

where $\mathbf{X}_0^n = (\mathbf{X}_0^n(1),\dots,\mathbf{X}_0^n(N))$, $\boldsymbol{\theta}_\mathscr{B} = (\theta_i)_{i\in\mathscr{B}}$, and we recall that by our convention $\prod_{i=m}^n Z_i = 1$ for $m > n$. This model can be further generalized covering the most general case where also the data between streams are dependent:

$$
p(\mathbf{X}_0^n|H_{\nu,\mathscr{B},\theta}) = \prod_{t=1}^{\nu} g(\mathbf{X}_t|\mathbf{X}_0^{t-1}) \times \prod_{t=\nu+1}^{n} f_{\mathscr{B},\theta}(\mathbf{X}_t|\mathbf{X}_0^{t-1}), \tag{2.6}
$$

where $\{g(\mathbf{X}_n|\mathbf{X}_0^{n-1})\}_{n\ge1}$ and $\{f_{\mathscr{B},\theta}(\mathbf{X}_n|\mathbf{X}_0^{n-1})\}_{n\ge1}$ are sequences of conditional densities of $\mathbf{X}_n = (X_n(1),\dots,X_n(N))$ given $\mathbf{X}_0^{n-1}$.

2.2.2 Models for the Change Point

2.2.2.1 Types of Changes

There are two different kinds of changes – additive and non-additive. Additive changes lead to the change of the mean value of the sequence of observations. Nonadditive changes are typically produced by a change in variance or covariance, i.e., these are spectral changes. Figure 2.2 illustrates typical additive and non-additive changes in the observed data. Specifically, Figure 2.2(a) shows the change in the mean value of the i.i.d. Gaussian sequence and Figure 2.2(b) shows the change in the correlation coefficient of the first-order autoregression process (AR(1)) driven by the i.i.d. Gaussian errors. It is clearly seen that these changes are qualitatively different. In principle, these two types of changes can occur simultaneously. Figure 2.2(c) illustrates this situation where the simultaneous change in the mean value and the variance of the i.i.d. Gaussian sequence is shown.

2.2.2.2 Models

The changepoint ν may be considered either as an unknown deterministic number or as a random variable.

If the changepoint is treated as a random variable, then the model has to be supplied with the *prior distribution* of the changepoint. There may be several changepoint mechanisms and, as

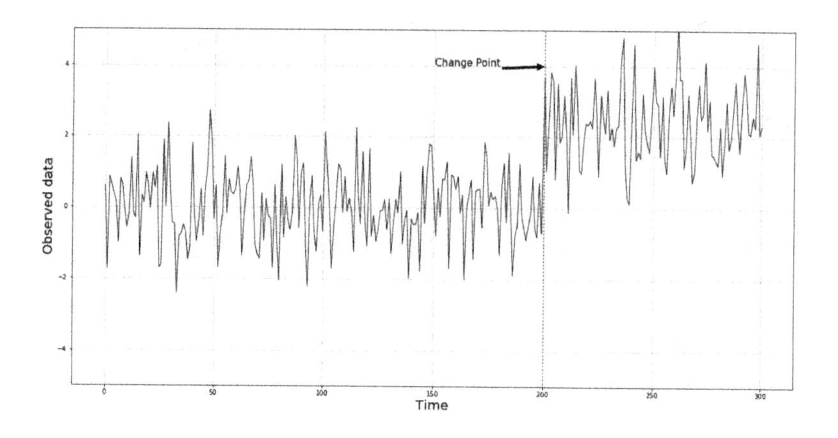

(a) Additive change: changepoint $\nu = 200$.

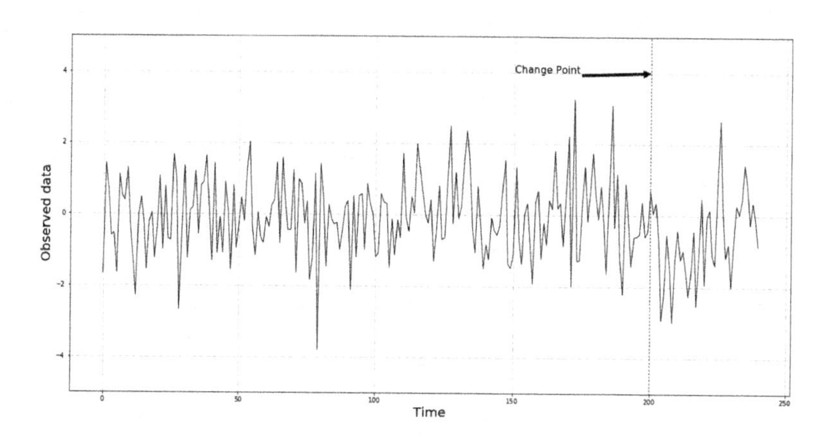

(b) Non-additive change: changepoint $\nu = 200$.

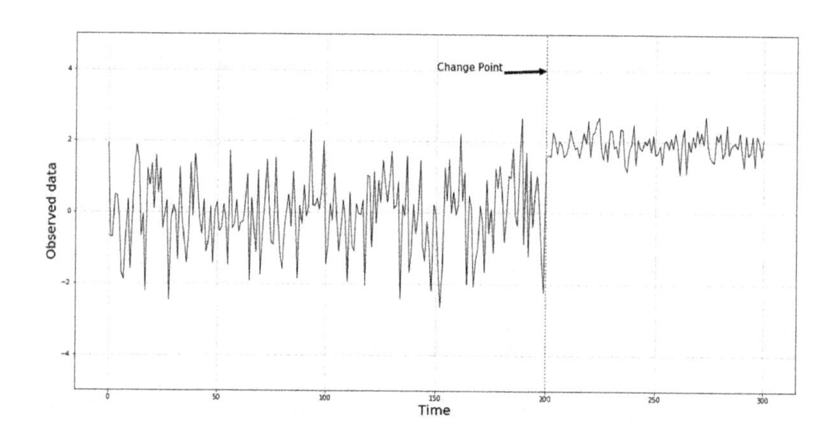

(c) A combination of additive and nonadditive changes: changepoint $\nu = 200$.

FIGURE 2.2
Illustration of additive and nonadditive changes.

a result, a random variable ν may be partially or completely dependent on the observations or independent of the observations. To account for these possibilities at once, let $q = \Pr(\nu < 0)$ and $\pi_k = \Pr(\nu = k|\mathbf{X}_1^k)$, $k \geq 0$, and observe that π_k, $k = 1,2,\dots$ are $\mathscr{F}_k$-adapted. That is, the probability of a change occurring at the time instant $\nu = k$ depends on $\mathbf{X}_1^k$, the observations' history accumulated up to and including the time $k \geq 1$. The probability $q + \pi_0 = \Pr(\nu \leq 0)$ represents the probability of the event that the change already took place before the observations became available. With the so-defined prior distribution, one can describe very general changepoint models, including those that assume ν to be a $\{\mathscr{F}_n\}$-adapted stopping time (see Moustakides [94]). Below we discuss Moustakides's concept by allowing the prior distribution to depend on some additional information available to "Nature".

To be more specific, let us consider the sequential change detection problem as a conflict between Nature (the Unknown) and the Statistician. At an unknown time ν Nature generates a change in the statistical properties of an observed process, while the Statistician uses the observations to decide whether the change took place or not. While the Statistician has access only to the observed data $\{X_n\}_{n\geq 1}$, Nature may have access to certain additional or even completely independent information to produce a change. To formulate this difference in available information, we assume that Nature obtains *sequentially* a process $\{Z_n\}_{n=-\infty}^{\infty}$ for making a decision about the change. The processes $\{Z_n\}_{n=-\infty}^{\infty}$ and $\{X_n\}_{n\geq 1}$ may be dependent. Moreover, in general, Nature may have access to more information than the Statistician, so we will assume that $\{X_n\}_{n\geq 1}$ is part of $\{Z_n\}$. The process $\{Z_n\}_{n=-\infty}^{\infty}$ can be acquired by Nature *before* the observations become available to the Statistician. This allows for modeling the case where the change occurs before the Statistician obtains any observations. For example, in engineering applications, this can happen if a system works for a long time before a "control process" starts.

Write $Y_n = (Z_n, X_n)$ and assume that there are two probability measures P_i, $i = \infty, 0$ that correspond to the process $Y = \{Y_n\}$ that produce conditional densities $p_i(Y_n|\mathbf{Y}^n)$ $(i = \infty, 0)$. For the sake of simplicity, we will focus on the case of a simple post-change hypothesis when the post-change parameter is known. If ν is a deterministic time of change, then the density of induced measure P_ν follows (2.1) with X replaced by Y. Consider the two filtrations $\{\mathscr{F}_n^Z\}_{n=-\infty}^{\infty}$ and $\{\mathscr{F}_n^X\}_{n=0}^{\infty}$ generated by Z and X, i.e., $\mathscr{F}_n^Z = \sigma\{Z_t; -\infty < t \leq n\}$; $\mathscr{F}_n^X = \sigma\{X_t; 0 < t \leq n\}$, with $\mathscr{F}_n^X$ being the trivial σ-algebra for $n \leq 0$ since the observations are available only for $n \geq 1$. Let ξ be a random variable measurable with respect to $\mathscr{F}_\infty^X$. Then, using the general model (2.1), we obtain

$$\mathsf{E}_\nu[\xi|\mathscr{F}_\nu^Z \cup \mathscr{F}_\nu^X] = \int \xi p_0(X_{\nu+1}^\infty|\mathbf{X}_1^\nu, \mathbf{Z}_{-\infty}^\nu)\mathrm{d}X_{\nu+1}^\infty = \mathsf{E}_0[\xi|\mathscr{F}_\nu^Z \cup \mathscr{F}_\nu^X]. \tag{2.7}$$

Inspired by the idea of Moustakides [94], we suggest the following general model for the changepoint mechanism: *A changepoint ν produced by Nature is a stopping time adapted to the filtration $\{\mathscr{F}_n^Z\}$ available to Nature.*

As usual, a sequential detection rule is a stopping time T adapted to the filtration $\{\mathscr{F}_n^X\}$ generated by observations.

Note that for every fixed ν the detection delay $T - \nu$ is a $\mathscr{F}_\infty^X$-measurable random variable. Therefore, in order to define a suitable performance measure that involves the delay to detection $(T-\nu)\mathbb{1}_{\{T>\nu\}}$, it is first useful to derive a convenient formula for the expectation of a randomly stopped sequence of $\mathscr{F}_\infty^X$-measurable nonnegative random variables.

Expectation of a randomly stopped sequence. Consider a process $\{\xi_n\}_{n=0}^\infty$, where for all $n \geq 0$ the random variables ξ_n are nonnegative and $\mathscr{F}_\infty^X$-measurable. We are interested in computing the expectation of the randomly stopped random variable ξ_ν, where ν is an $\{\mathscr{F}_n^Z\}$-adapted stopping time. Define $\xi_\nu \mathbb{1}_{\{\nu<\infty\}} = \xi_0 \mathbb{1}_{\{\nu\leq 0\}} + \sum_{t=1}^\infty \xi_t \mathbb{1}_{\{\nu=t\}}$. Taking expectation, applying (2.7) and using the fact that if ν is adapted to $\{\mathscr{F}_n^Z\}$ it is also adapted to $\{\mathscr{F}_n^Z \cup \mathscr{F}_n^X\}$, we obtain

$$\mathsf{E}_\nu[\xi_\nu \mathbb{1}_{\{\nu<\infty\}}] = \mathsf{E}_\infty[\mathsf{E}_0[\xi_0|\mathscr{F}_0^Z \cup \mathscr{F}_0^X]\mathbb{1}_{\{\nu\leq 0\}}] + \sum_{t=1}^\infty \mathsf{E}_\infty[\mathsf{E}_0[\xi_t|\mathscr{F}_t^Z \cup \mathscr{F}_t^X]\mathbb{1}_{\{\nu=t\}}]$$

$$= \mathsf{E}_\infty[\mathsf{E}_0[\xi_0|\mathscr{F}_0^Z]\mathbb{1}_{\{\nu\le 0\}}] + \sum_{t=1}^{\infty} \mathsf{E}_\infty[\mathsf{E}_\infty[\mathsf{E}_0[\xi_t|\mathscr{F}_t^Z \cup \mathscr{F}_t^X]|\mathscr{F}_t^Z]\mathbb{1}_{\{\nu=t\}}]. \quad (2.8)$$

When obtaining the second equality we used the fact that $\mathscr{F}_0^X$ is the trivial σ-algebra and that ν is $\{\mathscr{F}_t^Z\}$-adapted. Equality (2.8) will be extensively used for the development of a general performance measure. Note that by substituting $\xi_t = 1$ yields

$$\mathsf{P}_\nu(\nu < \infty) = \mathsf{P}_\infty(\nu < \infty). \quad (2.9)$$

2.3 Optimality Criteria

2.3.1 Preliminaries

Tartakovsky et al. [164, Sec. 6.3] suggested five changepoint problem settings — the Bayesian approach, the generalized Bayesian approach, the minimax approach, the uniform (pointwise) approach, and the approach related to multicyclic detection of a disorder in a stationary regime. The objective of this section is to briefly discuss several problem settings extending the discussion to the case where the point of change is modeled as a stopping time with respect to the information available to Nature.

We iterate that a *sequential change detection procedure* is a stopping time T on $X_1, X_2, \ldots$, i.e., $\{T \le n\} \in \mathscr{F}_n^X$, where $\mathscr{F}_n^X = \sigma(X_1, \ldots, X_n)$ is the sigma-algebra induced by the first n observations, with the convention $\mathscr{F}_n^X = \{\varnothing, \Omega\}$ for $n \le 0$. We assume that all random objects are defined on a complete probability space $(\Omega, \mathscr{F}, \mathsf{P})$. After observing $X_1, \ldots, X_T$ it is declared that the change is in effect. That may or may not be the case. If it is not, then $T \le \nu$, and it is said that a *false alarm* has been raised.

Tartakovsky et al. [164, Sec. 6.3] emphasize the importance of distinguishing between two different scenarios – single-run detection procedures and multicyclic procedures. In the single-run case, it is assumed that the change detection procedure is applied *only once*; the result is either a false alarm or a correct possibly delayed detection. What takes place beyond the stopping point T is of no concern. This *single-run paradigm* is shown in Figure 2.3. The observation process $\{X_n\}_{n\ge 1}$ undergoes a shift in the mean at some time instant ν, the changepoint. The grey line in Figure 2.3(b) gives an example of the corresponding detection statistic trajectory that exceeds the detection threshold prematurely before the change occurs. This is a false alarm situation, and T can be regarded as the random run length to a false alarm. Another possibility is depicted by the black line in Figure 2.3(b). This is an example where the detection statistic exceeds the detection threshold after the changepoint. Note that the detection delay $T - \nu$ is random.

In a variety of surveillance applications, the detection procedure should be applied *repeatedly*. This requires the specification of a renewal mechanism after each false or true alarm. The simplest renewal strategy is to restart from scratch, in which case the procedure becomes multicyclic with similar cycles in a statistical sense if the process is homogeneous. In the present book, we will not consider this scenario and refer to [164] where such an approach was considered in detail.

2.3.2 Measures of the False Alarm Risk

Figure 2.3 suggests that it is reasonable to measure the risk associated with a false alarm by the mean time to false alarm $\mathsf{E}_\infty[T]$ and the risk associated with a true change detection by the conditional average delay to detection $\mathsf{CADD}_\nu(T) = \mathsf{E}_\nu(T - \nu | T > \nu)$, $\nu = 0, 1, \ldots$. A good detection procedure should guarantee small values of the expected detection delay $\mathsf{CADD}_\nu(T)$ for all $\nu \ge 0$ when $\mathsf{E}_\infty[T]$ is fixed at a certain level. However, if the false alarm risk is measured in terms of

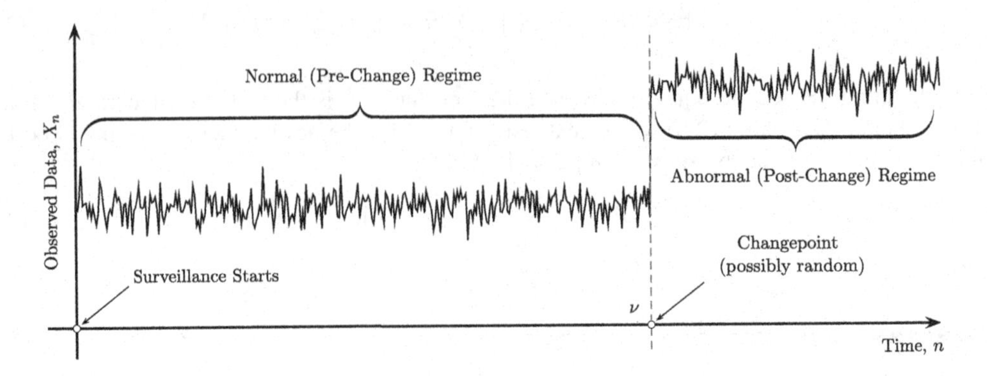

(a) An example of the behavior of the observed process $\{X_n\}_{n\geq 1}$.

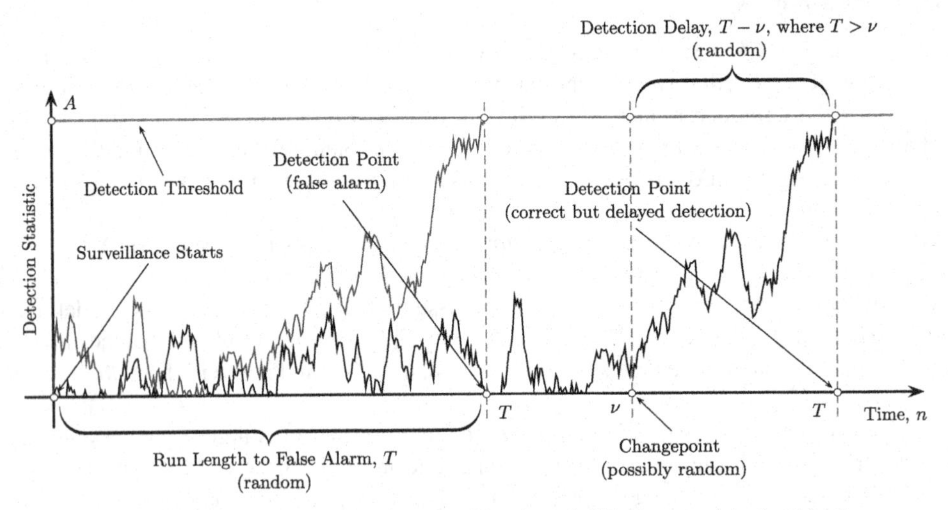

(b) Two possibilities in the detection process: false alarm (left) and correct detection (right).

FIGURE 2.3
Illustration of single-run sequential changepoint detection.

the mean time to false alarm, i.e., it is required that $\mathsf{E}_\infty[T] \geq \gamma$ for some $\gamma > 1$, then a procedure that minimizes the conditional average delay to detection $\mathsf{CADD}_\nu(T)$ uniformly over all ν does not exist. For this reason, we have to resort to different optimality criteria, e.g., to Bayes or minimax criteria. Having said that, we now introduce several possible measures of the false alarm risk.

2.3.2.1 Average Run Length to False Alarm

The expected value of the stopping time T under the no-change hypothesis H_∞, i.e., the mean time to false alarm $\mathsf{E}_\infty[T] = \mathsf{ARL2FA}(T)$ is usually referred to as the *Average Run Length to False Alarm*. One way of measuring the false alarm risk (or rate) is to keep $\mathsf{ARL2FA}(T)$ at a given level γ. This is perhaps the most popular way of measuring the false alarm rate in sequential change detection problems.

Thus, for $\gamma \geq 1$, introduce the class of changepoint detection rules

$$\mathbb{C}_\gamma = \{T : \mathsf{ARL2FA}(T) \geq \gamma\} \tag{2.10}$$

for which the average run length (ARL) to false alarm is not smaller than a prescribed level γ.

2.3.2.2 Weighted Probability of False Alarm

In Bayesian problems, the point of change ν is treated as random with a prior distribution $\pi_k = \Pr(\nu = k)$, $-\infty < k < +\infty$. For an event $\mathscr{A}$, define the probability measure $\mathsf{P}_\pi(\mathscr{A} \times \mathscr{K}) = \sum_{k\in\mathscr{K}} \pi_k \mathsf{P}_k(\mathscr{A})$. From the Bayesian point of view, it is reasonable to measure the false alarm risk with the *Weighted Probability of False Alarm* (PFA), defined as

$$\mathsf{PFA}_\pi(T) := \mathsf{P}_\pi(T \le \nu) = \sum_{k=-\infty}^{\infty} \pi_k \mathsf{P}_k(T \le k) = \sum_{k=0}^{\infty} \pi_k \mathsf{P}_\infty(T \le k). \tag{2.11}$$

Actually, the summation in (2.11) is over $k \in \mathbb{Z}_+ = \{0,1,2,\dots\}$ since $\mathsf{P}_\infty(T < 0) = 0$. Also, the last equality follows from the fact that $\mathsf{P}_k(T \le k) = \mathsf{P}_\infty(T \le k)$ because the event $\{T \le k\}$ depends on the first k observations which under measure P_k correspond to the no-change hypothesis H_∞.

Thus, for $\alpha \in (0,1)$, introduce the class of changepoint detection rules

$$\mathbb{C}_\pi(\alpha) = \{T : \mathsf{PFA}_\pi(T) \le \alpha\} \tag{2.12}$$

for which the weighted PFA does not exceed a prescribed level α.

2.3.2.3 Global Probability of False Alarm

The purpose of this section is to discuss serious issues related to the constraint imposed on the so-called *global* PFA.

Let $\mathsf{GPFA}(T) = \sup_{k\in\mathbb{Z}_+} \mathsf{P}_k(T \le k)$ denote the maximal (worst-case) probability of false alarm to which we will refer as the *Global Probability of False Alarm* (GPFA). Recall that $\mathsf{P}_k(T \le k) = \mathsf{P}_\infty(T \le k)$ since the event $\{T \le k\}$ is measurable with respect to the sigma-algebra $\mathscr{F}_k^X = \sigma(X_1,\dots,X_k)$ and, by our convention, if the change occurs at $\nu = k$, then the measure P_∞ changes to P_k at $k+1$. Hence,

$$\mathsf{GPFA}(T) = \sup_{k\in\mathbb{Z}_+} \mathsf{P}_\infty(T \le k) = \mathsf{P}_\infty(T < \infty).$$

For $\alpha \in (0,1)$, introduce the class of changepoint detection rules

$$\mathbb{C}_\infty(\alpha) = \{T : \mathsf{GPFA}(T) \le \alpha\} \tag{2.13}$$

for which the global PFA does not exceed a prescribed level α.

Obviously, the constraint $\mathsf{P}_\infty(T < \infty) \le \alpha$ is very strong – it is the strongest among all possible constraints on the false alarm risk. In fact, if $\mathsf{P}_\infty(T < \infty) < 1$ (i.e., a rule T does not terminate w.p. 1 under H_∞), then $\mathsf{E}_\infty[T] = \infty$. Thus, the ARL to false alarm is unbounded due to the high price that has to be paid for a such strong constraint. This fact definitely causes issues for practical applications and even methodological difficulties, as discussed by Tartakovsky [157] and Tartakovsky et al. [164]. Indeed, the minimax criteria considered in Subsection 2.3.5 cannot be used in this case since the maximal average detection delay $\sup_k \mathsf{E}_k(T-k|T>k) = \infty$. Moreover, even though that the asymptotically optimal (as $\alpha \to 0$) Bayesian solution obtained in [157] formally exists, still, it is questionable whether or not it makes sense to exploit this detection rule in any applications. We believe that the answer is negative, providing a quote from [157] with a small modification:

While in the Bayesian setting it is possible to devise the nearly optimal change detection rule in class $\mathbb{C}_\infty(\alpha)$ to handle the GPFA bound, our personal opinion is that this constraint is too strong to be useful in applications ....

See Subsection 2.3.4 for further details.

We will avoid considering class $\mathbb{C}_\infty(\alpha)$ in this book.

2.3.2.4 Local Probabilities of False Alarm

As have been already mentioned, the constraint $\mathsf{E}_\infty[T] \geq \gamma$ imposed on the ARL to false alarm (i.e., class of rules (2.10)) is the most popular. However, the requirement of having large values of the $\mathsf{ARL2FA}(T) = \mathsf{E}_\infty[T]$ generally does not guarantee small values of the maximal local conditional probability of false alarm (LCPFA) $\sup_{\ell \geq 0} \mathsf{P}_\infty(T \leq \ell + m | T > \ell)$ or the maximal local (unconditional) probability of false alarm (LPFA) $\sup_{\ell \geq 0} \mathsf{P}_\infty(\ell < T \leq \ell + m)$ in a time interval (window) with fixed length $m \geq 1$. Indeed, we have the following result.

Lemma 2.1. *Let m be a positive integer, $m < \gamma$. If $T \in \mathbb{C}_\gamma$, then there exists an ℓ, possibly depending on γ, such that*

$$\mathsf{P}_\infty(T \leq \ell + m | T > \ell) < m/\gamma. \tag{2.14}$$

Proof. Without loss of generality we may assume that $\mathsf{P}_\infty(T > \ell) > 0$, since otherwise $\mathsf{P}_\infty(T > \ell) = 0$ for all ℓ and $\mathsf{E}_\infty[T] = 0$, which contradicts the inequality $\mathsf{E}_\infty[T] \geq \gamma$. We have

$$\mathsf{E}_\infty[T] = \sum_{\ell=0}^{\infty} \mathsf{P}_\infty(T > \ell) = \sum_{i=0}^{m-1} \sum_{k=0}^{\infty} \mathsf{P}_\infty(T > i + km) = \sum_{i=0}^{m-1} \sum_{k=0}^{\infty} \mathsf{P}_\infty(T > i)\mathsf{P}_\infty(T > i + km | T > i), \tag{2.15}$$

where we used the equality

$$\mathsf{P}_\infty(T > i + km) = \mathsf{P}_\infty(T > i)\mathsf{P}_\infty(T > i + km | T > i).$$

Now, inequality (2.14) can be proved by contradiction as follows. Assume that

$$\mathsf{P}_\infty(T \leq \ell + m | T > \ell) \geq m/\gamma \quad \text{for all } \ell \in \mathbb{Z}_+,$$

that is,

$$\mathsf{P}_\infty(T > \ell + m | T > \ell) < 1 - m/\gamma \quad \text{for all } \ell \in \mathbb{Z}_+. \tag{2.16}$$

By the assumption (2.16),

$$\mathsf{P}_\infty(T > i + km | T > i) < 1 - km/\gamma < (1 - m/\gamma)^k,$$

so it follows from equality (2.15) that

$$\mathsf{E}_\infty[T] < \sum_{i=0}^{m-1} \mathsf{P}_\infty(T > i) \sum_{k=0}^{\infty} (1 - m/\gamma)^k = (\gamma/m) \sum_{i=0}^{m-1} \mathsf{P}_\infty(T > i) < \gamma,$$

which contradicts the assumption $T \in \mathbb{C}_\gamma$. □

Hence, the condition $\mathsf{E}_\infty[T] \geq \gamma$ only guarantees, for any $m \geq 1$, the existence of some ℓ, that possibly depends on γ, for which the inequality (2.14) holds, but not that the LCPFA $\mathsf{P}_\infty(T \leq \ell + m | T > \ell)$ is small for all $\ell \in \mathbb{Z}_+$ when the ARL to false alarm $\mathsf{E}_\infty[T]$ is large.

At the same time, for many practical applications, it is desirable to control the maximal LCPFA $\sup_{\ell \in \mathbb{Z}_+} \mathsf{P}_\infty(T \leq \ell + m | T > \ell)$ or the maximal LPFA $\sup_{\ell \in \mathbb{Z}_+} \mathsf{P}_\infty(\ell < T \leq \ell + m)$ at a certain (usually low) level β. Thus, we introduce the classes of detection rules

$$\mathbb{C}(m, \beta) = \left\{ T : \sup_{\ell \in \mathbb{Z}_+} \mathsf{P}_\infty(T \leq \ell + m | T > \ell) \leq \beta \right\} \tag{2.17}$$

and

$$\widetilde{\mathbb{C}}(m, \beta) = \left\{ T : \sup_{\ell \in \mathbb{Z}_+} \mathsf{P}_\infty(\ell < T \leq \ell + m) \leq \beta \right\}, \tag{2.18}$$

for which the corresponding maximal local probabilities of raising a false alarm inside a sliding window of size $m \geq 1$ do not exceed a certain *a priori* chosen level $0 < \beta < 1$.

The important fact is that, for a given $0 < \beta < 1$, the constraints

$$\sup_{\ell \in \mathbb{Z}_+} \mathsf{P}_\infty(T \leq \ell + m | T > \ell) \leq \beta \quad \text{for some given } m \geq 1 \tag{2.19}$$

and

$$\sup_{\ell \in \mathbb{Z}_+} \mathsf{P}_\infty(\ell < T \leq \ell + m) \leq \beta \quad \text{for some given } m \geq 1 \tag{2.20}$$

are stronger than the ARL constraint $\mathsf{E}_\infty[T] \geq \gamma$ $(\gamma \geq 1)$, i.e., if $T \in \mathbb{C}(m,\beta)$ $(T \in \widetilde{\mathbb{C}}(m,\beta))$, then this implies that $T \in \mathbb{C}_\gamma$ for some $\gamma = \gamma(\beta, m)$, but the opposite is not true (see Lemma 2.1). The following lemma justifies this fact.

Lemma 2.2. *Let m be a positive integer.*
(i) *If $T \in \mathbb{C}(m,\beta)$, then T necessarily belongs to class $\mathbb{C}_\gamma$ for some $\gamma = \gamma(m,\beta)$, in particular for*

$$\gamma(m,\beta) = \frac{1}{2\beta} \sum_{i=0}^{m-1} \mathsf{P}_\infty(T > i). \tag{2.21}$$

(ii) *If $T \in \widetilde{\mathbb{C}}(m,\beta)$, then T necessarily belongs to class $\mathbb{C}_\gamma$ for some $\gamma = \gamma(m,\beta)$, in particular for*

$$\gamma(m,\beta) = \frac{m}{\beta} c(1-c) \tag{2.22}$$

with any $c \in (0,1)$.

Proof. Proof of part (i). Write $N = 1/m\beta_0$. Let the constraint (2.19) hold with $\beta = m\beta_0$. Then

$$\mathsf{P}_\infty(T > i + km | T > i) \geq 1 - km\beta_0 = 1 - k/N \quad \text{for all } i \geq 0,$$

and using (2.15) we obtain

$$\begin{aligned}
\mathsf{E}_\infty[T] &= \sum_{i=0}^{m-1} \sum_{k=0}^{\infty} \mathsf{P}_\infty(T > i) \mathsf{P}_\infty(T > i + km | T > i), \\
&\geq \sum_{i=0}^{m-1} \mathsf{P}_\infty(T > i) \sum_{k=0}^{N} (1 - k/N) = \left[1 + N - \frac{N}{2}\left(\frac{1}{N} + 1\right)\right] \sum_{i=0}^{m-1} \mathsf{P}_\infty(T > i) \\
&= \frac{1}{2}(N+1) \sum_{i=0}^{m-1} \mathsf{P}_\infty(T > i) \geq \frac{1}{2m\beta_0} \sum_{i=0}^{m-1} \mathsf{P}_\infty(T > i) = \frac{1}{2\beta} \sum_{i=0}^{m-1} \mathsf{P}_\infty(T > i).
\end{aligned}$$

Hence, (2.19) implies $\mathsf{E}_\infty[T] \geq \gamma$ with $\gamma = \gamma(m,\beta)$ defined in (2.21) and the proof of (i) is complete.

Proof of part (ii). Let $M = c/\beta$ with $0 < c < 1$. Then

$$\mathsf{P}_\infty(T < Mm) \leq \sum_{i=1}^{M} \mathsf{P}_\infty\{(i-1)m < T \leq im)\} \leq M\beta,$$

so $\mathsf{P}_\infty(T \geq Mm) \geq 1 - M\beta = 1 - c$ and, by Chebyshev's inequality,

$$\mathsf{E}_\infty[T] \geq Mm \mathsf{P}_\infty(T \geq Mm) \geq \frac{m}{\beta} c(1-c).$$

Therefore, constraint (2.20) implies $\mathsf{E}_\infty[T] \geq \gamma$ with $\gamma = \gamma(m,\beta)$ defined in (2.22) and the proof of (ii) is complete. □

Another reason for considering the constraints (2.19) and (2.20) is that the appropriateness of the ARL to false alarm $\mathsf{E}_\infty[T]$ as an exhaustive measure of the false alarm risk is questionable unless the P_∞-distribution of T is geometric, at least approximately. The geometric distribution is characterized entirely by a single parameter, which uniquely determines $\mathsf{E}_\infty[T]$ and is uniquely determined by $\mathsf{E}_\infty[T]$. As a result, if T is geometric, one can evaluate $\mathsf{P}_\infty(T \le \ell+m|T>\ell)$ for any $\ell \in \mathbb{Z}_+$ and both constraints are quite similar and can be recalculated in each other.

For the i.i.d. model (2.3), Pollak and Tartakovsky [112] showed that under mild assumptions the P_∞-distribution of the stopping times associated with detection schemes from a certain class is asymptotically exponential with parameter $1/\mathsf{E}_\infty[T]$ as $\mathsf{E}_\infty[T] \to \infty$. The class includes all the most popular procedures, including CUSUM and SR. Hence, for the i.i.d. model (2.3), the ARL to false alarm is an acceptable measure of the false alarm rate. However, for a general non-i.i.d. model this is not necessarily true, which suggests that alternative measures of the false alarm rate are in order.

We stress once again that, by Lemma 2.2, in general $\sup_\ell \mathsf{P}_\infty(T \le \ell+m|T>\ell) \le \beta$ and $\sup_\ell \mathsf{P}_\infty(\ell < T \le \ell+m) \le \beta$ are *more stringent* conditions than $\mathsf{E}_\infty[T] \ge \gamma$. Hence, in general, $\mathbb{C}(m,\beta) \subset \mathbb{C}_\gamma$ and $\widetilde{\mathbb{C}}(m,\beta) \subset \mathbb{C}_\gamma$.

2.3.3 An Expected Delay to Detection in a General Case

In this section, we use the formula (2.8) for the expectation of a randomly stopped sequence to provide suitable expressions for the average detection delay. Recall that ν is the stopping time generated by Nature to impose the change and T is the stopping time employed by the Statistician to detect this change. As discussed at the beginning of Subsection 2.3.2, the natural performance measure associated with the speed of detection is the conditional average delay to detection $\mathsf{CADD}_\nu(T) = \mathsf{E}_\nu(T-\nu|T>\nu)$ (for a fixed change point $\nu \in \mathbb{Z}_+$), which in the case where ν is a random variable (stopping time in our case) can be generalized to

$$\mathscr{J}_\nu(T) = \mathsf{E}_\nu\left(T-\nu|T>\nu; \nu<\infty\right) = \frac{\mathsf{E}_\nu[(T-\nu)^+ \mathbb{1}_{\{\nu<\infty\}}]}{\mathsf{E}_\nu[\mathbb{1}_{\{T>\nu\}}\mathbb{1}_{\{\nu<\infty\}}]}, \tag{2.23}$$

where we assume that $\mathsf{P}_\infty(\nu<\infty)>0$.

Since sequences of random variables $\{(T-k)^+\}$ and $\{\mathbb{1}_{\{T>k\}}\}$ are nonnegative and $\{\mathscr{F}_\infty^X\}$-measurable, using (2.8) and (2.23) we obtain

$$\mathscr{J}_\nu(T) = \frac{\mathsf{E}_\infty[\mathsf{E}_0(T|\mathscr{F}_0^Z)\mathbb{1}_{\{\nu\le 0\}}] + \sum_{k=1}^\infty \mathsf{E}_\infty[\mathsf{E}_\infty[\mathsf{E}_0[(T-k)^+|\mathscr{F}_k^Z \cup \mathscr{F}_k^X]|\mathscr{F}_k^Z]\mathbb{1}_{\{\nu=k\}}]}{\mathsf{E}_\infty[\mathsf{E}_0[\mathbb{1}_{\{T>0\}}|\mathscr{F}_0^Z]\mathbb{1}_{\{\nu\le 0\}}] + \sum_{k=1}^\infty \mathsf{E}_\infty[\mathsf{E}_\infty[\mathsf{E}_0[\mathbb{1}_{\{T>k\}}|\mathscr{F}_k^Z \cup \mathscr{F}_k^X]|\mathscr{F}_k^Z]\mathbb{1}_{\{\nu=k\}}]}.$$

Assume that $T>0$ a.s.[1] Now, using the fact that since the event $\{T>k\}$ is $\mathscr{F}_k^X$-measurable it is also $\mathscr{F}_k^Z \cup \mathscr{F}_k^X$-measurable, we obtain

$$\mathscr{J}_\nu(T) = \frac{\mathsf{E}_\infty[\mathsf{E}_0(T|\mathscr{F}_0^Z)\mathbb{1}_{\{\nu\le 0\}}] + \sum_{k=1}^\infty \mathsf{E}_\infty[\mathsf{E}_\infty[\mathsf{E}_0[(T-k)^+|\mathscr{F}_k^Z \cup \mathscr{F}_k^X]|\mathscr{F}_k^Z]\mathbb{1}_{\{\nu=k\}}]}{\mathsf{E}_\infty[\mathbb{1}_{\{\nu\le 0\}}] + \sum_{k=1}^\infty \mathsf{E}_\infty[\mathsf{P}_\infty(T>k|\mathscr{F}_k^Z)\mathbb{1}_{\{\nu=k\}}]}. \tag{2.24}$$

Typically, the changepoint mechanism is not known exactly. Then it makes sense to define an uncertainty class $\mathscr{T}$ of possible stopping times for ν and to consider the worst possible candidate from this class, i.e., to maximize $\mathscr{J}_\nu(T)$ over $\nu \in \mathscr{T}$, which leads to the following performance measure

$$\mathscr{J}(T) = \sup_{\nu\in\mathscr{T}} \frac{\mathsf{E}_\infty[\mathsf{E}_0(T|\mathscr{F}_0^Z)\mathbb{1}_{\{\nu\le 0\}}] + \sum_{k=1}^\infty \mathsf{E}_\infty[\mathsf{E}_\infty[\mathsf{E}_0[(T-k)^+|\mathscr{F}_k^Z \cup \mathscr{F}_k^X]|\mathscr{F}_k^Z]\mathbb{1}_{\{\nu=k\}}]}{\mathsf{E}_\infty[\mathbb{1}_{\{\nu\le 0\}}] + \sum_{k=1}^\infty \mathsf{E}_\infty[\mathsf{P}_\infty(T>k|\mathscr{F}_k^Z)\mathbb{1}_{\{\nu=k\}}]}. \tag{2.25}$$

[1] This assumption excludes stopping times randomized at 0, i.e., such that $\mathsf{P}(T=0)=p$ with $0<p<1$. In certain situations, it makes sense to consider this more general case.

The next proposition provides the worst case performance for change points from the class of all $\{\mathscr{F}_n^Z\}$-adapted stopping times suggested by Moustakides [94].

Proposition 2.1. *Let $\mathscr{T}$ be the class of $\{\mathscr{F}_n^Z\}$-adapted stopping times. Then the worst case performance measure with respect to the change times ν from $\mathscr{T}$ is*

$$\mathscr{J}_{\mathrm{M}}(T) = \sup_{\nu \in \mathscr{T}} \mathsf{E}_\nu(T-\nu | T > \nu; \nu < \infty) = \sup_{k \in \mathbb{Z}_+} \operatorname{ess\,sup} \mathsf{E}_k(T-k | T > k, \mathscr{F}_k^Z). \tag{2.26}$$

Proof. Using (2.24) we obtain

$$\begin{aligned}
\mathscr{J}(T) &\le \max\left\{\frac{\mathsf{E}_\infty[\mathsf{E}_0(T|\mathscr{F}_0^Z)\mathbb{1}_{\{\nu\le 0\}}]}{\mathsf{E}_\infty[\mathbb{1}_{\{\nu\le 0\}}]}, \sup_{k\ge 1}\frac{\mathsf{E}_\infty[\mathsf{E}_\infty[\mathsf{E}_0[(T-k)^+|\mathscr{F}_k^Z\cup\mathscr{F}_k^X]|\mathscr{F}_k^Z]\mathbb{1}_{\{\nu=k\}}]}{\mathsf{E}_\infty[\mathsf{P}_\infty(T>k|\mathscr{F}_k^Z)\mathbb{1}_{\{\nu=k\}}]}\right\} \\
&\le \sup_{k\ge 0}\operatorname{ess\,sup}\frac{\mathsf{E}_\infty[\mathsf{E}_0[(T-k)^+|\mathscr{F}_k^Z\cup\mathscr{F}_k^X]|\mathscr{F}_k^Z]}{\mathsf{P}_\infty(T>k|\mathscr{F}_k^Z)} \\
&= \sup_{k\ge 0}\operatorname{ess\,sup}\frac{\mathsf{E}_k[(T-k)^+|\mathscr{F}_k^Z]}{\mathsf{P}_\infty(T>k|\mathscr{F}_k^Z)} = \sup_{k\ge 0}\operatorname{ess\,sup}\mathsf{E}_k[T-k|T>k;\mathscr{F}_k^Z].
\end{aligned}$$

□

2.3.4 Bayesian Criteria with Respect to the Expected Delay to Detection

The characteristic feature of the Bayes criterion is the assumption that the changepoint is a random variable possessing a prior distribution. This is instrumental in certain applications [137, 138, 171]. This is also of interest since the limiting versions of Bayesian solutions lead to useful procedures, which are optimal or asymptotically optimal in more practical problems.

Let $\{\pi_k\}_{-\infty<k<+\infty}$ be the prior distribution of the changepoint ν, $\pi_k = \Pr(\nu = k)$, *independent* of the observations. While we allow the changepoint ν to take on negative values, there is no need to specify the detailed distribution for negative integers since $T \ge 0$ w.p. 1. Thus, the detailed distribution $\Pr(\nu = k)$ for $k = -1, -2, \dots$ is not important and the only value we need is the cumulative probability $q = \sum_{k=-\infty}^{-1} \Pr(\nu = k)$. So without loss of generality in the rest of the book we will assume that $\pi_{-1} = \Pr(\nu < 0) = q$, i.e., for negative values the prior is concentrated at a single point -1 and the prior distribution on nonnegative integers is defined as

$$\pi_k = (1-q)\tilde{\pi}_k \quad \text{for } k = 0, 1, 2 \dots,$$

where $\tilde{\pi}_k = \Pr(\nu = k | \nu \ge 0)$, $k \ge 0$. The probability $\Pr(\nu \le 0) = q + (1-q)\tilde{\pi}_0$ is interpreted as the probability that the change has already occurred before the observations became available.

Recall that, for any measurable event $\mathscr{A}$, the probability measure P_π is defined as

$$\mathsf{P}_\pi(\mathscr{A} \times \mathscr{K}) = \sum_{k \in \mathscr{K}} \pi_k \mathsf{P}_k(\mathscr{A}),$$

where the π in the subscript emphasizes the dependence on the prior distribution.

A reasonable way to benchmark the detection delay is through the *Average Delay to Detection* (ADD), which is defined as

$$\mathsf{ADD}_\pi(T) = \mathsf{E}_\pi(T-\nu | T > \nu; \nu < \infty) = \frac{\mathsf{E}_\pi[(T-\nu)^+; \nu < \infty]}{\mathsf{P}_\pi(T > \nu; \nu < \infty)}, \tag{2.27}$$

where E_π denotes the expectation with respect to P_π.

In what follows without loss of generality we shall assume that $\pi_\infty = 0$. In this case, conditioning on $\{\nu < \infty\}$ in (2.27) can be removed:

$$\mathsf{ADD}_\pi(T) = \mathsf{E}_\pi(T - \nu | T > \nu) = \frac{\mathsf{E}_\pi[(T-\nu)^+]}{\mathsf{P}_\pi(T > \nu)}, \tag{2.28}$$

Note that if Nature and the Statistician access independent information, i.e., the sequences $\{Z_n\}$ and $\{X_n\}$ are independent under both measures $\mathsf{P}_i, i = 0, \infty$, then the measure defined in (2.24) takes the special form

$$\mathscr{J}_{\mathrm{S}}(T) = \frac{\mathsf{E}_0[T]\Pr(\nu \le 0) + \sum_{k=1}^\infty \pi_k \mathsf{E}_k[(T-k)^+]}{\Pr(\nu \le 0) + \sum_{k=1}^\infty \pi_k \mathsf{P}_\infty(T > k)},$$

which, obviously, coincides with the ADD defined in (2.27). This measure was introduced by Shiryaev [131].

As discussed in Subsection 2.3.2.2, in Bayesian problems the false alarm risk is usually associated with the weighted PFA, defined in (2.11).

Let $\mathbb{C}_\pi(\alpha)$ be the class of detection rules, defined in (2.12), for which the PFA does not exceed a preset level $\alpha \in (0,1)$. Then under the Bayesian approach the goal is to find a stopping time $T_{\mathrm{opt}} \in \mathbb{C}_\pi(\alpha)$ such that

$$\mathscr{J}_{\mathrm{S}}(T_{\mathrm{opt}}) = \inf_{T \in \mathbb{C}_\pi(\alpha)} \mathscr{J}_{\mathrm{S}}(T) \quad \text{for any } \alpha \in (0,1). \tag{2.29}$$

For the i.i.d. model (2.3) and under the assumption that the changepoint ν has a *geometric* prior distribution, this problem was solved by Shiryaev [130, 131, 136]. Specifically, Shiryaev assumed that ν is distributed according to the zero-modified geometric distribution $\mathrm{Geometric}(q,\rho)$[2]

$$\Pr(\nu < 0) = \pi_{-1} = q \text{ and } \Pr(\nu = k) = (1-q)\rho(1-\rho)^k \quad \text{for } k = 0,1,2,\dots, \tag{2.30}$$

where $q \in [0,1)$ and $\rho \in (0,1)$. Note that in our previous notation $\Pr(\nu \le 0) = q + (1-q)\rho$, $\pi_k = (1-q)\rho(1-\rho)^k$, $k \ge 0$, and $\tilde{\pi}_k = \rho(1-\rho)^k$, $k \ge 0$.

Observe now that if $\alpha \ge 1-q$, then there is a trivial solution to the optimization problem (2.29) since we can simply stop right away without any observation. Indeed, this strategy produces $\mathsf{ADD}_\pi = 0$ and $\mathsf{PFA}_\pi = \Pr(\nu \ge 0) = 1-q$, which satisfies the constraint $\mathsf{PFA}_\pi(T) \le \alpha$. Therefore, to avoid trivialities we have to assume that $\alpha < 1-q$. In this case, Shiryaev [130, 131, 136] proved that the optimal detection procedure is based on comparing the posterior probability of a change currently being in effect with a certain detection threshold: the procedure stops as soon as $\Pr(\nu < n | \mathscr{F}_n^X)$ exceeds the threshold. We refer to this strategy as the *Shiryaev rule*. To guarantee its strict optimality the detection threshold should be set so as to guarantee that the PFA is exactly equal to the selected level α.

It is more convenient to express Shiryaev's procedure through the statistic

$$R_{n,\rho} = \frac{q}{(1-q)\rho}\prod_{j=1}^n \left(\frac{\mathscr{L}_j}{1-\rho}\right) + \sum_{k=1}^n \prod_{j=k}^n \left(\frac{\mathscr{L}_j}{1-\rho}\right), \tag{2.31}$$

where $\mathscr{L}_j = f_1(X_j)/f_0(X_j)$ is the likelihood ratio (LR) for the j-th data point X_j. Indeed, by using the Bayes rule, one can show that

$$\Pr(\nu < n | \mathscr{F}_n) = \frac{R_{n,\rho}}{R_{n,\rho} + 1/\rho}, \tag{2.32}$$

[2] Shiryaev considered a slightly different distribution $\Pr(\nu = 0) = q$ and $\Pr(\nu = k) = (1-q)\rho(1-\rho)^{k-1}$ for $k = 1,2,\dots$, assuming that X_k is the first post-change observation when $\nu = k$. This difference in definitions does not affect the results.

whence it is readily seen that thresholding the posterior probability $\Pr(\nu < n|\mathscr{F}_n)$ is the same as thresholding the process $\{R_{n,\rho}\}_{n\geq 1}$. Therefore, the Shiryaev detection rule has the form

$$T_S(A) = \inf\{n \geq 1 \colon R_{n,\rho} \geq A\}, \tag{2.33}$$

and if $A = A_\alpha$ can be selected in such a way that the PFA is exactly equal to α, i.e., $\mathsf{PFA}^\pi(T_S(A_\alpha)) = \alpha$, then it is strictly optimal in class $\mathbb{C}_\pi(\alpha)$,

$$\inf_{T\in\mathbb{C}_\pi(\alpha)} \mathscr{J}_S(T) = \mathscr{J}_S(T_S(A_\alpha)) \quad \text{for any } \; 0 < \alpha < 1-q.$$

Note that Shiryaev's statistic $R_{n,\rho}$ can be rewritten in the recursive form

$$R_{n,\rho} = (1+R_{n-1,\rho})\frac{\mathscr{L}_n}{1-\rho}, \quad n \geq 1, \;\; R_{0,\rho} = \frac{q}{(1-q)\rho}. \tag{2.34}$$

Note also that expressions (2.31) and (2.32) remain true under the geometric prior distribution (2.30) even in the general non-i.i.d. case (2.2) with $\mathscr{L}_n = f_{1,n}(X_n|\mathbf{X}_1^{n-1})/f_{0,n}(X_n|\mathbf{X}_1^{n-1})$. However, in order for the recursion (2.34) to hold in this case, $\{\mathscr{L}_n\}_{n\geq 1}$ should be independent of the changepoint ν, which may not be the case, as discussed in Subsection 2.2.1.

As $\rho \to 0$, where ρ is the parameter of the geometric prior distribution (2.30), the Shiryaev detection statistic (2.34) converges to the so-called *Shiryaev–Roberts* (SR) detection statistic. The latter is the basis for the *SR rule*.

Further details on properties of the Shiryaev and SR detection rules, including asymptotic optimality (as the PFA goes to zero) in the general non-i.i.d. case, will be presented in Chapter 3 and Chapter 4 where general asymptotic Bayesian and minimax changepoint detection theories that cover practically arbitrarily non-i.i.d. models and prior distributions are developed.

Next, consider an alternative setup of minimizing $\mathsf{ADD}_\pi(T)$ in the class of rules $\mathbb{C}_\infty(\alpha)$ defined in (2.13), i.e., that upper-bounds the global PFA $\mathsf{P}_\infty(T < \infty)$. It turns out that the first-order asymptotically optimal (as $\alpha \to 0$) Bayes rule exists under very general conditions and has the following form [157]:

$$T_A = \inf\left\{n \geq 1 : \bar{\Lambda}_n \geq A\right\}, \quad A > 1,$$

where

$$\bar{\Lambda}_n = \sum_{k=-1}^{\infty} \pi_k \prod_{i=k+1}^{\infty} \frac{f(X_i|\mathbf{X}^{i-1})}{g(X_i|\mathbf{X}^{i-1})}$$

and $A = A_\alpha = 1/\alpha$ guarantees $\mathsf{GPFA}(T_A) \leq \alpha$. For the geometric prior (2.30), this rule can be written as

$$T_A = \inf\left\{n \geq 1 : R_{n,\rho} \geq \frac{1}{\alpha\rho(1-\rho)^n} - \rho^{-1}\right\}.$$

Comparing to the Shiryaev rule (2.33) one can see that now the stopping boundary increases exponentially fast with n, which is an unavoidable penalty for the very strong global PFA constraint in place of the weighted PFA constraint. As a result, the conditional average detection delay $\mathsf{CADD}_k(T_A) = \mathsf{E}_k(T_A - k|T_A > k)$ grows quite fast with the changepoint k, approaching infinity as $k \to \infty$. This is a serious drawback, which makes this procedure impractical, as was already discussed in Subsection 2.3.2.3.

2.3.5 Minimax Criteria with Respect to the Expected Delay to Detection

In minimax problem formulations, it is assumed that the changepoint is an unknown not necessarily random number. Even if it is random, its distribution is unknown. The minimax approach has multiple optimality criteria that we consider below.

Lorden [84] was the first who developed the asymptotic minimax theory of change detection in class $\mathbb{C}_\gamma = \{T : \mathsf{E}_\infty[T] \ge \gamma\}$ introduced in (2.10), i.e., for the false alarm risk associated with the ARL to false alarm $\mathsf{ARL2FA}(T) = \mathsf{E}_\infty[T]$, which has to be lower-bounded by a given number $\gamma \ge 1$ (see Subsection 2.3.2.1 for a discussion). Recall also the false alarm scenario shown in Figure 2.3(b). As far as the risk associated with detection delay is concerned, Lorden suggested to evaluate the risk from the delay to detection by the double worst-worst-case average delay to detection defined as

$$\mathsf{ESADD}(T) = \sup_{k \in \mathbb{Z}_+} \operatorname{ess\,sup} \mathsf{E}_k[(T-k)^+ | \mathscr{F}_k^X] \tag{2.35}$$

when the changepoint $\nu = k$. In other words, the conditional ADD is first maximized over all possible trajectories of observations up to the changepoint and then over the changepoint.

Lorden's minimax optimization problem is to find a stopping time $T_{\rm opt} \in \mathbb{C}_\gamma$ such that

$$\mathsf{ESADD}(T_{\rm opt}) = \inf_{T \in \mathbb{C}_\gamma} \mathsf{ESADD}(T) \quad \text{for every } \gamma > 1. \tag{2.36}$$

Let us now discuss connection of Lorden's risk (2.35) with more general Moustakides's risk (2.26). Assume first conditional post-change independence from Nature's information Z:

$$f(X_{n+1}^\infty | X_1^n, Z_{-\infty}^n) = f(X_{n+1}^\infty | X_1^n) \quad \text{for all } n \ge 0.$$

This assumption implies that for any $\mathscr{F}_\infty^X$-measurable random variable Y

$$\mathsf{E}[Y | \mathscr{F}_n^X \cup \mathscr{F}_n^Z] = \mathsf{E}[Y | \mathscr{F}_n^X]. \tag{2.37}$$

Then we have the following simplifications for the risk $\mathscr{J}(T)$ defined in (2.25):

$$\begin{aligned}
\mathscr{J}(T) &= \sup_{\nu \in \mathscr{T}} \frac{\mathsf{E}_\infty[\mathsf{E}_0[T]\mathbb{1}_{\{\nu \le 0\}}] + \sum_{k=1}^\infty \mathsf{E}_\infty[\mathsf{E}_0[(T-k)^+ | \mathscr{F}_k^X]\mathbb{1}_{\{\nu = k\}}]}{\mathsf{E}_\infty[\mathbb{1}_{\{\nu \le 0\}}] + \sum_{k=1}^\infty \mathsf{E}_\infty[\mathsf{P}_\infty(T > k | \mathscr{F}_k^X)\mathbb{1}_{\{\nu = k\}}]} \\
&= \sup_{\nu \in \mathscr{T}} \frac{\mathsf{E}_0[T]\mathsf{P}_\infty(\nu \le 0) + \sum_{k=1}^\infty \mathsf{E}_\infty[\mathsf{E}_0[(T-k)^+ | \mathscr{F}_k^X]\mathsf{P}_\infty(\nu = k | \mathscr{F}_k^X)]}{\mathsf{P}_\infty(\nu \le 0) + \sum_{k=1}^\infty \mathsf{E}_\infty[\mathsf{P}_\infty(T > k | \mathscr{F}_k^X)\mathsf{P}_\infty(\nu = k | \mathscr{F}_k^X)]} \\
&= \sup_{k \in \mathbb{Z}_+} \operatorname{ess\,sup} \mathsf{E}_k[(T-k)^+ | \mathscr{F}_k^X],
\end{aligned}$$

which is nothing but Lorden's risk (2.35).

Note that if Nature and the Statistician have an access to the same information, i.e., $\mathscr{F}_n^Z = \mathscr{F}_n^X$, then equality (2.37) also holds. More generally, if Z_n contains X_n plus some additional *independent* information I_n, that is, if $\{I_n\}$ is independent of $\{X_n\}$ and Nature uses $Z_n = (X_n, I_n)$, then equality (2.37) holds as well. Yet another scenario is when we adopt the model (2.4) with independent pre-change and post-change processes. In this case, (2.37) holds true.

For the i.i.d. scenario (2.3), Lorden [84] showed that Page's Cumulative Sum (CUSUM) procedure [104] is first-order asymptotically minimax as $\gamma \to \infty$. For any $\gamma > 1$, this problem was solved by Moustakides [93], who showed that CUSUM is exactly optimal. See also Ritov [123] for a different decision-theoretic argument.

Though strict $\mathsf{ESADD}(T)$-optimality of the CUSUM procedure is a strong result, it is more natural to construct a procedure that minimizes the conditional average delay to detection $\mathsf{CADD}_\nu(T) = \mathsf{E}_\nu(T - \nu | T > \nu)$ for all $\nu \ge 0$ simultaneously. As no such uniformly optimal procedure is possible, Pollak [109] suggested to evaluate the risk from the delay to detection by the maximal conditional average delay to detection

$$\mathsf{SADD}(T) = \sup_{k \in \mathbb{Z}_+} \mathsf{E}_k(T - k | T > k). \tag{2.38}$$

Pollak's minimax optimization problem seeks to find a stopping time $T_{\rm opt} \in \mathbb{C}_\gamma$ such that

$$\mathsf{SADD}(T_{\rm opt}) = \inf_{T \in \mathbb{C}_\gamma} \mathsf{SADD}(T) \quad \text{for every } \gamma > 1. \tag{2.39}$$

Again, as in Section 2.3.4, assume that Nature and the Statistician access independent information, i.e., the sequences $\{Z_n\}$ and $\{X_n\}$ are independent. If the prior distribution of the change point $\pi = \{\pi_k\}$ is unknown, then it is of interest considering the worst possible scenario (the least favorable prior):

$$\mathscr{J}_{\mathrm{P}}(T) = \sup_{\pi} \frac{\mathsf{E}_0[T]\Pr(\nu \le 0) + \sum_{k=1}^{\infty} \mathsf{E}_k[(T-k)^+]\pi_k}{\Pr(\nu \le 0) + \sum_{k=1}^{\infty} \mathsf{P}_\infty(T>k)\pi_k} = \sup_{k \in \mathbb{Z}_+} \frac{\mathsf{E}_k[(T-k)^+]}{\mathsf{P}_\infty(T>k)} = \sup_{k \in \mathbb{Z}_+} \mathsf{E}_k(T-k|T>k),$$

which yields Pollaks's risk $\mathsf{SADD}(T)$ defined in (2.38).

Unlike Lorden's minimax problem (2.36), Pollak's minimax problem (2.39) is still not solved in general. Some light as to the possible solution in the i.i.d. case is shed in [97, 116, 166, 164].

2.3.6 Pointwise Uniform Optimality Criterion

While the Bayesian and minimax formulations are reasonable and can be justified in many applications, it would be most desirable to guarantee small values of the conditional average detection delay $\mathsf{CADD}_\nu(T) = \mathsf{E}_\nu(T-\nu|T \ge \nu)$ uniformly for all $\nu \ge 0$ when the false alarm risk is fixed at a certain level. However, if the false alarm risk is measured in terms of the ARL to false alarm, i.e., if it is required that $\mathsf{ARL2FA}(T) \ge \gamma$ for some $\gamma > 1$, then a procedure that minimizes $\mathsf{CADD}_\nu(T)$ for all ν does not exist, as we previously discussed. More importantly, as discussed in Subsection 2.3.2.4, large values of the $\mathsf{ARL2FA}(T) = \mathsf{E}_\infty[T]$ generally do not imply small values of the maximal local probability of false alarm $\mathsf{LCPFA}(T) = \sup_{\ell \ge 0} = \mathsf{P}_\infty(T \le \ell + m|T > \ell)$ in a time interval of a fixed length $m \ge 1$ (see Lemma 2.1), while the opposite is always true (see Lemma 2.2). Therefore, the constraint $\mathsf{LCPFA}(T) \le \beta$ is more stringent than $\mathsf{E}_\infty[T] \ge \gamma$. In Subsection 2.3.2.4, we discussed yet another reason for considering the LCPFA constraint instead of the ARL to false alarm constraint.

For these reasons, we now focus on the class of detection rules $\mathbb{C}(m,\beta)$, defined in (2.17), i.e., such rules that the $\mathsf{LCPFA}(T)$ is upper-bounded with the prespecified level $\beta \in (0,1)$. The goal is to find a stopping time $T_{\mathrm{opt}} \in \mathbb{C}(m,\beta)$ such that

$$\mathsf{CADD}_\nu(T_{\mathrm{opt}}) = \inf_{T \in \mathbb{C}(m,\beta)} \mathsf{CADD}_\nu(T) \quad \text{for all } \nu \in \mathbb{Z}_+ \text{ and any } 0 < \beta < 1. \tag{2.40}$$

Such an optimal procedure may exist, while its structure is not known unless β is small.

2.3.7 Criteria Maximizing Probability of Detection

The optimality criteria (2.29), (2.36), (2.39), (2.40) considered in the previous sections require minimization of expected delay to detection at an infinite time horizon and do not consider a probability of detection of a change in a given fixed time interval. Often, however, practitioners are interested in such probabilities under a given false alarm rate even if the change lasts infinitely long.[3] Besides, in many applications, the length of a change is finite, e.g., in problems of detecting transient changes with known and unknown durations [52, 53, 121, 150, 151]. Then stopping outside of the interval $(\nu, \nu+M]$ of a given duration M may not be quite appropriate. In such cases, it is reasonable to find detection rules that minimize the probability of missed detection in a certain fixed time interval and to consider the following optimality criteria. We recall the definition of classes $\mathbb{C}_\gamma$, $\mathbb{C}_\pi(\alpha)$, and $\mathbb{C}(m,\beta)$ given in (2.10), (2.12), and (2.17), respectively.

The Bayesian optimality criterion in class $\mathbb{C}_\pi(\alpha)$. Given the interval (time-window) of the length M ($M \ge 1$), define the (average) probability of detection

$$\mathsf{P}_\pi(T \le \nu + M|T > \nu) = \frac{\sum_{k=-1}^{\infty} \pi_k \mathsf{P}_k(k < T \le k+M)}{1 - \mathsf{PFA}_\pi(T)} \tag{2.41}$$

[3] In practice, this means that the length of a change is substantially larger than an average detection delay.

in this interval. The goal is to find such rule $T_{\rm opt} \in \mathbb{C}_\pi(\alpha)$ that for every $0 < \alpha < 1$

$$\mathsf{P}_\pi(T_{\rm opt} \le \nu + M | T_{\rm opt} > \nu) = \sup_{T \in \mathbb{C}_\pi(\alpha)} \mathsf{P}_\pi(T \le \nu + M | T > \nu) \quad \text{for some fixed } M \ge 1. \tag{2.42}$$

Note that summation in (2.41) is over $k \ge 0$ if $\mathsf{P}(T = 0) = 0$, i.e., when the randomized at 0 stopping times are excluded.

Assume now that M is not fixed but random with the given distribution $\pi_i^M = \Pr(M = i)$, $i = 1, 2, \dots$ and that M is independent of the change point ν. In particular, this assumption is reasonable when M is an unknown duration of a change, which is a nuisance parameter, i.e., the fact of change disappearance does not have to be detected. Then the probability of detection in (2.41) is re-written as

$$\mathsf{P}_{\pi,\pi^M}(T \le \nu + M | T > \nu) = \frac{\sum_{k=-1}^{\infty} \sum_{i=1}^{\infty} \pi_k \pi_i^M \mathsf{P}_k(k < T \le k+i)}{1 - \mathsf{PFA}_\pi(T)}. \tag{2.43}$$

The optimality criterion (2.42) gets modified as

$$\mathsf{P}_{\pi,\pi^M}(T_{\rm opt} \le \nu + M | T_{\rm opt} > \nu) = \sup_{T \in \mathbb{C}_\pi(\alpha)} \mathsf{P}_{\pi,\pi^M}(T \le \nu + M | T > \nu). \tag{2.44}$$

Maximin optimality criteria in class $\mathbb{C}_\gamma$. Given the interval of the length M, define minimal probabilities of detection

$$\inf_{\nu \ge 0} \operatorname{ess\,inf} \mathsf{P}_\nu(\nu < T \le \nu + M | \mathscr{F}_\nu, T > \nu)$$

and

$$\inf_{\nu \ge 0} \mathsf{P}_\nu(\nu < T \le \nu + M | T > \nu).$$

The goal is to find such rule $T_{\rm opt} \in \mathbb{C}_\gamma$ that for every $\gamma \ge 1$

$$\inf_{\nu \ge 0} \operatorname{ess\,inf} \mathsf{P}_\nu(T_{\rm opt} \le \nu + M | \mathscr{F}_\nu, T_{\rm opt} > \nu) = \sup_{T \in \mathbb{C}_\gamma} \inf_{\nu \ge 0} \operatorname{ess\,inf} \mathsf{P}_\nu(T \le \nu + M | \mathscr{F}_\nu, T > \nu) \tag{2.45}$$

and

$$\inf_{\nu \ge 0} \mathsf{P}_\nu(T_{\rm opt} \le \nu + M | T_{\rm opt} > \nu) = \sup_{T \in \mathbb{C}_\gamma} \inf_{\nu \ge 0} \mathsf{P}_\nu(T \le \nu + M | T > \nu). \tag{2.46}$$

If M is not fixed but random with the given distribution $\pi_i^M = \Pr(M = i)$, then the criteria (2.45) and (2.46) are modified as

$$\inf_{\nu \ge 0} \operatorname{ess\,inf} \bar{\mathsf{P}}_\nu(T_{\rm opt} \le \nu + M | \mathscr{F}_\nu, T_{\rm opt} > \nu) = \sup_{T \in \mathbb{C}_\gamma} \inf_{\nu \ge 0} \operatorname{ess\,inf} \bar{\mathsf{P}}_\nu(T \le \nu + M | \mathscr{F}_\nu, T > \nu) \tag{2.47}$$

and

$$\inf_{\nu \ge 0} \bar{\mathsf{P}}_\nu(T_{\rm opt} \le \nu + M | T_{\rm opt} > \nu) = \sup_{T \in \mathbb{C}_\gamma} \inf_{\nu \ge 0} \bar{\mathsf{P}}_\nu(T \le \nu + M | T > \nu), \tag{2.48}$$

where

$$\begin{aligned} \bar{\mathsf{P}}_\nu(T \le \nu + M | \mathscr{F}_\nu, T > \nu) &= \sum_{i=1}^{\infty} \pi_i^M \mathsf{P}_\nu(T \le \nu + i | \mathscr{F}_\nu, T > \nu, M = i), \\ \bar{\mathsf{P}}_\nu(T \le \nu + i | T > \nu) &= \sum_{i=1}^{\infty} \pi_i^M \mathsf{P}_\nu(T \le \nu + i | T > \nu, M = i). \end{aligned} \tag{2.49}$$

Maximin optimality criteria in class $\mathbb{C}(m,\beta)$. The goal is to find such rule $T_{\rm opt} \in \mathbb{C}(m,\beta)$ that for every $0 < \beta < 1$ and some $m \ge 1$

$$\inf_{\nu \ge 0} \operatorname{ess\,inf} \mathsf{P}_\nu(T_{\rm opt} \le \nu + M | \mathscr{F}_\nu, T_{\rm opt} > \nu) = \sup_{T \in \mathbb{C}(m,\beta)} \inf_{\nu \ge 0} \operatorname{ess\,inf} \mathsf{P}_\nu(T \le \nu + M | \mathscr{F}_\nu, T > \nu) \tag{2.50}$$

and

$$\inf_{\nu\geq 0} \mathsf{P}_\nu(T_{\rm opt} \leq \nu + M | T_{\rm opt} > \nu) = \sup_{T\in\mathbb{C}(m,\beta)} \inf_{\nu\geq 0} \mathsf{P}_\nu(T \leq \nu + M | T > \nu). \tag{2.51}$$

Again, if M is random with the distribution $\pi_i^M = \Pr(M = i)$, then the criteria (2.50) and (2.51) get modified similar to (2.47) and (2.48) by replacing the measure P_ν with $\bar{\mathsf{P}}_\nu$.

2.3.8 Asymptotic Optimality Criteria

Unfortunately, for general non-i.i.d. stochastic models it is very difficult, if at all possible, to find optimal solutions to optimization problems formulated in previous sections. To obtain reasonable results it is necessary to consider asymptotically optimal rules when the false alarm risk approaches zero. For example, in the Bayesian setup, instead of the optimization problem (2.29) for each PFA $\alpha \in (0,1)$ one may require to find an asymptotically optimal change detection rule $T_{\rm aopt}$ that satisfies

$$\text{Find } T_{\rm aopt} \in \mathbb{C}_\pi(\alpha) \text{ such that } \lim_{\alpha\to 0} \frac{\inf_{T\in\mathbb{C}_\pi(\alpha)} \mathsf{ADD}_\pi(T)}{\mathsf{ADD}_\pi(T_{\rm aopt})} = 1.$$

This is the so-called first-order asymptotic optimality criterion.

More generally, it is of interest to study the behavior of higher moments of the detection delay $\mathsf{E}_\pi[(T-\nu)^r | T > \nu]$ for some $r > 1$, i.e.,

$$\text{Find } T_{\rm aopt} \in \mathbb{C}_\pi(\alpha) \text{ such that } \lim_{\alpha\to 0} \frac{\inf_{T\in\mathbb{C}_\pi(\alpha)} \mathsf{E}_\pi[(T-\nu)^r | T > \nu]}{\mathsf{E}_\pi[(T_{\rm aopt}-\nu)^r | T > \nu]} = 1.$$

Our main focus in this book is on such kinds of asymptotic problems for single and multiple data streams. Specific asymptotic optimality criteria are formulated in each chapter.

3

Bayesian Quickest Change Detection in a Single Population

3.1 Introduction

Suppose $X_1, X_2, \ldots$ are random variables observed sequentially, which may change statistical properties at an unknown point in time $\nu \in \{0, 1, 2, \ldots\}$, so that $X_1, \ldots, X_\nu$ are generated by one stochastic model and $X_{\nu+1}, X_{\nu+2}, \ldots$ by another model. The value of the change point ν is unknown and the fact of change must be detected as soon as possible controlling for a risk associated with false detections.

More specifically, let $\mathbf{X}_0^n = (X_0, X_1, \ldots, X_n)$ denote a sample of size n with an attached initial value X_0 and the density of this sample is given by the general non-i.i.d. model (2.2), where $\{f_\theta(X_n|\mathbf{X}_0^{n-1})\}_{n\geq 0}$ is a sequence of post-change conditional densities of X_n given $\mathbf{X}_0^{n-1}$ with the unknown parameter $\theta \in \Theta$ and $\{g(X_n|\mathbf{X}_0^{n-1})\}_{n\geq 0}$ is a sequence of pre-change densities, which are known. As discussed in Chapter 2, the goal of the quickest changepoint detection problem is to develop a detection rule that stops as soon as possible after the real change occurs under a given risk of false alarms.

In early stages, the work focused on the i.i.d. case where $f_\theta(X_n|\mathbf{X}_0^{n-1}) = f_\theta(X_n)$ and $g(X_n|\mathbf{X}_0^{n-1}) = g(X_n)$, i.e., when the observations are i.i.d. according to a distribution with density $f_\theta(X_n)$ in the pre-change mode and with density $g(X_n)$ in the post-change mode. In the early 1960s, Shiryaev [131] developed a Bayesian sequential changepoint detection theory when $\theta = \theta_1$ is known. This theory implies that the detection procedure based on thresholding the posterior probability of the change being active before the current time is strictly optimal, minimizing the expected delay to detection in the class of procedures with a given weighted probability of false alarm if the prior distribution of the change point is geometric. At the beginning of the 1970s, Lorden [84] showed that Page's CUSUM procedure [104] is first-order asymptotically optimal in a minimax sense, minimizing the maximal expected delay to detection in the class of procedures with the prescribed average run length to false alarm (ARL2FA) as ARL2FA approaches infinity. In the mid-1980s, Moustakides [93] established exact minimaxity of the CUSUM procedure for any value of the ARL2FA. Pollak [109] suggested modifying the conventional Shiryaev–Roberts statistic (see [130, 131, 124]) by randomizing the initial condition to make it an equalizer. His version of the Shiryaev–Roberts statistic starts from a random point sampled from the quasi-stationary distribution of the Shiryaev–Roberts statistic. He proved that, for a large ARL2FA, this randomized procedure is asymptotically third-order minimax within an additive vanishing term. The articles [97, 116] indicate that the Shiryaev–Roberts–Pollak procedure is not exactly minimax for all values of the ARL2FA by showing that a generalized Shiryaev–Roberts procedure that starts from a specially designed deterministic point performs slightly better. Shiryaev [130, 131] was the first who established exact optimality of the Shiryaev–Roberts detection procedure in the problem of detecting changes occurring at a far time horizon after many re-runs among multi-cyclic procedures with the prescribed mean time between false alarms for detecting a change in the drift of the Brownian motion. Pollak and Tartakovsky [113] extended Shiryaev's result to the discrete-time i.i.d. (not necessarily Gaussian) case. Third-order asymptotic optimality of generalized Shiryaev–Roberts procedures with random and deterministic head-starts was established in [165]. Another trend related to evaluation of performance of CUSUM and EWMA detection procedures was initiated by the SPC (statistical process control) community (see, e.g., [15, 19, 27, 58, 59, 92, 174, 181, 191, 185, 186]).

In many practical applications, the i.i.d. assumption is too restrictive. The observations may be either non-identically distributed or correlated or both, i.e., non-i.i.d. Lai [75] generalized Lorden's asymptotic theory [84] for the general non-i.i.d. case establishing asymptotic optimality of the CUSUM procedure under very general conditions in the point-wise, minimax, and Bayesian settings. He also suggested a window-limited version of the CUSUM procedure, which is computationally less demanding than a conventional CUSUM, but still preserves asymptotic optimality properties. Tartakovsky and Veeravalli [171], Baron and Tartakovsky [7], and Tartakovsky [160] generalized Shiryaev's Bayesian theory for the general non-i.i.d. case and for a wide class of prior distributions. In particular, it was proved that the Shiryaev detection rule is asymptotically optimal – it minimizes not only the expected delay to detection but also higher moments of the detection delay as the weighted probability of a false alarm vanishes. Fuh and Tartakovsky [47] specified the results in [171, 160] for finite-state hidden Markov models (HMM), finding sufficient conditions under which the Shiryaev and Shiryaev–Roberts rules are first-order asymptotically optimal, assuming that both pre-change and post-change distributions are completely specified, i.e., the post-change parameter θ is known. Fuh [44] proved first-order asymptotic minimaxity of the CUSUM procedure as the ARL2FA goes to infinity. Pergamenchtchikov and Tartakovsky [108] established point-wise and minimax asymptotic optimality properties of the Shiryaev–Roberts rule for the general non-i.i.d. stochastic model in the class of rules with the prescribed local conditional probability of false alarm (in the given time interval) as well as presented sufficient conditions for ergodic Markov processes.

In a variety of applications, however, a pre-change distribution is known but the post-change distribution is rarely known completely. A more realistic situation is parametric uncertainty when the parameter θ of the post-change distribution is unknown since a putative value of θ is rarely representative. When the post-change parameter is unknown, so that the post-change hypothesis "$\mathsf{H}_k^{\vartheta} : \nu = k, \theta = \vartheta$", $\vartheta \in \Theta$ is composite, and it is desirable to detect quickly a change in a broad range of possible values, the natural modification of the CUSUM, Shiryaev and Shiryaev–Roberts procedures is based either on maximizing over ϑ or weighting over a mixing measure $W(\vartheta)$ the corresponding statistics tuned to $\theta = \vartheta$. The maximization leads to the generalized likelihood ratio (GLR)-based procedures and weighting to mixtures. Lorden [84] was the first established first-order asymptotic minimaxity of the GLR-CUSUM procedure for the i.i.d. exponential families as the ARL2FA goes to infinity (see also Dragalin [32] for refined results). Siegmund and Yakir [142] established third-order asymptotic minimaxity of the randomized mixture Shiryaev–Roberts–Pollak procedure for the exponential family with respect to the maximal Kullback–Leibler information. Lai [75] established point-wise and minimax asymptotic optimality of the window-limited mixture CUSUM and GLR-CUSUM procedures for general non-i.i.d. models. Further detailed overview and references can be found in the monographs [11, 164].

A variety of applications where sequential changepoint detection is important are discussed, e.g., in [9, 10, 11, 4, 24, 68, 88, 92, 115, 141, 152, 170, 162, 148, 159, 168, 169, 164, 167].

In this chapter, we provide the asymptotic Bayesian theory of change detection for the composite post-change hypothesis where the post-change parameter is unknown. We assume that the observations can have a very general structure, i.e., can be dependent and non-identically distributed. The key assumption in the general asymptotic theory is a stability property of the log-likelihood ratio process between the "change" and "no-change" hypotheses, which can be formulated in terms of a Law of Large Numbers and rates of convergence, e.g., as the r-complete convergence of the properly normalized log-likelihood ratio and its adaptive version in the vicinity of the true parameter value.

3.2 The Shiryaev and Shiryaev–Roberts Mixture Rules

Let P_∞ denote the probability measure corresponding to the sequence of observations $\{X_n\}_{n\geq 1}$ when there is never a change ($\nu = \infty$) and, for $k = 0, 1, \dots$ and $\vartheta \in \Theta$, let $\mathsf{P}_{k,\vartheta}$ denote the measure corresponding to the sequence $\{X_n\}_{n\geq 1}$ when $\nu = k < \infty$ and $\theta = \vartheta$ (i.e., $X_{\nu+1}$ is the first post-change observation), where $\theta \in \Theta$ is a parameter (possibly multidimensional). Let $\mathsf{E}_{k,\vartheta}$ and E_∞ denote expectations under $\mathsf{P}_{k,\vartheta}$ and P_∞, respectively.

In what follows, we consider the general non-i.i.d. model defined in (2.2), i.e.,

$$p_{\nu,\theta}(\mathbf{X}_0^n) = \prod_{i=0}^{\nu} g(X_i|\mathbf{X}_0^{i-1}) \times \prod_{i=\nu+1}^{n} f_\theta(X_i|\mathbf{X}_0^{i-1}). \tag{3.1}$$

The likelihood ratio (LR) of the hypothesis "$\mathsf{H}_k^\vartheta : \nu = k, \theta = \vartheta$" that the change occurs at $\nu = k$ with the post-change parameter $\theta = \vartheta$ against the no-change hypothesis "$\mathsf{H}_\infty : \nu = \infty$" based on the sample $\mathbf{X}^n = (X_1, \dots, X_n)$ is given by the product

$$LR_\vartheta(k,n) = \prod_{i=k+1}^{n} \frac{f_\vartheta(X_i|\mathbf{X}^{i-1})}{g(X_i|\mathbf{X}^{i-1})}, \quad n > k$$

and we set $LR_\vartheta(k,n) = 1$ for $n \leq k$. Recall that in general densities $f_{\vartheta,i}(X_i|\mathbf{X}^{i-1})$ and $g_i(X_i|\mathbf{X}^{i-1})$ may depend on i, and we omit i for brevity.

Assume that the change point ν is a random variable independent of the observations with prior distribution $\pi_k = \mathsf{P}(\nu = k)$, $k = 0, 1, 2, \dots$ with $\pi_k > 0$ for $k \in \{0, 1, 2, \dots\} = \mathbb{Z}_+$. We will also assume that a change point may take negative values, which means that the change has occurred by the time the observations became available. However, as we discussed in the previous chapter, the detailed structure of the distribution $\mathsf{P}(\nu = k)$ for $k = -1, -2, \dots$ is not important. The only value which matters is the total probability $q = \mathsf{P}(\nu \leq -1)$ of the change being in effect before the observations become available, so we set $\mathsf{P}(\nu \leq -1) = \mathsf{P}(\nu = -1) = \pi_{-1} = q$.

Let $\mathscr{L}_\vartheta(n) = f_\vartheta(X_n|\mathbf{X}^{n-1})/g(X_n|\mathbf{X}^{n-1})$. As discussed in [171, 160], for detecting a change from $\{g(X_n|\mathbf{X}^{n-1})\}$ to $\{f_\vartheta(X_n|\mathbf{X}^{n-1})\}$ it is natural to use the Shiryaev statistic

$$S_\vartheta^\pi(n) = \frac{1}{\mathsf{P}(\nu \geq n)} \left(q \prod_{t=1}^{n} \mathscr{L}_\vartheta(t) + \sum_{k=0}^{n-1} \pi_k \prod_{t=k+1}^{n} \mathscr{L}_\vartheta(t) \right), \quad n \geq 1, \quad S_\vartheta^\pi(0) = q/(1-q) \tag{3.2}$$

tuned to ϑ. Hereafter we set $\prod_{t=j}^{i} \mathscr{L}_\vartheta(t) = 1$ for $i < j$. The corresponding stopping time (time of detection) is

$$T_A^\vartheta = \inf\{n \geq 1 : S_\vartheta^\pi(n) \geq A\} \tag{3.3}$$

where A ($A > 0$) is a threshold controlling for the false alarm risk. In what follows, we refer to T_A^ϑ as the Shiryaev detection rule.

When the value of the parameter is unknown there are two conventional approaches to overcome uncertainty – either to maximize or average over ϑ. The second approach is usually referred to as *Mixtures*. To be more specific, introduce a mixing measure $W(\theta)$, $\int_\Theta \mathrm{d}W(\theta) = 1$, which can be interpreted as a prior distribution if needed. Define the average (mixed) LR

$$\Lambda_W(k,n) = \int_\Theta LR_\vartheta(k,n)\, \mathrm{d}W(\vartheta), \quad k < n \tag{3.4}$$

and the statistic

$$S_W^\pi(n) = \int_\Theta S_\vartheta^\pi(n)\, \mathrm{d}W(\vartheta) = \frac{1}{\mathsf{P}(\nu \geq n)} \left(q\Lambda_W(0,n) + \sum_{k=0}^{n-1} \pi_k \Lambda_W(k,n) \right), \quad n \geq 1$$
$$S_W^\pi(0) = q/(1-q), \tag{3.5}$$

where $S^{\pi}_{\vartheta}(n)$ is the Shiryaev statistic tuned to the parameter $\theta = \vartheta$ defined in (3.2). We will call $S^{\pi}_{W}(n)$ the *Mixture Shiryaev* (MS) statistic and will write $S_W(n)$, omitting dependence on the prior π for brevity.

In the sequel, we study the MS detection rule that stops and raises an alarm as soon as the statistic $S_W(n)$ reaches a positive level A, i.e., the MS rule is nothing but the stopping time

$$T_A^W = \inf\{n \geq 1 : S_W(n) \geq A\}, \tag{3.6}$$

where $A > 0$ is a threshold controlling for the false alarm risk.

Another popular statistic for detecting a change from $\{g(X_n|\mathbf{X}^{n-1})\}$ to $\{f_{\vartheta}(X_n|\mathbf{X}^{n-1})\}$, which has certain optimality properties [113, 165, 116, 164], is the generalized Shiryaev–Roberts (SR) statistic

$$R_{\vartheta}(n) = \omega LR_{\vartheta}(0,n) + \sum_{k=0}^{n-1} LR_{\vartheta}(k,n) = \omega \prod_{t=1}^{n} \mathscr{L}_{\vartheta}(t) + \sum_{k=1}^{n} \prod_{t=k}^{n} \mathscr{L}_{\vartheta}(t), \quad n \geq 1 \tag{3.7}$$

with a non-negative head-start $R_{\vartheta}(0) = \omega$, $\omega \geq 0$. The corresponding stopping time is

$$\widetilde{T}_B^{\vartheta} = \inf\{n \geq 1 : R_{\vartheta}(n) \geq B\}, \tag{3.8}$$

where B ($B > 0$) is a threshold controlling for the false alarm risk. In what follows, we refer to $\widetilde{T}_B^{\vartheta}$ as the SR detection rule.

The mixture counterpart, which we will refer to as the *Mixture Shiryaev–Roberts* (MSR) statistic, is

$$R_W(n) = \int_{\Theta} R_{\vartheta}(n)\, \mathrm{d}W(\vartheta) = \omega \Lambda_W(0,n) + \sum_{k=0}^{n-1} \Lambda_W(k,n), \quad n \geq 1, \; R_W(0) = \omega, \tag{3.9}$$

and the corresponding MSR detection rule is given by the stopping time

$$\widetilde{T}_B^W = \inf\{n \geq 1 : R_W(n) \geq B\}, \tag{3.10}$$

where $B > 0$ is a threshold controlling for the false alarm risk.

In Section 3.4, we show that the MS detection rule T_A^W is first-order asymptotically optimal, minimizing moments of the stopping time distribution for the low risk of false alarms under very general conditions. In Section 3.5, we establish asymptotic properties of the MSR rule, showing that it is also asymptotically optimal when the prior distribution becomes asymptotically flat, but not in general.

3.3 Asymptotic Problems

Let $\mathsf{P}^{\pi}_{\theta}(\mathscr{A} \times \mathscr{K}) = \sum_{k \in \mathscr{K}} \pi_k \mathsf{P}_{k,\theta}(\mathscr{A})$ denote the "weighted" probability measure and $\mathsf{E}^{\pi}_{\theta}$ the corresponding expectation.

For $r \geq 1$, $\nu = k \in \mathbb{Z}_+$, and $\theta \in \Theta$, introduce the risk associated with the conditional r-th moment of the detection delay

$$\mathscr{R}^r_{k,\theta}(T) = \mathsf{E}_{k,\theta}\left[(T-k)^r \,|\, T > k\right]. \tag{3.11}$$

In a Bayesian setting, the average risk associated with the moments of delay to detection is

$$\begin{aligned} \overline{\mathscr{R}}^r_{\pi,\theta}(T) &:= \mathsf{E}^{\pi}_{\theta}[(T-\nu)^r | T > \nu] = \frac{\sum_{k=-1}^{\infty} \pi_k \mathsf{E}_{k,\theta}\left[(T-k)^+\right]^r}{1 - \mathsf{PFA}_{\pi}(T)} \\ &= \frac{\sum_{k=-1}^{\infty} \pi_k \mathscr{R}^r_{k,\theta}(T) \mathsf{P}_{\infty}(T > k)}{1 - \mathsf{PFA}_{\pi}(T)}, \end{aligned} \tag{3.12}$$

where

$$\mathsf{PFA}_\pi(T) = \mathsf{P}_\theta^\pi(T \le \nu) = \sum_{k=0}^{\infty} \pi_k \mathsf{P}_\infty(T \le k) \tag{3.13}$$

is the weighted probability of false alarm (PFA) that corresponds to the risk associated with a false alarm, and we set $\mathscr{R}^r_{-1,\theta}(T) = \mathscr{R}^r_{0,\theta}(T) = \mathsf{E}_{0,\theta}[T]$ but not $\mathscr{R}^r_{-1,\theta}(T) = \mathsf{E}_{-1,\theta}[T+1]$ here and in what follows. Therefore, if we are interested only in stopping times that are positive and finite w.p. 1 (under P_∞), the risk $\overline{\mathscr{R}}^r_{\pi,\theta}(T)$ can be written as

$$\overline{\mathscr{R}}^r_{\pi,\theta}(T) = \frac{\Pr(\nu \le 0)\mathsf{E}_{0,\theta}[T] + \sum_{k=1}^{\infty} \pi_k \mathscr{R}^r_{k,\theta}(T)\mathsf{P}_\infty(T > k)}{\Pr(\nu \le 0) + \sum_{k=1}^{\infty} \pi_k \mathsf{P}_\infty(T > k)}. \tag{3.14}$$

Sometimes, it will be convenient to replace summation in (3.12) over $k \in \mathbb{Z}_+$, i.e., to write (with a certain abuse of notation)

$$\overline{\mathscr{R}}^r_{\pi,\theta}(T) = \frac{\sum_{k=0}^{\infty} \pi_k \mathscr{R}^r_{k,\theta}(T)\mathsf{P}_\infty(T > k)}{1 - \mathsf{PFA}_\pi(T)},$$

setting $\pi_0 = \Pr(\nu \le 0)$ instead of $\pi_0 = \Pr(\nu = 0)$.

Note that in (3.12) and (3.13) we used the fact that $\mathsf{P}_{k,\theta}(T \le k) = \mathsf{P}_\infty(T \le k)$ since the event $\{T \le k\}$ depends on the observations $X_1, \dots X_k$ generated by the pre-change probability measure P_∞ since by our convention X_k is the last pre-change observation if $\nu = k$.

We are interested in the Bayesian optimization problem

$$\inf_{\{T:\mathsf{PFA}_\pi(T) \le \alpha\}} \overline{\mathscr{R}}^r_{\pi,\theta}(T) \quad \text{for all } \theta \in \Theta. \tag{3.15}$$

However, in general this problem is not manageable for every value of the PFA $\alpha \in (0,1)$. So we will focus on the asymptotic problem assuming that the PFA α approaches zero. Specifically, in Section 3.4, we will be interested in proving that the MS rule is first-order asymptotically optimal, i.e.,

$$\lim_{\alpha \to 0} \frac{\inf_{T \in \mathbb{C}_\pi(\alpha)} \overline{\mathscr{R}}^r_{\pi,\theta}(T)}{\overline{\mathscr{R}}^r_{\pi,\theta}(T_A^W)} = 1 \quad \text{for all } \theta \in \Theta, \tag{3.16}$$

where $\mathbb{C}_\pi(\alpha) = \{T : \mathsf{PFA}_\pi(T) \le \alpha\}$ is the class of detection rules for which the PFA does not exceed a prescribed number $\alpha \in (0,1)$. In addition, we will prove that the MS rule is uniformly first-order asymptotically optimal in a sense of minimizing the conditional risk (3.11) for all change point values $\nu = k \in \mathbb{Z}_+$, i.e.,

$$\lim_{\alpha \to 0} \frac{\inf_{T \in \mathbb{C}_\pi(\alpha)} \mathscr{R}^r_{k,\theta}(T)}{\mathscr{R}^r_{k,\theta}(T_A^W)} = 1 \quad \text{for all } \theta \in \Theta \text{ and all } k \in \mathbb{Z}_+. \tag{3.17}$$

In Section 3.6, we consider a "purely" Bayes problem with the average (integrated) risk, which is the sum of the PFA and the cost of delay proportional to the r-th moment of the detection delay and prove that the MS rule is asymptotically optimal when the cost of delay to detection approaches 0.

Asymptotic properties of the MSR rule $\widetilde{T}_B$ will be also established.

For a fixed $\theta \in \Theta$, introduce the log-likelihood ratio (LLR) process $\{\lambda_\theta(k,n)\}_{n \ge k+1}$ between the hypotheses $\mathsf{H}_{k,\theta}$ $(k = 0, 1, \dots)$ and H_∞:

$$\lambda_\theta(k,n) = \sum_{j=k+1}^{n} \log \frac{f_\theta(X_j|\mathbf{X}^{j-1})}{g(X_j|\mathbf{X}^{j-1})}, \quad n > k$$

($\lambda_\theta(k,n) = 0$ for $n \le k$).

Let $k \in \mathbb{Z}_+$ and $r > 0$. We say that a sequence of the normalized LLRs $\{n^{-1}\lambda_\theta(k,k+n)\}_{n\geq 1}$ converges *r-completely* to a number I_θ under the probability measure $\mathsf{P}_{k,\theta}$ as $n \to \infty$ if

$$\sum_{n=1}^{\infty} n^{r-1} \mathsf{P}_{k,\theta}\left\{\left|n^{-1}\lambda_\theta(k,k+n) - I_\theta\right| > \varepsilon\right\} < \infty \quad \text{for all } \varepsilon > 0, \tag{3.18}$$

and we say that $\{n^{-1}\lambda_\theta(k,k+n)\}_{n\geq 1}$ converges to I_θ *uniformly r-completely* as $n \to \infty$ if

$$\sum_{n=1}^{\infty} n^{r-1} \sup_{0\leq k<\infty} \mathsf{P}_{k,\theta}\left\{\left|n^{-1}\lambda_\theta(k,k+n) - I_\theta\right| > \varepsilon\right\} < \infty \quad \text{for all } \varepsilon > 0. \tag{3.19}$$

(see Section B.2 in Appendix B for details regarding r-complete convergence).

3.4 Asymptotic Optimality of the Mixture Shiryaev Rule

3.4.1 Assumptions

To study asymptotic optimality we need certain constraints imposed on the prior distribution $\pi = \{\pi_k\}$ and on the asymptotic behavior of the decision statistics as the sample size increases (i.e., on the general stochastic model (3.1)).

The following two conditions are imposed on the prior distribution:

$\mathbf{CP}_1$. *For some* $0 \leq \mu < \infty$,

$$\lim_{n\to\infty} \frac{1}{n} \left|\log \sum_{k=n+1}^{\infty} \pi_k\right| = \mu. \tag{3.20}$$

$\mathbf{CP}_2$. *If* $\mu = 0$, *then in addition*

$$\sum_{k=0}^{\infty} \pi_k |\log \pi_k|^r < \infty \quad \text{for some } r \geq 1. \tag{3.21}$$

The class of prior distributions satisfying conditions $\mathbf{CP}_1$ and $\mathbf{CP}_2$ will be denoted by $\mathbf{C}(\mu)$. Also, without special mentioning we always assume that $\{\pi_k\}$ is fully supported, i.e., $\pi_k > 0$ for all $k \in \mathbb{Z}_+ = \{0,1,\dots\}$ and that $\pi_\infty = 0$.

Note that if $\mu > 0$, then the prior distribution has an exponential right tail. Distributions such as geometric and discrete versions of gamma and logistic distributions, i.e., models with bounded hazard rate, belong to this class. In this case, condition (3.21) holds automatically. If $\mu = 0$, the distribution has a heavy tail, i.e., belongs to the model with a vanishing hazard rate. However, we cannot allow this distribution to have a too heavy tail, which will generate very large time intervals between change points. This is guaranteed by condition $\mathbf{CP}_2$. Note that condition $\mathbf{CP}_1$ excludes light-tail distributions with unbounded hazard rates (e.g., Gaussian-type or Weibull-type with the shape parameter $\kappa > 1$) for which the time-intervals with a change point are very short. In this case, prior information dominates information obtained from the observations, the change can be easily detected at early stages, and the asymptotic analysis is impractical. A typical heavy-tailed prior distribution that satisfies both conditions $\mathbf{CP}_1$ with $\mu = 0$ and $\mathbf{CP}_2$ for all $r \geq 1$ is a discrete Weibull-type distribution with the shape parameter $0 < \kappa < 1$. Note also that constraint (3.21) is often guaranteed by finiteness of the r-th moment, $\sum_{k=0}^{\infty} k^r \pi_k < \infty$.

See Subsection 3.4.2 for a further discussion of the reason for imposing constraints (3.20)–(3.21) that limit possible prior distributions, which are appropriate for asymptotic study.

For $\delta > 0$, define $\Gamma_{\delta,\theta} = \{\vartheta \in \Theta : |\vartheta - \theta| < \delta\}$. In the sequel, we will exclude from consideration parameter values of θ that have W-measure null. Specifically, we assume that $W(\theta)$ is quite arbitrarily satisfying the condition:

For any $\delta > 0$, the measure $W(\vartheta)$ is positive on $\Gamma_{\delta,\theta}$ for any $\theta \in \Theta$, i.e.,

$$W\{\vartheta \in \Theta : |\vartheta - \theta| < \delta\} > 0 \quad \text{for any } \delta > 0 \text{ and any } \theta \in \Theta. \tag{3.22}$$

For $0 < I_\theta < \infty$, define

$$p_{N,k}(\varepsilon, \theta) = \mathsf{P}_{k,\theta}\left\{\frac{1}{N} \max_{1 \le n \le N} \lambda_\theta(k, k+n) \ge (1+\varepsilon) I_\theta\right\} \tag{3.23}$$

and

$$\Upsilon_{k,r}(\theta, \varepsilon) = \lim_{\delta \to 0} \sum_{n=1}^{\infty} n^{r-1} \mathsf{P}_{k,\theta}\left\{\frac{1}{n} \inf_{\vartheta \in \Gamma_{\delta,\theta}} \lambda_\vartheta(k, k+n) < I_\theta - \varepsilon\right\}. \tag{3.24}$$

Regarding the general model for the observations (3.1), we assume that the following two conditions are satisfied:

$\mathbf{C}_1$. *There exist positive and finite numbers I_θ, $\theta \in \Theta$, such that $n^{-1}\lambda_\theta(k, k+n)$ converges to I_θ in $\mathsf{P}_{k,\theta}$-probability and*

$$\lim_{N \to \infty} p_{N,k}(\varepsilon, \theta) = 0 \quad \text{for any } k \in \mathbb{Z}_+,\ \varepsilon > 0 \text{ and } \theta \in \Theta; \tag{3.25}$$

$\mathbf{C}_2$. *For every $\theta \in \Theta$, any $\varepsilon > 0$, and for some $r \ge 1$*

$$\sum_{k=0}^{\infty} \pi_k \Upsilon_{k,r}(\varepsilon, \theta) < \infty. \tag{3.26}$$

Note that condition $\mathbf{C}_1$ holds whenever $\lambda_\theta(k, k+n)/n$ converges almost surely to I_θ under $\mathsf{P}_{\theta,k}$,

$$\frac{1}{n}\lambda_\theta(k, k+n) \xrightarrow[n \to \infty]{\mathsf{P}_{k,\theta}\text{-a.s.}} I_\theta \quad \text{for all } \theta \in \Theta \tag{3.27}$$

(cf. Lemma B.1 in Appendix B).

3.4.2 Heuristics

We begin with a heuristic argument to obtain first-order asymptotic approximations for the moments of the detection delay when the threshold in the Shiryaev procedure is large (i.e., the PFA is small). This argument also explains the reason why the conditions $\mathbf{CP}_1$ and $\mathbf{CP}_2$ on the prior distribution are imposed.

Assume first that the change occurs at $\nu \le 0$. It is easy to see that the logarithm of the MS statistic defined in (3.5) can be written as

$$\log S_W(n) = |\log \mathsf{P}(\nu \ge n)| + \log \pi_0 + \log \Lambda_W(0, n) + Y_n,$$

where

$$Y_n = \log\left(1 + q/\pi_0 + \sum_{j=1}^{n-1} \frac{\pi_j \Lambda_W(j, n)}{\pi_0 \Lambda_W(0, n)}\right).$$

Under conditions $\mathbf{C}_1$ and $\mathbf{C}_2$ the normalized LLR $n^{-1}\log \Lambda_W(0, n)$ converges under $\mathsf{P}_{0,\theta}$ to I_θ, so we can expect that for a large n

$$\log \Lambda_W(0, n) \approx I_\theta\, n + \xi_n, \tag{3.28}$$

where ξ_n/n converges to 0. Also, Y_n, $n \ge 1$ are "slowly changing" and converge to a random variable Y_∞. Thus, ignoring the overshoot of $\log S_W(T_A^W)$ over $\log A$ and taking into account that, by condition (3.20), $|\log \mathsf{P}(\nu \ge T_A^W)| \approx \mu T_A^W$ for a large A, we obtain

$$\log A \approx \log S_W(T_A^W) \approx \mu T_A^W + \log \pi_0 + I_\theta T_A^W + \xi_{T_A^W} + Y_\infty.$$

Taking expectation on both sides and ignoring the term $\mathsf{E}_{0,\theta}[Y_\infty]$ yields

$$(\mu + I_\theta)\mathsf{E}_{0,\theta}[T_A^W] \approx \log(A/\pi_0). \tag{3.29}$$

A similar argument leads to the following approximate formula (for a large A) for the expected delay $\mathsf{E}_{k,\theta}[(T_A^W - k)^+]$ when the change occurs at $\nu = k$:

$$\mathsf{E}_{k,\theta}[(T_A^W - k)^+] \approx \frac{\log(A/\pi_k)}{I_\theta + \mu}, \quad k \in \mathbb{Z}_+,$$

where we kept π_k since it may become very small for certain prior distributions for large k. This gives us the following approximation for the average delay to detection:

$$\sum_{k=0}^{\infty} \pi_k \mathsf{E}_{k,\theta}[(T_A^W - k)^+] \approx \frac{\log A + \sum_{k=0}^{\infty} \pi_k |\log \pi_k|}{I_\theta + \mu},$$

which can be further generalized for the r-th moment of the delay as

$$\sum_{k=0}^{\infty} \pi_k \mathsf{E}_{k,\theta}[(T_A - k)^+]^r \approx \frac{(\log A)^r + \sum_{k=0}^{\infty} \pi_k |\log \pi_k|^r}{(I_\theta + \mu)^r}. \tag{3.30}$$

In subsequent sections, these approximations are justified rigorously.

There are two key points we would like to address. First, the term $\sum_{k=0}^{\infty} \pi_k |\log \pi_k|^r$ can be ignored only if it is substantially smaller than $(\log A)^r$. This explains the need for condition (3.21) when $\mu = 0$, which limits the rate of decay of π_k to 0 for large k forbidding it to be too slow. Also, if μ is positive but small, then $\sum_{k=0}^{\infty} \pi_k |\log \pi_k|^r$ is large. Since we are interested in asymptotics for a large A, in order to ignore the term $\sum_{k=0}^{\infty} \pi_k |\log \pi_k|^r$, we have to assume that

$$\sum_{k=0}^{\infty} \pi_k |\log \pi_k|^r \ll (\log A)^r.$$

This means that when we consider the asymptotic case of $A \to \infty$ ($\alpha \to 0$), the rate with which μ goes to 0 should be matched with A (α). In other words, we have to assume that $\mu = \mu(A)$, and therefore, $\pi_k = \pi_k(A)$ depend on A in such a way that

$$\frac{\sum_{k=0}^{\infty} \pi_k(A) |\log \pi_k(A)|^r}{(\log A)^r} = o(1) \quad \text{as } A \to \infty.$$

Even more importantly, the above argument allows us to understand the need for condition (3.20) with the factor n but not with some non-linear function $\phi(n)$. Whenever we require the normalized LLR $n^{-1}\lambda_\theta(k, k+n)$ with the normalizing factor n to converge to a number I_θ, information comes from the observations with an average rate $I_\theta n$, and the amount of prior information is determined by the term $|\log \mathsf{P}(\nu > n)|$. Hence, if we assume that $|\log \mathsf{P}(\nu > n)| \sim \mu n$ for large n (condition (3.20)), then prior information contributes as much information as additional μ "observations"; see (3.29) where the left-hand side corresponds to the total average information in T_A^W observations plus prior information. This explains the factor $(I_\theta + \mu)^{-r}$ in (3.30), which balances both prior information and information from observations. If $I_\theta \gg \mu$, then the prior distribution contributes very little information and most information comes from observations. However, if the prior distribution has a light tail, then $|\log \mathsf{P}(\nu > n)|$ increases with n faster than the linear function. In this case, prior information dominates information from observations, which is evidently a degenerate case. In other words, in asymptotic problems considered in this chapter, the class of prior distributions should include only distributions with a bounded hazard rate, which is expressed by condition (3.20). If, however, we assume that the proper normalizing factor for the LLR is not n but $\phi(n)$ (say the power function $\phi(n) = n^\beta$), then the "hazard rate" should fit into this assumption, i.e., condition (3.20) is replaced by $|\mathsf{P}(\nu > n)|/\phi(n) \to \mu$.

Unfortunately, in spite of the simplicity of the basic ideas and the approximate calculations, the rigorous argument in proving the result like (3.30) is rather tedious.

3.4.3 Asymptotic Optimality

In order to establish asymptotic optimality we first obtain, under condition $\mathbf{C}_1$, an asymptotic lower bound for moments of the detection delay $\overline{\mathscr{R}}^r_{\pi,\theta}(T) = \mathsf{E}^\pi_\theta[(T-\nu)^r | T > \nu]$ and $\mathscr{R}^r_{k,\theta} = \mathsf{E}_{k,\theta}[(T-k)^r | T > k]$ of any detection rule T from class $\mathbb{C}_\pi(\alpha)$, and then we show that under condition $\mathbf{C}_2$ this bound is attained for the MS rule T^W_A when $A = A_\alpha$ is properly selected.

Asymptotic lower bounds for all positive moments of the detection delay are specified in the following theorem. Condition (3.25) (and hence, the a.s. convergence condition (6.89)) is sufficient for this purpose.

Theorem 3.1. *Let, for some $\mu \ge 0$, the prior distribution belong to class $\mathbf{C}(\mu)$. Assume that for some positive and finite function $I(\theta) = I_\theta$, $\theta \in \Theta$ condition $\mathbf{C}_1$ holds. Then, for all $r > 0$ and all $\theta \in \Theta$*

$$\liminf_{\alpha\to 0} \frac{\inf_{T\in\mathbb{C}_\pi(\alpha)} \mathscr{R}^r_{k,\theta}(T)}{|\log\alpha|^r} \ge \frac{1}{(I_\theta+\mu)^r} \quad \text{for all } k \in \mathbb{Z}_+ \tag{3.31}$$

and

$$\liminf_{\alpha\to 0} \frac{\inf_{T\in\mathbb{C}_\pi(\alpha)} \overline{\mathscr{R}}^r_{\pi,\theta}(T)}{|\log\alpha|^r} \ge \frac{1}{(I_\theta+\mu)^r}. \tag{3.32}$$

Proof. For $\varepsilon \in (0,1)$ and $\delta > 0$, define

$$N_\alpha = N_\alpha(\varepsilon,\delta,\theta) = \frac{(1-\varepsilon)|\log\alpha|}{I_\theta+\mu+\delta}$$

and let $\Pi_k = \mathsf{P}(\nu > k)$ $(\Pi_{-1} = 1-q)$. By the Chebyshev inequality,

$$\mathscr{R}^r_{k,\theta}(T) \ge \mathsf{E}_{k,\theta}[(T-k)^+]^r \ge N^r_\alpha \mathsf{P}_{k,\theta}(T-k > N_\alpha) \ge N^r_\alpha\left[\mathsf{P}_{k,\theta}(T>k) - \mathsf{P}_{k,\theta}(k < T \le k+N_\alpha)\right],$$

where $\mathsf{P}_{k,\theta}(T>k) = \mathsf{P}_\infty(T>k)$, so that

$$\inf_{T\in\mathbb{C}_\pi(\alpha)} \mathscr{R}^r_{k,\theta}(T) \ge N^r_\alpha\left[\inf_{T\in\mathbb{C}_\pi(\alpha)} \mathsf{P}_\infty(T>k) - \sup_{T\in\mathbb{C}_\pi(\alpha)} \mathsf{P}_{k,\theta}(k<T\le k+N_\alpha)\right]. \tag{3.33}$$

Thus, to prove the lower bound (3.32) we need to show that, for arbitrarily small ε and δ and all fixed $k \in \mathbb{Z}_+$,

$$\lim_{\alpha\to 0} \inf_{T\in\mathbb{C}_\pi(\alpha)} \mathsf{P}_\infty(T>k) = 1 \tag{3.34}$$

and

$$\lim_{\alpha\to 0} \sup_{T\in\mathbb{C}_\pi(\alpha)} \mathsf{P}_{k,\theta}(k<T\le k+N_\alpha) = 0, \tag{3.35}$$

since in this case inequality (3.33) yields the asymptotic inequality

$$\inf_{T \in \mathbb{C}_\pi(\alpha)} \mathscr{R}^r_{k,\theta}(T) \geq \left[\frac{(1-\varepsilon)|\log \alpha|}{I_\theta + \mu + \delta}\right]^r (1+o(1)),$$

which holds for arbitrarily small ε and δ.

First, note that

$$\alpha \geq \sum_{i=k}^{\infty} \pi_i \mathsf{P}_\infty(T \leq i) \geq \mathsf{P}_\infty(T \leq k) \sum_{i=k}^{\infty} \pi_i,$$

and hence,

$$\inf_{T \in \mathbb{C}_\pi(\alpha)} \mathsf{P}_\infty(T > k) \geq 1 - \alpha/\Pi_{k-1}, \quad k \in \mathbb{Z}_+, \tag{3.36}$$

which approaches 1 as $\alpha \to 0$ for any fixed $k \in \mathbb{Z}_+$. Thus, (3.34) follows.

Now, introduce

$$U_{\alpha,k}(T) = e^{(1+\varepsilon)I_\theta N_\alpha} \mathsf{P}_\infty(k < T \leq k + N_\alpha),$$
$$p_{\alpha,k}(\theta) = \mathsf{P}_{k,\theta}\left(\frac{1}{N_\alpha} \max_{1 \leq n \leq N_\alpha} \lambda_\theta(k, k+n) \geq (1+\varepsilon) I_\theta\right).$$

Changing the measure $\mathsf{P}_\infty \to \mathsf{P}_{k,\theta}$, we obtain that for any $C > 0$ and $\varepsilon \in (0,1)$

$$\begin{aligned}
\mathsf{P}_\infty\{k < T \leq k + N_\alpha\} &= \mathsf{E}_{k,\theta}\left\{\mathbb{1}_{\{k<T\leq k+N_\alpha\}} \exp\{-\lambda_\theta(k,T)\}\right\} \\
&\geq \mathsf{E}_{k,\theta}\left\{\mathbb{1}_{\{k<T\leq k+N_\alpha, \lambda_{k,T}(\theta)<C\}} \exp\{-\lambda_\theta(k,T)\}\right\} \\
&\geq e^{-C}\mathsf{P}_{k,\theta}\left(\{k < T \leq k+N_\alpha\} \cap \left\{\max_{k<n\leq k+N_\alpha} \lambda_\theta(k,n) < C\right\}\right) \\
&\geq e^{-C}\left[\mathsf{P}_{k,\theta}\{k < T \leq k+N_\alpha\} - \mathsf{P}_k\left\{\max_{1\leq n\leq N_\alpha} \lambda_\theta(k,k+n) \geq C\right\}\right],
\end{aligned} \tag{3.37}$$

where the last inequality follows trivially from the fact that for any events $\mathscr{A}$ and $\mathscr{B}$ (with $\mathscr{B}^c$ being a complement to $\mathscr{B}$), $\mathsf{P}(\mathscr{A} \cap \mathscr{B}) \geq \mathsf{P}(\mathscr{A}) - \mathsf{P}(\mathscr{B}^c)$. Setting $C = (1+\varepsilon)I_\theta N_\alpha$, it follows from (3.37) that

$$\mathsf{P}_{k,\theta}(k < T \leq k + N_\alpha) \leq U_{\alpha,k}(T) + p_{\alpha,k}(\theta). \tag{3.38}$$

Due to (3.36)

$$\sup_{T \in \mathbb{C}_\pi(\alpha)} \mathsf{P}_\infty(T \leq k) \leq \alpha/\Pi_{k-1}, \quad k \geq 1, \tag{3.39}$$

so that

$$\begin{aligned}
U_{\alpha,k}(T) &\leq e^{(1+\varepsilon)I_\theta N_\alpha} \mathsf{P}_\infty(T \leq k + N_\alpha) \leq \alpha\, e^{(1+\varepsilon)I_\theta N_\alpha}/\Pi_{k-1+N_\alpha} \\
&\leq \exp\left\{(1+\varepsilon)I_\theta N_\alpha - |\log \alpha| + (k-1+N_\alpha)\frac{|\log \Pi_{k-1+N_\alpha}|}{k-1+N_\alpha}\right\}.
\end{aligned} \tag{3.40}$$

Using the fact that by condition (3.20), for all sufficiently large N_α (small α), there exists a (small) δ such that

$$\frac{|\log \Pi_{k-1+N_\alpha}|}{k-1+N_\alpha} \leq \mu + \delta,$$

we obtain that for a sufficiently small α

$$\sup_{T \in \mathbb{C}_\pi(\alpha)} U_{\alpha,k}(T) \leq \exp\{(1+\varepsilon)I_\theta N_\alpha - |\log \alpha| + (k-1+N_\alpha)(\mu+\delta)\}$$

$$= \exp\left\{-\frac{I_\theta \varepsilon^2 + (\mu+\delta)\varepsilon}{I_\theta + \mu + \delta}|\log \alpha| + (\mu+\delta)(k-1)\right\}, \tag{3.41}$$

where the last term is less or equal to

$$\exp\left\{-\varepsilon^2|\log \alpha| + (\mu+\delta)(k-1)\right\} := \overline{U}_{\alpha,k}(\varepsilon,\delta)$$

for all $\varepsilon \in (0,1)$. Thus, for all $\varepsilon \in (0,1)$,

$$U_{\alpha,k}(T) \le \overline{U}_{\alpha,k}(\varepsilon,\delta). \tag{3.42}$$

Since $\overline{U}_{\alpha,k}(\varepsilon,\delta)$ does not depend on the stopping time T and the value of $\overline{U}_{\alpha,k}(\varepsilon,\delta)$ goes to 0 as $\alpha \to 0$ for any fixed $k \in \mathbb{Z}_+$ and any $\varepsilon > 0$ and $\delta > 0$, it follows that

$$\sup_{T \in \mathbb{C}_\pi(\alpha)} U_{\alpha,k}(T) \to 0 \quad \text{as } \alpha \to 0 \text{ for any fixed } k \in \mathbb{Z}_+. \tag{3.43}$$

Also, by condition $\mathbf{C}_1$, $p_{\alpha,k}(\theta) \to 0$ for all $k \in \mathbb{Z}_+$, and therefore, (3.35) holds. This completes the proof of the lower bound (3.31).

Next, we get to proving the lower bound (3.32). Similarly to (3.33), by the Chebyshev inequality,

$$\inf_{T \in \mathbb{C}_\pi(\alpha)} \overline{\mathscr{R}}^r_{\pi,\theta}(T) \ge N_\alpha^r \left[1 - \alpha - \sup_{T \in \mathbb{C}_\pi(\alpha)} \mathsf{P}^\pi_\theta(0 < T - \nu \le N_\alpha)\right]. \tag{3.44}$$

Thus, to prove the lower bound (3.32) we need to show that, for arbitrarily small ε and δ,

$$\lim_{\alpha \to 0} \sup_{T \in \mathbb{C}_\pi(\alpha)} \mathsf{P}^\pi_\theta(0 < T - \nu \le N_\alpha) = 0. \tag{3.45}$$

Write

$$K_\alpha = K_\alpha(\varepsilon,\mu,\delta) = \left\lfloor \frac{\varepsilon^3|\log \alpha|}{\mu+\delta} \right\rfloor.$$

We have

$$\begin{aligned}
\sup_{T \in \mathbb{C}_\pi(\alpha)} \mathsf{P}^\pi_\theta(0 < T - \nu \le N_\alpha) &= \sum_{k=0}^{\infty} \pi_k \sup_{T \in \mathbb{C}_\pi(\alpha)} \mathsf{P}_{k,\theta}(k < T \le k + N_\alpha) \\
&= \sum_{k=0}^{K_\alpha} \pi_k \sup_{T \in \mathbb{C}_\pi(\alpha)} \mathsf{P}_{k,\theta}(k < T \le k + N_\alpha) + \sum_{k=K_\alpha+1}^{\infty} \pi_k \sup_{T \in \mathbb{C}_\pi(\alpha)} \mathsf{P}_{k,\theta}(k < T \le k + N_\alpha) \\
&\le \sum_{k=0}^{K_\alpha} \pi_k p_{\alpha,k}(\theta) + \sum_{k=0}^{K_\alpha} \pi_k \sup_{T \in \mathbb{C}_\pi(\alpha)} U_{\alpha,k}(T) + \sum_{k=K_\alpha+1}^{\infty} \pi_k \qquad (3.46) \\
&\le \Pi_{K_\alpha} + \sum_{k=0}^{K_\alpha} \pi_k p_{\alpha,k}(\theta) + \max_{0 \le k \le K_\alpha} \sup_{T \in \mathbb{C}_\pi(\alpha)} U_{\alpha,k}(T) \\
&\le \Pi_{K_\alpha} + \sum_{k=0}^{K_\alpha} \pi_k p_{\alpha,k}(\theta) + \overline{U}_{\alpha,K_\alpha}, \qquad (3.47)
\end{aligned}$$

where inequality (3.46) follows from inequality (3.38) and inequality (3.47) from (3.42). If $\mu > 0$, by condition (3.20), $\log \Pi_{K_\alpha} \sim -\mu K_\alpha$ as $\alpha \to 0$, so $\Pi_{K_\alpha} \to 0$. If $\mu = 0$, this probability goes to 0 as $\alpha \to 0$ as well since, by condition (3.21),

$$\Pi_{K_\alpha} < \sum_{k=K_\alpha}^{\infty} \pi_k |\log \pi_k| \xrightarrow[\alpha \to 0]{} 0.$$

Clearly, $\overline{U}_{\alpha,K_\alpha} \to 0$. By condition $\mathbf{C}_1$ and Lebesgue's dominated convergence theorem, $\sum_{k=0}^{K_\alpha} \pi_k p_{\alpha,k}(\theta) \to 0$, and therefore, all three terms go to zero as $\alpha \to 0$ for all $\varepsilon, \delta > 0$, so that (3.45) follows and the proof of the lower bound (3.32) is complete. □

The following lemma provides the upper bound for the PFA of the MS rule.

Lemma 3.1. *For all $A > q/(1-q)$ and any prior distribution of ν, the PFA of the MS rule T_A^W satisfies the inequality*

$$\mathsf{PFA}_\pi(T_A^W) \le 1/(1+A), \tag{3.48}$$

so that for $\alpha < 1-q$

$$A = A_\alpha = (1-\alpha)/\alpha \quad \text{implies} \quad \mathsf{PFA}_\pi(T_{A_\alpha}^W) \le \alpha. \tag{3.49}$$

Proof. Clearly,

$$\mathsf{PFA}_\pi(T_A^W) = \mathsf{E}^\pi[\mathsf{P}(T_A^W \le \nu|\mathscr{F}_{T_A^W}); T_A^W < \infty].$$

Using the Bayes rule and the fact that $\prod_{i=j+1}^n \mathscr{L}_i(\theta) = 1$ for $j \ge n$, we obtain

$$\mathsf{P}(\nu = k|\mathscr{F}_n) = \frac{\pi_k \Lambda_W(k,n)}{q\Lambda_W(0,n) + \sum_{j=0}^{n-1} \pi_j \Lambda_W(j,n) + \mathsf{P}(\nu \ge n)},$$

so that

$$\mathsf{P}(\nu \ge n|\mathscr{F}_n) = \sum_{k=n}^\infty \mathsf{P}(\nu = k|\mathscr{F}_n) = \frac{\mathsf{P}(\nu \ge n)}{q\Lambda_W(0,n) + \sum_{j=0}^{n-1} \pi_j \Lambda_W(j,n) + \mathsf{P}(\nu \ge n)} = \frac{1}{S_W(n)+1}.$$

Therefore, taking into account that $S_W(T_A^W) \ge A$ on $\{T_A^W < \infty\}$, we have

$$\mathsf{PFA}_\pi(T_A^W) = \mathsf{E}^\pi[1/(1+S_W(T_A^W)); T_A^W < \infty] \le 1/(1+A)$$

and the inequality (3.48) follows. Implication (3.49) is obvious. □

The following proposition establishes first-order asymptotic approximations for the moments of the detection delay of the MS rule when threshold A goes to infinity regardless of the PFA constraint.

Proposition 3.1. *Let $r \ge 1$ and let the prior distribution of the change point belong to class $\mathbf{C}(\mu)$. Assume that for some $0 < I_\theta < \infty$, $\theta \in \Theta$, right-tail and left-tail conditions $\mathbf{C}_1$ and $\mathbf{C}_2$ are satisfied. Then, for all $0 < m \le r$ and all $\theta \in \Theta$ as $A \to \infty$*

$$\mathscr{R}^m_{k,\theta}(T_A^W) \sim \left(\frac{\log A}{I_\theta + \mu}\right)^m \quad \text{for all } k \in \mathbb{Z}_+ \tag{3.50}$$

and

$$\overline{\mathscr{R}}^m_{\pi,\theta}(T_A^W) \sim \left(\frac{\log A}{I_\theta + \mu}\right)^m. \tag{3.51}$$

In order to prove this proposition we need the following lemma. Write

$$N_A = N_A(\theta,\mu,\varepsilon) = 1 + \left\lfloor \frac{\log(A/\pi_k)}{I_\theta + \mu - \varepsilon} \right\rfloor.$$

Lemma 3.2. *Let $r \ge 1$ and let the prior distribution of the change point satisfy condition (3.20). Then for a sufficiently large A, any $0 < \varepsilon < I_\theta + \mu$ and all $k \in \mathbb{Z}_+$,*

$$\begin{aligned} \mathsf{E}_{k,\theta}[(T_A^W - k)^+]^r \le & \left(1 + \frac{\log(A/\pi_k)}{I_\theta + \mu - \varepsilon}\right)^r \\ & + r2^{r-1} \sum_{n=N_A}^\infty n^{r-1} \mathsf{P}_{k,\theta}\left(\frac{1}{n} \inf_{\vartheta \in \Gamma_{\delta,\theta}} \lambda_\vartheta(k,k+n) < I_\theta - \varepsilon\right). \end{aligned} \tag{3.52}$$

Proof. For $k \in \mathbb{Z}_+$, define the stopping times

$$\tau_A^{(k)} = \inf\{n \ge 1 : \log \Lambda_W(k, k+n) + |\log \mathsf{P}(\nu \ge k+n)| \ge \log(A/\pi_k)\}.$$

Obviously, for any $n > k$,

$$\log S_W(n) \ge \log\left(\frac{\pi_k}{\mathsf{P}(\nu \ge n)} \Lambda_W(k,n)\right) = \log \Lambda_W(k,n) + \log \pi_k - \log \mathsf{P}(\nu \ge n),$$

and hence, for every $A > 0$, $(T_A^W - k)^+ \le \tau_A^{(k)}$ and $\mathsf{E}_{k,\theta}[(T_A^W - k)^+]^r \le \mathsf{E}_{k,\theta}[(\tau_A^{(k)})^r]$.

Setting $\tau = \tau_A^{(k)}$ and $N = N_A$ in inequality (A.1) in Lemma A.1 (see Appendix A) we obtain that for any $k \in \mathbb{Z}_+$ the following inequality holds:

$$\mathsf{E}_{k,\theta}\left[(T_A^W - k)^+\right]^r \le \mathsf{E}_{k,\theta}\left[\left(\tau_A^{(k)}\right)^r\right] \le N_A^r + r2^{r-1} \sum_{n=N_A}^{\infty} n^{r-1} \mathsf{P}_{k,\theta}\left(\tau_A^{(k)} > n\right). \tag{3.53}$$

It is easily seen that for all $k \in \mathbb{Z}_+$ and $n \ge N_A$

$$\begin{aligned}
\mathsf{P}_{k,\theta}\left(\tau_A^{(k)} > n\right) &\le \mathsf{P}_{k,\theta}\left\{\frac{\log \Lambda_W(k,k+n)}{n} < \frac{1}{n}\log\left(\frac{A}{\pi_k}\right) - \frac{|\log \mathsf{P}(\nu \ge k+n)|}{n}\right\} \\
&\le \mathsf{P}_{k,\theta}\left\{\frac{\log \Lambda_W(k,k+n)}{n} < I_\theta + \mu - \varepsilon - \frac{|\log \mathsf{P}(\nu \ge k+n)|}{n}\right\}.
\end{aligned}$$

Since, by condition $\mathbf{CP}_1$, $N_A^{-1}|\log \mathsf{P}(\nu \ge k + N_A)| \to \mu$ as $A \to \infty$, for a sufficiently large value of A there exists a small κ such that

$$\left|\mu - \frac{|\log \mathsf{P}(\nu \ge k + N_A)|}{N_A}\right| < \kappa.$$

Hence, for all sufficiently large A,

$$\mathsf{P}_{k,\theta}\left(\tau_A^{(k)} > n\right) \le \mathsf{P}_{k,\theta}\left(\frac{1}{n}\log \Lambda_W(k,k+n) < I_\theta - \varepsilon + \kappa\right).$$

Also,

$$\log \Lambda_W(k,k+n) \ge \inf_{\vartheta \in \Gamma_{\delta,\theta}} \lambda_\vartheta(k,k+n) + \log W(\Gamma_{\delta,\theta}),$$

where $\Gamma_{\delta,\theta} = \{\vartheta \in \Theta : |\vartheta - \theta| < \delta\}$. Thus, for all sufficiently large n and A, for which $\kappa + |\log W(\Gamma_{\delta,\theta})|/n < \varepsilon/2$, we have

$$\begin{aligned}
\mathsf{P}_{k,\theta}\left(\tau_A^{(k)} > n\right) &\le \mathsf{P}_{k,\theta}\left(\frac{1}{n}\inf_{\vartheta \in \Gamma_{\delta,\theta}} \lambda_\vartheta(k,k+n) < I_\theta - \varepsilon + \kappa + \frac{1}{n}|\log W(\Gamma_{\delta,\theta})|\right) \\
&\le \mathsf{P}_{k,\theta}\left(\frac{1}{n}\inf_{\vartheta \in \Gamma_{\delta,\theta}} \lambda_\vartheta(k,k+n) < I_\theta - \varepsilon/2\right).
\end{aligned} \tag{3.54}$$

Using (3.53) and (3.54), we obtain inequality (3.52). □

Proof of Proposition 3.1. Using Lemma 3.2 and inequality $\mathsf{P}_\infty(T_A^W > k) > 1 - [A\Pi_{k-1}]^{-1}$ (see (3.36)), we obtain

$$\mathscr{R}_{k,\theta}^r(T_A^W) = \frac{\mathsf{E}_{k,\theta}\left[\left(T_A^W - k\right)^+\right]^r}{\mathsf{P}_\infty(T_A^W > k)} \le \frac{\left(1 + \left\lfloor \frac{\log(A/\pi_k)}{I_\theta + \mu - \varepsilon} \right\rfloor\right)^r + r2^{r-1}\Upsilon_{k,r}(\theta,\varepsilon)}{1 - 1/(A\Pi_{k-1})}, \tag{3.55}$$

where $\Upsilon_{k,r}(\theta,\varepsilon)$ is defined in (3.24). Since, by condition $\mathbf{C}_2$, $\Upsilon_{k,r}(\theta,\varepsilon) < \infty$ for all $k \in \mathbb{Z}_+$ and $\theta \in \Theta$, this implies the asymptotic upper bound

$$\mathscr{R}^r_{k,\theta}(T_A^W) \le \left(\frac{\log A}{I_\theta + \mu}\right)^r (1+o(1)), \quad A \to \infty, \tag{3.56}$$

which along with the lower bound

$$\mathscr{R}^r_{k,\theta}(T_A^W) \ge \left(\frac{\log A}{I_\theta + \mu}\right)^r (1+o(1)), \quad A \to \infty \tag{3.57}$$

proves the asymptotic relation (3.50). Note that the lower bound (3.57) follows immediately from the lower bound (3.31) in Theorem 3.1 by replacing α with $1/(A+1)$ since it follows from (3.48) that $T_A^W \in \mathbb{C}_\pi(1/(A+1))$.

We now get to proving (3.51). Since the MS rule T_A^W belongs to class $\mathbb{C}_\pi(1/(A+1))$, replacing α by $1/(A+1)$ in the asymptotic lower bound (3.32), we obtain that under the right-tail condition $\mathbf{C}_1$ the following asymptotic lower bound holds for all $r > 0$ and $\theta \in \Theta$:

$$\overline{\mathscr{R}}^r_{\pi,\theta}(T_A^W) \ge \left(\frac{\log A}{I_\theta + \mu}\right)^r (1+o(1)), \ A \to \infty. \tag{3.58}$$

Thus, to prove (3.51) it suffices to show that, under the left-tail condition $\mathbf{C}_2$, for $0 < m \le r$ and $\theta \in \Theta$

$$\overline{\mathscr{R}}^m_{\pi,\theta}(T_A^W) \le \left(\frac{\log A}{I_\theta + \mu}\right)^m (1+o(1), \ A \to \infty. \tag{3.59}$$

Using Lemma 3.2, we obtain that for any $0 < \varepsilon < I_\theta + \mu$

$$\mathsf{E}^\pi_\theta[(T_A^W - \nu)^+]^r = \sum_{k=0}^\infty \pi_k \mathsf{E}_{k,\theta}\left[(T_A^W - k)^+\right]^r \le \sum_{k=0}^\infty \pi_k \left(1 + \frac{\log(A/\pi_k)}{I_\theta + \mu - \varepsilon}\right)^r + r2^{r-1}\sum_{k=0}^\infty \pi_k \Upsilon_{k,r}(\theta,\varepsilon), \tag{3.60}$$

where we set $\pi_0 = \Pr(\nu \le 0)$. This inequality together with the inequality $1 - \mathsf{PFA}_\pi(T_A^W) \ge A/(1+A)$ yields

$$\begin{aligned}\overline{\mathscr{R}}^r_{\pi,\theta}(T_A^W) &= \frac{\sum_{k=0}^\infty \pi_k \mathsf{E}_{k,\theta}\left[(T_A^W - k)^+\right]^r}{1 - \mathsf{PFA}_\pi(T_A^W)} \\ &\le \frac{\sum_{k=0}^\infty \pi_k \left(1 + \frac{\log(A/\pi_k)}{I_\theta+\mu-\varepsilon}\right)^r + r2^{r-1}\sum_{k=0}^\infty \pi_k \Upsilon_{k,r}(\theta,\varepsilon)}{A/(1+A)}.\end{aligned} \tag{3.61}$$

By condition $\mathbf{C}_2$, $\sum_{k=0}^\infty \pi_k \Upsilon_{k,r}(\theta,\varepsilon) < \infty$ for any $\varepsilon > 0$ and any $\theta \in \Theta$ and, by condition (3.21), $\sum_{k=0}^\infty \pi_k |\log \pi_k|^r < \infty$, which implies that, as $A \to \infty$, for all $0 < m \le r$ and all $\theta \in \Theta$

$$\overline{\mathscr{R}}^m_{\pi,\theta}(T_A^W) \le \left(\frac{\log A}{I_\theta + \mu - \varepsilon}\right)^r (1+o(1)).$$

Since ε can be arbitrarily small, the upper bound (3.59) follows and the proof of the asymptotic expansion (3.51) is complete. □

The following theorem, whose proof is immediate since everything was prepared, is the main result in the general non-i.i.d. case. It shows that the MS detection rule is asymptotically optimal to the first order under mild conditions for the observations and prior distributions.

Theorem 3.2. *Let $r \geq 1$ and let the prior distribution of the change point belong to class $\mathbf{C}(\mu)$. Assume that for some $0 < I_\theta < \infty$, $\theta \in \Theta$, right-tail and left-tail conditions $\mathbf{C}_1$ and $\mathbf{C}_2$ are satisfied. If $A = A_\alpha$ is so selected that $\mathsf{PFA}_\pi(T^W_{A_\alpha}) \leq \alpha$ and $\log A_\alpha \sim |\log \alpha|$ as $\alpha \to 0$, in particular $A = A_\alpha = (1-\alpha)/\alpha$, where $0 < \alpha < 1-q$, then $T^W_{A_\alpha}$ is first-order asymptotically optimal as $\alpha \to 0$ in class $\mathbb{C}_\pi(\alpha)$, minimizing moments of the detection delay up to order r, i.e., for all $0 < m \leq r$ and all $\theta \in \Theta$ as $\alpha \to 0$*

$$\inf_{T \in \mathbb{C}_\pi(\alpha)} \mathscr{R}^m_{k,\theta}(T) \sim \left(\frac{|\log \alpha|}{I_\theta + \mu}\right)^m \sim \mathscr{R}^m_{k,\theta}(T^W_{A_\alpha}) \quad \text{for all } k \in \mathbb{Z}_+ \tag{3.62}$$

and

$$\inf_{T \in \mathbb{C}_\pi(\alpha)} \overline{\mathscr{R}}^m_{\pi,\theta}(T) \sim \left(\frac{|\log \alpha|}{I_\theta + \mu}\right)^m \sim \overline{\mathscr{R}}^m_{\pi,\theta}(T^W_{A_\alpha}). \tag{3.63}$$

Proof. Setting $A = A_\alpha = (1-\alpha)/\alpha$ in (3.50) and (3.51) yields as $\alpha \to 0$

$$\mathscr{R}^m_{k,\theta}(T^W_{A_\alpha}) \sim \left(\frac{|\log \alpha|}{I_\theta + \mu}\right)^m, \quad \overline{\mathscr{R}}^m_{\pi,\theta}(T^W_{A_\alpha}) \sim \left(\frac{|\log \alpha|}{I_\theta + \mu}\right)^m, \tag{3.64}$$

which along with the lower bounds (3.31) and (3.32) in Theorem 3.1 completes the proof of (3.62) and (3.63). Obviously, approximations (3.64) are also correct if threshold A_α is so selected that $T^W_{A_\alpha} \in \mathbb{C}_\pi(\alpha)$ and $\log A_\alpha \sim |\log \alpha|$ as $\alpha \to 0$. The proof is complete. □

Theorem 3.2 covers a very wide class of non-i.i.d. models for the observations as well as a large class of prior distributions. However, condition (3.20) does not include the case where μ is strictly positive, but may go to zero, $\mu \to 0$. Indeed, as discussed in detail in [160] the distributions with an exponential right tail that satisfy condition (3.20) with $\mu > 0$ do not converge as $\mu \to 0$ to heavy-tailed distributions for which $\mu = 0$. As a result, the assertions of Theorem 3.2 do not hold with $\mu = 0$ if μ approaches 0 with an arbitrarily rate. The rate has to be matched somehow with α. This issue was already discussed in Section 3.4.2.

For this reason, we now consider the case where the prior distribution $\pi^\alpha = \{\pi^\alpha_k\}$ of the change point depends on the PFA constraint α and generalize condition $\mathbf{CP}_1$ as follows:

$\mathbf{CP}^{(\alpha)}_1$. *For some $0 \leq \mu_\alpha < \infty$ and $0 \leq \mu < \infty$,*

$$\lim_{n\to\infty} \frac{1}{n} \left| \log \sum_{k=n+1}^{\infty} \pi^\alpha_k \right| = \mu_\alpha \quad \text{and} \quad \lim_{\alpha \to 0} \mu_\alpha = \mu. \tag{3.65}$$

The class of prior distributions satisfying this condition will be denoted by $\mathbf{C}^{(\alpha)}(\mu)$.

For establishing asymptotic optimality properties of change detection procedures we will assume in addition that the following two conditions hold:

$\mathbf{CP}^{(\alpha)}_2$. *If $\mu_\alpha > 0$ for all α and $\mu = 0$, then μ_α approaches zero at such rate that for some $r \geq 1$*

$$\lim_{\alpha \to 0} \frac{\sum_{k=0}^\infty \pi^\alpha_k |\log \pi^\alpha_k|^r}{|\log \alpha|^r} = 0. \tag{3.66}$$

$\mathbf{CP}^{(\alpha)}_3$. *For all $k \in \mathbb{Z}_+$*

$$\lim_{\alpha \to 0} \frac{|\log \pi^\alpha_k|}{|\log \alpha|} = 0. \tag{3.67}$$

Note that if $\mu > 0$, then condition (3.66) holds since $\lim_{\alpha \to 0} \sum_{k=0}^\infty \pi^\alpha_k |\log \pi^\alpha_k|^r < \infty$ for all $r > 0$. If $\mu = 0$ (asymptotically heavy-tailed distribution), as before, we cannot allow this distribution to have a too heavy tail, which is guaranteed by condition $\mathbf{CP}_2^{(\alpha)}$. Note that if $\mu_\alpha = \mu$ in $\mathbf{CP}_1^{(\alpha)}$, i.e., it coincides with condition $\mathbf{CP}_1$, then condition $\mathbf{CP}_3^{(\alpha)}$ holds automatically and condition $\mathbf{CP}_2^{(\alpha)}$ holds when $\sum_{k=0}^\infty \pi_k|\log \pi_k|^r < \infty$ for some $r \ge 1$, so that we get back to the previous case.

In the next lemma, which is analogous to Theorem 3.1, we provide asymptotic lower bounds for moments of the detection delay in class $\mathbb{C}(\alpha) = \mathbb{C}_{\pi_\alpha}(\alpha)$ when the prior distribution $\pi^\alpha = \{\pi_k^\alpha\}$ depends on α and $\mu_\alpha \to \mu \ge 0$ as $\alpha \to 0$.

Lemma 3.3. *Let the prior distribution $\pi^\alpha = \{\pi_k^\alpha\}$ of the change point belong to class $\mathbf{C}^{(\alpha)}(\mu)$. Assume that for some $0 < I_\theta < \infty$, $\theta \in \Theta$, condition $\mathbf{C}_1$ holds. Then, for all $r > 0$ and $\theta \in \Theta$*

$$\liminf_{\alpha\to 0} \frac{\inf_{T\in\mathbb{C}(\alpha)} \mathscr{R}_{k,\theta}^r(T)}{|\log\alpha|^r} \ge \frac{1}{(I_\theta+\mu)^r} \quad \text{for all } k \in \mathbb{Z}_+ \tag{3.68}$$

and

$$\liminf_{\alpha\to 0} \frac{\inf_{T\in\mathbb{C}(\alpha)} \overline{\mathscr{R}}_{\pi^\alpha,\theta}^r(T)}{|\log\alpha|^r} \ge \frac{1}{(I_\theta+\mu)^r}. \tag{3.69}$$

Proof. The proof of the lower bound (3.68) is identical to the proof of the lower bound (3.31) in Theorem 3.1 if we replace Π_k by Π_k^α.

To establish inequality (3.69) note that similar to (3.47) we have

$$\sup_{T\in\mathbb{C}(\alpha)} \mathsf{P}_\theta^{\pi^\alpha}(0 < T-\nu \le N_\alpha) \le \mathsf{P}(\nu > K_\alpha) + \sum_{k=0}^{K_\alpha} \pi_k^\alpha p_{\alpha,k}(\theta) + \overline{U}_{\alpha,K_\alpha},$$

where

$$K_\alpha = K_\alpha(\varepsilon,\mu,\delta) = \left\lfloor \frac{\varepsilon^3|\log\alpha|}{\mu+\delta} \right\rfloor.$$

The last term on the right-hand side goes to zero as $\alpha \to 0$. By condition $\mathbf{C}_1$, the middle term also goes to 0. If condition (3.20) holds with $\mu > 0$, then $\log \mathsf{P}(\nu > K_\alpha) \sim -\varepsilon^3|\log\alpha| \to -\infty$ as $\alpha \to 0$, so the probability $\mathsf{P}(\nu > K_\alpha)$ goes to 0 as $\alpha \to 0$ and the same is true if $\mu_\alpha > 0$ for all α and $\mu = 0$ since, in this case, $\log \mathsf{P}(\nu > K_\alpha) \sim -\varepsilon^2|\log\alpha| \to -\infty$ for a sufficiently small ε and $\delta = \varepsilon$. If $\mu_\alpha = \mu = 0$ for all α, then $\mathsf{P}(\nu > K_\alpha) \to 0$ as well since

$$\sum_{k=K_\alpha}^{\infty} \pi_k^0 \xrightarrow[\alpha\to 0]{} 0.$$

Therefore, all three terms go to zero as $\alpha \to 0$ for all $\varepsilon, \delta > 0$, so that

$$\sup_{T\in\mathbb{C}(\alpha)} \mathsf{P}_\theta^\pi(\nu < T \le \nu + N_\alpha) \to 0 \quad \text{as } \alpha \to 0 \tag{3.70}$$

and asymptotically as $\alpha \to 0$

$$\inf_{T\in\mathbb{C}(\alpha)} \overline{\mathscr{R}}_\theta^r(T) \ge \left[\frac{(1-\varepsilon)|\log\alpha|}{I_\theta+\mu+\delta}\right]^r (1+o(1)),$$

where ε and δ can be arbitrarily small, which yields the lower bound (3.69). □

Using this lemma, we now establish first-order asymptotic optimality of the MS rule when $\mu = \mu_\alpha$ approaches zero as $\alpha \to 0$. To simplify the proof, we strengthen condition $\mathbf{C}_2$ in the following uniform version:

$\mathbf{C}_3$. *For every $\theta \in \Theta$, for any $\varepsilon > 0$, and for some $r \ge 1$*

$$\Upsilon_r(\theta,\varepsilon) := \sup_{k\in\mathbb{Z}_+} \Upsilon_{k,r}(\theta,\varepsilon) = \lim_{\delta\to 0} \sum_{n=1}^\infty n^{r-1} \sup_{k\in\mathbb{Z}_+} \mathsf{P}_{k,\theta}\left(\frac{1}{n} \inf_{\vartheta\in\Gamma_{\delta,\theta}} \lambda_\vartheta(k,k+n) < I_\theta - \varepsilon\right) < \infty. \tag{3.71}$$

Theorem 3.3. *Let $r \geq 1$. Assume that the prior distribution $\pi^\alpha = \{\pi_k^\alpha\}$ of the change point ν belongs to class $\mathbf{C}^\alpha(\mu = 0)$ and that μ_α approaches zero as $\alpha \to 0$ at such a rate that conditions $\mathbf{CP}_2^{(\alpha)}$ and $\mathbf{CP}_3^{(\alpha)}$ are satisfied. Assume further that for some $0 < I_\theta < \infty$, $\theta \in \Theta$, the right-tail condition $\mathbf{C}_1$ and the uniform left-tail condition $\mathbf{C}_3$ are satisfied. If $A = A_\alpha$ is so selected that $\mathsf{PFA}_\pi(T_{A_\alpha}^W) \leq \alpha$ and $\log A_\alpha \sim |\log \alpha|$ as $\alpha \to 0$, in particular $A_\alpha = (1-\alpha)/\alpha$, then the MS rule $T_{A_\alpha}^W$ is asymptotically optimal as $\alpha \to 0$ in class $\mathbb{C}(\alpha)$, minimizing moments of the detection delay up to order r: for all $0 < m \leq r$ and all $\theta \in \Theta$ as $\alpha \to 0$*

$$\inf_{T \in \mathbb{C}(\alpha)} \overline{\mathscr{R}}^m_{\pi^\alpha,\theta}(T) \sim \left(\frac{|\log \alpha|}{I_\theta}\right)^m \sim \overline{\mathscr{R}}^m_{\pi^\alpha,\theta}(T_{A_\alpha}^W) \tag{3.72}$$

and

$$\inf_{T \in \mathbb{C}(\alpha)} \mathscr{R}^m_{k,\theta}(T) \sim \left(\frac{|\log \alpha|}{I_\theta}\right)^m \sim \mathscr{R}^m_{k,\theta}(T_{A_\alpha}^W) \quad \text{for all } k \in \mathbb{Z}_+. \tag{3.73}$$

Proof. Setting $A = (1-\alpha)/\alpha$ in inequality (3.61), we obtain

$$\overline{\mathscr{R}}^r_{\pi^\alpha,\theta}(T_A^W) \leq (1-\alpha)^{-1} \sum_{k=0}^\infty \pi_k^\alpha \left(1 + \frac{\log((1-\alpha)/\alpha\pi_k^\alpha)}{I_\theta + \mu_\alpha - \varepsilon}\right)^r + r2^{r-1} \sup_{k \in \mathbb{Z}_+} \Upsilon_{k,r}(\theta, \varepsilon).$$

Using conditions $\mathbf{CP}_2^{(\alpha)}$ and $\mathbf{C}_3$ and taking into account that $\mu_\alpha \to 0$ as $\alpha \to 0$ yields

$$\overline{\mathscr{R}}^r_{\pi^\alpha,\theta}(T_{A_\alpha}^W) \leq \left(\frac{|\log \alpha|}{I_\theta - \varepsilon}\right)^r (1 + o(1)).$$

Since ε can be arbitrarily small, we obtain the asymptotic upper bound

$$\overline{\mathscr{R}}^r_{\pi^\alpha,\theta}(T_{A_\alpha}^W) \leq \left(\frac{|\log \alpha|}{I_\theta}\right)^r (1 + o(1)),$$

as $\alpha \to 0$, which along with the lower bound (3.69) (with $\mu = 0$) proves (3.72). Obviously, this upper bound holds if we take any $A = A_\alpha$ such that $\mathsf{PFA}_\pi(T_{A_\alpha}^W) \leq \alpha$ and $\log A_\alpha \sim |\log \alpha|$ as $\alpha \to 0$.

Next, substituting $A = A_\alpha = (1-\alpha)/\alpha$ in (3.55) (or more generally any A_α such that $\log A_\alpha \sim |\log \alpha|$ and $\mathsf{PFA}_\pi(T_{A_\alpha}^W) \leq \alpha$), we obtain

$$\mathscr{R}^r_{k,\theta}(T_{A_\alpha}^W) \leq \frac{\left(1 + \left\lfloor \log \frac{(1-\alpha)/\alpha\pi_k^\alpha}{I_\theta + \mu_\alpha - \varepsilon} \right\rfloor\right)^r + r2^{r-1} \Upsilon_{k,r}(\theta, \varepsilon)}{1 - 1/(A_\alpha \Pi_{k-1})},$$

which due to conditions $\mathbf{C}_3$ and $\mathbf{CP}_2^{(\alpha)}$ and the fact that $\mu_\alpha \to 0$ implies that, for all fixed $k \in \mathbb{Z}_+$ and all $\theta \in \Theta$ as $\alpha \to 0$,

$$\mathscr{R}^r_{k,\theta}(T_{A_\alpha}^W) \leq \left(\frac{|\log \alpha|}{I_\theta}\right)^r (1 + o(1)).$$

This upper bound together with the lower bound (3.68) (with $\mu = 0$) yields (3.73) and the proof is complete. □

3.5 Asymptotic Performance of the Mixture Shiryaev–Roberts Rule

Consider now the MSR detection rule $\widetilde{T}_B^W$ defined in (3.9) and (3.10).

The following lemma shows how to select threshold $B = B_\alpha$ in the MSR rule to embed it in class $\mathbb{C}_\pi(\alpha)$. Write

$$\bar{\nu} = \sum_{k=0}^{\infty} k\pi_k = (1-q)\sum_{k=1}^{\infty} k\mathsf{P}(\nu = k|\nu \ge 0) = (1-q)\mathsf{E}[\nu|\nu \ge 0].$$

Since we are not interested in negative values of ν we will refer to $\bar{\nu}$ as the mean of the prior distribution. Recall that ω ($\omega \ge 0$) is a head-start of the MSR statistic $R_W(n)$ (see (3.9)).

Lemma 3.4. *For all $B > 0$ and any prior distribution of ν with finite mean $\bar{\nu}$, the PFA of the MSR rule $\widetilde{T}_B^W$ satisfies the inequality*

$$\mathsf{PFA}_\pi(\widetilde{T}_B^W) \le \frac{\omega b + \bar{\nu}}{B}, \tag{3.74}$$

where $b = \sum_{k=1}^{\infty} \pi_k$, so that if

$$B = B_\alpha = (\omega b + \bar{\nu})/\alpha$$

then $\mathsf{PFA}_\pi(\widetilde{T}_{B_\alpha}) \le \alpha$, i.e., $\widetilde{T}_{B_\alpha} \in \mathbb{C}_\pi(\alpha)$.

Proof. Evidently, $\mathsf{E}_\infty[R_\vartheta(n)|\mathscr{F}_{n-1}] = 1 + R_\vartheta(n-1)$, and hence,

$$\mathsf{E}_\infty[R_W(n)|\mathscr{F}_{n-1}] = \int_\Theta \mathrm{d}W(\vartheta) + \int_\Theta R_\vartheta(n-1)\,\mathrm{d}W(\vartheta) = 1 + R_W(n-1).$$

So $\{R_W(n) - \omega - n\}_{n\ge 1}$ is a zero-mean $(\mathsf{P}_\infty, \mathscr{F}_n)$–martingale and the MSR statistic $R_W(n)$ is a $(\mathsf{P}_\infty, \mathscr{F}_n)$–submartingale with mean $\mathsf{E}_\infty[R_W(n)] = \omega + n$. Applying Doob's submartingale inequality, we obtain that for $j = 1, 2, \dots$

$$\mathsf{P}_\infty(\widetilde{T}_B^W \le j) = \mathsf{P}_\infty\left(\max_{1\le i\le j} R_W(i) \ge B\right) \le (\omega + j)/B \tag{3.75}$$

and $\mathsf{P}_\infty(\widetilde{T}_B^W \le 0) = 0$. Thus,

$$\mathsf{PFA}_\pi(\widetilde{T}_B) = \sum_{j=1}^{\infty} \pi_j \mathsf{P}_\infty(\widetilde{T}_B^W \le j) \le \frac{\omega\sum_{j=1}^{\infty}\pi_j + \sum_{j=1}^{\infty} j\pi_j}{B},$$

which proves inequality (3.74). Therefore, assuming $\bar{\nu} < \infty$, we obtain that setting $B = B_\alpha = (\omega b + \bar{\nu})/\alpha$ implies $\widetilde{T}_{B_\alpha} \in \mathbb{C}_\pi(\alpha)$ and the proof is complete. □

The following proposition establishes asymptotic operating characteristics of the MSR rule $\widetilde{T}_B^W$ as threshold $B \to \infty$ regardless of the PFA constraint.

Proposition 3.2. *Let $\bar{\nu} < \infty$ and $0 \le \omega < \infty$. Let $r \ge 1$. Assume that for some function $0 < I_\theta < \infty$, $\theta \in \Theta$, conditions $\mathbf{C}_1$ and $\mathbf{C}_2$ are satisfied. Then, for all $0 < m \le r$ and $\theta \in \Theta$*

$$\lim_{B\to\infty} \frac{\mathscr{R}^m_{k,\theta}(\widetilde{T}_B^W)}{(\log B)^m} = \frac{1}{I_\theta^m} \quad \textit{for all } k \in \mathbb{Z}_+ \tag{3.76}$$

and

$$\lim_{B\to\infty} \frac{\overline{\mathscr{R}}^m_{\pi,\theta}(\widetilde{T}_B^W)}{(\log B)^m} = \frac{1}{I_\theta^m}. \tag{3.77}$$

Proof. For $\varepsilon \in (0,1)$, let $M_B = M_B(\varepsilon,\theta) = (1-\varepsilon)I_\theta^{-1}\log B$. Similarly to (3.33) we obtain

$$\mathscr{R}^r_{k,\theta}(\widetilde{T}^W_B) \geq M^r_B\left[\mathsf{P}_\infty(\widetilde{T}^W_B > k) - \mathsf{P}_{k,\theta}(k < \widetilde{T}^W_B \leq k + M_B)\right] \tag{3.78}$$

and similarly to (3.38),

$$\mathsf{P}_{k,\theta}\left(0 < \widetilde{T}^W_B - k \leq M_B\right) \leq U_{A,k}(\widetilde{T}^W_B) + p_{B,k}(\theta), \tag{3.79}$$

where

$$U_{A,k}(\widetilde{T}^W_B) = e^{(1+\varepsilon)I_\theta M_B}\mathsf{P}_\infty\left(0 < \widetilde{T}^W_B - k \leq M_B\right),$$

$$p_{B,k}(\theta) = \mathsf{P}_{k,\theta}\left(\frac{1}{M_B}\max_{1\leq n\leq M_B}\lambda_\theta(k,k+n) \geq (1+\varepsilon)I_\theta\right).$$

Since

$$\mathsf{P}_\infty\left(0 < \widetilde{T}^W_B - k \leq M_B\right) \leq \mathsf{P}_\infty\left(\widetilde{T}^W_B \leq k + M_B\right) \leq (k+\omega+M_B)/B,$$

(see (3.75)) we have

$$U_{A,k}(\widetilde{T}^W_B) \leq \frac{k+\omega+(1-\varepsilon)I_\theta^{-1}\log B}{B^{\varepsilon^2}}. \tag{3.80}$$

Therefore, $U_{B,k}(\widetilde{T}^W_B) \to 0$ as $B \to \infty$ for any fixed k. Also, $p_{B,k}(\theta) \to 0$ by condition $\mathbf{C}_1$, so that $\mathsf{P}_{k,\theta}\left(0 < \widetilde{T}^W_B - k < M_B\right) \to 0$ for any fixed k. Since $\mathsf{P}_\infty(\widetilde{T}^W_B > k) > 1 - (\omega+k)/B$, it follows from (3.78) that for an arbitrarily $\varepsilon \in (0,1)$ as $B \to \infty$

$$\mathscr{R}^r_{k,\theta}(\widetilde{T}^W_B) \geq \left(\frac{(1-\varepsilon)\log B}{I_\theta}\right)^r (1+o(1)),$$

which yields the asymptotic lower bound (for any fixed $k \in \mathbb{Z}_+$ and $\theta \in \Theta$)

$$\mathscr{R}^r_{k,\theta}(\widetilde{T}^W_B) \geq \left(\frac{\log B}{I_\theta}\right)^r (1+o(1)), \quad B \to \infty. \tag{3.81}$$

To prove (3.76) it suffices to show that

$$\mathscr{R}^r_{k,\theta}(\widetilde{T}^W_B) \leq \left(\frac{\log B}{I_\theta}\right)^r (1+o(1)), \quad B \to \infty. \tag{3.82}$$

For $k \in \mathbb{Z}_+$, define the stopping times

$$\tilde{\tau}^{(k)}_B = \inf\{n \geq 1 : \log\Lambda_W(k,k+n) \geq \log B\}.$$

Obviously, for any $n > k$, $\log R_W(n) \geq \log\Lambda_W(k,n)$, and hence, for every $B > 0$, $(\widetilde{T}^W_B - k)^+ \leq \tilde{\tau}^{(k)}_B$. Analogously to (3.53), we have

$$\mathsf{E}_{k,\theta}\left[(\widetilde{T}^W_B - k)^+\right]^r \leq \mathsf{E}_{k,\theta}\left[\left(\tilde{\tau}^{(k)}_B\right)^r\right] \leq \widetilde{M}^r_B + r2^{r-1}\sum_{n=\widetilde{M}_B}^\infty n^{r-1}\mathsf{P}_{k,\theta}\left(\tilde{\tau}^{(k)}_B > n\right), \tag{3.83}$$

where $\widetilde{M}_B = \widetilde{M}_B(\varepsilon,\theta) = 1 + \lfloor\log(B)/(I_\theta - \varepsilon)\rfloor$. For a sufficiently large n similarly to (3.54) we have

$$\mathsf{P}_{k,\theta}\left(\tilde{\tau}^{(k)}_B > n\right) \leq \mathsf{P}_{k,\theta}\left(\frac{1}{n}\inf_{\vartheta\in\Gamma_{\delta,\theta}}\lambda_\vartheta(k,k+n) < I_\theta - \varepsilon\right). \tag{3.84}$$

Using (3.83) and (3.84), we obtain the inequality

$$\mathsf{E}_{k,\theta}\left[\left(\widetilde{T}_B^W - k\right)^+\right]^r \le \left(1 + \left\lfloor \frac{\log B}{I_\theta - \varepsilon} \right\rfloor\right)^r + r2^{r-1}\Upsilon_{k,r}(\theta,\varepsilon), \tag{3.85}$$

which along with the inequality $\mathsf{P}_\infty(\widetilde{T}_B^W > k) > 1 - (\omega + k)/B$ implies the inequality

$$\mathscr{R}^r_{k,\theta}(\widetilde{T}_B^W) = \frac{\mathsf{E}_{k,\theta}\left[\left(\widetilde{T}_B^W - k\right)^+\right]^r}{\mathsf{P}_\infty(\widetilde{T}_B^W > k)} \le \frac{\left(1 + \left\lfloor \frac{\log B}{I_\theta - \varepsilon} \right\rfloor\right)^r + r2^{r-1}\Upsilon_{k,r}(\theta,\varepsilon)}{1 - (\omega + k)/B}. \tag{3.86}$$

Since, by condition $\mathbf{C}_2$, $\Upsilon_{k,r}(\theta,\varepsilon) < \infty$ for all $k \in \mathbb{Z}_+$ and $\theta \in \Theta$, this implies the asymptotic upper bound (3.82) and completes the proof of the asymptotic approximation (3.76).

We now continue with proving (3.77). By the Chebyshev inequality,

$$\begin{aligned}
\overline{\mathscr{R}}^r_{\pi,\theta}(\widetilde{T}_B^W) &\ge \mathsf{E}^\pi_\theta[(\widetilde{T}_B^W - k)^+]^r \ge M_B^r \mathsf{P}^\pi_\theta(\widetilde{T}_B^W - \nu > M_B) \\
&\ge M_B^r \left[\mathsf{P}^\pi_\theta(\widetilde{T}_B^W > \nu) - \mathsf{P}^\pi_\theta(\nu < \widetilde{T}_B < \nu + M_B)\right] \\
&\ge M_B^r \left[1 - \frac{\bar{\nu} + \omega}{B} - \mathsf{P}^\pi_\theta\left(0 < \widetilde{T}_B^W - \nu < M_B\right)\right].
\end{aligned} \tag{3.87}$$

Let K_B be an integer number that approaches infinity as $B \to \infty$. Using (3.79) and (3.80), we obtain the following upper bound

$$\begin{aligned}
\mathsf{P}^\pi_\theta(0 < \widetilde{T}_B^W - \nu < M_B) &= \sum_{k=0}^{\infty} \pi_k \mathsf{P}_{k,\theta}\left(0 < \widetilde{T}_B^W - k < M_B\right) \\
&\le \mathsf{P}(\nu > K_B) + \sum_{k=0}^{\infty} \pi_k U_{B,k}(\widetilde{T}_B^W) + \sum_{k=0}^{K_B} \pi_k p_{B,k} \\
&\le \mathsf{P}(\nu > K_B) + \frac{\bar{\nu} + \omega + (1-\varepsilon) I_\theta^{-1} \log B}{B^{\varepsilon^2}} + \sum_{k=0}^{K_A} \pi_k p_{B,k},
\end{aligned} \tag{3.88}$$

where the first two terms go to zero as $B \to \infty$ since $\bar{\nu}$ and ω are finite (by Markov's inequality $\mathsf{P}(\nu > K_B) \le \bar{\nu}/K_B$) and the last term also goes to zero by condition $\mathbf{C}_1$ and Lebesgue's dominated convergence theorem. Thus, for all $0 < \varepsilon < 1$, $\mathsf{P}^\pi_\theta(0 < \widetilde{T}_B^W - \nu < M_B)$ approaches 0 as $B \to \infty$. Using inequality (3.87), we obtain that for any $0 < \varepsilon < 1$ as $B \to \infty$

$$\overline{\mathscr{R}}^r_{\pi,\theta}(\widetilde{T}_B^W) \ge (1-\varepsilon)^r \left(\frac{\log B}{I_\theta}\right)^r (1 + o(1)),$$

which yields the asymptotic lower bound (for any $r > 0$ and $\theta \in \Theta$)

$$\overline{\mathscr{R}}^r_{\pi,\theta}(\widetilde{T}_B^W) \ge \left(\frac{\log B}{I_\theta}\right)^r (1 + o(1)), \quad B \to \infty. \tag{3.89}$$

To obtain the upper bound it suffices to use inequality (3.85), which along with the fact that $\mathsf{P}^\pi(\widetilde{T}_B^W > \nu) > 1 - (\bar{\nu} + \omega)/B$ yields (for every $0 < \varepsilon < I_\theta$)

$$\overline{\mathscr{R}}^r_{\pi,\theta}(\widetilde{T}_B^W) = \frac{\sum_{k=0}^{\infty} \pi_k \mathsf{E}_k[(\widetilde{T}_B^W - k)^+]^r}{\mathsf{P}_\pi(\widetilde{T}_B^W > \nu)} \le \frac{\left(1 + \frac{\log B}{I_\theta - \varepsilon}\right)^r + r2^{r-1} \sum_{k=0}^{\infty} \pi_k \Upsilon_{k,r}(\theta,\varepsilon)}{1 - (\omega + \bar{\nu})/B}, \tag{3.90}$$

where we set $\pi_0 = \Pr(\nu \le 0)$. Since by condition $\mathbf{C}_2$,

$$\sum_{k=0}^{\infty} \pi_k \Upsilon_{k,r}(\theta,\varepsilon) < \infty \quad \text{for any } \varepsilon > 0 \text{ and } \theta \in \Theta,$$

we obtain that, for every $0 < \varepsilon < I_\theta$ as $B \to \infty$,

$$\overline{\mathscr{R}}^r_{\pi,\theta}(\widetilde{T}^W_B) \le \left(\frac{\log B}{I_\theta - \varepsilon}\right)^r (1+o(1)).$$

Since ε can be arbitrarily small, this implies the asymptotic (as $B \to \infty$) upper bound

$$\overline{\mathscr{R}}^r_{\pi,\theta}(\widetilde{T}^W_B) \le \left(\frac{\log B}{I_\theta}\right)^r (1+o(1)), \tag{3.91}$$

which along with the lower bound (3.89) completes the proof. □

It follows from Proposition 3.2 that if $B = B_\alpha$ is so selected that $\widetilde{T}_{B_\alpha} \in \mathbb{C}_\pi(\alpha)$ and $\log B_\alpha \sim |\log \alpha|$ as $\alpha \to 0$, in particular $B_\alpha = (\omega b + \bar{\nu})/\alpha$, then for all $0 < m \le r$ and $\theta \in \Theta$

$$\lim_{\alpha \to 0} \frac{\mathscr{R}^m_{k,\theta}(\widetilde{T}_{B_\alpha})}{|\log \alpha|^m} = \frac{1}{I^m_\theta} \quad \text{for all } k \in \mathbb{Z}_+ \tag{3.92}$$

and

$$\lim_{\alpha \to 0} \frac{\overline{\mathscr{R}}^m_{\pi,\theta}(\widetilde{T}_{B_\alpha})}{|\log \alpha|^m} = \frac{1}{I^m_\theta}. \tag{3.93}$$

Comparing asymptotic approximations (3.92) and (3.93) with (3.62) and (3.63) in Theorem 3.2 shows that the MSR rule is asymptotically optimal if, and only if, $\mu = 0$, i.e., for heavy-tailed prior distributions, which is intuitively expected since the MSR rule exploits uniform prior. However, if $\mu > 0$ but small it is expected that the MSR rule will preserve the asymptotic optimality property.

The next theorem addresses the case where the head-start $\omega = \omega_\alpha$ of the MSR statistic and the mean value $\bar{\nu} = \bar{\nu}_\alpha$ of the prior distribution approach infinity as $\alpha \to 0$ with a certain rate. In this case, $\mu_\alpha \to 0$ as $\alpha \to 0$ and the MSR rule is asymptotically optimal.

Theorem 3.4. *Assume that $\omega_\alpha \to \infty$ and $\bar{\nu}_\alpha \to \infty$ with such a rate that the following condition holds:*

$$\lim_{\alpha \to 0} \frac{\log(\omega_\alpha + \bar{\nu}_\alpha)}{|\log \alpha|} = 0. \tag{3.94}$$

Assume further that for some $0 < I_\theta < \infty$ and $r \ge 1$ conditions $\mathbf{C}_1$ and $\mathbf{C}_3$ are satisfied. If threshold B_α is so selected that $\mathsf{PFA}_\pi(T_{B_\alpha}) \le \alpha$ and $\log B_\alpha \sim |\log \alpha|$ as $\alpha \to 0$, in particular $B_\alpha = (\omega_\alpha b_\alpha + \bar{\nu}_\alpha)/\alpha$, then for all $0 < m \le r$ and $\theta \in \Theta$, as $\alpha \to 0$

$$\overline{\mathscr{R}}^m_{\pi^\alpha,\theta}(\widetilde{T}_{B_\alpha}) \sim \left(\frac{|\log \alpha|}{I_\theta}\right)^m \sim \inf_{T \in \mathbb{C}(\alpha)} \overline{\mathscr{R}}^m_{\pi^\alpha,\theta}(T), \tag{3.95}$$

and

$$\mathscr{R}^m_{k,\theta}(\widetilde{T}_{B_\alpha}) \sim \left(\frac{|\log \alpha|}{I_\theta}\right)^m \sim \inf_{T \in \mathbb{C}(\alpha)} \mathscr{R}^m_{k,\theta}(T) \quad \textit{for all } k \in \mathbb{Z}_+. \tag{3.96}$$

Therefore, the MSR rule $\widetilde{T}_{B_\alpha}$ is asymptotically optimal as $\alpha \to 0$ in class $\mathbb{C}(\alpha)$, minimizing moments of the detection delay up to order r.

Proof. Previous results make the proof elementary. Note first that $\mu_\alpha \to 0$ as $\alpha \to 0$ since by the assumption of the theorem $\bar{\nu}_\alpha \to \infty$. Hence, the prior distribution π^α belongs to class $\mathbb{C}^{(\alpha)}(\mu = 0)$ and, by Lemma 3.3, the following asymptotic lower bounds hold as $\alpha \to 0$ for any $r \geq 1$:

$$\inf_{T \in \mathbb{C}(\alpha)} \overline{\mathscr{R}}^r_{\pi^\alpha,\theta}(T) \geq \left(\frac{|\log \alpha|}{I_\theta}\right)^r (1 + o(1)), \tag{3.97}$$

$$\inf_{T \in \mathbb{C}(\alpha)} \mathscr{R}^r_{k,\theta}(T) \geq \left(\frac{|\log \alpha|}{I_\theta}\right)^r (1 + o(1)) \quad \text{for all } k \in \mathbb{Z}_+. \tag{3.98}$$

Substitution $B = B_\alpha = (b_\alpha \omega_\alpha + \bar{\nu}_\alpha)/\alpha$ in (3.90) (or more generally such B_α that $\log B_\alpha \sim |\log \alpha|$) yields the upper bound

$$\overline{\mathscr{R}}^r_{\pi^\alpha,\theta}(\widetilde{T}_{B_\alpha}) \leq \frac{\left(1 + \frac{\log((\omega_\alpha + \bar{\nu}_\alpha)/\alpha)}{I_\theta - \varepsilon}\right)^r + r2^{r-1} \sup_{k \in \mathbb{Z}_+} \Upsilon_{k,r}(\theta, \varepsilon)}{1 - \alpha},$$

which implies the asymptotic upper bound

$$\overline{\mathscr{R}}^r_{\pi^\alpha,\theta}(\widetilde{T}_{B_\alpha}) \leq \left(\frac{|\log \alpha|}{I_\theta}\right)^r (1 + o(1)), \quad \alpha \to 0,$$

since, by condition (3.94), $\log[(\omega_\alpha + \bar{\nu}_\alpha)/\alpha] \sim |\log \alpha|$ and by condition $\mathbf{C}_3$, $\sup_{k \in \mathbb{Z}_+} \Upsilon_{k,r}(\theta, \varepsilon) < \infty$ for all $\theta \in \Theta$. This upper bound along with the lower bound (3.97) proves (3.95).

Finally, the asymptotic upper bound

$$\mathscr{R}^r_{k,\theta}(\widetilde{T}_{B_\alpha}) \leq \left(\frac{|\log \alpha|}{I_\theta}\right)^r (1 + o(1)), \quad \alpha \to 0,$$

for all $k \in \mathbb{Z}_+$ and $\theta \in \Theta$ follows immediately from (3.86) with $B = B_\alpha = (c_\alpha \omega_\alpha + \bar{\nu}_\alpha)/\alpha$ (or more generally with such B_α that $\log B_\alpha \sim |\log \alpha|$), which along with the lower bound (3.98) proves (3.96). □

3.6 Asymptotic Optimality with Respect to the Integrated Risk

Instead of the constrained optimization problem (3.15) consider now the unconstrained, "purely" Bayes problem with the loss function

$$L_r(T, \nu) = \mathbb{1}_{\{T \leq \nu\}} + c\,(T - \nu)^r \mathbb{1}_{\{T > \nu\}},$$

where $c > 0$ is the cost of delay per unit of time and $r \geq 1$. The unknown parameter θ is now assumed random and the weight function $W(\vartheta)$ is interpreted as the prior distribution of θ. The expected loss (integrated or average risk) associated with the detection rule T is given by

$$\rho^{c,r}_{\pi,W}(T) = \mathsf{PFA}_\pi(T) + c \int_\Theta \mathsf{E}^\pi_\vartheta[(T - \nu)^+]^r \, \mathrm{d}W(\vartheta).$$

Below we show that the MS rule T^W_A with a certain threshold $A = A_{c,r}$ that depends on the cost c is asymptotically optimal, minimizing the integrated risk $\rho^{c,r}_{\pi,W}(T)$ over all stopping times as the cost vanishes, $c \to 0$.

Define

$$\mathscr{R}^r_{\pi,W}(T) = \int_\Theta \overline{\mathscr{R}}^r_{\pi,\vartheta}(T) \, \mathrm{d}W(\vartheta).$$

Observe that, if we ignore the overshoot, then $\mathsf{PFA}_\pi(T_A^W) \approx 1/(1+A)$ and that using approximation (3.51) we may expect that for a large A

$$\mathscr{R}^r_{\pi,W}(T_A^W) \approx \int_\Theta \left(\frac{\log A}{I_\vartheta + \mu}\right)^r \, \mathrm{d}W(\vartheta) = (\log A)^r D_{\mu,r}$$

where

$$D_{\mu,r} = \int_\Theta \left(\frac{1}{I_\vartheta + \mu}\right)^r \, \mathrm{d}W(\vartheta).$$

So for large A the integrated risk of the MS rule is approximately equal to

$$\rho^{c,r}_{\pi,W}(T_A^W) = \mathsf{PFA}_\pi(T_A^W) + c[1 - \mathsf{PFA}_\pi(T_A^W)]\mathscr{R}^r_{\pi,W}(T_A^W) \approx G_{c,r}(A),$$

where

$$G_{c,r}(A) = \frac{1}{A} + D_{\mu,r} c (\log A)^r. \tag{3.99}$$

The threshold value $A = A_{c,r}$ that minimizes $G_{c,r}(A)$, $A > 0$, is a solution of the equation

$$r D_{\mu,r} A (\log A)^{r-1} = 1/c. \tag{3.100}$$

In particular, for $r = 1$ we obtain $A_{c,1} = 1/(cD_{\mu,1})$. Thus, it is reasonable to conjecture that threshold $A_{c,r}$ optimizes the performance of the MS rule for a small c, and hence, makes this rule asymptotically optimal as $c \to 0$.

The following lemma establishes an asymptotic lower bound for the average risk in the class of all (Markov) stopping times $\mathscr{M}$.

Lemma 3.5. *Let, for some $\mu \ge 0$, the prior distribution belong to class $\mathbf{C}(\mu)$. Assume that for some positive and finite numbers I_θ ($\theta \in \Theta$) condition $\mathbf{C}_1$ holds. Then for all $r > 0$*

$$\liminf_{c \to 0} \frac{\inf_{T \in \mathscr{M}} \rho^{c,r}_{\pi,W}(T)}{c|\log c|^r} \ge D_{\mu,r}. \tag{3.101}$$

Proof. Recall that the function $G_{c,r}(A)$ is defined in (3.99). It is easily seen that $\min_{A>0} G_{c,r}(A) = G_{c,r}(A_{c,r})$, where $A_{c,r}$ satisfies the equation

$$r D_{\mu,r} A (\log A)^{r-1} - 1/c = 0,$$

and goes to infinity as $c \to 0$ so that $\log A_{c,r} \sim |\log c|$ and that

$$\lim_{c \to 0} \frac{G_{c,r}(A_{c,r})}{D_{\mu,r} c |\log c|^r} = 1.$$

Thus, it suffices to prove that

$$\frac{\inf_{T \ge 0} \rho^{c,r}_{\pi,W}(T)}{G_{c,r}(A_{c,r})} \ge 1 + o(1) \quad \text{as } c \to 0. \tag{3.102}$$

If (3.102) is wrong, then there is a stopping rule T_c such that

$$\frac{\rho^{c,r}_{\pi,W}(T_c)}{G_{c,r}(A_{c,r})} < 1 + o(1) \quad \text{as } c \to 0. \tag{3.103}$$

Let $\alpha_c = \mathsf{PFA}_\pi(T_c)$. Since

$$\alpha_c \le \rho^{c,r}_{\pi,W}(T_c) < G_{c,r}(A_{c,r})(1 + o(1)) \to 0 \quad \text{as } c \to 0,$$

it follows that $\alpha_c \to 0$ as $c \to 0$. Using inequality (3.32), we obtain that as $\alpha_c \to 0$

$$\overline{\mathscr{R}}^r_{\pi,W}(T_c) \geq D_{\mu,r} |\log \alpha_c|^r (1+o(1)),$$

and hence, as $c \to 0$,

$$\rho^{c,r}_{\pi,W}(T_c) = \alpha_c + c(1-\alpha_c)\overline{\mathscr{R}}^r_{\pi,W}(T_c) \geq \alpha_c + c D_{\mu,r} |\log \alpha_c|^r (1+o(1)).$$

Thus,

$$\frac{\rho^{c,r}_{\pi,W}(T_c)}{G_{c,r}(A_{c,r})} \geq \frac{G_{c,r}(1/\alpha_c) + c D_{\mu,r} |\log \alpha_c|^r o(1)}{\min_{A>0} G_{c,r}(A)} \geq 1 + o(1),$$

which contradicts (3.103). Hence, (3.102) follows and the proof of (3.101) is complete. □

In the next theorem, we establish that the MS rule $T_{A_{c,r}}$ with threshold $A_{c,r}$ that satisfies (3.100) is indeed asymptotically optimal as $c \to 0$ under conditions $\mathbf{C}_1$ and $\mathbf{C}_2$ when the set Θ is compact.

Theorem 3.5. *Let the prior distribution of the change point belong to class* $\mathbf{C}(\mu)$. *Assume that for some* $0 < I_\theta < \infty$, $\theta \in \Theta$, *right-tail and left-tail conditions* $\mathbf{C}_1$ *and* $\mathbf{C}_2$ *are satisfied and that* Θ *is a compact set. Let* $A = A_{c,r}$ *be the solution of the equation* (3.100). *Then, as* $c \to 0$,

$$\inf_{T \in \mathscr{M}} \rho^{c,r}_{\pi,W}(T) \sim D_{\mu,r}\, c\, |\log c|^r \sim \rho^{c,r}_{\pi,W}(T^W_{A_{c,r}}). \tag{3.104}$$

Proof. Since Θ is compact it follows from the asymptotic approximation (3.51) in Theorem 3.2 that as $A \to \infty$

$$\int_\Theta \overline{\mathscr{R}}^r_{\pi,\vartheta}(T^W_A)\, \mathrm{d}W(\vartheta) = \int_\Theta \left(\frac{\log A}{I_\vartheta + \mu}\right)^r \mathrm{d}W(\vartheta)(1+o(1)) = D_{\mu,r}(\log A)^r(1+o(1)).$$

Since $\mathsf{PFA}_\pi(T^W_A) < 1/A$, we obtain the following asymptotic approximation for the integrated risk

$$\rho^{c,r}_{\pi,W}(T^W_A) \sim D_{\mu,r}\, c\, (\log A)^r \quad \text{as } A \to \infty. \tag{3.105}$$

Next, it is easily seen that, for any $r \geq 1$ and $\mu \in (0,\infty)$, threshold $A_{c,r}$ goes to infinity as $c \to 0$ with such a rate that $\log A_{c,r} \sim |\log c|$. As a result, we obtain that

$$\rho^{c,r}_{\pi,W}(T^W_{A_{c,r}}) \sim D_{\mu,r}\, c\, |\log c|^r \quad \text{as } c \to 0.$$

Comparing it with the asymptotic lower bound (3.101) in Lemma 3.5 yields (3.104). □

Finally, the results analogous to Theorems 3.3 and 3.4 in the case where the prior distribution has an exponential tail, i.e., $\mu > 0$, but $\mu = \mu_c \to 0$ as $c \to 0$ also hold for the integrated risk. Specifically, let $D_r = D_{\mu=0,r}$, i.e.,

$$D_r = \int_\Theta \left(\frac{1}{I^r_\vartheta}\right) \mathrm{d}W(\vartheta).$$

Note that the values of the mean of the prior distribution of the change point $\bar{\nu} = \sum_{j=1}^\infty j\pi^c_j = \bar{\nu}_c$, the head-start of the MSR statistic $\omega = \omega_c$, and the value of $b = \sum_{j=1}^\infty \pi^c_j = b_c$ are the functions of the cost c. The following theorem spells out details.

Theorem 3.6. *Assume that for some* $0 < I_\theta < \infty$, $\theta \in \Theta$, *right-tail and left-tail conditions* $\mathbf{C}_1$ *and* $\mathbf{C}_3$ *are satisfied and that* Θ *is compact.*
(i) *If the prior distribution* $\pi^c = \{\pi^c_k\}$ *satisfies condition* (3.20) *with* $\mu = \mu_c \to 0$ *as* $c \to 0$ *at such a rate that*

$$\lim_{c \to 0} \frac{\sum_{k=0}^\infty \pi^c_k |\log \pi^c_k|^r}{|\log c|^r} = 0 \tag{3.106}$$

and threshold $A = A_{c,r}$ of the MS rule T_A^W is the solution of the equation

$$rD_r A(\log A)^{r-1} = 1/c, \tag{3.107}$$

then, as $c \to 0$,

$$\inf_{T\geq 0} \rho_{\pi^c,W}^{c,r}(T) \sim D_r c\,|\log c|^r \sim \rho_{\pi^c,W}^{c,r}(T_{A_{c,r}}^W). \tag{3.108}$$

Therefore, the MS rule $T_{A_{c,r}}^W$ is asymptotically optimal as $c \to 0$.
(ii) *If the head-start ω_c and the mean of the prior distribution $\bar{\nu}_c$ approach infinity at such a rate that*

$$\lim_{c\to 0} \frac{\log(\omega_c + \bar{\nu}_c)}{|\log c|} = 0 \tag{3.109}$$

and if $B = B_{c,r}$ of the MSR rule $\widetilde{T}_B^W$ is the solution of the equation

$$rD_r B(\log B)^{r-1} = (\omega_c b_c + \bar{\nu}_c)/c, \tag{3.110}$$

then, as $c \to 0$,

$$\inf_{T\in\mathscr{M}} \rho_{\pi^c,W}^{c,r}(T) \sim D_r c\,|\log c|^r \sim \rho_{\pi^c,W}^{c,r}(\widetilde{T}_{B_{c,r}}^W). \tag{3.111}$$

Therefore, the MSR rule $\widetilde{T}_{A_{c,r}}^W$ is asymptotically optimal as $c \to 0$.

Proof. The proof of part (i). Using (3.60), we obtain

$$\int_\Theta \mathsf{E}_\vartheta^{\pi^c}[(T_A^W - \nu)^+]^r \,\mathrm{d}W(\vartheta) \leq \sum_{k=0}^\infty \pi_k^c \int_\Theta \left(1 + \frac{\log(A/\pi_k^c)}{I_\vartheta + \mu_c - \varepsilon}\right)^r \mathrm{d}W(\vartheta) + r2^{r-1}\int_\Theta \Upsilon_r(\vartheta,\varepsilon)\,\mathrm{d}W(\vartheta),$$

where the last term is finite since Θ is compact and $\Upsilon_r(\vartheta,\varepsilon_1) < \infty$ for all $\vartheta \in \Theta$ due to condition $\mathbf{C}_3$ and where

$$\sum_{k=0}^\infty \pi_k^c |\log \pi_k^c|^r = o(|\log c|^r) \quad \text{as } c \to 0$$

by assumption (3.106). Recall that, as established in the proof of Theorem 3.5 above, $\log A_{c,r} \sim |\log c|$ as $c \to 0$ when $A_{c,r}$ satisfies (3.107) and that $\mu_c \to 0$. Therefore, as $c \to 0$,

$$\int_\Theta \mathsf{E}_\vartheta^{\pi^c}[(T_{A_{c,r}}^W - \nu)^+]^r \,\mathrm{d}W(\vartheta) \leq \int_\Theta \left(\frac{|\log c|}{I_\vartheta - \varepsilon}\right)^r \mathrm{d}W(\vartheta)(1+o(1)).$$

Since ε can be arbitrarily small, we obtain

$$c\int_\Theta \mathsf{E}_\vartheta^{\pi^c}[(T_{A_{c,r}}^W - \nu)^+]^r \,\mathrm{d}W(\vartheta) \leq D_r c\,|\log c|^r(1+o(1)) \quad \text{as } c \to 0.$$

Since $\mathsf{PFA}_\pi(T_{A_{c,r}}^W) < 1/A_{c,r} = o(c|\log c|^r)$ it follows that

$$\rho_{\pi^c,W}^{c,r}(T_{A_{c,r}}^W) \leq D_r c\,|\log c|^r(1+o(1)) \quad \text{as } c \to 0. \tag{3.112}$$

The lower bound

$$\inf_{T\geq 0} \rho_{\pi^c,W}^{c,r}(T) \geq D_r c\,|\log c|^r(1+o(1)) \quad \text{as } c \to 0 \tag{3.113}$$

can be deduced using the argument essentially similar to that used in the proof of the lower bound (3.101) in Lemma 3.5 with $G_{c,r}(A) = 1/A + cD_r(\log A)^r$.

Using the asymptotic upper bound (3.112) and the lower bound (3.113) simultaneously, we obtain (3.108), which completes the proof of part (i).

The proof of part (ii). In order to prove (3.111) it suffices to prove the asymptotic upper bound

$$\rho_{\pi^c,W}^{c,r}(\widetilde{T}_{B_{c,r}}^W) \le D_r c|\log c|^r(1+o(1)) \quad \text{as } c \to 0. \tag{3.114}$$

Define

$$\widetilde{G}_{c,r}(B) = (\omega_c b_c + \bar{\nu}_c)/B + cD_r(\log B)^r.$$

Threshold $B_{c,r}$ that satisfies equation (3.110) minimizes $\widetilde{G}_{c,r}(B)$, and it is easily seen that $\log B_{c,r} \sim |\log c|$ as $c \to 0$ since, by assumption (3.109), $\omega_c b_c + \bar{\nu}_c = o(|\log c|)$ as $c \to 0$.

Using inequality (3.85), we obtain

$$\int_\Theta \mathsf{E}_\vartheta^{\pi^c}[(\widetilde{T}_B^W - \nu)^+]^r \, \mathrm{d}W(\vartheta) \le \int_\Theta \left(1 + \frac{\log B}{I_\vartheta - \varepsilon}\right)^r \mathrm{d}W(\vartheta) + r2^{r-1} \int_\Theta \Upsilon_r(\vartheta, \varepsilon) \, \mathrm{d}W(\vartheta),$$

where the last term is finite since Θ is compact and $\Upsilon_r(\vartheta,\varepsilon) < \infty$ for all $\vartheta \in \Theta$ due to condition $\mathbf{C}_3$. Therefore, for an arbitrarily small ε as $c \to 0$,

$$\int_\Theta \mathsf{E}_\vartheta^{\pi^c}[(\widetilde{T}_B^W - \nu)^+]^r \, \mathrm{d}W(\vartheta) \le \int_\Theta \left(\frac{|\log c|}{I_\vartheta - \varepsilon}\right)^r \mathrm{d}W(\vartheta)(1+o(1)),$$

which implies that, as $c \to 0$,

$$c \int_\Theta \mathsf{E}_\vartheta^{\pi^c}[(\widetilde{T}_B^W - \nu)^+]^r \, \mathrm{d}W(\vartheta) \le D_r c|\log c|^r(1+o(1)).$$

Since $\mathsf{PFA}_\pi(\widetilde{T}_{A_{c,r}}^W) \le (\omega_c b_c + \bar{\nu}_c)/B_{c,r} = o(c|\log c|^r)$, we obtain (3.114), which along with the lower bound (3.113) proves (3.111). The proof of part (ii) is complete. □

3.7 The Case of a Simple Post-Change Hypothesis

Consider now a particular case where the post-change hypothesis is simple, i.e., the post-change parameter $\theta = \vartheta$ is known. Alternatively, this situation also occurs when a putative value ϑ of the post-change parameter is of special interest, representing a nominal (or minimal) change. Then it is reasonable to turn the mixture rules T_A^W and $\widetilde{T}_B^W$ in rules T_A^ϑ and $\widetilde{T}_B^\vartheta$ that are tuned to ϑ. Obviously, this is equivalent to taking the degenerate weight function $W(\theta)$ concentrated at $\theta = \vartheta$. These detection procedures are nothing but the Shiryaev and SR rules defined in (3.2), (3.3) and (3.7), (3.8), respectively.

These rules have first-order asymptotic optimality properties at the point $\theta = \vartheta$ (and only at this point) with respect to $\mathscr{R}_{k,\vartheta}(T)$ and $\overline{\mathscr{R}}_{\pi,\vartheta}(T)$ when the right-tail condition $\mathbf{C}_1$ is satisfied for $\theta = \vartheta$ and the following left-tail condition holds:

$\widetilde{\mathbf{C}}_2$. *For every* $\varepsilon > 0$ *and for some* $r \ge 1$

$$\widetilde{\Upsilon}_r(\varepsilon) := \sum_{n=1}^{\infty} n^{r-1} \sup_{k \in \mathbb{Z}_+} \mathsf{P}_{k,\vartheta}\left(\frac{1}{n}\lambda_\vartheta(k,k+n) < I_\vartheta - \varepsilon\right) < \infty. \tag{3.115}$$

Indeed, since the Shiryaev rule is a particular case of the MS rule when the weight function $W(\theta)$ is concentrated at the point $\theta = \vartheta$ inequality (3.52) for T_A^ϑ in Lemma 3.2 has the form

$$\begin{aligned} \mathsf{E}_{k,\vartheta}\left[\left(T_A^\vartheta - k\right)^+\right]^r &\le \left(1 + \frac{\log(A/\pi_k)}{I_\vartheta + \mu - \varepsilon}\right)^r \\ &+ r2^{r-1} \sum_{n=N_A}^{\infty} n^{r-1} \mathsf{P}_{k,\vartheta}\left(\frac{1}{n}\lambda_\vartheta(k,k+n) < I_\vartheta - \varepsilon\right) \end{aligned} \tag{3.116}$$

(for a sufficiently large A, any $0 < \varepsilon < I_\vartheta + \mu$ and all $k \in \mathbb{Z}_+$). Hence, it suffices to replace condition $\mathbf{C}_2$ in previous theorems by condition $\widetilde{\mathbf{C}}_2$.

More specifically, the following two theorems hold true for the Shiryaev and the SR changepoint detection rules T_A^ϑ and $\widetilde{T}_B^\vartheta$.

Theorem 3.7. *Let $r \ge 1$ and let the prior distribution of the change point belong to class $\mathbf{C}(\mu)$. Assume that for $\theta = \vartheta$ and some $0 < I_\vartheta < \infty$ the right-tail condition $\mathbf{C}_1$ is satisfied and the left-tail condition $\widetilde{\mathbf{C}}_2$ holds as well.*

(i) *If threshold $A = A_\alpha$ in the Shiryaev rule is so selected that $\mathsf{PFA}_\pi(T_{A_\alpha}^\vartheta) \le \alpha$ and $\log A_\alpha \sim |\log \alpha|$ as $\alpha \to 0$, e.g., as $A = (1-\alpha)/\alpha$, then the Shiryaev rule $T_{A_\alpha}^\vartheta$ is first-order asymptotically optimal as $\alpha \to 0$ in class $\mathbb{C}_\pi(\alpha)$, minimizing moments of the detection delay up to order r: for all $0 < m \le r$ as $\alpha \to 0$*

$$\inf_{T \in \mathbb{C}_\pi(\alpha)} \mathscr{R}_{k,\vartheta}^m(T) \sim \left(\frac{|\log \alpha|}{I_\vartheta + \mu}\right)^m \sim \mathscr{R}_{k,\vartheta}^m(T_{A_\alpha}^\vartheta) \quad \textit{for all } k \in \mathbb{Z}_+ \tag{3.117}$$

and

$$\inf_{T \in \mathbb{C}_\pi(\alpha)} \overline{\mathscr{R}}_{\pi,\vartheta}^m(T) \sim \left(\frac{|\log \alpha|}{I_\vartheta + \mu}\right)^m \sim \overline{\mathscr{R}}_{\pi,\vartheta}^m(T_{A_\alpha}^\vartheta). \tag{3.118}$$

(ii) *If threshold $B = B_\alpha$ in the SR rule is so selected that $\mathsf{PFA}_\pi(\widetilde{T}_{B_\alpha}^\vartheta) \le \alpha$ and $\log B_\alpha \sim |\log \alpha|$ as $\alpha \to 0$, e.g., as $B_\alpha = (\omega b + \bar{\nu})/\alpha$, then for all $0 < m \le r$ as $\alpha \to 0$*

$$\mathscr{R}_{k,\vartheta}^m(\widetilde{T}_{B_\alpha}^\vartheta) \sim \left(\frac{|\log \alpha|}{I_\vartheta}\right)^m \quad \textit{for all } k \in \mathbb{Z}_+ \tag{3.119}$$

and

$$\overline{\mathscr{R}}_{\pi,\vartheta}^m(\widetilde{T}_{B_\alpha}^\vartheta) \sim \left(\frac{|\log \alpha|}{I_\vartheta}\right)^m. \tag{3.120}$$

Therefore, the SR rule $\widetilde{T}_{B_\alpha}^\vartheta$ is first-order asymptotically optimal as $\alpha \to 0$ in class $\mathbb{C}_\pi(\alpha)$, minimizing moments of the detection delay up to order r when the prior distribution π is heavy-tailed, i.e., when π belongs to class $\mathbf{C}(\mu = 0)$.

Theorem 3.8. *Let $r \ge 1$. Suppose that for $\theta = \vartheta$ and some $0 < I_\vartheta < \infty$ the right-tail condition $\mathbf{C}_1$ is satisfied and the left-tail condition $\widetilde{\mathbf{C}}_2$ holds as well.*

(i) *Assume that the prior distribution $\pi^\alpha = \{\pi_k^\alpha\}$ of the change point ν satisfies condition (3.20) with $\mu = \mu_\alpha \to 0$ as $\alpha \to 0$ and that μ_α approaches zero at such rate that*

$$\lim_{\alpha \to 0} \frac{\sum_{k=0}^{\infty} \pi_k^\alpha |\log \pi_k^\alpha|^r}{|\log \alpha|^r} = 0. \tag{3.121}$$

If $A = A_\alpha$ is so selected that $\mathsf{PFA}_\pi(T_{A_\alpha}) \le \alpha$ and $\log A_\alpha \sim |\log\alpha|$ as $\alpha \to 0$, e.g., as $A_\alpha = (1-\alpha)/\alpha$, then the Shiryaev rule $T^\vartheta_{A_\alpha}$ is asymptotically optimal as $\alpha \to 0$ in class $\mathbb{C}(\alpha)$, minimizing moments of the detection delay up to order r: for all $0 < m \le r$ as $\alpha \to 0$

$$\inf_{T\in\mathbb{C}(\alpha)} \mathscr{R}^m_{k,\vartheta}(T) \sim \left(\frac{|\log\alpha|}{I_\vartheta}\right)^m \sim \mathscr{R}^m_{k,\vartheta}(T^\vartheta_{A_\alpha}) \quad \text{for all } k \in \mathbb{Z}_+, \tag{3.122}$$

and

$$\inf_{T\in\mathbb{C}(\alpha)} \overline{\mathscr{R}}^m_{\pi^\alpha,\vartheta}(T) \sim \left(\frac{|\log\alpha|}{I_\vartheta}\right)^m \sim \overline{\mathscr{R}}^m_{\pi^\alpha,\vartheta}(T^\vartheta_{A_\alpha}) \tag{3.123}$$

(ii) *Let $\omega = \omega_\alpha$ and $\bar\nu = \bar\nu_\alpha$ go to infinity as $\alpha \to 0$ with such rate that*

$$\lim_{\alpha\to 0} \frac{\log(\omega_\alpha + \bar\nu_\alpha)}{|\log\alpha|} = 0. \tag{3.124}$$

If threshold $B = B_\alpha$ in the SR rule is so selected that $\mathsf{PFA}_\pi(\widetilde{T}^\vartheta_{B_\alpha}) \le \alpha$ and $\log B_\alpha \sim |\log\alpha|$ as $\alpha \to 0$, e.g., as $B = (\omega_\alpha b_\alpha + \bar\nu_\alpha)/\alpha$, then the SR rule $\widetilde{T}^\vartheta_{B_\alpha}$ is asymptotically optimal as $\alpha \to 0$ in class $\mathbb{C}(\alpha)$, minimizing moments of the detection delay up to order r: for all $0 < m \le r$ as $\alpha \to 0$

$$\inf_{T\in\mathbb{C}(\alpha)} \mathscr{R}^m_{k,\vartheta}(T) \sim \left(\frac{|\log\alpha|}{I_\vartheta}\right)^m \sim \mathscr{R}^m_{k,\vartheta}(\widetilde{T}^\vartheta_{B_\alpha}) \quad \text{for all } k \in \mathbb{Z}_+ \tag{3.125}$$

and

$$\inf_{T\in\mathbb{C}(\alpha)} \overline{\mathscr{R}}^m_{\pi^\alpha,\vartheta}(T) \sim \left(\frac{|\log\alpha|}{I_\vartheta}\right)^m \sim \overline{\mathscr{R}}^m_{\pi,\vartheta}(\widetilde{T}^\vartheta_{B_\alpha}). \tag{3.126}$$

The full proofs may be found in Tartakovsky [160]. In fact, Theorems 3.7 and 3.8 follow from Theorems 3.2–3.4 in an almost obvious manner.

Remark 3.1. Theorems 3.7 and 3.8 imply that the Shiryaev procedure T^ϑ_A is asymptotically optimal whenever the LLR converges to a constant I_ϑ uniformly r-completely, i.e., when for some $0 < I_\vartheta < \infty$ and all $\varepsilon > 0$

$$\sum_{n=1}^\infty n^{r-1} \sup_{k\in\mathbb{Z}_+} \mathsf{P}_{k,\vartheta}\left(\left|\frac{1}{n}\lambda_\vartheta(k,k+n) - I_\vartheta\right| > \varepsilon\right) < \infty. \tag{3.127}$$

Indeed, condition (3.127) implies both conditions $\mathbf{C}_1$ (for $\theta = \vartheta$) and $\widetilde{\mathbf{C}}_2$.

3.8 The Case of Independent Observations

We now study the case where observations are independent, but not necessarily identically distributed, i.e., $g_i(X_i|\mathbf{X}_0^{i-1}) = g_i(X_i)$ and $f_{\theta,i}(X_i|\mathbf{X}_0^{i-1}) = f_{\theta,i}(X_i)$ in (3.1). More generally, we may assume that the increments $\Delta\lambda_\theta(i)$ of the LLR $\lambda_\theta(k,n) = \sum_{i=k+1}^n \Delta\lambda_\theta(i)$ are independent, which is always the case if the observations are independent. This slight generalization is important for certain examples with dependent observations that lead to the LLR with independent increments. See Example 3.5.

The results of previous sections show that the lower bounds (3.31), (3.32) for moments of the detection delay hold whenever the LLR process $\lambda_\theta(k,k+n)$ obeys the SLLN (3.27), since in this case condition $\mathbf{C}_1$ is satisfied. However, in general, almost sure convergence (3.27) is not sufficient for obtaining the upper bounds, and therefore, for asymptotic optimality of the MS procedure. In fact, this condition does not even guarantee finiteness of the average delay to detection

$\mathsf{E}_\theta^\pi[T_A^W - \nu | T_A^W > \nu]$, and to obtain meaningful results we need to strengthen the SLLN into the left-tail r-complete version **C2**. However, in the case of independent observations, sufficient conditions for asymptotic optimality can be substantially relaxed. Specifically, the right-tail condition $\mathbf{C}_1$ and the following left-tail condition

$$\lim_{n\to\infty}\lim_{\delta\to 0} \mathsf{P}_{k,\theta}\left(\frac{1}{n}\int_{\Gamma_{\theta,\delta}} \lambda_\vartheta(\ell,\ell+n)\,\mathrm{d}W(\vartheta) < I_\theta - \varepsilon\right) = 0 \tag{3.128}$$
$$\text{for all } \ell \ge k, \text{ all } k \ge 0, \text{ and all } \varepsilon > 0$$

turn out to be sufficient for asymptotic optimality with respect to all positive moments of the detection delay.

The idea of relaxing condition $\mathbf{C}_2$ by condition (3.128) is based on splitting integration, when obtaining the upper bound for the expectation $\mathsf{E}_{k,\theta}[(T_A^W - k)^+]^r$, into a sequence of intervals (cycles) of the size $N_A \approx \log A/(I_\theta + \mu)$ and then showing that $\mathsf{P}_{k,\theta}(T_A^W - k > \ell N_A) \le \delta^\ell$, $\ell = 1,2,\dots$ for some small δ under condition (3.128), using independence of the LLR increments.

Theorem 3.9. *Assume that the LLR process* $\{\lambda_\theta(k,k+n)\}_{n\ge 1}$ *has independent, not necessarily identically distributed increments under* $\mathsf{P}_{k,\theta}$, $k \in \mathbb{Z}_+$. *Suppose that the right-tail condition* $\mathbf{C}_1$ *and the left-tail condition* (3.128) *are satisfied.*

(i) *Let the prior distribution of the change point be* Geometric(ρ). *Then relations* (3.62) *and* (3.63) *in Theorem 3.2(ii) for moments of the detection delay of the MS rule hold true for all* $m > 0$ *with* $\mu = |\log(1-\rho)|$. *Therefore, the MS rule* $T_{A_\alpha}^W$ *minimizes asymptotically as* $\alpha \to 0$ *all positive moments of the detection delay in class* $\mathbb{C}_\pi(\alpha)$. *Also, relations* (3.92) *and* (3.93) *in Theorem 3.2(ii) for moments of the detection delay of the MSR rule hold true for all* $m > 0$.

(ii) *Let the prior distribution be geometric with parameter* $\rho = \rho_\alpha$ *that depends on* α *and goes to zero as* $\alpha \to 0$ *at such a rate that*

$$|\log \rho_\alpha| = o(|\log \alpha|) \quad \text{as } \alpha \to 0. \tag{3.129}$$

Then relations (3.72) *and* (3.73) *in Theorem 3.3 for moments of the detection delay of the MS rule as well as relations* (3.95) *and* (3.96) *in Theorem 3.4 for moments of the detection delay of the MSR rule hold for all* $m > 0$. *Therefore, both detection rules* $T_{A_\alpha}^W$ *and* $\widetilde{T}_{B_\alpha}^W$ *are asymptotically optimal with respect to all positive moments of the detection delay.*

Proof. The proof of part (i). Introduce the following notation: $A_\rho = A/\rho$, $a_\rho = \log A_\rho - N_A\mu$, $\mu = |\log(1-\rho)|$,

$$N_A = N(A,\rho,\theta,\varepsilon) = 1 + \left\lfloor \frac{\log A_\rho}{I_\theta + \mu - \varepsilon} \right\rfloor.$$

Since under condition $\mathbf{C}_1$ the lower bounds (3.31) and (3.32) hold, in order to prove the assertion (i) for the MS rule, we need to prove the asymptotic (as $\alpha \to 0$) upper bounds (for any $r \ge 1$ under condition (3.128))

$$\mathscr{R}_{k,\theta}^r(T_{A_\alpha}^W) \le \left(\frac{|\log\alpha|}{I_\theta + \mu}\right)^r (1 + o(1)), \tag{3.130}$$

$$\mathscr{R}_{\pi,\theta}^r(T_{A_\alpha}^W) \le \left(\frac{|\log\alpha|}{I_\theta + \mu}\right)^r (1 + o(1)) \tag{3.131}$$

as long as $\log A_\alpha \sim |\log\alpha|$.

To this end, note that inequality (A.2) in Lemma A.1 (see Appendix A) yields the inequality

$$\mathsf{E}_{k,\theta}\left[(T_A^W - k)^+\right]^r \le N_A^r\left(1 + r2^{r-1}\sum_{\ell=1}^\infty \ell^{r-1}\mathsf{P}_{k,\theta}(T_A^W - k > \ell N_A)\right) \tag{3.132}$$

and that for the geometric prior we have

$$\widetilde{S}_W(n) := \rho^{-1} S_W(n) = \sum_{m=0}^{n-1} (1-\rho)^{m-n} \Lambda_W(m,n).$$

Consider the intervals (cycles) of the length N_A and let $K_n(k,N_A) = K_n = k + nN_A$ and

$$Z_{W,\theta}(K_{n-1},K_n) := \int_{\Gamma_{\delta,\theta}} \lambda_\vartheta(K_{n-1},K_n)\, \mathrm{d}W(\vartheta) = \sum_{i=K_{n-1}+1}^{K_n} \int_{\Gamma_{\delta,\theta}} \Delta\lambda_\vartheta(i)\, \mathrm{d}W(\vartheta).$$

Since for any $n \ge 1$

$$\log \widetilde{S}_W(K_n) \ge \mu N_A + \log \Lambda_W(K_{n-1},K_n) \ge \mu N_A + \log \int_{\Gamma_{\delta,\theta}} \exp\{\lambda_\vartheta(K_{n-1},K_n)\}\, \mathrm{d}W(\vartheta)$$

and, by Jensen's inequality,

$$\begin{aligned}\int_{\Gamma_{\delta,\theta}} \exp\{\lambda_\vartheta(K_{n-1},K_n)\}\, \mathrm{d}W(\vartheta) &\ge W(\Gamma_{\delta,\theta}) \exp\left\{\int_{\Gamma_{\delta,\theta}} \lambda_\vartheta(K_{n-1},K_n) \frac{\mathrm{d}W(\vartheta)}{W(\Gamma_{\delta,\theta})}\right\} \\ &= W(\Gamma_{\delta,\theta}) \exp\left\{\frac{Z_{W,\theta}(K_{n-1},K_n)}{W(\Gamma_{\delta,\theta})}\right\},\end{aligned}$$

we obtain

$$\begin{aligned}\mathsf{P}_{k,\theta}\left(T_A^W - k > \ell N_A\right) &= \mathsf{P}_{k,\theta}\left(\log \widetilde{S}_W(n) < \log A_\rho \text{ for } n = 1,\dots,k+\ell N_A\right) \\ &\le \mathsf{P}_{k,\theta}\left(\log \widetilde{S}_W(k+nN_A) < \log A_\rho \text{ for } n = 1,\dots,\ell\right) \\ &\le \mathsf{P}_{k,\theta}\left(\frac{Z_{W,\theta}(K_{n-1},K_n)}{W(\Gamma_{\delta,\theta})} < a_\rho - \log W(\Gamma_{\delta,\theta}) \text{ for } n = 1,\dots,\ell\right).\end{aligned}$$

Since the increments of the LLR are independent, the random variables $Z_{W,\theta}(K_{n-1},K_n)$, $n = 1,\dots,\ell$, are independent, and hence,

$$\begin{aligned}&\mathsf{P}_{k,\theta}\left(T_A^W - k > \ell N_A\right) \le \mathsf{P}_{k,\theta}\left(\frac{Z_{W,\theta}(K_{n-1},K_n)}{W(\Gamma_{\delta,\theta})} < a_\rho + |\log W(\Gamma_{\delta,\theta})| \text{ for } n = 1,\dots,\ell\right) \\ &= \prod_{n=1}^{\ell} \mathsf{P}_{k,\theta}\left(\frac{Z_{W,\theta}(K_{n-1},K_n)}{W(\Gamma_{\delta,\theta})} < a_\rho + |\log W(\Gamma_{\delta,\theta})|\right) \\ &\le \prod_{n=1}^{\ell} \mathsf{P}_{k,\theta}\left(\frac{1}{N_A W(\Gamma_{\delta,\theta})} Z_{W,\theta}(K_{n-1},K_n) < I_\theta - \varepsilon + |\log W(\Gamma_{\delta,\theta})|/N_A\right).\end{aligned}$$

Thus, for a large A such that $|\log W(\Gamma_{\delta,\theta})|/N_A < \varepsilon/2$, we obtain the inequality

$$\mathsf{P}_{k,\theta}\left(T_A^W - k > \ell N_A\right) \le \prod_{n=1}^{\ell} \mathsf{P}_{k,\theta}\left(\frac{1}{N_A W(\Gamma_{\delta,\theta})} \int_{\Gamma_{\delta,\theta}} \lambda_\vartheta(K_{n-1},K_n)\, \mathrm{d}W(\vartheta) < I_\theta - \varepsilon/2\right).$$

By condition (3.128), for a sufficiently large A there exists a small δ_A such that

$$\mathsf{P}_{k,\theta}\left(\frac{1}{N_A W(\Gamma_{\delta,\theta})} \int_{\Gamma_{\delta,\theta}} \lambda_\vartheta(K_{n-1},K_n)\, \mathrm{d}W(\vartheta) < I_\theta - \varepsilon/2\right) \le \delta_A, \quad n \ge 1.$$

Therefore, for any $\ell \geq 1$,

$$\mathsf{P}_{k,\theta}\left(T_A^W - k > \ell N_A\right) \leq \delta_A^\ell. \tag{3.133}$$

Combining this inequality with (3.132) and using the fact that $L_{r,A} = \sum_{\ell=1}^{\infty} \ell^{r-1} \delta_A^\ell \to 0$ as $A \to \infty$ for any $r > 0$, we obtain

$$\begin{aligned}
\mathscr{R}_{k,\theta}^r(T_{A_\alpha}^W) &= \frac{\mathsf{E}_{k,\theta}\left[(T_{A_\alpha}^W - k)^+\right]^r}{\mathsf{P}_\infty(T_{A_\alpha}^W > k)} \\
&\leq \frac{\left(1 + \frac{\log A_\alpha}{I_\theta + \mu - \varepsilon}\right)^r + r2^{r-1} L_{r,A_\alpha}}{1 - 1/[1-\rho)^{k-1} A_\alpha]} \\
&= \left(\frac{|\log \alpha|}{I_\theta + \mu - \varepsilon}\right)^r (1 + o(1)) \quad \text{as } \alpha \to 0.
\end{aligned} \tag{3.134}$$

Since $\varepsilon \in (0, I_\theta)$ is an arbitrarily number, this implies the upper bound (3.130), and hence, (3.62).

Next, using inequalities (3.132) and (3.133) and the fact that $1 - \mathsf{PFA}_\pi(A_\alpha) \geq 1 - \alpha$, we obtain

$$\begin{aligned}
\overline{\mathscr{R}}_{\pi,\theta}^r(T_{A_\alpha}^W) &= \frac{\sum_{k=1}^{\infty} \rho(1-\rho)^k \mathsf{E}_{k,\theta}\left[(T_{A_\alpha}^W - k)^+\right]^r}{1 - \mathsf{PFA}_\pi(A_\alpha)} \\
&\leq \frac{\left(1 + \frac{\log A_\alpha}{I_\theta + \mu - \varepsilon}\right)^r + r2^{r-1} L_{r,A_\alpha}}{1 - \alpha} \\
&= \left(\frac{|\log \alpha|}{I_\theta + \mu - \varepsilon}\right)^r (1 + o(1)) \quad \text{as } \alpha \to 0,
\end{aligned} \tag{3.135}$$

which implies the upper bound (3.131), and hence, (3.63).

The proof of asymptotic equalities (3.92) and (3.93) for the MSR rule is essentially analogous. It suffices to note that the inequality (3.132) holds for the MSR rule $\widetilde{T}_B$ with $N_B = 1 + \lfloor \log B/(I_\theta - \varepsilon) \rfloor$, i.e.,

$$\mathsf{E}_{k,\theta}\left[(\widetilde{T}_B^W - k)^+\right]^r \leq N_B^r \left(1 + r2^{r-1} \sum_{\ell=1}^{\infty} \ell^{r-1} \mathsf{P}_{k,\theta}(\widetilde{T}_B^W - k > \ell N_B)\right), \tag{3.136}$$

as well as that $R_W(n) = \lim_{\rho \to 0} \widetilde{S}_W(n) \geq \log \widetilde{S}_W(K_n) \geq \log \Lambda_W(K_{n-1}, K_n)$. So setting $a_\rho = \log B$ in previous inequalities for $\mathsf{P}_{k,\theta}(T_A^W - k > \ell N_A)$, similarly to (3.133) we obtain that $\mathsf{P}_{k,\theta}\left(\widetilde{T}_B^W - k > \ell N_B\right) \leq \delta_B^\ell$, where $\delta_B \to 0$ as $B \to \infty$. Since $\mathsf{P}_\infty(\widetilde{T}_{B_\alpha}^W > k) \geq 1 - k/B_\alpha$ and $\log B_\alpha \sim |\log \alpha|$, we obtain

$$\begin{aligned}
\mathscr{R}_{k,\theta}^r(\widetilde{T}_{B_\alpha}^W) &= \frac{\mathsf{E}_{k,\theta}\left[(\widetilde{T}_{B_\alpha}^W - k)^+\right]^r}{\mathsf{P}_\infty(\widetilde{T}_{B_\alpha}^W > k)} \\
&\leq \frac{\left(1 + \frac{\log B_\alpha}{I_\theta - \varepsilon}\right)^r + r2^{r-1} L_{r,B_\alpha}}{1 - k/B_\alpha} \\
&= \left(\frac{|\log \alpha|}{I_\theta - \varepsilon}\right)^r (1 + o(1)) \quad \text{as } \alpha \to 0,
\end{aligned} \tag{3.137}$$

which holds for an arbitrarily $\varepsilon \in (0, I_\theta)$. Thus, for any $r \geq 1$

$$\mathscr{R}_{k,\theta}^r(\widetilde{T}_{B_\alpha}^W) \leq \left(\frac{|\log \alpha|}{I_\theta}\right)^r (1 + o(1)) \quad \text{as } \alpha \to 0. \tag{3.138}$$

But according to the inequality (3.81) (with $B = B_\alpha$, $\log B_\alpha \sim |\log \alpha|$) under condition $\mathbf{C}_1$ the right hand-side is also the lower bound. This implies (3.92).

For the risk $\overline{\mathscr{R}}^r_{\pi,\theta}(\widetilde{T}^W_{B_\alpha})$, similarly to (3.135)

$$\overline{\mathscr{R}}^r_{\pi,\theta}(\widetilde{T}^W_{B_\alpha}) = \frac{\sum_{k=1}^{\infty} \rho(1-\rho)^k \mathsf{E}_{k,\theta}\left[(\widetilde{T}^W_{B_\alpha}-k)^+\right]^r}{1-\mathsf{PFA}_\pi(\widetilde{T}_{B_\alpha})}$$
$$\leq \frac{\left(1+\frac{\log B_\alpha}{I_\theta-\varepsilon}\right)^r + r2^{r-1} L_{r,B_\alpha}}{1-1/(\rho B_\alpha)}$$
$$= \left(\frac{|\log \alpha|}{I_\theta-\varepsilon}\right)^r (1+o(1)) \quad \text{as } \alpha \to 0, \tag{3.139}$$

where $\varepsilon \in (0, I_\theta)$ is arbitrarily, so that for all $r \geq 1$

$$\overline{\mathscr{R}}^r_{\pi,\theta}(\widetilde{T}^W_{B_\alpha}) \leq \left(\frac{|\log \alpha|}{I_\theta}\right)^r (1+o(1)) \quad \text{as } \alpha \to 0. \tag{3.140}$$

Applying this upper bound together with the lower bound (3.89) (which holds due to condition $\mathbf{C}_1$) proves (3.63).

Proof of part (ii). We provide the proof only for the MSR rule since for the MS rule it is practically the same. Asymptotic relations (3.96) follow immediately from the upper bound (3.140) and the lower bound (3.68) in Lemma 3.3 where we should set $\mu = 0$. Substituting $\rho = \rho_\alpha$ in (3.139) yields

$$\overline{\mathscr{R}}^r_{\pi^\alpha,\theta}(\widetilde{T}^W_{B_\alpha}) \leq \frac{\left(1+\frac{\log B_\alpha}{I_\theta-\varepsilon}\right)^r + r2^{r-1} L_{r,B_\alpha}}{1-1/(\rho_\alpha B_\alpha)} = \left(\frac{|\log \alpha|}{I_\theta-\varepsilon}\right)^r (1+o(1)) \quad \text{as } \alpha \to 0,$$

where we used the fact that $1/(\rho_\alpha B_\alpha) \to 0$ due to condition (3.129). Hence, the upper bound (3.140) for $\overline{\mathscr{R}}^r_{\pi^\alpha,\theta}(\widetilde{T}^W_{B_\alpha})$ holds, which along with the lower bound (3.69) in Lemma 3.3 (with $\mu = 0$) proves (3.95). □

Note that condition (3.128) is dominated by the condition

$$\lim_{n\to\infty} \lim_{\delta\to 0} \mathsf{P}_{k,\theta}\left(\frac{1}{n} \inf_{\vartheta\in\Gamma_{\theta,\delta}} \lambda_\vartheta(\ell,\ell+n)) < I_\theta - \varepsilon\right) = 0$$
$$\text{for all } \ell \geq k, \text{ all } k \geq 0, \text{ and all } \varepsilon > 0, \tag{3.141}$$

that is, all assertions of Theorem 3.9 hold under condition (3.141).

In the case of the simple post-change hypothesis when the parameter $\theta = \vartheta$ is known conditions (3.128) and (3.141) become

$$\lim_{n\to\infty} \mathsf{P}_{k,\vartheta}\left(\frac{1}{n}\lambda_\vartheta(\ell,\ell+n) < I_\vartheta - \varepsilon\right) = 0 \quad \text{for all } \varepsilon > 0, \text{ all } \ell \geq k \text{ and all } k \in \mathbb{Z}_+. \tag{3.142}$$

Since both conditions $\mathbf{C}_1$ (with $\theta = \vartheta$) and (3.142) are satisfied whenever $n^{-1}\lambda_\theta(k,k+n) \to I_\vartheta$ a.s. under $\mathsf{P}_{k,\vartheta}$ the SLLN is sufficient for asymptotic optimality of the Shiryaev procedure $T^\vartheta_{A_\alpha}$.

3.9 The i.i.d. Case

Let the post-change hypothesis be simple, i.e., $\theta = \vartheta$, where ϑ is known. In the i.i.d. case, where the increments $\{\Delta\lambda_t\}_{t>k}$ of the LLR $\lambda_\vartheta(k,k+n) = \sum_{t=k+1}^{n} \Delta\lambda_t$ are i.i.d. for all $k \in \mathbb{Z}_+$, i.e., the LLR is a random walk, both right-tail and left-tail conditions $\mathbf{C}_1$ and (3.142) are satisfied since by the SLLN $n^{-1}\lambda_\vartheta(0,n) \to I_\vartheta$ as $n \to \infty$ $\mathsf{P}_{0,\theta}$-a.s. Therefore, we have the following corollary of Theorem 3.9.

Corollary 3.1. *If the prior distribution is geometric and if the LLR $\{\lambda_\vartheta(k,k+n)\}_{n\in\mathbb{Z}_+}$ is a random walk with positive and finite drift $I_\vartheta = \mathsf{E}_{k,\vartheta}[\lambda_\vartheta(k,k+1)]$ under $\mathsf{P}_{k,\theta}$ for all $k \in \mathbb{Z}_+$, then the Shiryaev procedure $T^\vartheta_{A_\alpha}$ asymptotically (as $\alpha \to 0$) minimizes all positive moments of the detection delay in class $\mathbb{C}_\pi(\alpha)$. This is the case, in particular, when conditioned on $\nu = k$ the observations $X_1,\dots,X_k$ are i.i.d. with pre-change density $g(x)$ and $X_{k+1},X_{k+2},\dots$ are i.i.d. with post-change density $f_\vartheta(x)$, with $I_\vartheta = \mathscr{K}_\vartheta$ being the Kullback–Leibler information number*

$$\mathscr{K}_\vartheta = \mathsf{E}_{0,\vartheta}[\lambda_\vartheta(0,1)] = \int \log\left(\frac{f_\vartheta(x)}{g(x)}\right) f_\vartheta(x)\,\mathrm{d}x, \tag{3.143}$$

assuming that it is positive and finite.

Using the nonlinear renewal theory for perturbed random walks (see Appendix F) one can also derive higher order approximations for the PFA and the expected detection delay of the Shiryaev rule. Specifically, assume that the prior distribution of the change point is Geometric(q,ρ) and introduce the following additional notation: $S^\rho_\vartheta(n) = S^\pi_\vartheta(n)/\rho$, where $S^\pi_\vartheta(n)$ is defined in (3.2);

$$\lambda^\rho_n = \lambda_\vartheta(0,n) + |\log(1-\rho)|n = \sum_{t=1}^n \Delta\lambda^\rho_t, \quad \Delta\lambda^\rho_t = \Delta\lambda_t + |\log(1-\rho)|;$$

$$\tau_a = \inf\{n \ge 1 : \lambda^\rho_n \ge a\}. \tag{3.144}$$

Next, let $\kappa_a = \lambda^\rho_{\tau_a} - a$ on $\{\tau_a < \infty\}$ denote the overshoot, let

$$\mathsf{G}(y) = \mathsf{G}(y,\rho,\mathscr{K}_\vartheta) = \lim_{a\to\infty} \mathsf{P}_{0,\vartheta}\{\kappa_a \le y\} \tag{3.145}$$

denote the limiting distribution of the overshoot and let

$$\varkappa(\rho,\mathscr{K}_\vartheta) = \lim_{a\to\infty} \mathsf{E}_{0,\vartheta}[\kappa_a] = \int_0^\infty y\,\mathrm{d}\mathsf{G}(y)$$

denote the limiting average overshoot. Define also

$$\zeta(\rho,\mathscr{K}_\vartheta) = \lim_{a\to\infty} \mathsf{E}_{0,\vartheta}[e^{-\kappa_a}] = \int_0^\infty e^{-y}\,\mathrm{d}\mathsf{G}(y)$$

and

$$C(\rho,\mathscr{K}_\vartheta) = \mathsf{E}_0\left[\log\left(1 + S^\rho_\vartheta(0) + \sum_{t=1}^\infty e^{-\Delta\lambda^\rho_t}\right)\right], \tag{3.146}$$

where $S^\rho_\vartheta(0) = q/[(1-q)\rho]$.

Under assumptions that $\lambda^\rho(0,1)$ is nonarithmetic with respect to P_∞ and P_0 and that the Kullback–Leibler number is finite, $\mathscr{K}_\vartheta < \infty$, the following approximation for the PFA holds:

$$\mathsf{PFA}_\pi(T^\vartheta_A) = \frac{\zeta(\rho,\mathscr{K}_\vartheta)}{A}(1+o(1)) \quad \text{as } A \to \infty. \tag{3.147}$$

The proof of (3.147) is based on Theorem F.1, the First Nonlinear Renewal Theorem. If, in addition, the second moment $\mathsf{E}_0|\lambda(0,1)|^2$ is finite, then as $A \to \infty$

$$\mathsf{E}_0[T^\vartheta_A] = \frac{1}{\mathscr{K}_\vartheta + |\log(1-\rho)|}\left[\log(A/\rho) - C(\rho,\mathscr{K}_\vartheta) + \varkappa(\rho,\mathscr{K}_\vartheta)\right] + o(1). \tag{3.148}$$

The approximation (3.148) is established by using Theorem F.2, the Second Nonlinear Renewal Theorem. The details can be found in Tartakovsky et al. [164] (Theorem 7.1.5, page 327).

3.10 Window-Limited Change Detection Rules

In the general non-i.i.d. case, the computational complexity and memory requirements of the MS and MSR detection rules can be high when the post-change distributions depend on the point of change. In these cases, it is reasonable to exploit window-limited versions of mixture detection rules where the summation over potential change points k is restricted to a sliding window of size $\ell = \ell_1 - \ell_0$. The idea of using a window-limited generalized likelihood ratio procedure for stochastic dynamic systems described by linear state-space models belongs to Willsky and Jones [184], and a general (mostly minimax) quickest changepoint detection theory for window-limited CUSUM-type procedures based on the maximization over k restricted to $n - \ell_1 \le k \le n$ was developed by Lai [74, 75] who suggested a method of selection of ℓ_1 (depending on the given false alarm rate) to make the detection procedures asymptotically optimal. The role of ℓ_1 is to reduce the memory requirements and computational complexity of stopping rules. The values of ℓ_0 bigger than 0 can be used to protect against outliers, but $\ell_0 = 0$ is reasonable in most cases, which we assume in what follows.

To be more specific, in the window-limited versions of T_A^W and $\widetilde{T}_B^W$, the statistics $S_W(n)$ and $R_W(n)$ are replaced by the window-limited statistics

$$\widehat{S}_W(n) = S_W(n) \quad \text{for } n \le \ell; \quad \widehat{S}_W(n) = \frac{1}{\mathsf{P}(\nu \ge n)} \sum_{k=n-\ell}^{n-1} \pi_k \Lambda_W(k,n) \quad \text{for } n > \ell$$

and

$$\widehat{R}_W(n) = R_W(n) \quad \text{for } n \le \ell; \quad \widehat{R}_W(n) = \sum_{k=n-\ell}^{n-1} \Lambda_W(k,n) \quad \text{for } n > \ell.$$

Thus, the window-limited MS and MSR rules are defined as

$$T_A^{W,\ell} = \inf\left\{n \ge 1 : \widehat{S}_W(n) \ge A\right\}, \tag{3.149}$$

$$\widetilde{T}_B^{W,\ell} = \inf\left\{n \ge 1 : \widehat{R}_W(n) \ge B\right\}, \tag{3.150}$$

respectively.

Obviously, $T_A^W \le T_A^{W,\ell}$ and $\widetilde{T}_B^W \le \widetilde{T}_B^{W,\ell}$, so applying Lemmas 3.1 and 3.4, we obtain

$$\mathsf{PFA}_\pi(T_A^{W,\ell}) \le \frac{1}{A+1}, \quad \mathsf{PFA}(\widetilde{T}_B^{W,\ell}) \le \frac{\omega b + \bar{\nu}}{B}. \tag{3.151}$$

Thus, again

$$A = A_\alpha = \frac{1-\alpha}{\alpha} \quad \text{implies} \quad \mathsf{PFA}(T_{A_\alpha}^{W,\ell}) \le \alpha$$

and

$$B = B_\alpha = \frac{\omega b + \bar{\nu}}{\alpha} \quad \text{implies} \quad \mathsf{PFA}(\widetilde{T}_{B_\alpha}^{W,\ell}) \le \alpha.$$

Obviously, in order to preserve asymptotic optimality properties these window-limited versions have to exploit relatively large sizes of the windows ℓ. Specifically, $\ell = \ell(\alpha)$ must approach infinity as $\alpha \to 0$ at a certain rate. We now show that the window-limited MS and MSR detection rules are first-order asymptotically optimal if

$$\lim_{\alpha \to 0} \frac{\ell(\alpha)}{|\log \alpha|} = \infty.$$

For the sake of simplicity, we will consider the geometric prior distribution Geometric(ρ),

$$\Pr(\nu = k) = \pi_k = \rho(1-\rho)^k \quad \text{for } k = 0, 1, 2, \dots,$$

since a proof for an arbitrarily prior is much more technical. Write $\mu = |\log(1-\rho)|$.

Theorem 3.10. *Let $r \geq 1$ and let the prior distribution of the change point be* Geometric(ρ). *Assume that for some $0 < I_\theta < \infty$, $\theta \in \Theta$, right-tail and left-tail conditions $\mathbf{C}_1$ and $\mathbf{C}_2$ are satisfied.*

(i) *Let $\ell = \ell_A$ be positive integers such that*

$$\lim_{A\to\infty} \frac{\ell_A}{\log A} = \infty. \tag{3.152}$$

Then, for all $0 < m \leq r$ and all $\theta \in \Theta$ as $A \to \infty$

$$\mathscr{R}^m_{k,\theta}(T_A^{W,\ell}) \sim \left(\frac{\log A}{I_\theta + \mu}\right)^m \quad \textit{for all } k \in \mathbb{Z}_+ \tag{3.153}$$

and

$$\overline{\mathscr{R}}^m_{\pi,\theta}(T_A^{W,\ell}) \sim \left(\frac{\log A}{I_\theta + \mu}\right)^m. \tag{3.154}$$

(ii) *Let $\ell = \ell_\alpha$ be positive integers such that*

$$\lim_{\alpha\to 0} \frac{\ell_\alpha}{|\log \alpha|} = \infty. \tag{3.155}$$

If $A = A_\alpha$ is so selected that $\mathsf{PFA}_\pi(T_{A_\alpha}^{W,\ell}) \leq \alpha$ and $\log A_\alpha \sim |\log \alpha|$ as $\alpha \to 0$, in particular as $A = A_\alpha = (1-\alpha)/\alpha$, then $T_{A_\alpha}^{W,\ell}$ is first-order asymptotically optimal as $\alpha \to 0$ in class $\mathbb{C}_\pi(\alpha)$, minimizing moments of the detection delay up to order r: for all $0 < m \leq r$ and all $\theta \in \Theta$ as $\alpha \to 0$

$$\inf_{T\in\mathbb{C}_\pi(\alpha)} \mathscr{R}^m_{k,\theta}(T) \sim \left(\frac{|\log \alpha|}{I_\theta + \mu}\right)^m \sim \mathscr{R}^m_{k,\theta}(T_{A_\alpha}^{W,\ell}) \quad \textit{for all } k \in \mathbb{Z}_+ \tag{3.156}$$

and

$$\inf_{T\in\mathbb{C}_\pi(\alpha)} \overline{\mathscr{R}}^m_{\pi,\theta}(T) \sim \left(\frac{|\log \alpha|}{I_\theta + \mu}\right)^m \sim \overline{\mathscr{R}}^m_{\pi,\theta}(T_{A_\alpha}^{W,\ell}). \tag{3.157}$$

Proof. The proof of part (i). For any $0 < \varepsilon < I_\theta + \mu$, let

$$N_A = N_A(\theta, \varepsilon, \rho) = 1 + \left\lfloor \frac{\log A}{I_\theta + \mu - \varepsilon} \right\rfloor.$$

Since $T_A^W \leq T_A^{W,\ell}$ and by Proposition 3.1 under right-tail and left-tail conditions $\mathbf{C}_1$ and $\mathbf{C}_2$ for all $\theta \in \Theta$ as $A \to \infty$

$$\mathscr{R}^r_{k,\theta}(T_A^W) \sim \left(\frac{\log A}{I_\theta + \mu}\right)^r \quad \text{for all } k \in \mathbb{Z}_+; \quad \overline{\mathscr{R}}^r_{\pi,\theta}(T_A^W) \sim \left(\frac{\log A}{I_\theta + \mu}\right)^r,$$

it suffices to show that

$$\limsup_{A\to\infty} \frac{\mathscr{R}^r_{k,\theta}(T_A^{W,\ell})}{\log A} = \left(\frac{1}{I_\theta + \mu}\right)^r \quad \text{for all } k \in \mathbb{Z}_+, \tag{3.158}$$

$$\limsup_{A\to\infty} \frac{\overline{\mathscr{R}}^r_{\pi,\theta}(T_A^{W,\ell})}{\log A} = \left(\frac{1}{I_\theta+\mu}\right)^r. \tag{3.159}$$

Consider a multicyclic scheme with cycles of the length N_A. Let $K_n(k,N_A) = k + nN_A$ and

$$Z_{W,\theta}(K_{n-1},K_n) = \int_{\Gamma_{\delta,\theta}} \lambda_\vartheta(K_{n-1},K_n)\, \mathrm{d}W(\vartheta).$$

Since by Jensen's inequality,

$$\begin{aligned}
\log\left[\int_{\Gamma_{\delta,\theta}} \exp\{\lambda_\vartheta(K_{n-1},K_n)\}\, \mathrm{d}W(\vartheta)\right] &= \log\left[W(\Gamma_{\delta,\theta}\int_{\Gamma_{\delta,\theta}} \exp\{\lambda_\vartheta(K_{n-1},K_n)\}\frac{\mathrm{d}W(\vartheta)}{W(\Gamma_{\delta,\theta})}\right] \\
&\ge \log\left[W(\Gamma_{\delta,\theta})\exp\left\{\frac{\int_{\Gamma_{\delta,\theta}}\lambda_\vartheta(K_{n-1},K_n)\,\mathrm{d}W(\vartheta)}{W(\Gamma_{\delta,\theta})}\right\}\right] \\
&= \frac{Z_{W,\theta}(K_{n-1},K_n)}{W(\Gamma_{\delta,\theta})} - |\log W(\Gamma_{\delta,\theta})|,
\end{aligned}$$

and for any $n \ge 1$ and $\ell \ge N_A$

$$\begin{aligned}
\widehat{S}_W(n) &\ge \frac{\pi_{n-N_A}}{\mathsf{P}(\nu\ge n)}\Lambda_W(n-N_A,n) = \rho(1-\rho)^{-N_A}\Lambda_W(n-N_A,n) \\
&\ge \rho(1-\rho)^{-N_A}\int_{\Gamma_{\delta,\theta}} \exp\{\lambda_\vartheta(n-N_A,n)\}\, \mathrm{d}W(\vartheta),
\end{aligned}$$

it follows that

$$\log\widehat{S}_W(k+nN_A) \ge \log\rho + N_A|\log(1-\rho)| - |\log W(\Gamma_{\delta,\theta})| + \frac{Z_{W,\theta}(K_{n-1},K_n)}{W(\Gamma_{\delta,\theta})}.$$

Therefore, we obtain that for $\ell = \ell_A > N_A(1+o(1))$

$$\begin{aligned}
\mathsf{P}_{k,\theta}\left(T_A^{W,\ell} - k > iN_A\right) &= \mathsf{P}_{k,\theta}\left(\log\widehat{S}_W(n) < \log A_\rho \text{ for } n = 1,\dots,k+iN_A\right) \\
&\le \mathsf{P}_{k,\theta}\left(\log\widehat{S}_W(k+nN_A) < \log A_\rho \text{ for } n = 1,\dots,i\right) \\
&\le \mathsf{P}_{k,\theta}\left(\frac{Z_{W,\theta}(K_{n-1},K_n)}{W(\Gamma_{\delta,\theta})} < a_\rho + |\log W(\Gamma_{\delta,\theta})| \text{ for } n = 1,\dots,i\right),
\end{aligned}$$

where $A_\rho = A/\rho$, $a_\rho = \log A_\rho - \mu N_A$. Clearly, for a sufficiently large N_A

$$\begin{aligned}
&\mathsf{P}_{k,\theta}\left(\frac{Z_{W,\theta}(K_{n-1},K_n)}{W(\Gamma_{\delta,\theta})} < a_\rho + |\log W(\Gamma_{\delta,\theta})| \text{ for } n = 1,\dots,i\right) \\
&\le \mathsf{P}_{k,\theta}\left(\frac{1}{N_A W(\Gamma_{\delta,\theta})} Z_{W,\theta}(K_{n-1},K_n) < I_\theta - \varepsilon + \frac{|\log W(\Gamma_{\delta,\theta})|\rho}{N_A} \text{ for } n = 1,\dots,i\right) \\
&\le \mathsf{P}_{k,\theta}\left(\frac{1}{N_A}\inf_{\vartheta\in\Gamma_{\theta,\delta}} \lambda_\vartheta(K_{n-1},K_n) < I_\theta - \varepsilon + \frac{|\log W(\Gamma_{\delta,\theta})|\rho}{N_A} \text{ for } n = 1,\dots,i\right) \\
&\le \mathsf{P}_{k,\theta}\left(\frac{1}{N_A}\inf_{\vartheta\in\Gamma_{\theta,\delta}} \lambda_\vartheta(K_{i-1},K_i) < I_\theta - \varepsilon + \frac{|\log W(\Gamma_{\delta,\theta})|\rho}{N_A}\right)
\end{aligned}$$

and, hence, for a sufficiently large A such that $|\log\rho W(\Gamma_{\delta,\theta})|/N_A \le \varepsilon/2$ and any $\varepsilon > 0$

$$\mathsf{P}_{k,\theta}\left(T_A^{W,\ell} - k > iN_A\right) \le \mathsf{P}_{k,\theta}\left(\frac{1}{N_A}\inf_{\vartheta\in\Gamma_{\theta,\delta}} \lambda_\vartheta(K_{i-1},K_i) < I_\theta - \varepsilon/2\right). \tag{3.160}$$

Next, since by condition (3.152), $\ell_A > N_A(1+o(1))$, setting $\tau = T_A^{W,\ell} - k > iN_A$ and $N = N_A$ in inequality (A.2) in Lemma A.1 (see Appendix A) yields (for any $k \in \mathbb{Z}_+$)

$$\mathsf{E}_{k,\theta}\left[(T_A^{W,\ell} - k)^+\right]^r \le N_A^r\left[1 + r2^{r-1}\sum_{i=1}^{\infty} i^{r-1}\mathsf{P}_{k,\theta}\left(T_A^{W,\ell} - k > i\right)\right]. \tag{3.161}$$

Inequalities (3.160) and (3.161) imply

$$\begin{aligned}\mathsf{E}_{k,\theta}\left[(T_A^{W,\ell} - k)^+\right]^r &\le N_A^r\left[1 + r2^{r-1}\sum_{i=1}^{\infty} i^{r-1}\mathsf{P}_{k,\theta}\left(\frac{1}{N_A}\inf_{\vartheta\in\Gamma_{\theta,\delta}}\lambda_\vartheta(K_{i-1},K_i) < I_\theta - \varepsilon/2\right)\right]\\ &\le \left(1 + \frac{\log A}{I_\theta + \mu - \varepsilon}\right)^r\left[1 + r2^{r-1}\Upsilon_{k,r}(\theta,\varepsilon/2)\right].\end{aligned} \tag{3.162}$$

Using this inequality and the facts that

$$\mathsf{P}_\infty(T_A^{W,\ell} > k) \ge \mathsf{P}_\infty(T_A^{W} > k) \ge 1 - [(1-\rho)^{k-1}A]^{-1} \tag{3.163}$$

and that, by condition $\mathbf{C}_2$, $\Upsilon_{k,r}(\theta,\varepsilon/2) < \infty$ for all $k \in \mathbb{Z}_+$, we obtain

$$\begin{aligned}\mathscr{R}_{k,\theta}^r(T_A^{W,\ell}) &= \frac{\mathsf{E}_{k,\theta}\left[(T_A^{W,\ell} - k)^+\right]^r}{\mathsf{P}_\infty(T_A^{W,\ell} > k)}\\ &\le \frac{\left(1 + \frac{\log A}{I_\theta + \mu - \varepsilon}\right)^r\left[1 + r2^{r-1}\Upsilon_{k,r}(\theta,\varepsilon/2)\right]}{1 - 1/[(1-\rho)^{k-1}A]}\\ &= \left(\frac{\log A}{I_\theta + \mu - \varepsilon}\right)^r(1+o(1)) \quad \text{as } A \to \infty.\end{aligned} \tag{3.164}$$

Since $\varepsilon \in (0, I_\theta)$ is an arbitrarily number, this implies the asymptotic upper bound (3.158), and hence, (3.153).

Now, using inequalities (3.162) and $1 - \mathsf{PFA}_\pi(T_A^{W,\ell}) \ge A/(A+1)$ (see (3.151)), we obtain

$$\begin{aligned}\overline{\mathscr{R}}_{\pi,\theta}^r(T_A^{W,\ell}) &= \frac{\sum_{k=0}^{\infty}\rho(1-\rho)^k\mathsf{E}_{k,\theta}\left[(T_A^{W,\ell} - k)^+\right]^r}{1 - \mathsf{PFA}_\pi(T_A^{W,\ell})}\\ &\le \frac{\left(1 + \frac{\log A}{I_\theta + \mu - \varepsilon}\right)^r\left[1 + r2^{r-1}\sum_{k=0}^{\infty}\rho(1-\rho)^k\Upsilon_{k,r}(\theta,\varepsilon/2)\right]}{A/(A+1)}\\ &= \left(\frac{\log A}{I_\theta + \mu - \varepsilon}\right)^r(1+o(1)) \quad \text{as } \alpha \to 0,\end{aligned} \tag{3.165}$$

where $\varepsilon \in (0, I_\theta + \mu)$ is an arbitrarily number. This implies the upper bound (3.159), and hence, (3.154).

The proof of part (ii). Selecting in (3.153) and (3.154) A_α such that $\log A_\alpha \sim |\log\alpha|$ and $\mathsf{PFA}_\pi(T_{A_\alpha}^{W,\ell}) \le \alpha$ (in particular, we may take $A_\alpha = (1-\alpha)/\alpha$) we obtain

$$\mathscr{R}_{k,\theta}^r(T_{A_\alpha}^{W,\ell}) \sim \left(\frac{|\log\alpha|}{I_\theta + \mu}\right)^r, \quad \overline{\mathscr{R}}_{\pi,\theta}^r(T_{A_\alpha}^{W,\ell}) \sim \left(\frac{|\log\alpha|}{I_\theta + \mu}\right)^r, \tag{3.166}$$

which along with the lower bounds (3.31) and (3.32) in Theorem 3.1 completes the proof of (3.156) and (3.157). □

A similar argument shows that for the window-limited MSR rule $\widetilde{T}_{B_\alpha}^{W,\ell}$ asymptotic approximations (3.166) hold with $\mu = 0$ if threshold $B_\alpha = 1/(\alpha\rho)$ or more generally if $\log B_\alpha \sim |\log\alpha|$ as $\alpha \to 0$ and $\mathsf{PFA}_\pi(\widetilde{T}_{B_\alpha}^{W,\ell}) \le \alpha$. Thus, as the MSR rule, the window-limited MSR rule is not optimal when $\mu = |\log(1-\rho)| > 0$ but it is asymptotically optimal when $\rho \to 0$. The following theorem establishes sufficient conditions for asymptotic optimality of the window-limited MSR rule when $\rho = \rho_\alpha \to 0$ as $\alpha \to 0$.

Theorem 3.11. *Let $r \ge 1$ and let the prior distribution $\pi = \pi^\alpha$ of the change point be* Geometric(ρ) *with the parameter $\rho = \rho_\alpha$ that goes to 0 with such a rate that*

$$\lim_{\alpha\to 0} \frac{\log \rho_\alpha}{\log \alpha} = 0. \tag{3.167}$$

Assume that for some $0 < I_\theta < \infty$, $\theta \in \Theta$, the right-tail condition $\mathbf{C}_1$ and the uniform left-tail condition $\mathbf{C}_3$ are satisfied. Further, let the size of the window $\ell = \ell_\alpha$ be such that

$$\lim_{\alpha\to 0} \frac{\ell_\alpha}{|\log \alpha|} = \infty. \tag{3.168}$$

(i) *If threshold $A = A_\alpha$ in the window-limited MS rule $T_A^{W,\ell}$ is so selected that $\mathsf{PFA}_\pi(T_{A_\alpha}^{W,\ell}) \le \alpha$ and $\log A_\alpha \sim |\log\alpha|$ as $\alpha \to 0$, in particular $A_\alpha = (1-\alpha)/\alpha$, then $T_{A_\alpha}^{W,\ell}$ is first-order asymptotically optimal as $\alpha \to 0$ in class $\mathbb{C}_{\pi^\alpha}(\alpha) = \mathbb{C}(\alpha)$, minimizing moments of the detection delay up to order r, i.e., for all $0 < m \le r$ and all $\theta \in \Theta$ as $\alpha \to 0$*

$$\inf_{T\in\mathbb{C}(\alpha)} \mathscr{R}_{k,\theta}^m(T) \sim \left(\frac{|\log\alpha|}{I_\theta}\right)^m \sim \mathscr{R}_{k,\theta}^m(T_{A_\alpha}^{W,\ell}) \quad \text{for all } k \in \mathbb{Z}_+ \tag{3.169}$$

and

$$\inf_{T\in\mathbb{C}(\alpha)} \overline{\mathscr{R}}_{\pi^\alpha,\theta}^m(T) \sim \left(\frac{|\log\alpha|}{I_\theta}\right)^m \sim \overline{\mathscr{R}}_{\pi^\alpha,\theta}^m(T_{A_\alpha}^{W,\ell}). \tag{3.170}$$

(ii) *If threshold $B = B_\alpha$ in the window-limited MSR rule $\widetilde{T}_B^{W,\ell}$ is so selected that $\mathsf{PFA}_{\pi^\alpha}(\widetilde{T}_{B_\alpha}^{W,\ell}) \le \alpha$ and $\log B_\alpha \sim |\log\alpha|$ as $\alpha \to 0$, in particular $B_\alpha = 1/(\rho_\alpha\alpha)$, then $\widetilde{T}_{B_\alpha}^{W,\ell}$ is first-order asymptotically optimal as $\alpha \to 0$ in class $\mathbb{C}(\alpha)$, minimizing moments of the detection delay up to order r, i.e., for all $0 < m \le r$ and all $\theta \in \Theta$ as $\alpha \to 0$*

$$\inf_{T\in\mathbb{C}(\alpha)} \mathscr{R}_{k,\theta}^m(T) \sim \left(\frac{|\log\alpha|}{I_\theta}\right)^m \sim \mathscr{R}_{k,\theta}^m(\widetilde{T}_{B_\alpha}^{W,\ell}) \quad \text{for all } k \in \mathbb{Z}_+ \tag{3.171}$$

and

$$\inf_{T\in\mathbb{C}(\alpha)} \overline{\mathscr{R}}_{\pi^\alpha,\theta}^m(T) \sim \left(\frac{|\log\alpha|}{I_\theta}\right)^m \sim \overline{\mathscr{R}}_{\pi^\alpha,\theta}^m(\widetilde{T}_{B_\alpha}^{W,\ell}). \tag{3.172}$$

Proof. The proof of (i). Substituting $\mu = |\log(1-\rho_\alpha)|$ and $A = A_\alpha$ in (3.162) yields the inequality

$$\mathsf{E}_{k,\theta}\left[(T_{A_\alpha}^{W,\ell} - k)^+\right]^r \le \left(1 + \frac{\log A_\alpha}{I_\theta + |\log(1-\rho_\alpha)| - \varepsilon}\right)^r \left[1 + r2^{r-1} \sup_{j\in\mathbb{Z}_+} \Upsilon_{j,r}(\theta, \varepsilon/2)\right]. \tag{3.173}$$

Since by (3.163),

$$\mathsf{P}_\infty(T_A^{W,\ell} > k) \ge 1 - \frac{1}{(1-\rho_\alpha)^{k-1} A_\alpha},$$

since $\log A_\alpha \sim |\log\alpha|$ and, by condition $\mathbf{C}_3$, $\sup_{k\in\mathbb{Z}_+} \Upsilon_{k,r}(\theta, \varepsilon/2) < \infty$, we obtain

$$\mathscr{R}_{k,\theta}^r(T_A^{W,\ell}) = \frac{\mathsf{E}_{k,\theta}\left[(T_{A_\alpha}^{W,\ell} - k)^+\right]^r}{\mathsf{P}_\infty(T_{A_\alpha}^{W,\ell} > k)}$$

$$\leq \frac{\left(1+\frac{\log A_\alpha}{I_\theta+|\log(1-\rho_\alpha)|-\varepsilon}\right)^r \left[1+r2^{r-1}\sup_{j\in\mathbb{Z}_+}\Upsilon_{j,r}(\theta,\varepsilon/2)\right]}{1-1/[(1-\rho_\alpha)^{k-1}A_\alpha]}$$
$$= \left(\frac{|\log\alpha|}{I_\theta-\varepsilon}\right)^r (1+o(1)) \quad \text{as } \alpha\to 0.$$

Since $\varepsilon \in (0, I_\theta)$ can be arbitrarily small, this implies the asymptotic upper bound

$$\mathscr{R}^r_{k,\theta}(T^{W,\ell}_{A_\alpha}) \leq \left(\frac{|\log\alpha|}{I_\theta}\right)^r (1+o(1)) \quad \text{as } \alpha\to 0,$$

which along with the lower bound (3.68) (with $\mu=0$) in Lemma 3.3 proves (3.169).

Next, using inequalities (3.173) and $1-\mathsf{PFA}_{\pi^\alpha}(T^{W,\ell}_{A_\alpha}) \geq 1-\alpha$, we obtain

$$\overline{\mathscr{R}}^r_{\pi^\alpha,\theta}(T^{W,\ell}_{A_\alpha}) = \frac{\sum_{k=0}^\infty \rho_\alpha(1-\rho_\alpha)^k \mathsf{E}_{k,\theta}\left[(T^{W,\ell}_{A_\alpha}-k)^+\right]^r}{1-\mathsf{PFA}_{\pi^\alpha}(T^{W,\ell}_{A_\alpha})}$$
$$\leq \frac{\left(1+\frac{\log A_\alpha}{I_\theta+|\log(1-\rho_\alpha)|-\varepsilon}\right)^r \left[1+r2^{r-1}\sup_{k\in\mathbb{Z}_+}\Upsilon_{k,r}(\theta,\varepsilon/2)\right]}{1-\alpha}$$
$$= \left(\frac{|\log\alpha|}{I_\theta-\varepsilon}\right)^r (1+o(1)) \quad \text{as } \alpha\to 0,$$

where $\varepsilon \in (0, I_\theta)$ is an arbitrarily number. This upper bound along with the lower bound (3.69) (with $\mu=0$) in Lemma 3.3 proves (3.170).

The proof of part (ii). Write

$$N_B = N_B(\theta,\varepsilon) = 1+\left\lfloor \frac{\log B}{I_\theta-\varepsilon}\right\rfloor.$$

Noting that $\widehat{R}_W(n) \geq \Lambda_W(n-\ell,n)$ for any $n\geq 1$ and applying essentially the same argument that has led to inequality (3.161), we obtain that for a sufficiently large B

$$\begin{aligned}\mathsf{E}_{k,\theta}\left[(\widetilde{T}^{W,\ell}_B-k)^+\right]^r &\leq N_B^r\left[1+r2^{r-1}\sum_{i=1}^\infty i^{r-1}\mathsf{P}_{k,\theta}\left(\widetilde{T}^{W,\ell}_B-k>i\right)\right] \\ &\leq N_B^r\left[1+r2^{r-1}\sum_{i=1}^\infty i^{r-1}\mathsf{P}_{k,\theta}\left(\frac{1}{N_B}\inf_{\vartheta\in\Gamma_{\theta,\delta}}\lambda_\vartheta(K_{i-1},K_i)<I_\theta-\varepsilon/2\right)\right] \\ &\leq \left(1+\frac{\log B}{I_\theta-\varepsilon}\right)^r\left[1+r2^{r-1}\sup_{j\in\mathbb{Z}_+}\Upsilon_{j,r}(\theta,\varepsilon/2)\right]. \end{aligned} \tag{3.174}$$

Since $\mathsf{P}_\infty(\widetilde{T}^{W,\ell}_{B_\alpha}>k) \geq 1-k/B_\alpha$ and $\log B_\alpha \sim |\log\alpha|$, it follows that

$$\mathscr{R}^r_{k,\theta}(T^{W,\ell}_{B_\alpha}) = \frac{\mathsf{E}_{k,\theta}\left[(\widetilde{T}^{W,\ell}_{B_\alpha}-k)^+\right]^r}{\mathsf{P}_\infty(\widetilde{T}^{W,\ell}_{B_\alpha}>k)}$$
$$\leq \frac{\left(1+\frac{\log B_\alpha}{I_\theta-\varepsilon}\right)^r + r2^{r-1}\sup_{j\in\mathbb{Z}_+}\Upsilon_{j,r}(\theta,\varepsilon/2)}{1-k/B_\alpha}$$
$$= \left(\frac{|\log\alpha|}{I_\theta-\varepsilon}\right)^r (1+o(1)) \quad \text{as } \alpha\to 0,$$

which holds for an arbitrarily $\varepsilon \in (0, I_\theta)$. Thus, for any fixed $k \geq 0$ and $\theta \in \Theta$ the following asymptotic upper bound holds:

$$\mathscr{R}^r_{k,\theta}(\widetilde{T}^{W,\ell}_{B_\alpha}) \leq \left(\frac{|\log \alpha|}{I_\theta}\right)^r (1+o(1)) \quad \text{as } \alpha \to 0. \tag{3.175}$$

However, according to the inequality (3.68) in Lemma 3.3 where one has to set $\mu = 0$ under condition $\mathbf{C}_1$ the right hand-side is also the lower bound. This implies (3.171).

Finally, using inequality (3.174) and taking into account that $\mathsf{PFA}_{\pi^\alpha}(\widetilde{T}^{W,\ell}_{B_\alpha}) \geq 1 - (\rho_\alpha B_\alpha)$ and $\log B_\alpha \sim |\log \alpha|$, we obtain

$$\begin{aligned}
\overline{\mathscr{R}}^r_{\pi^\alpha,\theta}(\widetilde{T}^{W,\ell}_{B_\alpha}) &= \frac{\sum_{k=1}^\infty \rho_\alpha (1-\rho_\alpha)^k \mathsf{E}_{k,\theta}\left[(\widetilde{T}^{W,\ell}_{B_\alpha} - k)^+\right]^r}{1 - \mathsf{PFA}_{\pi^\alpha}(\widetilde{T}^{W,\ell}_{B_\alpha})} \\
&\leq \frac{\left(1 + \frac{\log B_\alpha}{I_\theta - \varepsilon}\right)^r + r2^{r-1} \sup_{j \in \mathbb{Z}_+} \Upsilon_{j,r}(\theta, \varepsilon/2)}{1 - 1/(\rho_\alpha B_\alpha)} \\
&= \left(\frac{|\log \alpha|}{I_\theta - \varepsilon}\right)^r (1+o(1)) \quad \text{as } \alpha \to 0,
\end{aligned}$$

where $\varepsilon \in (0, I_\theta)$ can be arbitrarily small, so that

$$\overline{\mathscr{R}}^r_{\pi^\alpha,\theta}(\widetilde{T}^{W,\ell}_{B_\alpha}) \leq \left(\frac{|\log \alpha|}{I_\theta}\right)^r (1+o(1)) \quad \text{as } \alpha \to 0.$$

Applying this upper bound together with the lower bound (3.69) in Lemma 3.3 (with $\mu = 0$) yields (3.172), and the proof is complete. □

3.11 Sufficient Conditions of Asymptotic Optimality for a Class of Markov Processes

Usually the right-tail condition $\mathbf{C}_1$ is not difficult to verify. In fact, for this purpose it suffices to establish the a.s. convergence (3.27). However, in practice, verifying the left-tail conditions $\mathbf{C}_2$ and $\mathbf{C}_3$ can cause some difficulty. In this section, we address this challenging problem for a class of ergodic Markov processes.

Note that the following condition implies conditions $\mathbf{C}_2$ and $\mathbf{C}_3$:

$\mathbf{C}_4(r)$. *Let the $\Theta \to \mathbb{R}_+$ function $I(\theta) = I_\theta$ be continuous and assume that for every compact set $\Theta_c \subseteq \Theta$, every $\varepsilon > 0$, and for some $r \geq 1$*

$$\begin{aligned}
&\Upsilon^*_r(\varepsilon, \Theta_c) := \sup_{\theta \in \Theta_c} \Upsilon_r(\varepsilon, \theta) = \\
&\lim_{\delta \to 0} \sum_{n=1}^\infty n^{r-1} \sup_{\theta \in \Theta_c} \sup_{k \in \mathbb{Z}_+} \mathsf{P}_{k,\theta}\left(\frac{1}{n} \inf_{\vartheta \in \Gamma_{\delta,\theta}} \lambda_\vartheta(k, k+n) < I_\theta - \varepsilon\right) < \infty.
\end{aligned} \tag{3.176}$$

Hence, condition $\mathbf{C}_4$ is sufficient for asymptotic optimality of the MS rule as well as for asymptotic results related to the MSR rule. Furthermore, if there exists a continuous $\Theta \times \Theta \to \mathbb{R}_+$ function $I(\vartheta, \theta)$ such that for any $\varepsilon > 0$, any compact $\Theta_c \subseteq \Theta$ and for some $r \geq 1$

$$\Upsilon^{**}_r(\varepsilon, \Theta_c) := \sum_{n=1}^\infty n^{r-1} \sup_{k \in \mathbb{Z}_+} \sup_{\theta \in \Theta_c} \mathsf{P}_{k,\theta}\left(\sup_{\vartheta \in \Theta_c} \left|\frac{1}{n}\lambda_\vartheta(k, k+n) - I(\vartheta, \theta)\right| > \varepsilon\right) < \infty, \tag{3.177}$$

then condition $\mathbf{C}_4$, and hence, conditions $\mathbf{C}_2$ and $\mathbf{C}_3$ are satisfied with $I_\theta = I(\theta,\theta)$ since

$$\mathsf{P}_{k,\theta}\left(\frac{1}{n}\inf_{|\vartheta-\theta|<\delta}\lambda_\vartheta(k,k+n) < I_\theta - \varepsilon\right) \le \mathsf{P}_{k,\theta}\left(\sup_{\vartheta\in\Theta_c}\left|\frac{1}{n}\lambda_\vartheta(k,k+n) - I(\vartheta,\theta)\right| > \varepsilon\right).$$

As we will see, conditions $\mathbf{C}_4$ and (3.177) are useful in checking applicability of theorems in particular examples.

In this section, we obtain sufficient conditions for $\mathbf{C}_4$ for a class of homogeneous Markov processes. We begin with certain preliminary results for homogeneous Markov processes regardless of the changepoint problem. These results will be used later in the changepoint detection problem.

Let $(X_n^\theta)_{n\ge1}$ be a time homogeneous Markov process with values in a measurable space $(\mathscr{X},\mathscr{B})$ defined by a family of the transition probabilities $(P^\theta(x,A))_{\theta\in\Theta}$ for some parameter set $\Theta\subseteq\mathbb{R}^p$. In the sequel, we denote by E_x^θ the expectation with respect to this probability. In addition, we assume that this process is geometrically ergodic, i.e., the following condition holds:

$\mathbf{A}_1$. *For any $\theta\in\Theta$ there exist a probability measure Q^θ on $(\mathscr{X},\mathscr{B})$ and the Lyapunov $\mathscr{X}\to[1,\infty)$ function $\mathbf{V}^\theta$ with $\mathsf{Q}^\theta(\mathbf{V}^\theta)<\infty$, such that for some positive and finite constants R and κ,*

$$\sup_{n\in\mathbb{Z}_+} e^{\kappa n}\sup_{x\in\mathbb{R}}\sup_{\theta\in\Theta}\sup_{0<F\le\mathbf{V}^\theta}\frac{1}{\mathbf{V}^\theta(x)}\left|\mathsf{E}_x^\theta[F(X_n)] - \mathsf{Q}^\theta(F)\right| \le R.$$

Now, for some $q>0$, we set

$$\upsilon_q^*(x) = \sup_{n\in\mathbb{Z}_+}\sup_{\theta\in\Theta}\mathsf{E}_x^\theta\left[\mathbf{V}^\theta(X_n)\right]^q. \tag{3.178}$$

Let ψ be a measurable $\Theta\times\mathscr{X}\times\mathscr{X}\to\mathbb{R}$ function such that the following integrals exist

$$\widetilde{\psi}(\theta,x) = \int_{\mathscr{X}}\psi(\theta,\mathrm{v},x)P^\theta(x,\mathrm{dv}) \quad\text{and}\quad \mathsf{Q}^\theta(\widetilde{\psi}) = \int_{\mathscr{X}}\widetilde{\psi}(u)\,\mathsf{Q}^\theta(\mathrm{d}u). \tag{3.179}$$

$\mathbf{A}_2$. *Assume that ψ satisfies the Hölder condition of the power $0<\gamma\le1$ with respect to the first variable, i.e., there exists a measurable positive $\mathscr{X}\times\mathscr{X}\to\mathbb{R}$ function h such that*

$$\sup_{u,\theta\in\Theta}\frac{|\psi(u,y,x)-\psi(\theta,y,x)|}{|u-\theta|^\gamma} \le h(y,x) \quad \textit{for any } x,y\in\mathscr{X}, \tag{3.180}$$

and the corresponding integrals $\widetilde{h}(\theta,x)$ and $\mathsf{Q}^\theta(\widetilde{h})$ exist for any $\theta\in\Theta$, where $\widetilde{h}(\theta,x)$ is defined as $\widetilde{\psi}$ in (3.179) and $|\cdot|$ is the Euclidean norm in $\mathbb{R}^p$.

$\mathbf{A}_3$. *Assume that the functions ψ and h are such that $|\widetilde{\psi}(\theta,x)|\le\mathbf{V}^\theta(x)$ and $|\widetilde{h}(\theta,x)|\le\mathbf{V}^\theta(x)$ for all $\theta\in\Theta$ and $x\in\mathscr{X}$.*

Now, for a function ψ for which there exist the integrals (3.179) we introduce the deviation processes

$$W_n^\psi(u,\theta) = n^{-1}\sum_{j=1}^n\psi(u,X_j,X_{j-1}) - \mathsf{Q}^\theta(\widetilde{\psi}) \quad\text{and}\quad \widetilde{W}_n^\psi(\theta) = W_n^\psi(\theta,\theta). \tag{3.181}$$

Similarly to (3.178), we define for some $q>0$

$$\psi_q^*(x) = \sup_{n\ge1}\sup_{\theta\in\Theta}\mathsf{E}_x^\theta|\psi(\theta,X_n,X_{n-1})|^q \quad\text{and}\quad h_q^*(x) = \sup_{n\ge1}\sup_{\theta\in\Theta}\mathsf{E}_x^\theta|h(X_n,X_{n-1})|^q. \tag{3.182}$$

Proposition 1 in [108] implies the following result.

Proposition 3.3. *Assume that conditions* $\mathbf{A}_1$–$\mathbf{A}_3$ *hold. Then for any* $q \geq 2$, *for which* $\upsilon_q^*(x) < \infty$, $\psi_q^*(x) < \infty$ *and* $h_q^*(x) < \infty$, *one has*

$$\sup_{n\geq 2} n^{q/2} \sup_{\theta\in\Theta} \sup_{x\in\mathscr{X}} \frac{\mathsf{E}_x^\theta |\widetilde{W}_n^\psi(\theta)|^q + \mathsf{E}_x^\theta |\widetilde{W}_n^h(\theta)|^q}{(1+\upsilon_q^*(x)+\psi_q^*(x)+h_q^*(x))} < \infty. \tag{3.183}$$

Conditions $\mathbf{A}_1$–$\mathbf{A}_3$ imply the following result which is essential for verifying the uniform left-sided condition $\mathbf{C}_3$.

Proposition 3.4. *Assume that conditions* $\mathbf{A}_1$–$\mathbf{A}_3$ *hold and* $\upsilon_q^*(x) < \infty$, $\psi_q^*(x) < \infty$ *and* $h_q^*(x) < \infty$ *for some* $q \geq 2$. *Then for any* $\varepsilon > 0$ *there exists* $\delta_0 > 0$ *such that*

$$\sup_{0<\delta\leq\delta_0} \sup_{n\geq 2} n^{q/2} \sup_{\theta\in\Theta} \sup_{x\in\mathscr{X}} \frac{\mathsf{P}_x^\theta\left(\sup_{|u-\theta|<\delta} |W_n^\psi(u,\theta)| > \varepsilon\right)}{(1+\upsilon_q^*(x)+\psi_q^*(x)+h_q^*(x))} < \infty. \tag{3.184}$$

Proof. First note that

$$\sup_{|u-\theta|<\delta} |W_n^\psi(u,\theta)| \leq \delta^\gamma \left(\mathsf{Q}^\theta(\widetilde{h}) + \widetilde{W}_n^h(\theta)\right) + |\widetilde{W}_n^\psi(\theta)|.$$

Therefore, for any

$$0 < \delta \leq \left(\frac{\varepsilon}{2(1+\mathsf{Q}^\theta(\widetilde{h}))}\right)^{1/\gamma} := \delta_0$$

we obtain

$$\mathsf{P}_x^\theta\left(\sup_{|u-\theta|<\delta} |W_n^\psi(u,\theta)| > \varepsilon\right) \leq \mathsf{P}_x^\theta\left(|\widetilde{W}_n^\psi(\theta)| > \varepsilon/2\right) + \mathsf{P}_x^\theta\left(|\widetilde{W}_n^h(\theta)| > 1\right).$$

Applying the bound (3.183) we obtain (3.184). □

We return to the change detection problem for Markov processes, assuming that the sequence of observations $(X_n)_{n\geq 1}$ is a Markov process, such that $(X_n)_{1\leq n\leq \nu}$ is a homogeneous process with the transition (from x to y) density $g(y|x)$. In the sequel, we denote by $\check{\mathsf{P}}$ the distribution of this process when $\nu = \infty$, i.e., when the Markov process $(X_n)_{n\geq 1}$ has transition density $g(y|x)$. The expectation with respect to this distribution we denote by $\check{\mathsf{E}}(\cdot)$. Moreover, $(X_n)_{n>\nu}$ is homogeneous positive ergodic with the transition density $f_\theta(y|x)$ and the stationary distribution Q^θ. The densities $g(y|x)$ and $f_\theta(y|x)$ are defined with respect to a sigma-finite positive measure μ on $\mathscr{B}$.

In this case, we can represent the LLR process $\lambda_\theta(k,n)$ as

$$\lambda_\theta(k,n) = \sum_{j=k+1}^{n} \psi(\theta, X_j, X_{j-1}), \quad \psi(\theta,y,x) = \log\frac{f_\theta(y|x)}{g(y|x)}, \tag{3.185}$$

and therefore,

$$\widetilde{\psi}(\theta,x) = \int_{\mathscr{X}} \psi(\theta,y,x) f_\theta(y|x)\, \mu(\mathrm{d}y). \tag{3.186}$$

If we assume in addition that the density $f_u(y|x)$ is continuously differentiable with respect to u in a compact set $\Theta_c \subseteq \Theta$, then the inequality (3.180) holds with $\gamma = 1$ and for any function $h(y,x)$ for which

$$\max_{u\in\Theta} \max_{1\leq j\leq p} |\partial\psi(u,y,x)/\partial u_j| \leq h(y,x). \tag{3.187}$$

$\mathbf{B}_1$. *Assume that there exists a set* $C \in \mathscr{B}$ *with* $\mu(C) < \infty$ *such that*

($\mathbf{B}$1.1) $f_* = \inf_{\theta \in \Theta_c} \inf_{x,y \in C} f_\theta(y|x) > 0$.

($\mathbf{B}$1.2) *For any $\theta \in \Theta_c$ there exists $\mathscr{X} \to [1,\infty)$ Lyapunov's function $\mathbf{V}^\theta$ such that*

- $\mathbf{V}^\theta(x) \ge \widetilde{\psi}(\theta,x)$ *and* $\mathbf{V}^\theta(x) \ge \widetilde{h}(\theta,x)$ *for any* $\theta \in \Theta_c$ *and* $x \in \mathscr{X}$,
- $\sup_{\theta \in \Theta} \sup_{x \in C} \mathbf{V}^\theta(x) < \infty$.
- *For some* $0 < \rho < 1$ *and* $D > 0$ *and for all* $x \in \mathscr{X}$ *and* $\theta \in \Theta$,

$$\mathsf{E}_x^\theta[\mathbf{V}^\theta(X_1)] \le (1-\rho)\mathbf{V}^\theta(x) + D\mathbb{1}_{\{C\}}(x). \tag{3.188}$$

$\mathbf{B}_2$. *Assume that there exists $q > 2$ such that*

$$\sup_{k\ge 1} \check{\mathsf{E}}[\psi_q^*(X_k)] < \infty, \quad \sup_{k\ge 1} \check{\mathsf{E}}[h_q^*(X_k)] < \infty \quad \text{and} \quad \sup_{k\ge 1} \check{\mathsf{E}}[\upsilon_q^*(X_k)] < \infty,$$

where the function $\upsilon_q^*(x)$ is defined in (3.178), *$\psi_q^*(x)$ and $h_q^*(x)$ are given in* (3.182).

Theorem 3.12. *Assume that conditions $\mathbf{B}_1$ and $\mathbf{B}_2$ hold for some compact set $\Theta_c \subseteq \Theta$. Then, for any $0 < r < q/2$, condition $\mathbf{C}_4(r)$ holds with $I_\theta = \mathsf{Q}^\theta(\widetilde{\psi})$.*

Proof. First note that it follows from Theorem D.1 in the appendix that the conditions $\mathbf{B}_1$ imply the property $\mathbf{A}_1$. So, by Proposition 3.4, there exists a positive constant $\mathbf{C}^*$ such that for any $x \in \mathscr{X}$

$$\mathsf{P}^\theta\left(\sup_{|u-\theta|<\delta} |W_n^\psi(u,\theta)| > \varepsilon | X_0 = x\right) \le \mathbf{C}^*\,\mathbf{U}^*(x)\, n^{-q/2},$$

where $\mathbf{U}^*(x) = 1 + \upsilon_q^*(x) + \psi_q^*(x) + h_q^*(x)$. Note now that

$$\mathsf{P}_{k,\theta}\left(\frac{1}{n}\inf_{|u-\theta|<\delta} \lambda_u(k,k+n) < I_\theta - \varepsilon\right) \le \mathsf{P}_{k,\theta}\left(\sup_{|u-\theta|<\delta}\left|\frac{1}{n}\lambda_u(k,k+n) - I_\theta\right| > \varepsilon\right).$$

In view of the homogeneous Markov property we obtain that for $I_\theta = \mathsf{Q}^\theta(\widetilde{\psi})$ the last probability can be represented as

$$\mathsf{P}_{k,\theta}\left(\sup_{|u-\theta|<\delta}\left|\frac{1}{n}\lambda_u(k,k+n) - I_\theta\right| > \varepsilon\right) = \check{\mathsf{E}}[\Psi^\theta(X_k)],$$

where $\Psi^\theta(x) = \mathsf{P}^\theta\left(\sup_{|u-\theta|<\delta} |W_n^\psi(u,\theta)| > \varepsilon | X_0 = x\right)$. Therefore,

$$\mathsf{P}_{k,\theta}\left(\sup_{|u-\theta|<\delta}\left|\frac{1}{n}\lambda_u(k,k+n) - I_\theta\right| > \varepsilon\right) \le \mathbf{C}^*\, n^{-r/2}\check{\mathsf{E}}[\mathbf{U}^*(X_k)].$$

Now condition $\mathbf{B}_2$ implies $\mathbf{C}_4(r)$ for any $0 < r < q/2$. □

Note that condition ($\mathbf{C}$1.1) does not always hold for the process $(X_n)_{n\ge 1}$ directly. Unfortunately, this condition does not hold for the practically important autoregression process of the order more than one. For this reason, we need to weaken this requirement. Specifically, assume that there exists $p \ge 2$ for which the process $(\widetilde{X}_{\iota,n})_{n \ge \widetilde{\nu}}$ for $\widetilde{\nu} = \nu/p - \iota$ defined as $\widetilde{X}_{\iota,n} = X_{np+\iota}$ satisfies the following properties:

$\mathbf{B}_1'$. *Assume that there exists a set $C \in \mathscr{B}$ with $\mu(C) < \infty$ such that*

($\mathbf{B}'$1.1) $\widetilde{f}_* = \inf_{1\le\iota\le p} \inf_{\theta \in \Theta_c} \inf_{x,y \in C} \widetilde{f}_{\iota,\theta}(y|x) > 0$, *where $\widetilde{f}_{\iota,\theta}(y|x)$ is the transition density for the process $(\widetilde{X}_{\iota,n})_{n\ge 1}$.*

($\mathbf{B}'1.2$) *For any $\theta \in \Theta_c$ there exists $\mathscr{X} \to [1,\infty)$ Lyapunov's function $\mathbf{V}^\theta$ such that*

$$\upsilon^* = \max_{1\le j\le p} \sup_{\theta\in\Theta_c} \sup_{x\in\mathscr{X}} \frac{\mathsf{E}_x^\theta[\mathbf{V}^\theta(X_j)]}{\mathbf{V}^\theta(x)} < \infty; \tag{3.189}$$

$$\upsilon_1^* = \sup_{\theta\in\Theta_c} \varkappa^\theta(\mathbf{V}^\theta) < \infty. \tag{3.190}$$

- $\mathbf{V}^\theta(x) \ge \widetilde{\psi}(\theta,x)$ *and* $\mathbf{V}^\theta(x) \ge \widetilde{h}(\theta,x)$ *for* $\theta \in \Theta_c$ *and* $x \in \mathscr{X}$ *and* $\sup_{\theta\in\Theta_c} \sup_{x\in C} \mathbf{V}^\theta(x) < \infty$.
- *For some $0<\rho<1$ and $D>0$ and for all $x \in \mathscr{X}$, $\theta \in \Theta_c$, and $0 \le \iota \le p-1$*

$$\mathsf{E}^\theta\left(\mathbf{V}^\theta(\widetilde{X}_{\iota,1})|\widetilde{X}_{\iota,0}=x\right) \le (1-\rho)\mathbf{V}^\theta(x) + D\mathbb{1}_{\{C\}}(x). \tag{3.191}$$

Theorem 3.13. *Suppose that conditions $\mathbf{B}_2$ and $\mathbf{B}_1'$ hold. Then, for any $0<r<q/2$, condition $\mathbf{C}_4(r)$ holds with $I_\theta = \mathsf{Q}^\theta(\widetilde{\psi})$.*

Proof. Note again that by Theorem D.1 (see Appendix) conditions $\mathbf{B}_1$ imply condition $\mathbf{A}_1$ for $(\widetilde{X}_{\iota,n})_{n\ge\widetilde{\nu}}$, i.e., for some positive constants $0<R_\iota<\infty$ and $\kappa_\iota>0$,

$$\sup_{n\ge0} e^{\kappa_\iota n} \sup_{x\in\mathbb{R}} \sup_{\theta\in\Theta_c} \sup_{0<F\le\mathbf{V}^\theta} \frac{1}{\mathbf{V}^\theta(x)} \left|D_m^\theta(x)\right| \le R_\iota,$$

where $D_m^\theta(x) = \mathsf{E}^\theta\left[\left(F(\widetilde{X}_{\iota,m}) - \varkappa^\theta(F)\right)|\widetilde{X}_{\iota,0}=x\right]$. So, for any $n \ge p$, we can write that $n = mp+\iota$ for some $0\le\iota\le p-1$ and we obtain

$$|\mathsf{E}_x^\theta(F(X_n) - \varkappa^\theta(F))| = |\mathsf{E}_x^\theta D_m^\theta(X_\iota)| \le R_\iota \mathbf{V}^\theta(X_\iota)e^{\kappa_\iota m}.$$

Now, the upper bound (3.189) implies

$$\sup_{n\ge p} e^{\kappa n} \sup_{x\in\mathbb{R}} \sup_{\theta\in\Theta_c} \sup_{0<F\le\mathbf{V}^\theta} \frac{1}{\mathbf{V}^\theta(x)} |\mathsf{E}_x^\theta(F(X_n) - \varkappa^\theta(F))| \le R_*,$$

where $R_* = \upsilon^* \max_{0\le\iota} R_\iota$ and $\kappa = \min_{0\le\iota\le p-1} \kappa_\iota/p$. Thus, using bound (3.190) we obtain condition $\mathbf{A}_1$ with $R = e^{\kappa_* p}(\upsilon^* + \upsilon_1^*) + R_*$. By an argument similar to the proof of Theorem 3.12 we obtain that condition $\mathbf{C}_4(r)$ holds with $I(\theta) = \mathsf{Q}^\theta(\widetilde{\psi})$ for any $0<r<q/2$. □

By an analogy with the i.i.d. case, in the Markov case, it is reasonable to call the number $I_\theta = \mathsf{Q}^\theta(\widetilde{\psi})$ as the Kullback–Leibler information number since $\mathsf{Q}^\theta(\widetilde{\psi})$ is equal to the expectation of the LLR increment when the LLR starts from the stationary distribution. In the sequel, we will use the notation $\mathscr{K}_\theta = \mathsf{Q}^\theta(\widetilde{\psi})$ and will refer to $\mathscr{K}_\theta$ as the *generalized* Kullback–Leibler information number.

Example 3.1 (*Detection of a change in the correlation coefficient of the AR*(1) *process*)**.** Let the observations represent the Markov sequence with the known correlation coefficient θ_0 before change and the unknown correlation coefficient θ after change, i.e.,

$$X_n = \left(\theta_0\mathbb{1}_{\{\nu\le n\}} + \theta\mathbb{1}_{\{\nu>n\}}\right)X_{n-1} + w_n, \quad n\ge1,$$

where X_0 is an initial value independent of the sequence $(w_n)_{n\ge1}$, and $(w_n)_{n\ge1}$ are i.i.d. random variables with $\mathsf{E}[w_1]=0$, $\mathsf{E}[w_1^2]=1$ and a known continuously differentiable density $\phi(x)$ such that for any $n\ge1$

$$\inf_{-n\le x\le n} \phi(x) > 0. \tag{3.192}$$

Since the observed process is Markov to establish asymptotic optimality of MS and MSR rules we will use the results established above.

Assume that θ belongs to a compact set $\Theta_c \subset]-1,1[\setminus\{\theta_0\}$. The ergodic distributions for $(X_n)_{n\le\nu}$ and $(X_n)_{n\ge\nu+1}$ are given by the random variables $\widetilde{w}_0$ and $\widetilde{w}_\theta$, respectively, which are defined as

$$\widetilde{w}_0 = \sum_{j=1}^{\infty} (\theta_0)^{j-1} w_j \quad \text{and} \quad \widetilde{w}_\theta = \sum_{j=1}^{\infty} \theta^{j-1} w_j. \tag{3.193}$$

The pre-change and post-change conditional densities are $g(X_n|X_{n-1}) = f_{\theta_0}(X_n|X_{n-1}) = \phi(X_n - \theta_0 X_{n-1})$ for all $1 \le n \le \nu$ and $f_\theta(X_n|X_{n-1}) = \phi(X_n - \theta X_{n-1})$ for $n > \nu$. In this case, for $\theta \in \Theta_c$

$$\psi(\theta,y,x) = \log\frac{f_\theta(y|x)}{f_{\theta_0}(y|x)} = \log\frac{\phi(y-\theta x)}{\phi(y-\theta_0 x)} \tag{3.194}$$

and

$$\widetilde{\psi}(\theta,x) = \int_{\mathbb{R}} \log\frac{\phi(y-\theta x)}{\phi(y-\theta_0 x)}\,\phi(y-\theta x)\,\mathrm{d}y. \tag{3.195}$$

Assume that there exist $c^* \ge 1$ and $\iota > 0$ such that $\mathsf{E}\,|w_1|^\iota < \infty$ and

$$\sup_{\theta\in\Theta_c}\ \sup_{y,x\in\mathbb{R}} \left(\frac{|\psi(\theta,y,x)| + |h'(\theta,y,x)|}{(1+|y|^\iota+|x|^\iota)} + \frac{\widetilde{\psi}(\theta,x)}{(1+|x|^\iota)}\right) \le c^*, \tag{3.196}$$

where $h'(\theta,y,x) = \partial h(\theta,y,x)/\partial\theta$.

Define the Lyapunov function as

$$\mathbf{V}(x) = c^*\,(1+|x|^\iota). \tag{3.197}$$

Obviously,

$$\lim_{|x|\to\infty} \frac{\mathsf{E}_x^\theta[\mathbf{V}(X_1)]}{\mathbf{V}(x)} = \lim_{|x|\to\infty} \frac{1+\mathsf{E}\,|\theta X + w_1|^\iota}{1+|x|^\iota} \le |\mathbf{u}_*|^\iota < 1.$$

Therefore, for any $|\mathbf{u}_*|^\iota < \rho < 1$ there exist $n \ge 1$ and $D > 0$ such that condition $\mathbf{B}_1$ holds with $C = [-n,n]$.

For verifying condition $\mathbf{B}_2$, suppose that there exists $q > 2$ such that

$$\int_{\mathbb{R}} |v|^r\,\psi(v)\,\mathrm{d}v < \infty, \qquad r = \iota q. \tag{3.198}$$

This condition implies that $\mathsf{E}\,|\widetilde{w}_0|^r < \infty$ and $\mathsf{E}\,|\widetilde{w}_1|^r < \infty$. Therefore,

$$\sup_{n\in\mathbb{Z}_+} \check{\mathsf{E}}\,|X_n|^r < \infty \tag{3.199}$$

and for any $x \in \mathbb{R}$

$$M^*(x) = \sup_{\theta\in\Theta_c}\ \sup_{n\ge1} \mathsf{E}_x^\theta\,|X_n|^r \le C^*(1+|x|^r).$$

So, in view of (3.199) we obtain $\sup_{n\ge1} \check{\mathsf{E}}[M^*(X_n)] < \infty$, which implies condition $\mathbf{B}_2$.

By Theorem 3.12, condition $\mathbf{C}_4(r)$ holds with $I_\theta = \mathscr{K}_\theta$, where

$$\mathscr{K}_\theta = \mathsf{Q}^\theta(\widetilde{\psi}) = \int_{\mathbb{R}} \left(\int_{\mathbb{R}} \log\frac{\phi(y-\theta x)}{\phi(y-\theta_0 x)}\,\phi(y-\theta x)\mathrm{d}y\right) \mathsf{Q}^\theta(\mathrm{d}x)$$

is the generalized Kullback–Leibler information number and where Q^θ is the distribution of $\widetilde{w}_\theta$ given in (3.193).

Consider now the particular Gaussian case where $w_n \sim \mathscr{N}(0,1), n \geq 1$ are i.i.d. standard normal random variables. Then $\psi(y) = \varphi(y) := (2\pi)^{-1/2} e^{-y^2/2}$,

$$\psi(\theta, y, x) = \frac{(y-\theta_0 x)^2 - (y-\theta x)^2}{2} \quad \text{and} \quad \widetilde{\psi}(\theta, x) = \frac{(\theta-\theta_0)^2 x^2}{2}.$$

Hence, we have

$$\sup_{\theta \in \Theta_c} \sup_{y,x \in (-\infty,\infty)} \frac{|\psi(\theta,y,x)|}{1+|y|^2+|x|^2} \leq 5/2 \quad \text{and} \quad \sup_{\theta \in \Theta_c} \sup_{x \in (-\infty,\infty)} \frac{\widetilde{\psi}(\theta,x)}{1+|x|^2} \leq 5/2, \tag{3.200}$$

i.e., conditions (3.196) are satisfied with $\iota = 2$ and $q^* = 5$. The random variable $\widetilde{w}_\theta$ is normal $\mathscr{N}\left(0,(1-\theta^2)^{-1}\right)$, i.e., the stationary distribution $\mathsf{Q}^\theta(x) = \mathsf{P}_{0,\theta}(X_\infty \leq x)$ of the Markov process X_n under $\mathsf{P}_{0,\theta}$ is normal $\mathscr{N}(0,(1-\theta^2)^{-1})$, and the generalized Kullback–Leibler information number can be calculated explicitly as

$$\mathscr{K}_\theta = \int_{-\infty}^{\infty} \left(\int_{-\infty}^{\infty} \psi(\theta,y,x)\, \varphi(y-\theta x) \mathrm{d}y \right) \mathsf{Q}^\theta(\mathrm{d}x) = \frac{(\theta-\theta_0)^2}{2(1-\theta^2)}. \tag{3.201}$$

Condition $\mathbf{C}_4(r)$ holds for all $r \geq 1$ and by Theorems 3.2 and 3.3, the MS detection procedure minimizes asymptotically all moments of the detection delay.

A typical behavior of the observed sequence $\{X_n\}_{n \geq 1}$ in the Gaussian case as well as of the MS and MSR statistics is shown in Figure 3.1 when the point of change $\nu = 200$. While the change is not visible to the naked eye (Figure 3.1(a)) it is easily detected by both change detections rules quite fast (Figure 3.1(b)). The large value of the changepoint was selected to illustrate false alarms.

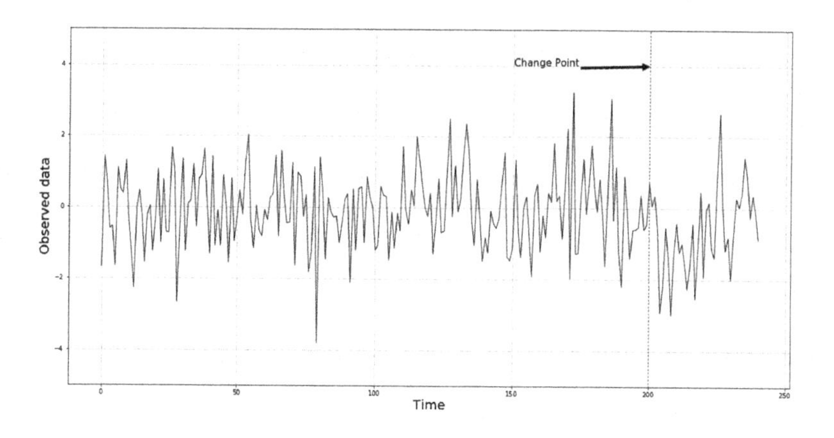

(a) Observations $\{X_n\}_{n\geq 1}$.

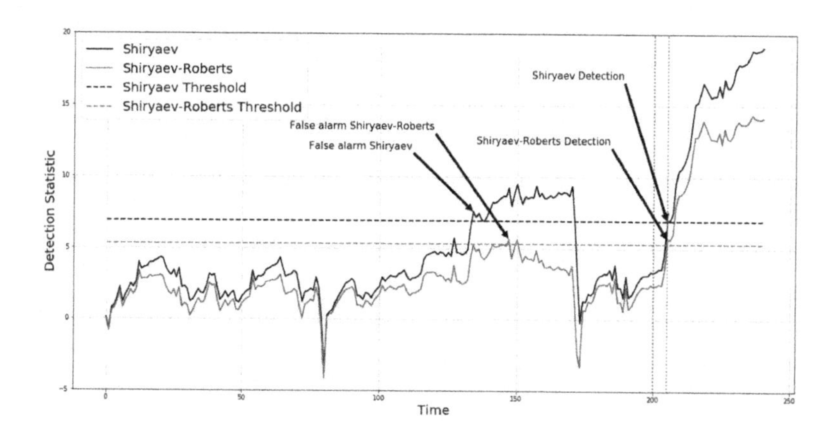

(b) MS and MSR statistics.

FIGURE 3.1
Behavior of observations $\{X_n\}_{n\geq 1}$ and the MS and MSR statistics in the Gaussian case: changepoint $\nu = 200$.

3.12 Asymptotic Optimality for Hidden Markov Models

Hidden Markov models (HMM) is a class of very important stochastic models that find extraordinary applications in a wide variety of fields such as speech recognition [67, 118]; handwriting recognition [63, 72]; computational molecular biology and bioinformatics, including DNA and protein modeling [26]; human activity recognition [193]; target detection and tracking [16, 173, 175]; and modeling, rapid detection, and tracking of malicious activity of terrorist groups [119, 120], to name a few. In this section, we focus on this class of models and specify the general results of Sections 3.4, 3.5 and 3.7 for HMMs, finding a set of general conditions under which the Shiryaev changepoint detection procedure is asymptotically optimal as the probability of false alarm goes to zero. Our approach for hidden Markov models is based on representation of the LLR in the form of a Markov random walk with the use of random matrices and r-complete convergence for Markov random walks. In addition, using uniform Markov renewal theory and Markov nonlinear renewal theory, we provide a high-order asymptotic approximation for the expected delay to detection and a first-order asymptotic approximation for the probability of false alarm of the Shiryaev detection rule.

Throughout this section we consider only the case of a known post-change parameter θ and the corresponding Shiryaev detection rule T_A^θ defined in (3.2), (3.3). We will omit the superscript and subscript θ in all objects in the rest of this section, including all probability measures and expectations. In particular, we will write T_A for T_A^θ, $LR(k,n)$ for the likelihood ratio $LR_\theta(k,n)$, and $\lambda(k,n)$ for the LLR $\lambda_\theta(k,n)$.

3.12.1 Markov Random Walk Representation of the LLR for HHM

We begin with describing a finite state HMM that is of interest in this section, which is a particular case of a general HMM defined in Definition D.4. Let $\mathbf{Z} = \{Z_n\}_{n\in\mathbb{Z}_+}$ be an ergodic (positive recurrent, irreducible and aperiodic) Markov chain on a finite state space $D = \{1,2,\cdots,d\}$ with transition probability matrix $[p(z,z')]_{z,z'=1,\cdots,d}$ and stationary distribution $\omega = \{\omega(z)\}_{z=1,\cdots,d}$. A random sequence $\{X_n\}_{n\in\mathbb{Z}_+}$, taking values in $\mathbb{R}^m$, is adjoined to the chain such that $\{(Z_n,X_n)\}_{n\in\mathbb{Z}_+}$ is a Markov chain on $D\times\mathbb{R}^m$ satisfying $\mathsf{P}\{Z_1\in A|Z_0=z,X_0=x_0\} = \mathsf{P}\{Z_1\in A|Z_0=z\}$ for $A\in\mathscr{B}(D)$, the Borel σ-algebra of D. Also, conditioning on the full $\mathbf{Z}$ sequence, we have

$$\mathsf{P}\{X_{n+1}\in B|Z_0,Z_1,\ldots;X_0,X_1,\ldots,X_n\} = \mathsf{P}\{X_{n+1}\in B|Z_{n+1},X_n\} \tag{3.202}$$

for each $n\in\mathbb{Z}_+$ and $B\in\mathscr{B}(\mathbb{R}^m)$, the Borel σ-algebra of $\mathbb{R}^m$. By $f(\cdot|Z_k,X_{k-1})$ denote the transition probability density of X_k given Z_k and X_{k-1} with respect to a σ-finite measure Q on $\mathbb{R}^m$ such that

$$\mathsf{P}\{Z_1\in A,X_1\in B|Z_0=z,X_0=x\} = \sum_{z'\in A}\int_{x'\in B} p(z,z')f(x'|z',x)Q(\mathrm{d}x'), \tag{3.203}$$

for $B\in\mathscr{B}(\mathbb{R}^m)$. We also assume that the Markov chain $\{(Z_n,X_n),n\ge 0\}$ has a stationary probability Γ_{st} with probability density function $\omega(z)f(\cdot|z)$.

We are interested in the changepoint detection problem for the d-state HMM $\{X_n\}_{n\in\mathbb{Z}_+}$ that satisfies (3.202) and (3.203). This HMM is of course a particular case of the general stochastic model described in (3.1). Specifically, for $j=\infty,0$, let $p_j(z,z')$ be the transition probability, $\omega_j(z)$ be the stationary probability, and $f_j(x'|z,x)$ be the transition probability density of the HMM in (3.202) and (3.203). In the changepoint problem, we suppose that the conditional density $f(x'|z,x)$ and the transition probability $p(z,z')$ change at an unknown time ν from (p_∞,f_∞) to (p_0,f_0).

Let $\mathbf{X}_0^n = (X_0,X_1,\ldots,X_n)$ be the sample obtained from the HMM $\{X_n,n\ge 0\}$. By (3.1), for $0\le k\le n$, the likelihood ratio of the hypothesis $\mathsf{H}_k:\nu=k$ against $\mathsf{H}_\infty:\nu=\infty$ for the sample $\mathbf{X}_0^n$ is given by

$$LR(k,n) := \frac{p_k(\mathbf{X}_0^n)}{p_\infty(\mathbf{X}_0^n)} = \prod_{i=k+1}^n \mathscr{L}_i, \tag{3.204}$$

where $\mathscr{L}_i = p_0(X_i|\mathbf{X}_0^{i-1})/p_\infty(X_i|\mathbf{X}_0^{i-1})$. Note that here we use $p_\infty(X_i|\mathbf{X}_0^{i-1})$ and $p_0(X_i|\mathbf{X}_0^{i-1})$ instead of $g(X_i|\mathbf{X}_0^{i-1})$ and $f(X_i|\mathbf{X}_0^{i-1})$ as in (3.1) for conditional pre-change and post-change densities since, in this section, $f(x'|z,x)$ stands for the transition probability density of the HMM. Denote

$$LR_n = LR(0,n) = \frac{\sum\limits_{z_0\in D,\ldots,z_n\in D} \omega_0(z_0)f_0(X_0|z_0)\prod\limits_{l=1}^n p_0(z_{l-1},z_l)f_0(X_l|z_l,X_{l-1})}{\sum\limits_{z_0\in D,\ldots,z_n\in D} \omega_\infty(z_0)f_\infty(X_0|z_0)\prod\limits_{l=1}^n p_\infty(z_{l-1},z_l)f_\infty(X_l|z_l,X_{l-1})}. \tag{3.205}$$

Recall that we assume that only the sample $\mathbf{X}_1^n = (X_1, \dots, X_n)$ can be observed and the initial value X_0 is used for producing the observed sequence $\{X_n, n = 1, 2, \dots\}$ with the desirable property. The initialization X_0 affects the initial value of the likelihood ratio, $LR_0 = \mathscr{L}_0$, which can be either random or deterministic. In turn, this influences the behavior of LR_n for $n \geq 1$. Using the sample $\mathbf{X}_0^n$ in (3.205) and (3.204) is convenient for Markov and hidden Markov models which can be initialized either randomly or deterministically. If X_0 cannot be observed (or properly generated), then we assume $\mathscr{L}_0 = LR_0 = 1$, which is equivalent to $f_0(X_0|z)/f_\infty(X_0|z) = 1$ for all $z \in D$ in (3.205). This is also the case when the change cannot occur before the observations become available, i.e., when $\Pr(\nu < 0) = q = 0$.

We now provide certain preliminary results required for the analysis of the Shiryaev rule. Since the Shiryaev detection statistic $S^\pi(n)$ (see (3.2)) involves $LR(k,n)$ we explore the structure of the likelihood ratio LR_n in (3.205) first and show that the LLR $\lambda(k,n) = \log LR(k,n)$ can be represented in the form of a Markov random walk. This is important for verification of conditions $\mathbf{C}_1$ and $\widetilde{\mathbf{C}}_2$ for HMM, and therefore, for establishing asymptotic optimality of the Shiryaev rule.

For this purpose, we use the method proposed by Fuh [44] to study a recursive CUSUM change-point detection procedure in HMM, representing (3.205) as the ratio of L_1-norms of products of Markovian random matrices. Specifically, given a column vector $\mathbf{u} = (u_1, \cdots, u_d)^\top \in \mathbb{R}^d$, where $\top$ denotes the transpose of the underlying vector in $\mathbb{R}^d$, define the L_1-norm of u as $\|\mathbf{u}\| = \sum_{i=1}^d |u_i|$. The likelihood ratio LR_n then can be represented as

$$LR_n = \frac{\|M_n^0 \cdots M_1^0 M_0^0 \omega_0\|}{\|M_n^\infty \cdots M_1^\infty M_0^\infty \omega_\infty\|}, \tag{3.206}$$

where

$$M_0^j = \begin{bmatrix} f_j(X_0|Z_0 = 1) & 0 & \cdots & 0 \\ \vdots & \ddots & \vdots & \vdots \\ 0 & 0 & \cdots & f_j(X_0|Z_0 = d) \end{bmatrix}, \tag{3.207}$$

$$M_k^j = \begin{bmatrix} p_j(1,1) f_j(X_k|Z_k = 1, X_{k-1}) & \cdots & p_j(d,1) f_j(X_k|Z_k = 1, X_{k-1}) \\ \vdots & \ddots & \vdots \\ p_j(1,d) f_j(X_k|Z_k = d, X_{k-1}) & \cdots & p_j(d,d) f_j(X_k|Z_k = d, X_{k-1}) \end{bmatrix} \tag{3.208}$$

for $j = 0, \infty$, $k = 1, \cdots, n$, and

$$\omega_j = \left(\omega_j(1), \cdots, \omega_j(d)\right)^\top.$$

Note that each component $p_j(z,z') f_j(X_k|Z_k = z', X_{k-1})$ in the matrix M_k^j defined in (3.208) corresponds to $Z_{k-1} = z$ and $Z_k = z'$, and X_k is a random variable with transition probability density $f_j(X_k|z', X_{k-1})$ for $j = 0, \infty$ and $k = 1, \cdots, n$. Therefore, the M_k^j are random matrices. Since $\{(Z_n, X_n), n \geq 0\}$ is a Markov chain by definition (3.202) and (3.203), this implies that $\{M_k^j, k = 1, \cdots, n\}$ is a sequence of Markov random matrices for $j = 0, \infty$ governed by a Markov chain $\{(Z_n, X_n), n \geq 0\}$. Hence, LR_n is the ratio of the L_1-norm of the products of Markov random matrices via representation (3.206). Note that $\omega_j(\cdot)$ is fixed in (3.206).

Let $\{(Z_n, X_n), n \geq 0\}$ be the Markov chain defined in (3.202) and (3.203). Denote $Y_n = (Z_n, X_n)$ and $\bar{D} = D \times \mathbb{R}^m$. As in [18], we define $Gl(d, \mathbb{R})$ as the set of invertible $d \times d$ matrices with real entries. For given $k = 0, 1, \cdots, n$ and $j = \infty, 0$, let M_k^j be the random matrix from $\bar{D} \times \bar{D}$ to $Gl(d, \mathbb{R})$, as defined in (3.207) and (3.208). For each n, let

$$\mathbf{M}_n = (\mathbf{M}_n^\infty, \mathbf{M}_n^0) = \left(M_n^\infty \circ \cdots \circ M_1^\infty \circ M_0^\infty, M_n^0 \circ \cdots \circ M_1^0 \circ M_0^0\right), \tag{3.209}$$

where $\circ$ denotes the product of two matrices. Then the system $\{(Y_n, \mathbf{M}_n), n \geq 0\}$ is called a product of Markov random matrices on $\bar{D} \times Gl(d, \mathbb{R}) \times Gl(d, \mathbb{R})$. Denote by P^y the probability distribution of $\{(Y_n, \mathbf{M}_n), n \geq 0\}$ with $Y_0 = y$, and by E^y the expectation under P^y.

We will now use the standard notation in random matrices (cf. Bougerol [17, 18]). Let $\mathbf{u} \in \mathbb{R}^d$ be a d-dimensional vector, $\bar{\mathbf{u}} := \mathbf{u}/||\mathbf{u}||$ the normalization of $\mathbf{u}$ ($||\mathbf{u}|| \neq 0$), and by $P(\mathbb{R}^d)$ denote the projection space of $\mathbb{R}^d$ which contains all elements $\bar{\mathbf{u}}$. For given $\bar{\mathbf{u}} \in P(\mathbb{R}^d)$ and $M \in Gl(d,\mathbb{R})$, denote $M \cdot \bar{\mathbf{u}} = \overline{M\mathbf{u}}$ and $\overline{\mathbf{M}_k\mathbf{u}} = \left(\overline{\mathbf{M}_k^\infty\mathbf{u}}, \overline{\mathbf{M}_k^0\mathbf{u}}\right)$, $k = 0, \cdots, n$. Let

$$W_0 = (Y_0, \overline{\mathbf{M}_0 u}),\ W_1 = (Y_1, \overline{\mathbf{M}_1\mathbf{u}}), \cdots, W_n = (Y_n, \overline{\mathbf{M}_n\mathbf{u}}). \tag{3.210}$$

Then, $\{W_n, n \geq 0\}$ is a Markov chain on the state space $\bar{D} \times P(\mathbb{R}^d) \times P(\mathbb{R}^d)$ with the transition kernel

$$\mathsf{P}((y,\bar{\mathbf{u}}), A \times B) := \mathsf{E}^y[I_{A\times B}(Y_1, \overline{M_1\mathbf{u}})] \tag{3.211}$$

for all $y \in \bar{D}$, $\bar{\mathbf{u}} := (\bar{\mathbf{u}}, \bar{\mathbf{u}}) \in P(\mathbb{R}^d) \times P(\mathbb{R}^d)$, $A \in \mathscr{B}(\bar{D})$, and $B \in \mathscr{B}(P(\mathbb{R}^d) \times P(\mathbb{R}^d))$, the σ-algebra of $P(\mathbb{R}^d) \times P(\mathbb{R}^d)$. For simplicity, we let $\mathsf{P}^{(y,\bar{\mathbf{u}})} := \mathsf{P}(\cdot,\cdot)$ and denote $\mathsf{E}^{(y,\bar{\mathbf{u}})}$ as the expectation under $\mathsf{P}^{(y,\bar{\mathbf{u}})}$. Since the Markov chain $\{(Z_n, X_n), n \geq 0\}$ has transition probability density and the random matrix $M_1(\theta)$ is driven by $\{(Z_n, X_n), n \geq 0\}$, the induced transition probability $\mathsf{P}(\cdot,\cdot)$ has a density p with respect to $m \times \bar{Q}$ for some probability measures m and $\bar{Q}$ (cf. [18]). Here the transition probability density p is defined in the sense of the transition probability for the Markov chain $\{(Z_n, X_n), n \geq 0\}$, as the last component is a deterministic function of (Z_n, X_n). According to Theorem 1(iii) in Fuh [44], under conditions $\mathbf{C}_1^h$ and $\mathbf{C}_2^h$ given below in Subsection 3.12.2, the stationary distribution of $\{W_n\}_{n\in\mathbb{Z}_+}$ exists. Denote it by Q_{st}.

The crucial observation is that the LLR can now be written as an additive functional of the Markov chain $\{W_n, n \geq 0\}$, that is,

$$\lambda(n) = \log LR_n = \sum_{t=1}^{n} \psi(W_{t-1}, W_t), \tag{3.212}$$

where

$$\psi(W_{t-1}, W_t) := \log \frac{||M_t^0 \circ \cdots \circ M_1^0 M_0^0 \omega_0|| / ||M_{t-1}^0 \circ \cdots \circ M_1^0 M_0^0 \omega_0||}{||M_t^\infty \circ \cdots \circ M_1^\infty M_0^\infty \omega_\infty|| / ||M_{t-1}^\infty \circ \cdots \circ M_1^\infty M_0^\infty \omega_\infty||}. \tag{3.213}$$

Hence the LLR is a Markov random walk with respect to the induced Markov process $\{W_n\}_{n\in\mathbb{Z}_+}$ generated by random matrices. See Definition D.2 in Appendix D for the definition of a Markov random walk.

3.12.2 Asymptotic Optimality

Recall the definition of the V-ergodicity of the Markov process given in Definition D.3 in Appendix D. Recall also that a Markov process $\{Y_n\}_{n\in\mathbb{Z}_+}$ on $\mathscr{Y}$ is Harris recurrent if there exist a recurrent set $\mathscr{A} \in \mathscr{B}(\mathscr{Y})$, a probability measure φ on $\mathscr{A}$ and an integer n_0 such that $\mathsf{P}\{Y_n \in \mathscr{A}$ for some $n \geq n_0 | Y_0 = z\} = 1$ for all $y \in \mathscr{Y}$, and there exists $c > 0$ such that

$$\mathsf{P}\{Y_n \in A | Y_0 = y\} \geq c\,\varphi(A) \tag{3.214}$$

for all $y \in \mathscr{A}$ and $A \subset \mathscr{A}$ (cf. Harris [57]). Note that under the irreducibility and aperiodicity assumption, V-uniform ergodicity implies Harris recurrence.

Let $\mathscr{W} = \bar{D} \times P(\mathbb{R}^d) \times P(\mathbb{R}^d)$ be the state space of the induced Markov chain $\{W_n\}_{n\in\mathbb{Z}_+}$ defined in (3.210). Denote $w := (x, \bar{\mathbf{u}}, \bar{\mathbf{u}})$ and $\bar{w} := (x_0, \omega_\infty, \omega_0)$, where x_0 is the initial state $X_0 = x_0$ taken from the stationary distribution with density $\omega_\infty f_\infty(\cdot|z)$.

Define the generalized Kullback–Leibler information number as

$$\mathscr{K}(\mathsf{P}_0, \mathsf{P}_\infty) = \mathsf{E}_{\mathsf{P}_0}^{\mathsf{Q}_{\mathrm{st}}}\left[\log \frac{||M_1^0 M_0^0 \omega_0||}{||M_1^\infty M_0^\infty \omega_\infty||}\right], \tag{3.215}$$

where P_∞ (P_0) denotes the probability of the Markov chain $\{W_n^\infty, n \ge 0\}$ ($\{W_n^0, n \ge 0\}$), and $\mathsf{E}_{\mathsf{P}_\infty}^{\mathsf{Q}_{\mathrm{st}}}$ ($\mathsf{E}_{\mathsf{P}_0}^{\mathsf{Q}_{\mathrm{st}}}$) refers to the expectation for P_∞ (P_0) under the invariant probability Q_{st}, i.e., when the initial value W_0 is distributed according to the stationary distribution Q_{st} of the Markov process $\{W_n\}$.

To prove first-order asymptotic optimality of the Shiryaev rule and to derive a high-order asymptotic approximation for the average detection delay for HMMs, the following two conditions are assumed.

$\mathbf{C}_1^h$. For each $j = \infty, 0$, the Markov chain $\{Z_n\}_{n \in \mathbb{Z}_+}$ defined in (3.202) and (3.203) is ergodic (positive recurrent, irreducible and aperiodic) on a finite state space $D = \{1, \cdots, d\}$. Moreover, the Markov chain $\{(Z_n, X_n)\}_{n \in \mathbb{Z}_+}$ is irreducible, aperiodic and V-uniformly ergodic for some $V \ge 1$ and finite on $\bar{D}$.

Note that $\mathbf{C}_1^h$ implies that $\{(Z_n, X_n)\}_{n \in \mathbb{Z}_+}$ has stationary probability Γ_{st}. Furthermore, we assume that Γ_{st} has probability density $\omega(z) f(\cdot|z)$ with respect to a σ-finite measure.

$\mathbf{C}_2^h$. Assume that $0 < \mathscr{K} < \infty$. Also, for each $j = \infty, 0$, assume that the random matrices M_0^j and M_1^j, defined in (3.207) and (3.208), are invertible P_j-a.s. and for some $r \ge 1$

$$\sup_{(z,x) \in D \times \mathbb{R}^m} \mathsf{E}^{(z,x)} \left| \frac{|X_1|^r V(Z_1, X_1)}{V(z,x)} \right| < \infty. \tag{3.216}$$

The ergodicity condition for Markov chains $\mathbf{C}_1^h$ is quite general and holds in many applications. The first part of the condition $\mathbf{C}_2^h$ is a standard constraint on the generalized Kullback–Leibler information number. Note that positiveness of the generalized Kullback–Leibler information number is not at all restrictive since it holds whenever the probability density functions of P_0 and P_∞ do not coincide almost everywhere. The finiteness condition is quite natural and holds in most cases. Moreover, the cases where it is infinite are easy to handle and can be viewed as degenerate from the asymptotic theory standpoint. The invertibility assumption in $\mathbf{C}_2^h$ follows from the standard products of random matrices assumption, cf. Bougerol [17]. Condition (3.216) is a moment condition under the V-norm. For a general discussion of the weight function V, which is usually referred to as the Lyapunov function, see Chapter 16 of Meyn and Tweedie [91].

Under conditions $\mathbf{C}_1^h$–$\mathbf{C}_2^h$ the Markov chain $\{W_n\}_{n \in \mathbb{Z}_+}$ has several useful properties that are summarized in the following proposition whose proof is the same as the proof of Proposition 2 of Fuh [45]. Denote $\chi(M) = \sup(\log \|M\|, \log \|M^{-1}\|)$.

Proposition 3.5. *Let a HMM defined in* (3.202) *and* (3.203) *satisfy conditions* $\mathbf{C}_1^h$–$\mathbf{C}_2^h$. *Then the induced Markov chain* $\{W_n\}_{n \in \mathbb{Z}_+}$, *defined in* (3.210) *and* (3.211), *is an aperiodic, irreducible and Harris recurrent Markov chain under* P_∞. *Moreover, it is also a V-uniformly ergodic Markov chain for some V on* $\mathscr{X}$. *We have* $\sup_w \{\mathsf{E}_\infty[V(W_1)|W_0 = w]/V(w)\} < \infty$, *and there exist* $a, C > 0$ *such that* $\mathsf{E}_\infty(\exp\{a\chi(M_1)\}|W_0 = w) \le C$ *for all* $w = (z_0, \omega, \omega) \in \mathscr{W}$.

Observe that by Theorems 3.7 and 3.8 uniform r-complete convergence of the normalized LLR processes $n^{-1}\lambda(k, k+n)$, $k = 0, 1, 2, \dots$ to the generalized Kullback–Leibler information number $I = \mathscr{K}$ under P_k is sufficient for asymptotic optimality of the Shiryaev procedure since if for all $\varepsilon > 0$

$$\sum_{n=1}^{\infty} n^{r-1} \sup_{k \in \mathbb{Z}_+} \mathsf{P}_k \left(\left| \frac{1}{n} \lambda(k, k+n) - \mathscr{K} \right| > \varepsilon \right) < \infty, \tag{3.217}$$

then both conditions $\mathbf{C}_1$ and $\widetilde{\mathbf{C}}_2$ hold (see Remark 3.1). Thus, it suffices to establish convenient conditions under which condition (3.217) holds for HMMs.

Denote by $\mathsf{P}_k^{(\omega,f)}$ the probability of the Markov chain $\{(Z_n, X_n)\}_{n \in \mathbb{Z}_+}$ that starts with the stationary initial distribution (ω, f) and conditioned on the change point $\nu = k$. Let $\mathsf{E}_k^{(\omega,f)}$ be the corresponding expectation. For brevity, we omit (ω, f) and simply write P_k and E_k.

Lemma 3.6. *Assume that conditions* $\mathbf{C}_1^h$–$\mathbf{C}_2^h$ *hold and that* $\mathsf{E}_0|\lambda(0,1)|^{r+1} < \infty$ *for some* $r \geq 1$. *Then, for all* $k \in \mathbb{Z}_+$,

$$\frac{1}{n}\lambda(k,k+n) \xrightarrow[n\to\infty]{\mathsf{P}_k} \mathscr{K} \quad uniformly - r - completely, \tag{3.218}$$

i.e., condition (3.217) *holds, and therefore, conditions* $\mathbf{C}_1$ *and* $\widetilde{\mathbf{C}}_2$ *hold as well.*

Proof. Recall that we use the notation $Y_n = (Z_n, X_n)$. By Proposition 3.5, Proposition 1 in [44], and the ergodic theorem for Markov chains, it is easy to see that the upper Lyapunov exponent for the Markov chain $\{W_n\}_{n\in\mathbb{Z}_+}$ under the probability P_0 is nothing but the relative entropy defined as

$$\begin{aligned} H(\mathsf{P}_0,\mathsf{P}_j) &= \mathsf{E}_{\mathsf{P}_0}\left[\log p_j(Y_1|Y_0,Y_{-1},\cdots)\right] \\ &= \mathsf{E}_{\mathsf{P}_0}\left[\log \sum_{x=1}^{d}\sum_{x'=1}^{d} f_j(Y_1|Y_0;X_1=x')p_j(x,x')\mathsf{P}_j(X_0=x|Y_0,Y_{-1},\cdots)\right] \\ &= \mathsf{E}_{\mathsf{P}_0}\left[\log \|M_1^j \mathsf{P}_j(X_0=x|Y_0,Y_{-1},\cdots)\|\right]. \end{aligned}$$

Therefore, the generalized Kullback–Leibler information number $\mathscr{K}$ can be defined as in (3.215).

Next, by Proposition 3.5, we can apply Lemma D.1 in Appendix D to obtain that (3.217) holds whenever $\mathsf{E}_0|\lambda(0,1)|^{r+1} < \infty$ and $\sup_{n\geq 0}\mathsf{E}^w|\Delta(W_n)|^{r+1} < \infty$ for some $r \geq 1$, where $\Delta : \mathscr{W} \to \mathbb{R}$ is a solution of the following Poisson equation

$$\mathsf{E}^w[\Delta(W_1)] - \Delta(w) = \mathsf{E}^w[\lambda(0,1)] - \mathsf{E}^{\mathsf{Q}_{\mathrm{st}}}[\lambda(0,1)] \tag{3.219}$$

for almost every $w \in \mathscr{W}$ with $\mathsf{E}^{\mathsf{Q}}_{\mathrm{st}}[\Delta(W_1)] = 0$. Here $\mathsf{E}^w(\cdot) = \mathsf{E}_0(\cdot|W_0 = w)$ is the conditional expectation when the change occurs at $\nu = 0$ conditioned on $W_0 = w$, i.e., when the Markov chain $\{W_n\}_{n\in\mathbb{Z}_+}$ is initialized from the point w, and $\mathsf{E}^{\mathsf{Q}_{\mathrm{st}}}(\cdot) = \int \mathsf{E}^w(\cdot)\,\mathrm{d}\mathsf{Q}_{\mathrm{st}}$. Note that it follows from Proposition 3.5 and Theorem 17.4.2 of Meyn and Tweedie [91] that a solution of (3.219) exists. Moreover, condition $\sup_{n\geq 0}\mathsf{E}^w|\Delta(W_n)|^{r+1} < \infty$ for some $r \geq 1$ follows from the boundedness property in Theorem 17.4.2 of Meyn and Tweedie [91] under conditions $\mathbf{C}_1^h$–$\mathbf{C}_2^h$. This completes the proof. □

Remark 3.2. The assertions of Lemma 3.6 hold true even if the Markov chain $\{(Z_n, X_n)\}_{n\in\mathbb{Z}_+}$ is initialized from any deterministic or random point with "nice" distribution. However, the proof in this case becomes more complicated.

Now, it follows from Theorem 3.7(ii) that if the prior distribution of the change point belongs to class $\mathbf{C}(\mu)$ and $A = A_\alpha$ is so selected that $\mathsf{PFA}_\pi(T_{A_\alpha}) \leq \alpha$ and $\log A_\alpha \sim |\log\alpha|$, then as $\alpha \to 0$

$$\inf_{T\in\mathbb{C}_\pi(\alpha)} \mathscr{R}_k^r(T) \sim \left(\frac{|\log\alpha|}{\mathscr{K}+\mu}\right)^r \sim \mathscr{R}_k^r(T_{A_\alpha}) \quad \text{for all } k \in \mathbb{Z}_+ \tag{3.220}$$

and

$$\inf_{T\in\mathbb{C}_\pi(\alpha)} \overline{\mathscr{R}}_\pi^r(T) \sim \left(\frac{|\log\alpha|}{\mathscr{K}+\mu}\right)^m \sim \overline{\mathscr{R}}_\pi^m(T_{A_\alpha}). \tag{3.221}$$

Also, it follows from Theorem 3.8 that if the prior distribution $\pi^\alpha = \{\pi_k^\alpha\}$ of the change point ν satisfies condition (3.20) with $\mu = \mu_\alpha \to 0$ as $\alpha \to 0$ and μ_α approaches zero at such a rate that condition (3.121) is satisfied, then asymptotic approximations (3.220) and (3.221) hold with $\mu = 0$. Thus, the Shiryaev rule T_{A_α} minimizes asymptotically to first order as $\alpha \to 0$ moments of the detection delay up to order r.

Similar asymptotic optimality results also hold true for the SR rule $\widetilde{T}_B$ when the prior is asymptotically flat, as can be expected. The details can be found in the paper by Fuh and Tartakovsky [47].

3.12.3 Higher Order Asymptotic Approximations for the Average Detection Delay and PFA

Note that when $r = 1$ in (3.220) we obtain the following first-order asymptotic approximation for the average detection delay of the Shiryaev rule as $A \to \infty$:

$$\mathsf{ADD}_k(T_A) = \left(\frac{\log A}{\mathcal{K} + \mu}\right)(1 + o(1)). \tag{3.222}$$

This approximation holds as long as conditions $\mathbf{C}_1^h$–$\mathbf{C}_2^h$ are satisfied and $\mathsf{E}_0|\lambda(0,1)| < \infty$.

In this section, we derive high-order approximations for the ADD up to a vanishing term $o(1)$ based on the Markov nonlinear renewal theory, assuming that the prior distribution of the change point is zero-modified Geometric(q,ρ) given in (2.30). We also derive a first-order asymptotic approximation for the PFA of the Shiryaev rule.

Define the statistic $S_n^\rho = S_\pi(n)/\rho$, which is given by the recursion

$$S_n^\rho = (1 + S_{n-1}^\rho)\Lambda_n^\rho, \quad n \ge 1, \quad S_0^\rho = q/[(1-q)\rho], \tag{3.223}$$

where

$$\Lambda_n^\rho = \frac{e^{\psi(W_{n-1},W_n)}}{1-\rho} = e^{\psi_\rho(W_{n-1},W_n)},$$
$$\psi_\rho(W_{i-1},W_i) = \psi(W_{i-1},W_i) + |\log(1-\rho)|.$$

Obviously, the Shiryaev rule can be written as

$$T_A = \inf\{n \ge 1 : S_n^\rho \ge A/\rho\}.$$

Note that we have

$$S_n^\rho = (1 + S_0^\rho)\prod_{i=1}^n \Lambda_i^\rho + \left(\prod_{i=1}^n \Lambda_i^\rho\right)\sum_{j=1}^{n-1}\prod_{s=1}^{j}(\Lambda_s^\rho)^{-1}$$
$$= \left(1 + S_0^\rho + \sum_{j=1}^{n-1}(1-\rho)^j e^{-\sum_{s=1}^j \psi(W_{s-1},W_s)}\right)\prod_{i=1}^n \Lambda_i^\rho,$$

and hence,

$$\log S_n^\rho = \sum_{t=1}^n \psi_\rho(W_{t-1},W_t) + \log\left(1 + S_0^\rho + V_n\right), \tag{3.224}$$

where

$$V_n = \sum_{j=1}^{n-1} e^{-\sum_{t=1}^j \psi_\rho(W_{t-1},W_t)}.$$

Let $a = \log(A/\rho)$. Obviously, using (3.224), the stopping time $T_A = T_a$ can be equivalently written as

$$T_a = \inf\{n \ge 1 : \lambda_n^\rho + \eta_n \ge a\}, \tag{3.225}$$

where $\eta_n = \log\left(1 + S_0^\rho + V_n\right)$ and

$$\lambda_n^\rho = \sum_{i=1}^n \psi_\rho(W_{i-1},W_i) = \lambda_n + n|\log(1-\rho)|, \quad n \ge 1.$$

Hereafter we will write $\lambda_n \equiv \lambda(0,n) = \sum_{i=1}^n \psi(W_{i-1},W_i)$, $n \ge 1$ for the LLR. In (3.225) the initial condition W_0 can be an arbitrarily fixed number or a random variable. Note that $\{\lambda_n\}_{n\ge1}$ is a Markov random walk with stationary mean $\mathsf{E}^{\mathsf{Q}_{\rm st}}[\lambda_1] = \mathscr{K}$, so $\{\lambda_n^\rho\}_{n\ge1}$ is a Markov random walk with stationary mean $\mathsf{E}^{\mathsf{Q}_{\rm st}}[\lambda_1^\rho] = \mathscr{K} + |\log(1-\rho)|$, where $\mathsf{E}^{\mathsf{Q}_{\rm st}}(\cdot) = \int \mathsf{E}_0(\cdot|W_0 = w)\mathrm{d}\mathsf{Q}_{\rm st}(w)$ denotes the expectation of the induced Markov chain $\{W_n\}_{n\ge0}$ under the invariant measure $\mathsf{Q}_{\rm st}$. Let $\chi_a = \lambda_n^\rho + \eta_{T_a} - a$ be a corresponding overshoot. Then we have

$$\mathsf{E}_0[\lambda_{T_a}^\rho|W_0 = w] = a + \mathsf{E}_0[\chi(a)|W_0 = w] - \mathsf{E}_0[\eta_{T_a}|W_0 = w]. \tag{3.226}$$

For $b > 0$, define

$$\tau_b = \inf\{n \ge 1 : \lambda_n^\rho \ge b\}, \tag{3.227}$$

and let $\kappa_b = \lambda_{\tau_b}^\rho - b$ (on $\{\tau_b < \infty\}$) denote the overshoot of the statistic λ_n^ρ crossing threshold b at time $n = \tau_b$. When $b = 0$, we denote τ_b in (3.227) as τ_+. For a given $w \in \mathscr{X}$, let

$$G(y,\rho,\mathscr{K}) = \lim_{b\to\infty} \mathsf{P}_1\{\kappa_b \le y|W_0 = w\} \tag{3.228}$$

be the limiting distribution of the overshoot. Note that this distribution does not depend on w.

To approximate the expected value of κ_b introduce the following notation. Let $\mathsf{P}_{0,+}(w,B) = \mathsf{P}_{0,+}\left(W_{\tau_+} \in B|W_0 = w\right)$ denote the transition probability associated with the Markov chain $\{W_n\}_{n\in\mathbb{Z}_+}$ generated by the ascending ladder variable $\lambda_{\tau_+}^\rho$. Under the V-uniform ergodicity condition and $\mathsf{E}_{\rm st}^{\mathsf{Q}}[\lambda_1] = \mathscr{K} > 0$, a similar argument as on page 255 of Fuh and Lai [46] yields that the transition probability $\mathsf{P}_{0,+}(w,\cdot)$ has an invariant measure Q_+. Let $\mathsf{E}^{\mathsf{Q}_+}$ denote expectation under W_0 having the distribution Q_+.

By Markov renewal theory (see Theorem E.8 in Appendix E),

$$\varkappa(\rho,\mathscr{K}) := \lim_{b\to\infty} \mathsf{E}_0(\kappa_b|W_0 = w) = \int_0^\infty y\,\mathrm{d}G(y,\rho,\mathscr{K}) = \frac{\mathsf{E}^{\mathsf{Q}_+}[\lambda_{\tau_+}^\rho]^2}{2\mathsf{E}^{\mathsf{Q}_+}[\lambda_{\tau_+}^\rho]}.$$

Let us also define

$$\zeta(\rho,\mathscr{K}) = \lim_{b\to\infty} \mathsf{E}_0(e^{-\kappa_b}|W_0 = w) = \int_0^\infty e^{-y}\mathrm{d}G(y,\rho,\mathscr{K})$$

and

$$C(\rho,\mathscr{K}) = \mathsf{E}_0\left\{\log\left[1 + S_0^\rho + \sum_{i=1}^\infty (1-\rho)^i e^{-\lambda_i}\right]\right\}. \tag{3.229}$$

Note that by (3.225),

$$\lambda_{T_a}^\rho = a - \eta_{T_a} + \chi_a \quad \text{on } \{T_a < \infty\},$$

where $\chi_a = \lambda_{T_a}^\rho + \eta_{T_a} - a$ is the overshoot of $\lambda_n^\rho + \eta_n$ crossing the boundary a at time T_a. Taking the expectations on both sides, using (3.226) and applying Wald's identity for Markov random walks (cf. Theorem D.2 in Appendix D), we obtain

$$\begin{aligned}(\mathscr{K} + |\log(1-\rho)|)\mathsf{E}_0(T_a|W_0 = w) + \int_{\mathscr{X}} \Delta(w')\mathrm{d}\mathsf{Q}_{\rm st}(w') - \Delta(w) &= \mathsf{E}_0(\lambda_{T_a}^\rho|W_0 = w) \\ = a - \mathsf{E}_0(\eta_{T_a}|W_0 = w) + \mathsf{E}_0(\chi_a|W_0 = w), &\end{aligned} \tag{3.230}$$

where $\Delta : \mathscr{W} \to \mathbb{R}$ solves the Poisson equation

$$\mathsf{E}_0\left[\Delta(W_1)|W_0 = w\right] - \Delta(w) = \mathsf{E}_0\left[\lambda_1^\rho|W_0 = w\right] - \mathsf{E}_{\rm st}^{\mathsf{Q}}[\lambda_1^\rho] \tag{3.231}$$

for almost every $w \in \mathscr{W}$ with $\mathsf{E}_{\rm st}^{\mathsf{Q}}[\Delta(W_1)] = 0$.

The crucial observations are that the sequence $\{\eta_n\}_{n\geq 1}$ is slowly changing and that η_n converges P_0-a.s. as $n \to \infty$ to the random variable

$$\eta = \log\left\{1 + S_0^\rho + \sum_{i=1}^{\infty}(1-\rho)^i e^{-\lambda_i}\right\} \tag{3.232}$$

with finite expectation $\mathsf{E}^{\mathsf{Q}_+}[\eta] = C(\rho, \mathscr{K})$, where $C(\rho, \mathscr{K})$ is defined in (3.229). By the first Markov Nonlinear Renewal Theorem (see Theorem F.6 in Appendix F), under mild conditions, the limiting distribution of the overshoot of a Markov random walk over threshold does not change by the addition of a slowly changing nonlinear term.

The mathematical details are given in Theorem 3.14 below. More importantly, Markov nonlinear renewal theory allows us to obtain an approximation for the probability of false alarm $\mathsf{PFA}_\pi(T_A)$ that takes the overshoot into account.

Theorem 3.14. *Let $\{Y_n = (Z_n, X_n)\}_{n\in\mathbb{Z}_+}$ be a HMM and assume that conditions $\mathbf{C}_1^h$–$\mathbf{C}_2^h$ hold. Let the prior distribution $\{\pi_k\}$ of the change point ν be the zero-modified geometric distribution* Geometric(q,ρ) *given in* (2.30), *and assume that λ_1^ρ is non-arithmetic with respect to P_∞ and P_0.*

(i) *As $A \to \infty$,*

$$\mathsf{PFA}_\pi(T_A) = \frac{\zeta(\rho, \mathscr{K})}{A}(1+o(1)). \tag{3.233}$$

(ii) *If, in addition, the second moment of the log-likelihood ratio is finite, $\mathsf{E}_0|\lambda_1|^2 < \infty$, then for $w \in \mathscr{W}$, as $A \to \infty$*

$$\begin{aligned}\mathsf{E}_0(T_A|W_0 = w) = \frac{1}{\mathscr{K} + |\log(1-\rho)|}\Bigg(&\log\frac{A}{\rho} - C(\rho,\mathscr{K}) \\ &+ \varkappa(\rho,\mathscr{K}) - \int_{\mathscr{S}} \Delta(\tilde{w})\,\mathrm{d}\mathsf{Q}_+(\tilde{w}) + \Delta(w)\Bigg) + o(1).\end{aligned} \tag{3.234}$$

Remark 3.3. The constants $\zeta(\rho,\mathscr{K})$, $\varkappa(\rho,\mathscr{K}) = \mathsf{E}^{\mathsf{Q}_+}(\lambda_{\tau_+}^\rho)^2/2\mathsf{E}^{\mathsf{Q}_+}[\lambda_{\tau_+}^\rho]$, and $C(\rho,\mathscr{K}) = \mathsf{E}^{\mathsf{Q}_+}[\eta]$ are the subject of Markov renewal and Markov nonlinear renewal theories. The constant $-\int_{\mathscr{S}} \Delta(\tilde{w})\mathrm{d}\mathsf{Q}_+(\tilde{w}) + \Delta(w)$ is due to Markovian dependence via Poisson equation (3.231).

Proof. The proof of part (i). Using the same argument as in the proof of Theorem 7.1.5(i) in [164, page 327], which is correct not only in the i.i.d. case but in a most general non-i.i.d. case too, we obtain

$$\mathsf{PFA}_\pi(T_A) = \frac{1}{\rho A}\mathsf{E}_\pi\left[e^{-\chi_a}; T_a > \nu\right](1+o(1)), \quad A \to \infty,$$

where $\chi_a = \lambda_{T_a}^\rho + \eta_{T_a} - a$. In order to evaluate the value of

$$\mathsf{E}_\pi\left[e^{-\chi_a}; T_a > \nu\right] = \sum_{k=0}^{\infty} \pi_k \mathsf{E}_k\left[e^{-\chi_a}; T_a > k\right]$$

we recall that the stopping time T_a can be represented as in (3.225), where $\lambda_n^\rho = \lambda_n + n|\log(1-\rho)|$ is a Markov random walk with the drift $\mathscr{K} + |\log(1-\rho)|$ and η_n, $n \geq 1$ are slowly changing under P_0. Since, by condition $\mathbf{C}_2^h$, $0 < \mathscr{K} < \infty$, and $\mathsf{P}_k(T_a > k) = \mathsf{P}_\infty(T_a > k) \to 1$ as $a \to \infty$, we can apply Theorem F.6 in Appendix F to obtain

$$\lim_{a\to\infty} \mathsf{E}_k\left[e^{-\chi_a}; T_a > k\right] = \int_0^\infty e^{-y}\,\mathrm{d}G(y,\rho,\mathscr{K}) = \zeta(\rho,\mathscr{K}),$$

which implies that $\lim_{a\to\infty} \mathsf{E}_\pi\left[e^{-\chi_a}; T_a > \nu\right] = \zeta(\rho,\mathscr{K})$ and completes the proof of (3.233).

The proof of part (ii). The probability P_0 and expectation E_0 in the proof below are taken under $W_0 = w$, i.e., $\mathsf{P}_0(\cdot|W_0 = w)$ and $\mathsf{E}_0(\cdot|W_0 = w)$. We omit conditioning on $W_0 = w$ for brevity. The proof of (3.234) is based on the second Markov nonlinear renewal theorem (see Theorem F.7 in Appendix F).

By (3.225), the stopping time $T_A = T_a$ is based on thresholding the sum of the Markov random walk λ_n^ρ and the perturbation term η_n. Note that $\eta_n \to \eta$ P_0-a.s. and $\mathsf{E}_0[\eta_n] \to \mathsf{E}_0[\eta]$ as $n \to \infty$, so η_n, $n \geq 1$ are slowly changing under P_0. In order to apply Theorem F.7 in our case, we have to check the following three conditions:

$$\sum_{n=1}^{\infty} \mathsf{P}_0\{\eta_n \leq -\varepsilon n\} < \infty \text{ for some } 0 < \varepsilon < \mathscr{K}; \tag{3.235}$$

$$\max_{0 \leq k \leq n} |\eta_{n+k}|,\ n \geq 1, \text{ are } \mathsf{P}_0\text{-uniformly integrable;} \tag{3.236}$$

$$\lim_{a\to\infty} a\, \mathsf{P}_0\left\{T_A \leq \frac{\varepsilon a}{\mathscr{K} + |\log(1-\rho)|}\right\} = 0 \quad \text{for some } 0 < \varepsilon < 1. \tag{3.237}$$

Condition (3.235) obviously holds because $\eta_n \geq 0$. Condition (3.236) holds because η_{2n}, $n \geq 1$, are P_0-uniformly integrable since $\eta_{2n} \leq \eta$ and $\mathsf{E}_0[\eta] < \infty$ and $\max_{0 \leq k \leq n} |\eta_{n+k}| = \eta_{2n}$.

To verify condition (3.237) introduce

$$N_{a,\varepsilon} = \frac{(1-\varepsilon)a}{\mathscr{K} + |\log(1-\rho)|}; \quad p(a,\varepsilon) = \mathsf{P}_0\left(\frac{1}{N_{a,\varepsilon}} \max_{1\leq n \leq N_{a,\varepsilon}} \lambda_n \geq (1+\varepsilon)\mathscr{K}\right).$$

Using inequalities (3.38) and (3.40) with $k = 0$, $T = T_a$, $\alpha = e^{-a}$, $\mu = |\log(1-\rho)|$, $N_\alpha = N_{a,\varepsilon}$ and the fact that for the zero-modified geometric prior $\log \Pi_{N_{a,\varepsilon}-1} = \log(1-q) + N_{a,\varepsilon} \log(1-\rho)$, we obtain that for all $a = \log A > 0$ and $0 < \varepsilon < 1$

$$\mathsf{P}_0(T_a \leq N_{a,\varepsilon}) \leq \frac{1}{1-q} \exp\{-\varepsilon^2 a\} + p(a,\varepsilon). \tag{3.238}$$

Hence, it suffices to show that under the second moment condition $\mathsf{E}_0|\lambda_1|^2 < \infty$ the probability $p(a,\varepsilon)$ goes to 0 as $a \to \infty$ faster than $1/a$, i.e., $\lim_{a\to\infty}[p(a,\varepsilon)a] = 0$. By Proposition 3.5, we can apply Lemma D.1 which yields

$$\sum_{n=1}^{\infty} \mathsf{P}_0\left\{\max_{1\leq k \leq n} (\lambda_k - k\mathscr{K}) \geq \varepsilon n\right\} < \infty \quad \text{for all } \varepsilon > 0$$

whenever $\mathsf{E}_0|\lambda_1|^2 < \infty$. This implies that the summand is $o(1/n)$ for a large n. Owing to the fact that

$$p(a,\varepsilon) = \mathsf{P}_0\left(\max_{1\leq n\leq N_{a,\varepsilon}} (\lambda_n - N_{a,\varepsilon}\mathscr{K}) \geq \varepsilon \mathscr{K} N_{a,\varepsilon}\right) \leq \mathsf{P}_0\left(\max_{1\leq n\leq N_{a,\varepsilon}} (\lambda_n - n\mathscr{K}) \geq \varepsilon \mathscr{K} N_{a,\varepsilon}\right)$$

we obtain that $p(a,\varepsilon) = o(1/a)$ as $a \to \infty$. This implies condition (3.237).

An appeal to the second Markov Nonlinear Renewal Theorem F.7 completes the proof. □

An analogous result can be also established for the SR procedure $\widetilde{T}_B$ defined in (3.7), (3.8). Since this procedure is a limit of the Shiryaev procedure as the parameter of the geometric prior distribution ρ goes to zero, it is intuitively obvious that the higher-order approximation for the conditional average detection delay $\mathsf{E}_0(\widetilde{T}_B|W_0 = w)$ is given by (3.234) with $\rho = 0$ and $A/\rho = B$. This is indeed the case as the following theorem shows. The argument is essentially similar to the proof of Theorem 3.14(ii).

Theorem 3.15. *Let $(Z_n, X_n)_{n\in\mathbb{Z}_+}$ be a hidden Markov model. Assume conditions $\mathbf{C}_1^h$–$\mathbf{C}_2^h$ hold. Let the prior distribution of the change point be the zero-modified geometric distribution* (2.30). *Assume that λ_1 is non-arithmetic with respect to P_∞ and P_0 and that $\mathsf{E}_0|\lambda_1|^2 < \infty$. Then for $w \in \mathscr{W}$, as $B \to \infty$*

$$\mathsf{E}_0(\widetilde{T}_B|W_0 = w) = \frac{1}{\mathscr{K}}\Bigg(\log B - \mathsf{E}^{\mathsf{Q}_+}[\tilde{\eta}] + \varkappa(\rho,\mathscr{K}) \\ - \int_{\mathscr{S}} \Delta(\tilde{w})\,\mathrm{d}\mathsf{Q}_+(\tilde{w}) + \Delta(w)\Bigg) + o(1), \tag{3.239}$$

where $\tilde{\eta} = \log(1+\ell+\sum_{j=1}^\infty e^{-\lambda_j})$ and $\ell = \lim_{\rho\to\infty}[q(\rho)/\rho]$.

The asymptotic approximations for $\mathsf{ADD}_0(T_A|W_0 = w) = \mathsf{E}_0[T_A|W_0 = w]$ can be extended to the case of an arbitrarily point of change $\nu = k$, i.e., for $\mathsf{ADD}_k(T_A|W_k = w) = \mathsf{E}_k[T_A - k|T_A > k, W_k = w]$. The difference compared to (3.234) is only in the constant $C(\rho,\mathscr{K})$, which now depends on k:

$$C_k(\rho,\mathscr{K}) = \mathsf{E}_k\left[\log\left(1 + S_k^\rho + \sum_{i=k+1}^{\infty}(1-\rho)^i e^{-\lambda_i}\right)\right].$$

Specifically, we have that as $A \to \infty$

$$\mathsf{ADD}_k(T_A|W_k = w) = \frac{1}{\mathscr{K} + |\log(1-\rho)|}\Bigg(\log\frac{A}{\rho} - C_k(\rho,\mathscr{K}) \\ + \varkappa(\rho,\mathscr{K}) - \int_{\mathscr{S}} \Delta(\tilde{w})\,\mathrm{d}\mathsf{Q}_+(\tilde{w}) + \Delta(w)\Bigg) + o(1). \tag{3.240}$$

Write $\mathsf{ADD}_k(T_A) = \mathsf{E}_k[T_A - k|T_A > k]$. Since $\{W_n\}_{n\in\mathbb{Z}_+}$ is an ergodic Markov process, averaging over the stationary distribution $\mathrm{d}\mathsf{Q}_{\mathrm{st}}(w)$ on the right-hand side in (3.240) yields

$$\mathsf{ADD}_k(T_A) = \frac{1}{\mathscr{K} + |\log(1-\rho)|}\left(\log\frac{A}{\rho} - C_k(\rho,\mathscr{K}) + \varkappa(\rho,\mathscr{K}) - \bar{\Delta}_+ + \bar{\Delta}\right) + o(1), \tag{3.241}$$

where

$$\bar{\Delta}_+ = \int_{\mathscr{S}} \Delta(v)\,\mathrm{d}\mathsf{Q}_+(v), \quad \bar{\Delta} = \int_{\mathscr{W}} \Delta(w)\,\mathrm{d}\mathsf{Q}_{\mathrm{st}}(w).$$

Also, if we are interested in the asymptotic approximation for the $\mathsf{ADD}(T_A) = \mathsf{E}_\pi[T_A - \nu|T_A > \nu] = \sum_{k=0}^\infty \pi_k \mathsf{E}_k[T_A - k|T_A > k]$, then formally averaging the right-hand side in (3.241) over k, we obtain

$$\mathsf{ADD}(T_A) = \frac{1}{\mathscr{K} + |\log(1-\rho)|}\left(\log\frac{A}{\rho} - \bar{C}(\rho,\mathscr{K}) + \varkappa(\rho,\mathscr{K}) - \bar{\Delta}_+ + \bar{\Delta}\right) + o(1), \tag{3.242}$$

where $\bar{C}(\rho,\mathscr{K}) = qC_0(\rho,\mathscr{K}) + (1-q)\sum_{k=0}^\infty \rho(1-\rho)^k C_k(\rho,\mathscr{K})$.

It should be noted that while the asymptotic approximations (3.233), (3.234), (3.239), (3.240)–(3.242) are of certain theoretical interest, their practical significance is minor. The reason is that Markov renewal-theoretic and nonlinear renewal-theoretic constants $\zeta(\rho,\mathscr{K})$, $\varkappa(\rho,\mathscr{K})$ and $C(\rho,\mathscr{K})$ usually cannot be computed either analytically or numerically. Also, currently, there are no numerical techniques for solving (two-dimensional) integral equations for performance metrics (average detection delays, PFA, etc.) of change detection rules similar to those developed in the i.i.d. case when the detection statistics are Markov (see, e.g., [97, 163, 164]). Therefore, development of such numerical techniques is an important problem that still is waiting for its solution. Without such techniques the only way we see is evaluation of the required constants by Monte Carlo. But such MC simulations do not give any advantage over the straightforward MC evaluation of operating characteristics.

3.12.4 The Case of Conditionally Independent Observations

In many applications, including several examples considered below, one is interested in a simplified HMM where the observations X_n, $n = 1,2,\dots$ are conditionally independent, conditioned on the Markov chain Z_n, i.e.,

$$f_j(X_n|Z_n,X_{n-1}) = f_j(X_n|Z_n), \quad j = 0,\infty.$$

In this particular case, the weight function $V = 1$, and hence, the conditions $\mathbf{C}_1^h$ and $\mathbf{C}_2^h$ in the above theorems can be simplified to the following conditions.

$\widehat{\mathbf{C}}_1^h$. For each $j = \infty, 0$, the Markov chain $\{Z_n\}_{n\in\mathbb{Z}_+}$, defined in (3.202)) and (3.203), is ergodic (positive recurrent, irreducible and aperiodic) on a finite state space $D = \{1,\cdots,d\}$ and has stationary probability ω_j.

$\widehat{\mathbf{C}}_2^h$. The generalized Kullback–Leibler information number is positive and finite, $0 < \mathscr{K} < \infty$. For each $j = \infty, 0$, the random matrices M_0^j and M_1^j, defined in (3.207) and (3.208), are invertible P_j almost surely and for some $r > 0$,

$$\left| \sum_{z_0,z_1\in\mathscr{Z}} \int_{x\in\mathbb{R}^d} \omega_j(z_0) p_j(z_0,z_1) x^{r+1} f_j(x|z_1) Q(\mathrm{d}x) \right| < \infty. \tag{3.243}$$

Example 3.2 (*Detection of changes in a two-state HMM with i.i.d. observations*). The following example, which deals with a two-state hidden Markov model with i.i.d. observations in each state, may be of interest, in particular, for rapid detection and tracking of sudden spurts and downfalls in activity profiles of terrorist groups that could be caused by various factors such as changes in the organizational dynamics of terrorist groups, counterterrorism activity, changing socio-economic and political contexts, etc. In [119], based on the analysis of real data from *Fuerzas Armadas Revolucionarias de Colombia* (FARC) terrorist group from Colombia (RDWTI) it was shown that two-state HMMs can be recommended for detecting and tracking sudden changes in activity profiles of terrorist groups. The HMM framework provides good explanation capability of past/future activity across a large set of terrorist groups with different ideological attributes.

Consider a binary-state case with i.i.d. observations in each state. Specifically, let θ be a parameter taking two possible values θ_0 and θ_1 and let $Z_n \in \{1,2\}$ be a two-state ergodic Markov chain with the transition matrix

$$[p_\theta(i,l)] = \begin{bmatrix} 1-p_\theta & p_\theta \\ \tilde{p}_\theta & 1-\tilde{p}_\theta \end{bmatrix}$$

and stationary initial distribution $\mathsf{P}_\theta(Z_0 = 2) = 1 - \mathsf{P}_\theta(Z_0 = 1) = \omega_\theta(2) = \tilde{p}_\theta/(p_\theta + \tilde{p}_\theta)$ for some $\{p_\theta, \tilde{p}_\theta\} \in [0,1]$. Further, assume that conditioned on Z_n the observations X_n are i.i.d. with densities $f_\theta(x|Z_n = l) = f_\theta^{(l)}(x)$ $(l = 1,2)$, where the parameter $\theta = \theta_0$ pre-change and $\theta = \theta_1$ post-change. In other words, in this scenario,

$$p(X_n|\mathbf{X}_0^{n-1}) = \begin{cases} p_{\theta_0}(X_n|\mathbf{X}_0^{n-1}) & \text{if } n \le \nu \\ p_{\theta_1}(X_n|\mathbf{X}_0^{n-1}) & \text{if } n > \nu \end{cases},$$

and therefore, the increment of the likelihood ratio $\mathscr{L}_n = LR_n/LR_{n-1}$ does not depend on the change point. As a result, the Shiryaev detection statistic obeys the recursion (3.249), so that in order to implement the Shiryaev procedure it suffices to develop an efficient computational scheme for the likelihood ratio $LR_n = p_{\theta_1}(\mathbf{X}_0^n)/p_{\theta_0}(\mathbf{X}_0^n)$. To obtain a recursion for LR_n, define the probabilities $P_{\theta,n} := \mathsf{P}_\theta(\mathbf{X}_0^n, Z_n = 2)$ and $\widetilde{P}_{\theta,n} := \mathsf{P}_\theta(\mathbf{X}_0^n, Z_n = 1)$. Straightforward argument shows that for $n \ge 1$

$$P_{\theta,n} = \left[P_{\theta,n-1}\, p_\theta(2,2) + \widetilde{P}_{\theta,n-1}\, p_\theta(1,2)\right] f_\theta^{(2)}(X_n); \tag{3.244}$$

$$\widetilde{P}_{\theta,n} = \left[P_{\theta,n-1}\, p_\theta(2,1) + \widetilde{P}_{\theta,n-1}\, p_\theta(1,1)\right] f_\theta^{(1)}(X_n) \tag{3.245}$$

with $P_{\theta,0} = \omega_\theta(2)$ and $\widetilde{P}_{\theta,0} = \omega_\theta(1) = 1 - \omega_\theta(2)$. Since $p_\theta(\mathbf{X}_0^n) = P_{\theta,n} + \widetilde{P}_{\theta,n}$ we obtain that

$$LR_n = \frac{P_{\theta_1,n} + \widetilde{P}_{\theta_1,n}}{P_{\theta_0,n} + \widetilde{P}_{\theta_0,n}}.$$

Therefore, to implement the Shiryaev (or the SR) rule we have to update and store $P_{\theta,n}$ and $\widetilde{P}_{\theta,n}$ for the two parameters values θ_0 and θ_1 using simple recursions (3.244) and (3.245).

Condition $\widehat{\mathbf{C}}_1^h$ obviously holds. Assume that the observations are Gaussian with unit variance and different mean values in pre- and post-change regimes as well as for different states, i.e., $f_\theta^{(l)}(x) = \varphi(x - \mu_\theta^{(l)})$ ($\theta = \theta_0, \theta_1$, $l = 1,2$), where $\varphi(x) = (2\pi)^{-1/2} \exp\{-y/2\}$ is density of the standard normal distribution. It is easily verified that the generalized Kullback–Leibler number is finite. Condition (3.243) in $\widehat{\mathbf{C}}_2^h$ has the form (for $\theta = \theta_0, \theta_1$)

$$\left| \sum_{i,l=1,2} \omega_\theta(i) p_\theta(i,l) \int_{-\infty}^{\infty} x^{r+1} \varphi(x - \mu_\theta^{(l)})\, \mathrm{d}x \right| < \infty,$$

which holds for all $r \geq 1$ due to the finiteness of all absolute positive moments of the normal distribution and the fact that

$$\left| \sum_{i,l=1,2} \omega_\theta(i) p_\theta(i,l) \int_{-\infty}^{\infty} x^{r+1} \varphi(x - \mu_\theta^{(l)})\, \mathrm{d}x \right| \\ \leq \sum_{i,l=1,2} \omega_\theta(i) p_\theta(i,l) \int_{-\infty}^{\infty} |x|^{r+1} \varphi(y - \mu_\theta^{(l)})\, \mathrm{d}x.$$

Therefore, the Shiryaev detection rule is nearly optimal, minimizing asymptotically all positive moments of the detection delay.

Example 3.3 (*A two-state HMM with conditionally independent observations*). Consider an example which is motivated by certain multisensor target track management applications [16] that are discussed later on in Chapter 8, Section 8.1.

Let $Z_n \in \{1,2\}$ be a two-state (hidden, unobserved) Markov chain with transition probabilities $\mathsf{P}_j(Z_n = 1|Z_{n-1} = 2) = p_j(2,1) = p$ and $\mathsf{P}_j(Z_n = 2|Z_{n-1} = 1) = p_j(1,2) = \tilde{p}$, $n \geq 1$ and initial stationary distribution $\mathsf{P}_j(Z_0 = 1) = \omega_j(1) = p/(p+\tilde{p})$ for both $j = \infty$ and $j = 0$. Under the pre-change hypothesis $\mathsf{H}_\infty : \nu = \infty$, the conditional density of the observation X_n is

$$p(X_n|\mathbf{X}_0^{n-1}, Z_n = l, \mathsf{H}_\infty) = g_l(X_n) \quad \text{for } l = 1,2,\ n \geq 1,$$

and under the hypothesis $\mathsf{H}_k : \nu = k$, the observations $X_{k+1}, X_{k+2}, \dots$ are i.i.d. with density $f(x)$. The pre-change joint density of the vector $\mathbf{X}_0^n$ is $p_\infty(\mathbf{X}_0^n) = \prod_{i=1}^n p_\infty(X_i|\mathbf{X}_0^{i-1})$, where

$$p_\infty(X_i|\mathbf{X}_0^{i-1}) = \sum_{l=1}^{2} g_l(X_i) \mathsf{P}(Z_i = l|\mathbf{X}_0^{i-1}), \quad i \geq 1, \tag{3.246}$$

and the posterior probability $\mathsf{P}(Z_i = l|\mathbf{X}_0^{i-1}) = P_{i|i-1}(l)$ is obtained by a Bayesian update as follows. By the Bayes rule, the posterior probability $\mathsf{P}(Z_i = l|\mathbf{X}_0^{i}) := P_i(l)$ is given by

$$P_i(l) = \frac{g_l(X_i) P_{i|i-1}(l)}{\sum_{s=1}^2 g_s(X_i) P_{i|i-1}(s)}. \tag{3.247}$$

The probability $P_{i|i-1}(Z_i)$ is used as the prior probability for the update (prediction term) and can be computed as

$$\begin{aligned} P_{i|i-1}(2) &= P_{i-1}(2)(1-p) + P_{i-1}(1)\tilde{p}, \\ P_{i|i-1}(1) &= P_{i-1}(1)(1-q) + P_{i-1}(2)p. \end{aligned} \tag{3.248}$$

The Shiryaev statistic $S_\pi(n)$ can be computed recursively as

$$S_\pi(n) = [S_\pi(n-1) + \pi_{n,n}]\,\mathscr{L}_n, \quad n \ge 1, \quad S_\pi(0) = q/(1-q), \tag{3.249}$$

where the likelihood ratio "increment" $\mathscr{L}_n = f(X_n)/p_\infty(X_n|\mathbf{X}_0^{n-1})$ can be effectively computed using (3.246), (3.247) and (3.248) and

$$\pi_{n,n} = \frac{\pi_n}{\sum_{k=n+1}^\infty \pi_k}.$$

Therefore, in this example, the computational cost of the Shiryaev rule is small, and it can be easily implemented on-line.

Condition $\widehat{\mathbf{C}}_1^h$ obviously holds. Condition (3.243) in $\widehat{\mathbf{C}}_2^h$ holds if

$$\int_{-\infty}^{\infty} |x|^{r+1} f(x)\,\mathrm{d}Q(x) < \infty, \quad \int_{-\infty}^{\infty} |x|^{r+1} g_l(x)\,\mathrm{d}Q(x) < \infty \;\text{ for } l = 1,2$$

since

$$\left| \sum_{i,l=1,2} \int_{-\infty}^{\infty} \omega_0(i) p_0(i,l) x^{r+1} f(x)\,\mathrm{d}Q(x) \right| \le \int_{-\infty}^{\infty} |x|^{r+1} f(x)\,\mathrm{d}Q(x)$$

and

$$\left| \sum_{i,l=1,2} \int_{-\infty}^{\infty} \omega_\infty(i) p_\infty(i,l) x^{r+1} g_l(x)\,\mathrm{d}Q(x) \right| \le \sum_{i,l=1,2} \omega_\infty(i) p_\infty(i,l) \int_{-\infty}^{\infty} |x|^{r+1} g_l(x)\,\mathrm{d}Q(x).$$

In this case, the generalized Kullback–Leibler number $\mathscr{K}$ is obviously finite. Therefore, the Shiryaev detection rule is nearly optimal, minimizing asymptotically moments of the detection delay up to order r. In particular, if $f(x)$ and $g_l(x)$ are Gaussian densities, then the Shiryaev procedure minimizes all positive moments of the delay to detection.

3.13 Additional Examples

Example 3.4 (*Detection of a change of variance in normal population with unknown mean*)**.** Let observations $X_n \sim \mathscr{N}(\theta, \sigma_\infty^2)$ be i.i.d. normal with variance σ_∞^2 before change and i.i.d. normal $\mathscr{N}(\theta, \sigma_0^2)$ with variance σ_0^2 after change with the same unknown mean θ. Formally, this problem is not in the class of problems considered in this chapter since both pre- and post-change densities depend on an unknown nuisance parameter θ, and hence, the pre-change hypothesis is also composite. However, this problem can be reduced to the problem of testing simple hypotheses using the principle of invariance, since it is invariant under the group of shifts $\{G_b(x) = x + b\}_{-\infty<b<\infty}$. The maximal invariant is $\mathbf{Y}^n = (Y_1, \dots, Y_n)$, $n \ge 2$, where $Y_k = X_k - X_1$, $Y_1 = 0$, and we can now consider a transformed sequence of observations $\{Y_n\}_{n\ge 2}$, which are not i.i.d. and not even independent anymore. Pre- and post-change densities of $\mathbf{Y}^n$ are equal to

$$\begin{aligned} p_i(\mathbf{Y}^n) &= \frac{1}{(2\pi\sigma_i^2)^{n/2}} \int_{-\infty}^{\infty} e^{-\frac{1}{2\sigma_i^2}\sum_{k=1}^n (Y_k+\theta)^2}\,\mathrm{d}\theta \\ &= \frac{1}{(2\pi\sigma_i^2)^{(n-1)/2}\sqrt{n}} e^{-\frac{1}{2\sigma_i^2}\sum_{k=1}^n (Y_k - \overline{Y}_n)^2}, \quad i = \infty, 0, \end{aligned}$$

where $\overline{Y}_n = n^{-1}\sum_{k=1}^n Y_k$. Here and in the rest of this example we use the notation p_∞ and p_0 for the pre-change density g and the post-change density f, respectively. Define $\overline{X}_n = n^{-1}\sum_{k=1}^n X_k$, $s_n^2 = (n-1)^{-1}\sum_{k=1}^n (X_k - \overline{X}_n)^2$, and $V_n = (n-1)s_n^2 - (n-2)s_{n-1}$. Noting that $\sum_{k=1}^n (Y_k - \overline{Y}_n)^2 = (n-1)s_n^2$, we obtain

$$p_i(Y_j|\mathbf{Y}^{j-1}) = \frac{1}{\sqrt{2\pi\sigma_i^2}}\sqrt{\frac{j-1}{j}}e^{-V_j/2\sigma_i^2}, \quad j \geq 2,$$

and therefore, the invariant LLR is

$$\lambda_\vartheta(k,k+n) = \sum_{j=k+1}^{k+n} \log\frac{p_0(Y_j|\mathbf{Y}^{j-1})}{p_\infty(Y_j|\mathbf{Y}^{j-1})} = \frac{\vartheta^2-1}{2\sigma_0^2}\sum_{j=k+1}^{k+n} V_j - (n-1)\log\vartheta, \quad n \geq 2$$

($\lambda_\vartheta(k,k+1) = 0$), where $\vartheta = \sigma_0/\sigma_\infty$. Taking into account that $\sum_{j=1}^{k+n} V_j = (k+n-1)s_{k+n}^2$, we have

$$\sum_{j=k+1}^{k+n} V_j = (k+n-1)s_{k+n}^2 - ks_{k+1}^2,$$

so using notation $S_{k,n}^2 = (k+n-1)s_{k+n}^2 - ks_{k+1}^2$ the LLR can be written as

$$\lambda_\vartheta(k,k+n) = \frac{\vartheta^2-1}{2\sigma_0^2}S_{k,n}^2 - (n-1)\log\vartheta, \quad n \geq 2. \tag{3.250}$$

Thus, we can now construct the invariant Shiryaev and SR rules based on the LLRs $\lambda_\vartheta(k,k+n)$, $n \geq 2$, defined in (3.250).

Note first that $S_{k,n}^2/n \to \sigma_0^2$ as $n \to \infty$ almost surely under $\mathsf{P}_{k,\vartheta}$, so that

$$n^{-1}\lambda_\vartheta(k,k+n) \xrightarrow[n\to\infty]{\mathsf{P}_{k,\vartheta}\text{-a.s.}} \frac{\vartheta^2-1}{2} - \log\vartheta = I_\vartheta,$$

and I_ϑ is positive for any $\vartheta \neq 1$ ($\vartheta > 0$). Thus, by Lemma B.1 in Appendix B, condition $\mathbf{C}_1$ holds with $I_\vartheta = (\vartheta^2-1)/2 - \log\vartheta$ and to apply the results of Section 3.7 it suffices to show that condition $\widetilde{\mathbf{C}}_2$ holds. We now show that this condition holds for all $r \geq 1$, i.e.,

$$\sup_{k\in\mathbb{Z}_+} \Upsilon_{k,r}(\theta,\varepsilon) < \infty \quad \text{for all } r \geq 1 \text{ and } \varepsilon > 0. \tag{3.251}$$

To this end, note that the statistic $S_{k,n}^2$ can be written as

$$S_{k,n}^2 = \sum_{i=k+1}^{k+n}(X_i - \overline{X}_{k+n}^{k+1})^2 + k(\overline{X}_{k+n} - \overline{X}_k)^2 + n(\overline{X}_{k+n} - \overline{X}_{k+n}^{k+1})^2,$$

where $\overline{X}_{k+n}^{k+1} = n^{-1}\sum_{i=k+1}^{k+n} X_i$. Denoting

$$k(\overline{X}_{k+n} - \overline{X}_k)^2 + n(\overline{X}_{k+n} - \overline{X}_{k+n}^{k+1})^2 = W_{k,n},$$

and $\gamma_{k,n} = (2\sigma_0^2\log\vartheta)/(\vartheta^2-1)n - W_{k,n}/n$ and using the fact that $W_{k,n} \geq 0$, we obtain that for some positive $\tilde{\varepsilon}$

$$\mathsf{P}_{k,\vartheta}\left(\frac{1}{n}\lambda_\vartheta(k,k+n) < I_\vartheta - \varepsilon\right) = \mathsf{P}_{k,\vartheta}\left(\frac{1}{n}\sum_{i=k+1}^{k+n}(X_i - \overline{X}_{k+n}^{k+1})^2 < \sigma_0^2 - \gamma_{k,n} - \tilde{\varepsilon}\right)$$

$$
\begin{aligned}
&\le \mathsf{P}_{k,\vartheta}\left(\frac{1}{n}\sum_{i=k+1}^{k+n}(X_i-\overline{X}_{k+n}^{k+1})^2<\sigma_0^2-\tilde{\varepsilon}\right)\\
&=\mathsf{P}_{0,\vartheta}\left(\frac{1}{n}\sum_{i=1}^{n}(X_i-\overline{X}_n)^2<\sigma_0^2-\tilde{\varepsilon}\right)\\
&=\mathsf{P}_{0,\vartheta}\left(\frac{1}{n}(n-1)s_n^2<\sigma_0^2-\tilde{\varepsilon}\right).
\end{aligned}
$$

Since $(n-1)s_n^2/\sigma_0^2$ has chi-squared distribution with $n-1$ degrees of freedom, $\mathsf{P}_{0,\vartheta}\left((n-1)s_n^2/n-\sigma_0^2<-\tilde{\varepsilon}\right)$ vanishes exponentially fast as $n\to\infty$ and it follows that for all $\tilde{\varepsilon}>0$ and all $r\ge 1$

$$
\begin{aligned}
&\sum_{n=1}^{\infty} n^{r-1}\sup_{k\in\mathbb{Z}_+}\mathsf{P}_{k,\vartheta}\left(\frac{1}{n}\lambda_\vartheta(k,k+n)<I_\vartheta-\varepsilon\right)\\
&\le\sum_{n=1}^{\infty} n^{r-1}\mathsf{P}_{0,\vartheta}\left(\frac{1}{n}(n-1)s_n^2<\sigma_0^2-\tilde{\varepsilon}\right)<\infty,
\end{aligned}
$$

i.e., (3.251) holds for all $r\ge 1$.

By Theorems 3.7, 3.8 and 3.10 the Shiryaev detection rule and the window-limited Shiryaev rule minimize asymptotically all positive moments of the detection delay.

Example 3.5 (*Detection of a change in the randomly switching correlation coefficient of the AR*(1) *process*)**.** Consider the extension of Example 3.1 to the HMM with switching correlation coefficient in the post-change mode. Specifically, let the observations follow the AR(1) model of the form

$$
X_n=(a_0\mathbb{1}_{\{n<\nu\}}+a(Z_n)\mathbb{1}_{\{n\ge\nu\}})X_{n-1}+\xi_n,\quad n\ge 1,\tag{3.252}
$$

where $Z_n\in\{1,\dots,d\}$ is a d-state unobservable ergodic Markov chain and, conditioned on $Z_n=l$, $a(Z_n=l)=a_l$, $l=1,\dots,d$. The noise sequence $\{\xi_n\}$ is the i.i.d. standard Gaussian sequence, $\xi_n\sim\mathcal{N}(0,1)$. Thus, the problem is to detect a change in the correlation coefficient of the Gaussian first-order AR process from the known value a_0 to a random value $a(Z_n)\in\{a_1,\dots,a_d\}$ with possible switches between the given levels $a_1,\dots,a_d$ for $n\ge\nu$.

We assume that the transition matrix $[p(i,l)]$ is positive definite, i.e., $\det[p(i,l)]>0$ (evidently, it does not depend on $j=0,\infty$) and that $|a_i|<1$ for $i=0,1,\dots,d$. The likelihood ratio for X_n given $\mathbf{X}_0^{n-1}$ and $Z_n=l$ between the hypotheses H_k and H_∞ is

$$
\frac{p(X_n|\mathbf{X}_0^{n-1},Z_n=l,\mathsf{H}_k)}{p(X_n|\mathbf{X}_0^{n-1},\mathsf{H}_\infty)}=\exp\left\{\frac{1}{2}\left[(X_n-a_0X_{n-1})^2-(X_n-a_lX_{n-1})^2\right]\right\},\ n\ge k,
$$

so that the likelihood ratio $\mathscr{L}_n=p_0(X_n|\mathbf{X}_0^{n-1})/p_\infty(X_n|\mathbf{X}_0^{n-1})$ can be computed as

$$
\mathscr{L}_n=\sum_{l=1}^{d}\exp\left\{\frac{1}{2}\left[(X_n-a_0X_{n-1})^2-(X_n-a_lX_{n-1})^2\right]\right\}\mathsf{P}(Z_n=l|\mathbf{X}_0^{n-1}),
$$

where using the Bayes rule, we obtain

$$
\begin{aligned}
\mathsf{P}(Z_n=l|\mathbf{X}_0^{n-1})&=\sum_{i=1}^{d}p(i,l)\mathsf{P}(Z_{n-1}=i|\mathbf{X}_0^{n-1}),\\
\mathsf{P}(Z_n=l|\mathbf{X}_0^{n})&=\frac{\mathsf{P}(Z_n=l|\mathbf{X}_0^{n-1})\exp\{-\frac{1}{2}(X_n-a_lX_{n-1})^2\}}{\sum_{i=1}^{d}\mathsf{P}(Z_n=i|\mathbf{X}_0^{n-1})\exp\{-\frac{1}{2}(X_n-a_iX_{n-1})^2\}}.
\end{aligned}
$$

The Markov chain (Z_n, X_n) is V-uniformly ergodic with the Lyapunov function $V(z,x) = c(1+x^2+z^2)$, where $c \geq 1$, so condition $\mathbf{C}_1^h$ is satisfied. The condition $\mathbf{C}_2^h$ also holds. Indeed, if the change occurs from a_0 to the ith component with probability 1, i.e., $\mathsf{P}(Z_n = i) = 1$ for $n \geq \nu$, then the Kullback–Leibler information number is equal to

$$I_i = \frac{(a_i - a_0)^2}{2(1-a_i^2)}$$

(see (3.201)). Hence,

$$0 < \mathscr{K} < \frac{(\max_{1\leq i\leq d} a_i - a_0)^2}{2(1-\max_{1\leq i\leq d} a_i^2)} < \infty.$$

The condition (3.216) in $\mathbf{C}_2^h$ holds for all $r \geq 1$ with $V(z,x) = c(1+z^2+x^2)$ since all moments of the Gaussian distribution are finite.

By Theorems 3.7, 3.8 and 3.10 the Shiryaev detection rule and the window-limited Shiryaev rule minimize asymptotically all positive moments of the detection delay.

Example 3.6 (*Detection of signals with unknown amplitudes in a multichannel system*)**.** Assume there is a multichannel system with N channels (or alternatively an N-sensor system) and one is able to observe the output vector $X_n = (X_n^1, \dots, X_n^N)$, $n = 1, 2, \dots$, where the observations in the ith channel are of the form

$$X_n^i = \theta_i S_n^i \mathbb{1}_{\{n>\nu\}} + \xi_n^i, \quad n \geq 1.$$

Here $\theta_i S_n^i$ is a deterministic signal with an unknown amplitude $\theta_i > 0$ that may appear at an unknown time ν in additive noise ξ_n^i. For the sake of simplicity, suppose that all signals appear at the same unknown time ν. Assume that noises $\{\xi_n^i\}_{n\in\mathbb{Z}_+}$, $i = 1, \dots, N$, are mutually independent p^i-th order Gaussian autoregressive processes AR(p^i), i.e.,

$$\xi_n^i = \sum_{j=1}^{p^i} \beta_j^i \xi_{n-j}^i + w_n^i, \quad n \geq 1, \tag{3.253}$$

where $\{w_n^i\}_{n\geq 1}$ are mutually independent i.i.d. normal $\mathscr{N}(0,1)$ sequences and the initial values $\xi_{1-p^i}^i, \xi_{2-p^i}^i, \dots, \xi_0^i$ are arbitrarily random or deterministic numbers, in particular we may set zero initial conditions $\xi_{1-p^i}^i = \xi_{2-p^i}^i = \dots = \xi_0^i = 0$. The coefficients $\beta_1^i, \dots, \beta_{p^i}^i$ are known and all roots of the equations $z^{p^i} - \beta_1^i z^{p^i-1} - \dots - \beta_{p^i}^i = 0$ are in the interior of the unit circle, so that the AR(p^i) processes are stable. Let $\varphi(x) = (2\pi)^{-1/2} e^{-x^2/2}$ denote density of the standard normal distribution. Define the p_n^i-th order residual

$$\widetilde{X}_n^i = X_n^i - \sum_{j=1}^{p_n^i} \beta_j^i X_{n-j}^i, \quad n \geq 1,$$

where $p_n^i = p^i$ if $n > p^i$ and $p_n^i = n$ if $n \leq p^i$. Write $\theta = (\theta_1, \dots, \theta_N)$ and $\Theta = (0,\infty) \times \dots \times (0,\infty)$ (N times). It is easy to see that the conditional pre-change and post-change densities are

$$g(X_n|\mathbf{X}^{n-1}) = \prod_{i=1}^N \varphi(\widetilde{X}_n^i), \quad f_\theta(X_n|\mathbf{X}^{n-1}) = \prod_{i=1}^N \varphi(\widetilde{X}_n^i - \theta_i \widetilde{S}_n^i), \quad \theta \in \Theta,$$

where $\widetilde{S}_n^i = S_n^i - \sum_{j=1}^{p_n^i} \beta_j^i S_{n-j}^i$. Obviously, due to the independence of the data across channels for all $k \in \mathbb{Z}_+$ and $n \geq 1$ the LLR has the form

$$\lambda_\vartheta(k, k+n) = \sum_{i=1}^N \left[\vartheta_i \sum_{j=k+1}^{k+n} \widetilde{S}_j^i \widetilde{X}_j^i - \frac{\vartheta_i^2 \sum_{j=k+1}^{k+n} (\widetilde{S}_j^i)^2}{2} \right].$$

Under measure $\mathsf{P}_{k,\theta}$ the random variables $\{\widetilde{X}_n^i\}_{n\ge k+1}$ are independent Gaussian random variables with mean $\mathsf{E}_{k,\theta}[\widetilde{X}_n^i] = \theta_i \widetilde{S}_n^i$ and unit variance, and hence, under $\mathsf{P}_{k,\theta}$ the normalized LLR can be written as

$$\frac{1}{n}\lambda_{\vartheta,\theta}(k,k+n) = \sum_{i=1}^N \frac{\vartheta_i\theta_i - \vartheta_i^2/2}{n} \sum_{j=k+1}^{k+n} (\widetilde{S}_j^i)^2 + \frac{1}{n}\vartheta_i \sum_{j=k+1}^{k+n} \widetilde{S}_j^i \eta_j^i, \tag{3.254}$$

where $\{\eta_j^i\}_{j\ge k+1}, i=1,\dots,N$, are mutually independent sequences of i.i.d. standard normal random variables.

Assume that

$$\lim_{n\to\infty} \frac{1}{n} \sup_{k\in\mathbb{Z}_+} \sum_{j=k+1}^{k+n} |\widetilde{S}_j^i|^2 = Q_i, \tag{3.255}$$

where $0 < Q_i < \infty$. This is typically the case in most signal processing applications, e.g., for harmonic signals $S_n^i = \sin(\omega_i n + \phi_n^i)$. Then for all $k \in \mathbb{Z}_+$ and $\theta \in \Theta$

$$\frac{1}{n}\lambda_\theta(k,k+n) \xrightarrow[n\to\infty]{\mathsf{P}_{k,\theta}-\text{a.s.}} \sum_{i=1}^N \frac{\theta_i^2 Q_i}{2} = I_\theta,$$

so that condition $\mathbf{C}_1$ holds. Furthermore, since all moments of the LLR are finite it is straightforward to show that conditions (3.176) and (3.177), and hence, conditions $\mathbf{C}_2$ and $\mathbf{C}_3$ hold for all $r \ge 1$. Indeed, using (3.254), we obtain that $I(\vartheta,\theta) = \sum_{i=1}^N (\vartheta_i\theta_i - \vartheta_i^2/2)Q_i$ and for any $\delta > 0$

$$\mathsf{P}_{k,\theta}\left(\sup_{\vartheta\in[\theta-\delta,\theta+\delta]} \left|\frac{1}{n}\lambda_\vartheta(k,k+n) - I(\vartheta,\theta)\right| > \varepsilon\right) = \mathsf{P}_{k,\theta}\left(|Y_{k,n}(\theta)| > \varepsilon\sqrt{n}\right),$$

where

$$Y_{k,n}(\theta) = \sum_{i=1}^N \frac{\theta_i}{\sqrt{n}} \sum_{j=k+1}^{k+n} \widetilde{S}_j^i \eta_j^i, \quad n \ge 1$$

is the sequence of normal random variables with mean zero and variance $\sigma_n^2 = n^{-1}\sum_{i=1}^N \theta_i^2 \sum_{j=k+1}^{k+n} (\widetilde{S}_j^i)^2$, which by (3.255) is asymptotic to $\sum_{i=1}^N \theta_i^2 Q_i$. Thus, for a sufficiently large n there exists $\delta_0 > 0$ such that $\sigma_n^2 \le \delta_0 + \sum_{i=1}^N \theta_i^2 Q_i$ and we obtain that for all large n

$$\begin{aligned}
&\mathsf{P}_{k,\theta}\left(\sup_{\vartheta\in[\theta-\delta,\theta+\delta]} \left|\frac{1}{n}\lambda_\vartheta(k,k+n) - I(\vartheta,\theta)\right| > \varepsilon\right) \\
&\le \mathsf{P}\left(|\hat{\eta}| > \frac{\delta_0 + \sum_{i=1}^N \theta_i^2 Q_i}{\sigma_n^2} \frac{\varepsilon\sqrt{n}}{\delta_0 + \sum_{i=1}^N \theta_i^2 Q_i}\right) \\
&\le \mathsf{P}\left(|\hat{\eta}| > \frac{\varepsilon\sqrt{n}}{\delta_0 + \sum_{i=1}^N \theta_i^2 Q_i}\right),
\end{aligned}$$

where $\hat{\eta} \sim \mathscr{N}(0,1)$ is a standard normal random variable. Hence, for all $r \ge 1$

$$\sum_{n=1}^\infty n^{r-1} \sup_{k\in\mathbb{Z}_+} \sup_{\vartheta\in\Theta_c} \mathsf{P}_{k,\theta}\Big(\sup_{\vartheta\in[\theta-\delta,\theta+\delta]} \left|\frac{1}{n}\lambda_\vartheta(k,k+n) - I(\vartheta,\theta)\right| > \varepsilon\Big) < \infty,$$

which implies (3.177) for all $r \ge 1$.

By Theorems 3.7, 3.8 and 3.10, the Shiryaev detection rule and the window-limited Shiryaev rule minimize asymptotically all positive moments of the detection delay. All asymptotic assertions for the MSR rule presented in Section 3.5 also hold with $I_\theta = \sum_{i=1}^N \theta_i^2 Q_i/2$. In particular, by Theorem 3.4, the MSR rule is also asymptotically optimal for all $r \ge 1$ if the prior distribution of the change point is either heavy-tailed or asymptotically flat.

Note that in this example despite the fact that the data are correlated in time the increments of the LLR are independent, so that it would be sufficient to use the results of Section 3.8, i.e., to verify a weaker condition (3.128).

Since by condition $\mathbf{C}_2$ the MS and MSR rules are asymptotically optimal for almost arbitrarily mixing distribution $W(\theta)$, in this example it is most convenient to select the conjugate prior, $W(\theta) = \prod_{i=1}^N F(\theta_i/v_i)$, where $F(y)$ is a standard normal distribution and $v_i > 0$, in which case the MS and MSR statistics can be computed explicitly.

Note that this example arises in certain interesting practical applications, as discussed in [164]. For example, surveillance systems (radar, acoustic, EO/IR) typically deal with detecting moving and maneuvering targets that appear at unknown times, and it is necessary to detect a signal from a randomly appearing target in clutter and noise with the smallest possible delay. In radar applications, often the signal represents a sequence of modulated pulses and clutter/noise can be modeled as a Markov Gaussian process or more generally as a pth order Markov process (see, e.g, [4, 122]). In underwater detection of objects with active sonars, reverberation creates very strong clutter that represents a correlated process in time [87], so that again the problem can be reduced to detection of a signal with an unknown intensity in correlated clutter. In applications related to detection of point and slightly extended objects with EO/IR sensors (on moving and still platforms such as space-based, airborne, ship-board, ground-based), sequences of images usually contain a cluttered background which is correlated in space and time, and it is a challenge to detect and track weak objects in correlated clutter [162].

Yet another challenging application area where the multichannel model is useful is cyber-security [159, 168, 167]. Malicious intrusion attempts in computer networks (spam campaigns, personal data theft, worms, distributed denial-of-service (DDoS) attacks, etc.) incur significant financial damage and are a severe harm to the integrity of personal information. It is therefore essential to devise automated techniques to detect computer network intrusions as quickly as possible so that an appropriate response can be provided and the negative consequences for the users are eliminated. In particular, DDoS attacks typically involve many traffic streams resulting in a large number of packets aimed at congesting the target's server or network. As a result, these attacks usually lead to abrupt changes in network traffic and can be detected by noticing a change in the average number of packets sent through the victim's link per unit time. Figure 3.2 illustrates how the multichannel anomaly Intrusion Detection System works for detecting a real UDP packet storm. The multichannel MSR algorithm with the AR(1) model and uniform prior $W(\theta_i)$ on a finite interval $[1,5]$ was used. The first plot shows packet rate. It is seen that there is a slight change in the mean, which is barely visible. The second plot shows the behavior of the multi-cyclic MSR statistic $W_n = \log R_W(n)$, which is restarted from scratch every time a threshold exceedance occurs. Threshold exceedances before the UDP DDoS attack starts (i.e., false alarms) are shown by grey dots and the true detections are marked by black dots.

3.14 Concluding Remarks

1. Since in general we do not assume a class of models for the observations such as Gaussian, Markov or HMM and build the decision statistics on the LLR process $\lambda_\theta(k,k+n)$, it is natural to impose conditions on the behavior of $\lambda_\theta(k,k+n)$, which is expressed by conditions $\mathbf{C}_1$, $\mathbf{C}_2$ and $\mathbf{C}_3$, related to the law of large numbers for the LLR and rates of convergence in the law of large numbers. The assertions of Theorems 3.2–3.4 hold if $n^{-1}\lambda_\theta(k,k+n)$ and $n^{-1}\log\Lambda_W(k,k+n)$ converge uniformly r-completely to I_θ under $\mathsf{P}_{k,\theta}$, i.e., when for all $\varepsilon > 0$ and $\theta \in \Theta$

$$\sum_{n=1}^{\infty} n^{r-1} \sup_{k\in\mathbb{Z}_+} \mathsf{P}_{k,\theta}\left(\left|\frac{1}{n}\lambda_\theta(k,k+n) - I_\theta\right| > \varepsilon\right) < \infty,$$

$$\sum_{n=1}^{\infty} n^{r-1} \sup_{k\in\mathbb{Z}_+} \mathsf{P}_{k,\theta}\left(\left|\frac{1}{n}\log\Lambda_W(k,k+n) - I_\theta\right| > \varepsilon\right) < \infty. \tag{3.256}$$

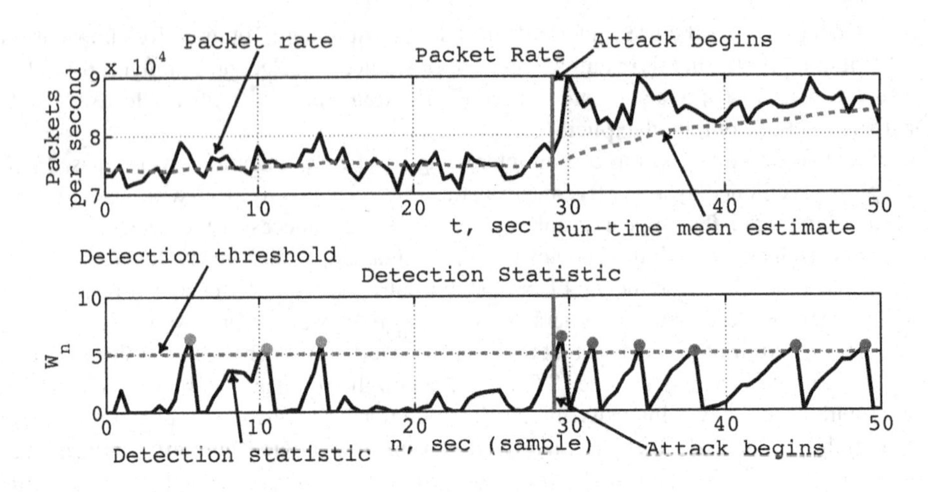

FIGURE 3.2
Detection of the UDP DDoS packet storm attack: upper picture — raw data (packet rate); bottom — log of the MSR statistic.

However, verifying the r-complete convergence condition (3.256) for the weighted LLR $\log \Lambda_W(k, k+n)$ is typically much more difficult than checking conditions $\mathbf{C}_2$ and $\mathbf{C}_3$ for the local values of the LLR in the vicinity of the true parameter value.

2. As expected, the results indicate that the MSR rule is not asymptotically optimal when the prior distribution of the change point has an exponential tail (i.e., $\mu > 0$), but it is asymptotically optimal for heavy-tailed prior distributions (i.e., $\mu = 0$) and also when $\mu \to 0$ with a certain rate.

3. The results show that first-order asymptotic optimality properties of the MS and MSR procedures hold for practically arbitrarily weight function $W(\theta)$, in particular for any prior that has strictly positive values on Θ. Therefore, the selection of $W(\theta)$ can be based solely on the computational aspects. The conjugate prior is typically the best choice when possible. However, if the parameter is vector and the parameter space is intricate, constructing mixture statistics may be difficult. In this case, discretizing the parameter space and selecting the prior $W(\theta = \theta_i)$ concentrated on discrete points θ_i, $i = 1, \dots, N$, suggested and discussed in [40] for the hypothesis testing problems, is perhaps the best option. Then one can easily compute the MS and MSR statistics (as long as the likelihood ratio $LR_\theta(k, k+n)$ can be computed) at the expense of losing optimality between the points θ_i since the resulting discrete versions are asymptotically optimal only at the points θ_i.

4. The asymptotic optimality results can be extended to the "asymptotically nonstationary" case where the normalized LLR $\lambda_\theta(k, k+n)/\phi(n)$ converges r-completely to a constant I_θ with the normalization factor $\phi(n)$ (in place of n), i.e., assuming that

$$\sum_{k=0}^{\infty} \pi_k \sum_{n=1}^{\infty} n^{r-1} \mathsf{P}_{k,\theta}\left(\left|\frac{\lambda_\theta(k, k+n)}{\phi(n)} - I_\theta\right| > \varepsilon\right) < \infty,$$

where $\phi(n)$ is some not too slowly increasing function of n, e.g., $\phi(n) = n^\beta$, $\beta > 0$. In this case, in the condition (3.20) the normalizing factor n should be replaced by $\phi(n)$, i.e., $|\mathsf{P}(\nu > n)|/\phi(n) \to \mu$ as $n \to \infty$ (in order to guarantee an allowable amount of information that comes from prior knowledge).

4

Nearly Optimal Pointwise and Minimax Change Detection in a Single Population

4.1 Introduction

In contrast to Chapter 3 where we investigated asymptotic optimality in the Bayesian setup in the class of detection rules with the given weighted probability of false alarm, this chapter is concerned with the problem of minimizing the moments of the detection delay, $\mathscr{R}^r_{\nu,\theta}(T) = \mathsf{E}_{\nu,\theta}\left[(T-\nu)^r \,|\, T>\nu\right]$, in pointwise (i.e., for all change points ν) and minimax (i.e., for a worst change point) settings among all procedures for which the local probability of a false alarm $\mathsf{P}_\infty(k \le T < k+m | T \ge k)$ in the fixed time window is upper-bounded by a given small number. Specifically, we show that the mixture version of the Shiryaev–Roberts rule (MSR) $\widetilde{T}^W_B$, defined in (3.9) and (3.10), is asymptotically optimal in the class of procedures with the prescribed maximal conditional probability of false alarm when it is small, minimizing moments of the detection delay pointwise (for all change points) as well as in a minimax sense (for the worst change point). An important feature of the MSR rule is applicability of a simple upper bound on the probability of false alarm. In this respect, the generalized CUSUM procedure based on maximization over unknown parameters is not as good.

4.2 Nearly Optimal Pointwise and Minimax Change Detection in the Class with Maximal PFA Constraint

4.2.1 Problem Setup

We again consider the general stochastic model for the observations given by (2.2). Recall that the MSR change detection rule is defined in (3.9) and (3.10). In this chapter, it is convenient to re-write the MSR stopping time $\widetilde{T}^W_A$ in the log scale as

$$\widetilde{T}^W_a = \inf\{n \ge 1 : \log R_W(n) \ge a\}, \quad a = \log A \in (-\infty, +\infty), \tag{4.1}$$

where

$$R_W(n) = \int_\Theta R_\theta(n)\, \mathrm{d}W(\theta), \quad n \ge 1, \; R_W(0) = 0 \tag{4.2}$$

is the MSR statistic and where

$$R_\theta(n) = \sum_{k=1}^{n} \prod_{i=k}^{n} \frac{f_{\theta,i}(X_i|\mathbf{X}_1^{i-1})}{g_i(X_i|\mathbf{X}_1^{i-1})} \tag{4.3}$$

is the SR statistic tuned to $\theta \in \Theta$. Compared to the generalized MSR statistic (3.9) that starts from an arbitrarily point $\omega \ge 0$, in this chapter, for the sake of simplicity we set $\omega = 0$. All the results can be easily generalized to the case $\omega > 0$.

For any $0<\beta<1$, $m\geq 1$ and $\ell\geq 1$, introduce the following class of change detection procedures that upper-bounds the *local conditional probability of false alarm* (LCPFA) $\mathsf{P}_\infty(T<k+m|T\geq k)=\mathsf{P}_\infty(k\leq T<k+m|T\geq k)$ in the time interval $[k,k+m-1]$ of the length m:

$$\mathbb{C}(\beta,\ell,m)=\left\{T\in\mathscr{M}:\sup_{1\leq k\leq \ell}\mathsf{P}_\infty(T<k+m|T\geq k)\leq\beta\right\}, \tag{4.4}$$

where $\mathscr{M}$ is a class of all Markov times.

For $r\geq 1$ and $\theta\in\Theta$, we consider the risk associated with the *conditional rth moment of the detection delay*

$$\mathscr{R}^r_{\nu,\theta}(T)=\mathsf{E}_{\nu,\theta}\left[(T-\nu)^r\,|\,T>\nu\right] \tag{4.5}$$

and the following optimization problems: the pointwise minimization

$$\inf_{T\in\mathbb{C}(\beta,\ell,m)}\mathscr{R}^r_{\nu,\theta}(T)\quad\text{for every }\nu\geq 0\text{ and }\theta\in\Theta \tag{4.6}$$

as well as the two minimax problems

$$\inf_{T\in\mathbb{C}(\beta,\ell,m)}\sup_{0\leq\nu<\infty}\mathscr{R}^r_{\nu,\theta}(T)\quad\text{for every }\theta\in\Theta \tag{4.7}$$

and

$$\inf_{T\in\mathbb{C}(\beta,\ell,m)}\sup_{\theta\in\Theta}\sup_{0\leq\nu<\infty}I^r_\theta\mathscr{R}^r_{\nu,\theta}(T). \tag{4.8}$$

The parameters ℓ and m and the function $I(\theta)=I_\theta$, which characterizes the distance between pre- and post-change distributions, will be specified later.

However, solving the optimization problems (4.6)–(4.8) for any LCPFA $\beta<1$ is practically impossible, especially for the general non-i.i.d. model. For this reason, we will focus on the asymptotic problems when the LCPFA β goes to 0. Our goal is to show that the MSR detection rule $\widetilde{T}^W_a$ with a suitable threshold $a=a_\beta$ is first-order asymptotically optimal in class $\mathbb{C}(\beta,\ell_\beta,m_\beta)$ when ℓ_β and m_β tend to infinity as $\beta\to 0$ with an appropriate rate.

4.2.2 Asymptotic Optimality of the MSR Detection Rule

4.2.2.1 The Non-i.i.d. Case

The Bayesian asymptotic optimality results of Chapter 3 are the key for solving these problems. So we begin with specification of the results from Chapter 3 in the form most convenient for this purpose.

In what follows, it suffices to assume that the prior probability distribution $\mathsf{P}(\nu=k)$ of the change point ν is geometric with the parameter $0<\rho<1$, i.e.,

$$\mathsf{P}(\nu=k)=\pi_k(\rho)=\rho\,(1-\rho)^k,\quad k=0,1,2,\dots, \tag{4.9}$$

so that $\mathsf{PFA}_\pi(T)=\mathsf{PFA}_\rho(T)=\sum_{k=0}^\infty\rho\,(1-\rho)^k\mathsf{P}_\infty(T\leq k)$.

Now, for some fixed $0<\rho,\alpha<1$, let $\mathbb{C}_\rho(\alpha)$ denote the class $\mathbb{C}_\pi(\alpha)$ when the prior distribution is Geometric(ρ), i.e.,

$$\mathbb{C}_\rho(\alpha)=\left\{T\in\mathscr{M}:\mathsf{PFA}_\rho(T)\leq\alpha\right\}. \tag{4.10}$$

As in Chapter 3, we assume that left-tail and right-tail conditions $\mathbf{C}_1$ and $\mathbf{C}_3$ hold for the LLR between the hypotheses H^θ_k $(k=0,1,\dots)$ and H_∞

$$\lambda_\theta(k,n)=\sum_{j=k+1}^n\log\frac{f_{\theta,j}(X_j|\mathbf{X}^{j-1})}{f_{0,j}(X_j|\mathbf{X}^{j-1})}. \tag{4.11}$$

Specifically,

$\mathbf{C}_1$. *Assume that there exists a positive and finite $\Theta \to \mathbb{R}$ function $I(\theta) = I_\theta > 0$ such that for every $k \geq 0$ and $\varepsilon > 0$*

$$\lim_{N\to\infty} \mathsf{P}_{k,\theta}\left\{\frac{1}{N}\max_{1\leq n\leq N} \lambda_\theta(k,k+n) \geq (1+\varepsilon)I_\theta\right\} = 0 \quad \text{for all } \theta \in \Theta. \tag{4.12}$$

$\mathbf{C}_2(r)$. *There exists a positive continuous $\Theta \to \mathbb{R}$ function $I(\theta) = I_\theta$ such that for every $\theta \in \Theta$, for any $\varepsilon > 0$, and for some $r \geq 1$*

$$\Upsilon_r(\varepsilon,\theta) := \lim_{\delta\to 0}\sum_{n=1}^{\infty} n^{r-1} \sup_{0\leq k<\infty} \mathsf{P}_{k,\theta}\left(\frac{1}{n}\inf_{|\vartheta-\theta|<\delta} \lambda_\vartheta(k,k+n) < I_\theta - \varepsilon\right) < \infty. \tag{4.13}$$

By Lemma B.1 in Appendix B, to check condition (4.12) it is sufficient to establish that $n^{-1}\lambda_\theta(k,k+n) \to I_\theta$ as $n\to\infty$ P_θ-a.s.

To check condition (4.13) it is sufficient to check the following condition:

$\mathbf{C}_2^*(r)$. *There exists a positive continuous $\Theta \to \mathbb{R}$ function $I(\theta)$ such that for every compact set $\Theta_c \subseteq \Theta$, for any $\varepsilon > 0$, and for some $r \geq 1$*

$$\Upsilon_r^*(\varepsilon,\Theta_c) := \sup_{\theta\in\Theta_c} \Upsilon_r(\varepsilon,\theta) < \infty. \tag{4.14}$$

As in the previous chapter, in order to exclude parameter values that have measure null we will assume that $W(\Gamma_{\delta,\theta}) > 0$ for all $\theta \in \Theta$, where $\Gamma_{\delta,\theta} = \{\vartheta \in \Theta : |\vartheta - \theta| < \delta\}$.

In what follows, we assume that the parameter $\rho = \rho_\alpha$ of the prior distribution is a function of α such that

$$\lim_{\alpha\to 0} \rho_\alpha = 0 \quad \text{and} \quad \lim_{\alpha\to 0} \frac{|\log\rho_\alpha|}{|\log\alpha|} = 0. \tag{4.15}$$

Let $k^* = k^*_\alpha$ be a function of α such that

$$\lim_{\alpha\to 0} k^*_\alpha = \infty \quad \text{and} \quad \lim_{\alpha\to 0} \alpha\,\rho_\alpha\, k^*_\alpha = 0. \tag{4.16}$$

Note that if in the MSR rule threshold $a = a(\alpha,\rho_\alpha)$ is selected as $a_\alpha = \log[(1-\rho_\alpha)/(\rho_\alpha\alpha)]$, where ρ_α satisfies condition (4.15), then $a_\alpha \sim |\log\alpha|$ and, by Lemma 3.4, $\mathsf{PFA}_\rho(T^W_{a_\alpha}) \leq \alpha$, i.e., this choice of the threshold guarantees that $\widetilde{T}^W_{a_\alpha} \in \mathbb{C}_{\rho_\alpha}(\alpha)$ for every $0 < \alpha < 1$.

The following proposition establishes first-order asymptotic optimality of the MSR rule in class $\mathbb{C}_{\rho_\alpha}(\alpha) = \mathbb{C}(\alpha)$.

Proposition 4.1. *Let $r \geq 1$. Assume that right-tail and left-tail conditions $\mathbf{C}_1$ and $\mathbf{C}_2(r)$ hold for some $0 < I_\theta < \infty$, the parameter $0 < \rho = \rho_\alpha < 1$ of the geometric prior distribution satisfies conditions* (4.15), *and conditions* (4.16) *hold for $k^* = k^*_\alpha$.*

(i) *If $a = a_\alpha$ is so selected that $\mathsf{PFA}_\rho(\widetilde{T}^W_{a_\alpha}) \leq \alpha$ and $\log a_\alpha \sim |\log\alpha|$ as $\alpha \to 0$, in particular $a_\alpha = \log[(1-\rho_\alpha)/\rho_\alpha\alpha]$, then, for all $\theta \in \Theta$ as $\alpha \to 0$,*

$$\inf_{T\in\mathbb{C}(\alpha)} \mathscr{R}^r_{\nu,\theta}(T) \sim \left(\frac{|\log\alpha|}{I_\theta}\right)^r \sim \mathscr{R}^r_{\nu,\theta}(\widetilde{T}^W_{a_\alpha}) \quad \textit{for all fixed } \nu \geq 0 \tag{4.17}$$

and

$$\inf_{T\in\mathbb{C}(\alpha)} \max_{0\leq\nu\leq k^*_\alpha} \mathscr{R}^r_{\theta,\nu}(T) \sim \left(\frac{|\log\alpha|}{I_\theta}\right)^r \sim \max_{0\leq\nu\leq k^*_\alpha} \mathscr{R}^r_{\theta,\nu}(\widetilde{T}^W_{a_\alpha}). \tag{4.18}$$

Thus, the MSR rule $\widetilde{T}^W_{a_\alpha}$ is first-order asymptotically pointwise optimal and minimax in class $\mathbb{C}(\alpha)$ with respect to the moments of the detection delay up to order r for all parameter values $\theta \in \Theta$.

(ii) *Assume further that for every $\varepsilon > 0$ and some $r \ge 1$*

$$\sup_{\theta\in\Theta_1} \Upsilon_r(\varepsilon,\theta) < \infty \quad \text{and} \quad \inf_{\theta\in\Theta_1} I_\theta > 0, \tag{4.19}$$

where Θ_1 is compact. If $a = a_\alpha$ is so selected that $\mathsf{PFA}_\rho(\widetilde{T}^W_{a_\alpha}) \le \alpha$ and $\log a_\alpha \sim |\log \alpha|$ as $\alpha \to 0$, in particular $a_\alpha = \log[(1-\rho_\alpha)/\rho_\alpha \alpha]$, then, as $\alpha \to 0$,

$$\inf_{T\in\mathbb{C}(\alpha)} \sup_{\theta\in\Theta_1} \max_{0\le\nu\le k^*_\alpha} I^r_\theta \mathscr{R}^r_{\nu,\theta}(T) \sim |\log\alpha|^r \sim \sup_{\theta\in\Theta_1} \max_{0\le\nu\le k^*_\alpha} I^r_\theta \mathscr{R}^r_{\nu,\theta}(\widetilde{T}^W_{a_\alpha}). \tag{4.20}$$

Thus, the MSR rule $\widetilde{T}^W_{a_\alpha}$ is first-order asymptotically double minimax.

Proof. Proof of part (i). Asymptotic approximations (4.17) in assertion (i) follow from (3.96) in Theorem 3.4 of Chapter 3.

To prove asymptotic approximations (4.18) recall that, by Doob's inequality (3.75)

$$\mathsf{P}_\infty(\widetilde{T}^W_a \le j) \le j e^{-a}, \quad j \ge 1, \tag{4.21}$$

and we obtain that

$$\mathsf{P}_\infty\left(\widetilde{T}^W_{a_\alpha} > \nu\right) \ge \mathsf{P}_\infty\left(\widetilde{T}^W_{a_\alpha} > k^*_\alpha\right) \ge 1 - k^*_\alpha e^{-a_\alpha} \quad \text{for any } 0 \le \nu \le k^*_\alpha. \tag{4.22}$$

Since $a_\alpha \sim |\log\alpha|$ it follows that $\mathsf{P}_\infty(\widetilde{T}^W_{a_\alpha} > \nu) \to 1$ as $\alpha \to 0$ for all $0 \le \nu \le k^*_\alpha$ and since $\mathscr{R}^r_{\nu,\theta}(\widetilde{T}^W_{a_\alpha}) = \mathsf{E}_{\nu,\theta}[(\widetilde{T}^W_{a_\alpha} - \nu)^+]^r / \mathsf{P}_\infty\left(\widetilde{T}^W_{a_\alpha} > \nu\right)$, inequality (3.86) implies (4.18) for $k^* = k^*_\alpha$ satisfying conditions (4.16). This completes the proof of part (i).

Proof of part (ii). Using the lower bound (3.68) in Lemma 3.3, we obtain that under condition $\mathbf{C}_1$ the following asymptotic lower bound holds:

$$\inf_{T\in\mathbb{C}(\alpha)} \sup_{\theta\in\Theta_1} \max_{0\le\nu\le k^*_\alpha} I^r_\theta \mathscr{R}^r_{\nu,\theta}(T) \ge |\log\alpha|^r(1+o(1)) \quad \text{as } \alpha \to 0. \tag{4.23}$$

To prove approximations in (4.20) it suffices to show that the right-hand side in (4.23) is attained for the risk $\sup_{\theta\in\Theta_1} \max_{0\le\nu\le k^*_\alpha} I^r_\theta \mathscr{R}^r_{\nu,\theta}(\widetilde{T}^W_{a_\alpha})$ of the MSR rule. Using inequality (3.85) in Chapter 3, we obtain that for an arbitrarily $0 < \varepsilon < I_\theta$

$$\sup_{\theta\in\Theta_1} I^r_\theta \sup_{\nu\ge 0} \mathsf{E}_{\nu,\theta}\left[(\widetilde{T}_a - \nu)^+\right]^r \le \sup_{\theta\in\Theta_1} I^r_\theta \left(1 + \left\lfloor \frac{a}{I_\theta - \varepsilon} \right\rfloor\right)^r + r2^{r-1} \sup_{\theta\in\Theta_1} I^r_\theta \Upsilon_r(\varepsilon,\theta).$$

By condition (4.19), the second term on the right side is finite, which immediately implies that

$$\sup_{\theta\in\Theta_1} I^r_\theta \sup_{\nu\ge 0} \mathsf{E}_{\nu,\theta}\left[(\widetilde{T}_a - \nu)^+\right]^r \le a^r(1+o(1)) \quad \text{as } a \to \infty.$$

Next, using inequality (4.22), we obtain

$$\mathscr{R}^r_{\nu,\theta}(\widetilde{T}^W_{a_\alpha}) = \frac{\mathsf{E}_{\nu,\theta}[(\widetilde{T}^W_{a_\alpha} - \nu)^+]^r}{\mathsf{P}_\infty\left(\widetilde{T}^W_{a_\alpha} > \nu\right)} \le \frac{\mathsf{E}_{\nu,\theta}[(\widetilde{T}^W_{a_\alpha} - \nu)^+]^r}{1 - k^*_\alpha e^{-a_\alpha}} \quad \text{for } \nu \le k^*_\alpha,$$

which along with the previous inequality and the fact that $a_\alpha \sim |\log\alpha|$ yields the upper bound

$$\sup_{\theta\in\Theta_1} \max_{0\le\nu\le k^*_\alpha} I^r_\theta \mathscr{R}^r_{\nu,\theta}(\widetilde{T}^W_{a_\alpha}) \le |\log\alpha|^r(1+o(1)) \quad \text{as } \alpha \to 0$$

and the proof is complete. □

We now proceed with the pointwise and minimax problems (4.6), (4.7) and (4.8) in the class of procedures with given LCPFA $\mathbb{C}(\beta,\ell,m)$ defined in (4.4). Note that the results of Proposition 4.1 for the Bayesian-type problem are essential, since asymptotic optimality in class $\mathbb{C}(\beta,\ell,m)$ is obtained by embedding this class in class $\mathbb{C}_\rho(\alpha)$ with specially selected parameters ρ and α.

For any $0<\beta<1$, $m\geq 1$ and $\ell\geq 1$, define

$$\alpha_1=\alpha_1(\beta,m)=\beta+(1-\rho_{1,\beta})^{m+1} \tag{4.24}$$

and

$$\alpha_2=\alpha_2(\beta,\ell,m)=\frac{\beta(1-\rho_{2,\beta})^{\ell+m}}{1+\beta}, \tag{4.25}$$

where $\rho_{2,\beta}=\check{\delta}_\beta\,\rho_{1,\beta}$ and the functions $0<\rho_{1,\beta}<1$ and $0<\check{\delta}_\beta<1$ are such that

$$\lim_{\beta\to 0}\left(\rho_{1,\beta}+\check{\delta}_\beta\right)=0 \quad\text{and}\quad \lim_{\beta\to 0}\frac{|\log\rho_{1,\beta}|+|\log\check{\delta}_\beta|}{|\log\beta|}=0. \tag{4.26}$$

For example, we can take

$$\rho_{1,\beta}=\frac{1}{1+|\log\beta|},\qquad \check{\delta}_\beta=\frac{\check{\delta}^*}{|\log\beta|}\quad\text{and}\quad 0<\check{\delta}^*<1. \tag{4.27}$$

To find asymptotic lower bounds for the problems (4.6) and (4.7) in addition to condition $\mathbf{C}_1$ we impose the following condition related to the growth of the window size m in the LCPFA:

$\mathbf{H}_1$. *The size of the window $m=m_\beta$ is a function of β, such that*

$$\lim_{\beta\to 0}\frac{|\log\alpha_{1,\beta}|}{|\log\beta|}=1, \tag{4.28}$$

where $\alpha_{1,\beta}=\alpha_1(\beta,m_\beta)$.

The following theorem establishes asymptotic lower bounds.

Theorem 4.1. *Assume that conditions $\mathbf{C}_1$ and $\mathbf{H}_1$ hold. Then, for every $\ell\geq 1$, $\nu\geq 0$, $\theta\in\Theta$ and $r\geq 1$*

$$\liminf_{\beta\to 0}\frac{\inf_{T\in\mathbb{C}(\beta,\ell,m_\beta)}\sup_{\nu\geq 0}\mathscr{R}^r_{\nu,\theta}(T)}{|\log\beta|^r}\geq\liminf_{\beta\to 0}\frac{\inf_{T\in\mathbb{C}(\beta,\ell,m_\beta)}\mathscr{R}^r_{\nu,\theta}(T)}{|\log\beta|^r}\geq\frac{1}{I^r_\theta}. \tag{4.29}$$

Proof. First, recall that by Lemma 3.3 in Chapter 3 under condition $\mathbf{C}_1$ the lower bounds hold (for any $r\geq 1$, $\nu\geq 0$ and $\theta\in\Theta$):

$$\liminf_{\alpha\to 0}\frac{\inf_{T\in\mathbb{C}(\alpha)}\sup_{\nu\geq 0}\mathscr{R}^r_{\nu,\theta}(T)}{|\log\alpha|^r}\geq\liminf_{\alpha\to 0}\frac{\inf_{T\in\mathbb{C}(\alpha)}\mathscr{R}^r_{\nu,\theta}(T)}{|\log\alpha|^r}\geq\frac{1}{I^r_\theta}. \tag{4.30}$$

Second, we show that for any $0<\beta<1$, $m\geq|\log(1-\beta)|/[|\log(1-\rho_{1,\beta})|]-1:=m^*_\beta$ and $\ell\geq 1$, the following inclusion holds:

$$\mathbb{C}(\beta,\ell,m)\subseteq\mathbb{C}_{\rho_{1,\beta}}(\alpha_1), \tag{4.31}$$

where $\alpha_1=\alpha_1(\beta,m)$ is defined in (4.24). Indeed, let T be from $\mathbb{C}(\beta,\ell,m)$. Then, using definition of class $\mathbb{C}(\beta,\ell,m)$, we obtain that $\mathsf{P}_\infty(T\leq m)\leq\beta$. Therefore, taking in (4.9) $\rho=\rho_{1,\beta}$, we obtain

$$\sum_{k=0}^{\infty}\pi_k(\rho_{1,\beta})\mathsf{P}_\infty(T\leq k)=\sum_{k=0}^{m}\pi_k(\rho_{1,\beta})\mathsf{P}_\infty(T\leq k)+\sum_{k=m+1}^{\infty}\pi_k(\rho_{1,\beta})\mathsf{P}_\infty(T\leq k)$$

$$\leq \beta + \sum_{k=m+1}^{\infty} \pi_k(\rho_{1,\beta}) = \beta + (1-\rho_{1,\beta})^{m+1} = \alpha_1,$$

i.e., $T \in \mathbb{C}_{\rho_{1,\beta}}(\alpha_1)$ where $\alpha_1 < 1$ for $m > m^*_\beta$ and any $\beta \in (0,1)$.

Inclusion (4.31) implies that for all $\nu \geq 0$ and $\theta \in \Theta$ and for a sufficiently small $\beta > 0$

$$\inf_{T \in \mathbb{C}(\beta,\ell,m)} \mathscr{R}^r_{\nu,\theta}(T) \geq \inf_{T \in \mathbb{C}_{\rho_{1,\beta}}(\alpha_{1,\beta})} \mathscr{R}^r_{\nu,\theta}(T).$$

Now, lower bounds (4.29) follow from lower bounds (4.30) and condition $\mathbf{H}_1$. □

To establish asymptotic optimality properties of the MSR rule with respect to the risks $\mathscr{R}^r_{\nu,\theta}(T)$ (for all $\nu \geq 0$) and $\sup_{\nu \geq 0} \mathscr{R}^r_{\nu,\theta}(T)$ in class $\mathbb{C}(\beta,\ell,m)$ we need the uniform left-tail r-complete convergence condition $\mathbf{C}_2(r)$ as well as the following condition:

$\mathbf{H}_2$. *Parameters $\ell = \ell_\beta$ and $m = m_\beta$ are functions of β, such that*

$$\lim_{\beta \to 0} \frac{|\log \alpha_{2,\beta}|}{|\log \beta|} = 1, \tag{4.32}$$

where $\alpha_{2,\beta} = \alpha_2(\beta, \ell_\beta, m_\beta)$.

The conditions (4.28) and (4.32) hold, for example, if

$$m_\beta = \lfloor |\log \beta| / \rho_{1,\beta} \rfloor \quad \text{and} \quad \ell_\beta = \check{\varkappa} m_\beta, \tag{4.33}$$

where $\check{\varkappa} > 0$ is fixed.

Note that

$$\lim_{\beta \to 0} \left(|\log \alpha_{2,\beta}| + (\ell_\beta + m_\beta) \log(1-\rho_{2,\beta}) \right) = \infty. \tag{4.34}$$

Denote by $\widetilde{T}^W_\beta$ the MSR rule $\widetilde{T}^W_{a_\beta}$ with threshold a_β given by

$$a_\beta = \frac{1-\alpha_{2,\beta}}{\rho_{2,\beta}\alpha_{2,\beta}}. \tag{4.35}$$

Theorem 4.2. *If conditions $\mathbf{H}_1$ and $\mathbf{H}_2$ hold, then, for any $0 < \beta < 1$, the MSR rule $\widetilde{T}^W_\beta$ with threshold a_β given by (4.35) belongs to class $\mathbb{C}(\beta, \ell_\beta, m_\beta)$. Assume in addition that conditions $\mathbf{C}_1$ and $\mathbf{C}_2$ are satisfied. Then for every $\nu \geq 0$ and $\theta \in \Theta$*

$$\inf_{T \in \mathbb{C}(\beta,\ell_\beta,m_\beta)} \mathscr{R}^r_{\nu,\theta}(T) \sim \left(\frac{|\log \beta|}{I_\theta} \right)^r \sim \mathscr{R}^r_{\nu,\theta}(\widetilde{T}^W_\beta) \tag{4.36}$$

and for all $\theta \in \Theta$

$$\inf_{T \in \mathbb{C}(\beta,\ell_\beta,m_\beta)} \max_{0 \leq \nu \leq \ell_\beta + m_\beta} \mathscr{R}^r_{\nu,\theta}(T) \sim \left(\frac{|\log \beta|}{I_\theta} \right)^r \sim \max_{0 \leq \nu \leq \ell_\beta + m_\beta} \mathscr{R}^r_{\nu,\theta}(\widetilde{T}^W_\beta). \tag{4.37}$$

Therefore, the MSR rule $\widetilde{T}^W_\beta$ is first-order asymptotically pointwise optimal and minimax in class $\mathbb{C}(\beta, \ell_\beta, m_\beta)$, minimizing moments of the detection delay up to order r for all parameter values $\theta \in \Theta$.

Proof. By Lemma 3.4, the MSR rule $\widetilde{T}^W_{a_\alpha(\rho)} \in \mathbb{C}_\rho(\alpha)$ for any $0<\alpha,\rho<1$. Now, definition (4.35) implies that $\widetilde{T}^W_\beta \in \mathbb{C}_{\rho_{2,\beta}}(\alpha_{2,\beta})$. Note that for any $0<\beta<1$, $m \ge |\log(1-\beta)|/[|\log(1-\rho_{1,\beta})|]-1$ and $\ell \ge 1$, the following inclusion holds:

$$\mathbb{C}_{\rho_{2,\beta}}(\alpha_2) \subseteq \mathbb{C}\left(\beta,\ell,m_\beta\right). \tag{4.38}$$

Indeed, by definition of class $\mathbb{C}_\rho(\alpha)$, we have that for any $0<\alpha,\rho<1$ and any $i\ge 1$

$$\alpha \ge \rho\sum_{k=i}^{\infty}(1-\rho)^k \mathsf{P}_\infty(T\le k) \ge \rho \mathsf{P}_\infty(T\le i)\sum_{k=i}^{\infty}(1-\rho)^k = \mathsf{P}_\infty(T\le i)(1-\rho)^i,$$

which implies that

$$\sup_{T\in\mathbb{C}_\rho(\alpha)} \mathsf{P}_\infty(T\le i) \le \alpha(1-\rho)^{-i}.$$

Let $T\in\mathbb{C}_{\rho_{2,\beta}}(\alpha_2)$. Then, taking into account the latter inequality with $i=\ell+m$ and using definition of α_2, we obtain that

$$\begin{aligned}
\max_{1\le k\le \ell_\beta} \mathsf{P}_\infty(T<k+m_\beta|T\ge k) &\le \max_{1\le k\le \ell_\beta}\frac{\mathsf{P}_\infty\left(T<k+m_\beta\right)}{\mathsf{P}_\infty\left(T\ge k\right)} \\
&\le \frac{\mathsf{P}_\infty\left(T<\ell_\beta+m_\beta\right)}{1-\mathsf{P}_\infty\left(T<\ell_\beta+m_\beta\right)} \\
&\le \frac{\alpha_2\left(1-\rho_{2,\beta}\right)^{-(\ell_\beta+m_\beta)}}{1-\alpha_2\left(1-\rho_{2,\beta}\right)^{-(\ell_\beta+m_\beta)}} = \beta,
\end{aligned}$$

i.e., T belongs to $\mathbb{C}(\beta,\ell_\beta,m_\beta)$.

Using inclusion (4.38), we obtain that the stopping time $\widetilde{T}^W_\beta$ belongs to $\mathbb{C}\left(\beta,\ell_\beta,m_\beta\right)$ for any $0<\beta<1$.

Next, in view of definition of a_β in (4.35) and of the form of the function $\rho_{2,\beta}$ in (4.26) we obtain, using condition $\mathbf{H}_2$, that $\lim_{\beta\to 0}\log a_\beta/|\log\beta| = 1$. Thus, by (3.76) in Proposition 3.2,

$$\lim_{\beta\to\infty}\frac{1}{|\log\beta|^r}\mathscr{R}^r_{\nu,\theta}(\widetilde{T}^W_\beta) = \frac{1}{I^r_\theta} \quad \text{for all } \nu\ge 0 \text{ and all } \theta\in\Theta.$$

Comparing this equality to the lower bound (4.29) implies (4.36).

To prove (4.37) it suffices to show that

$$\limsup_{\beta\to 0}\frac{\max\limits_{0\le\nu\le\ell_\beta+m_\beta}\mathscr{R}^r_{\nu,\theta}(\widetilde{T}^W_\beta)}{|\log\beta|^r} \le \frac{1}{I^r_\theta}. \tag{4.39}$$

Note that

$$\max_{0\le\nu\le\ell_\beta+m_\beta}\mathscr{R}^r_{\nu,\theta}(\widetilde{T}^W_\beta) \le \frac{\max\limits_{0\le\nu\le\ell_\beta+m_\beta}\mathsf{E}_{\nu,\theta}\left[(\widetilde{T}^W_\beta-\nu)^+]^r\right]}{\min\limits_{0\le\nu\le\ell_\beta+m_\beta}\mathsf{P}_\infty\left(\widetilde{T}^W_\beta>\nu\right)},$$

where

$$\min_{0\le\nu\le\ell_\beta+m_\beta}\mathsf{P}_\infty\left(\widetilde{T}^W_\beta>\nu\right) = \mathsf{P}_\infty\left(\widetilde{T}^W_\beta>\ell_\beta+m_\beta\right)\to 1 \quad \text{as } \beta\to 0.$$

Also, by inequality (3.85) in Chapter 3, for an arbitrarily $0<\varepsilon<I_\theta$

$$\sup_{\nu\ge 0}\mathsf{E}_{\nu,\theta}\left[(\widetilde{T}^W_\beta-\nu)^+\right]^r \le \left(1+\frac{a_\beta}{I_\theta-\varepsilon}\right)^r + r2^{r-1}\Upsilon_r(\varepsilon,\theta).$$

As a result, we obtain

$$\max_{0\le\nu\le\ell_\beta+m_\beta} \mathscr{R}^r_{\nu,\theta}(\widetilde{T}^W_\beta) \le \frac{\sup_{\nu\ge0} \mathsf{E}_{\nu,\theta}\left[(\widetilde{T}^W_\beta-\nu)^{+}]^r\right]}{\mathsf{P}_\infty\left(\widetilde{T}^W_\beta > \ell_\beta+m_\beta\right)} = \left(\frac{|\log\beta|}{I_\theta}\right)^r (1+o(1)) \quad \text{as } \beta\to 0.$$

This obviously yields the upper bound (4.39) and the proof is complete. □

Remark 4.1. For large values of threshold a regardless of the constraint on the LCPFA β the following approximation holds

$$\mathscr{R}^r_{\nu,\theta}(\widetilde{T}^W_a) \sim \left(\frac{a}{I_\theta}\right)^r \quad \text{as } a\to\infty. \tag{4.40}$$

The following theorem establishes asymptotic minimax properties of the MSR rule in the double minimax problem (4.8), i.e., with respect to the risk $\sup_{\theta\in\Theta_1}\max_{0\le\nu\le\ell_\beta+m_\beta} I^r_\theta \mathscr{R}^r_{\nu,\theta}(T)$ for compact subsets Θ_1 of Θ. Its proof follows almost immediately from Proposition 4.1(ii) and Theorem 4.2, and for this reason it is omitted.

Theorem 4.3. *Assume that for some $0<I_\theta<\infty$ conditions $\mathbf{C}_1$ and* (4.19) *are satisfied and that conditions $\mathbf{H}_1$ and $\mathbf{H}_2$ are satisfied as well. If Θ_1 is a compact subset of Θ, then*

$$\inf_{T\in\mathbb{C}(\beta,\ell_\beta,m_\beta)} \sup_{\theta\in\Theta_1} \max_{0\le\nu\le\ell_\beta+m_\beta} I^r_\theta \mathscr{R}^r_{\nu,\theta}(T) \sim |\log\beta|^r \sim \sup_{\theta\in\Theta_1} \max_{0\le\nu\le\ell_\beta+m_\beta} I^r_\theta \mathscr{R}^r_{\nu,\theta}(\widetilde{T}^W_\beta). \tag{4.41}$$

Remark 4.2. Let

$$T^*_B(\theta) = \inf\{n\ge 1: R_n(\theta)\ge B\}$$

be the stopping time of the SR detection procedure tuned to θ, where $R_n(\theta)$ is the SR statistic defined in (4.3). If the least favorable value of the parameter θ_* that maximizes the risk

$$\sup_{\theta\in\Theta_1} \max_{0\le\nu\le\ell_\beta+m_\beta} I^r_\theta \mathscr{R}^r_{\nu,\theta}(T^*_{B_\beta}(\theta)) = \max_{0\le\nu\le\ell_\beta+m_\beta} I^r_{\theta_*} \mathscr{R}^r_{\nu,\theta_*}(T^*_{B_\beta}(\theta_*))$$

can be found (at least approximately within a small term), then the SR rule $T^*_B(\theta_*)$ is asymptotically double minimax. This rule is easier to implement and it should have even smaller maximal risk than the MSR rule.

4.2.2.2 The Case of LLR with Independent Increments

As in Section 3.8 of Chapter 3, we now consider a particular (still quite general) case where the LLR process has independent increments. The following theorem is similar to Theorem 3.9.

Theorem 4.4. *Assume that the LLR process $\{\lambda_\theta(k,k+n)\}_{n\ge1}$ has independent, not necessarily identically distributed increments under $\mathsf{P}_{k,\theta}$, $k\ge 0$. Suppose that conditions $\mathbf{H}_1$, $\mathbf{H}_2$, $\mathbf{C}_1$, and* (3.128) *are satisfied. Then asymptotic relations* (4.36) *and* (4.37) *hold for all $r>0$, i.e., the MSR rule $\widetilde{T}^W_\beta$ is first-order asymptotically uniformly pointwise optimal and minimax in class $\mathbb{C}(\beta,\ell_\beta,m_\beta)$ with respect to all positive moments of the detection delay.*

Proof. The proof follows almost immediately from Theorem 3.9 and Theorem 4.2. □

4.3 Examples

Recall that conditions $\mathbf{C}_1$ and $\mathbf{C}_2$ has been already verified in a number of examples in Chapter 3 (see, e.g., Examples 3.5, 3.6 in Section 3.13). Here we present several additional examples that illustrate the general theory developed in previous sections. We refer to the corresponding conditions $\mathbf{A}_1$–$\mathbf{A}_3$, $\mathbf{B}_1$–$\mathbf{B}_2$ and $\mathbf{B}'_1$ in Section 3.11.

Example 4.1 (*Change in the parameters of the multivariate linear difference equation*). Consider the multivariate model in $\mathbb{R}^p$ given by

$$X_n = \left(\check{A}_n \mathbb{1}_{\{n\le\nu\}} + A_n \mathbb{1}_{\{n>\nu\}}\right) X_{n-1} + w_n\,, \tag{4.42}$$

where $\check{A}_n$ and A_n are $p\times p$ random matrices and $(w_n)_{n\ge1}$ is an i.i.d. sequence of Gaussian random vectors $\mathcal{N}(0,Q_0)$ in $\mathbb{R}^p$ with the positive definite $p\times p$ matrix Q_0. Assume also that $\check{A}_n = A_0 + B_n$, $A_n = \theta + B_n$ and $(B_n)_{n\ge1}$ are i.i.d. Gaussian random matrices $\mathcal{N}(0,Q_1)$, where the $p^2\times p^2$ matrix $Q_1 = \mathsf{E}[B_1\otimes B_1]$ is positive definite. Assume, in addition, that all eigenvalues of the matrix

$$\mathsf{E}[\check{A}_1\otimes\check{A}_1] = A_0\otimes A_0 + Q_1$$

are less than 1 in module. Define

$$\Theta = \{\theta\in\mathbb{R}^{p^2} : \max_{1\le p^4} \mathbf{e}_j(\theta\otimes\theta + Q_1) < 1\}\}\setminus\{A_0\}, \tag{4.43}$$

where $\mathbf{e}_j(A)$ is the jth eigenvalue of the matrix A, and assume further that the matrix $\theta\in\Theta$. In this case, the processes $\{X_n\}_{n\ge1}$ (in the case $\nu=\infty$) and $\{X_n\}_{n>\nu}$ (in the case $\nu<\infty$) are ergodic with the stationary distributions given by the vectors [71]

$$\check{\varsigma} = \sum_{i=1}^{\infty}\prod_{j=1}^{i-1}\check{A}_j w_l \quad\text{and}\quad \varsigma_\theta = \sum_{i=1}^{\infty}\prod_{j=1}^{i-1} A_j w_l$$

i.e., the corresponding invariant measures $\check{\varkappa}$ and $\varkappa^\theta$ on $\mathbb{R}^p$ are defined as $\check{\varkappa}(A) = \mathsf{P}(\check{\varsigma}\in\Gamma)$ and $\varkappa^\theta(A) = \mathsf{P}(\varsigma_\theta\in\Gamma)$ for any $\Gamma\in\mathscr{B}(\mathbb{R}^p)$.

Define the Lyapunov function as

$$\mathbf{V}^\theta(x) = \upsilon_*(1 + x^\top T(\theta)x), \quad T(\theta) = \left(I_{p^4} - \theta^\top\otimes\theta^\top - Q_1^\top\right)^{-1}\mathrm{Vect}(\check{I}_p), \tag{4.44}$$

where $\upsilon_*\ge1$, $\check{I}_m$ is the identity matrix of order m. As shown in [38], in this case, for any $x\in\mathbb{R}^p$ the quadratic form $x^\top T(\theta)x\ge|x|^2$. Hence all eigenvalues of the matrix $T(\theta)$ are greater than 1. Let now $\Theta_c\subset\Theta$ be some compact set. For some fixed $N_*>1$, define the set

$$C = \{x\in\mathbb{R}^p : \max_{\theta\in\Theta_c} x^\top T(\theta)x\le N_*\}. \tag{4.45}$$

Direct calculations yield

$$\mathsf{E}^\theta\left[\mathbf{V}^\theta(X_1)|X_0=x\right] = \mathbf{V}^\theta(x)\left(1 - \frac{|x|^2 - \mathrm{tr}\,T(\theta)Q_0}{\mathbf{V}^\theta(x)}\right). \tag{4.46}$$

Taking into account that the function $T(\theta)$ is continuous, we obtain that for any non-zero vector $x\in\mathbb{R}^p$ and $\theta\in\Theta_c$

$$1\le\mathbf{e}_{min}\le\frac{x^\top T(\theta)x}{|x|^2}\le\mathbf{e}_{max}<\infty,$$

where

$$\mathbf{e}_{min} = \min_{\theta\in\Theta_c} \inf_{x\in\mathbb{R}^p, x\neq 0} \frac{x^\top T(\theta)x}{|x|^2} \quad \text{and} \quad \mathbf{e}_{max} = \max_{\theta\in\Theta_c} \sup_{x\in\mathbb{R}^p, x\neq 0} \frac{x^\top T(\theta)x}{|x|^2}.$$

From here it follows that, for $x \in C^c$, $|x|^2 > N_*/\mathbf{e}_{max}$, and therefore,

$$\frac{|x|^2 - \operatorname{tr} T(\theta)Q_0}{\mathbf{V}^\theta(x)} \geq \frac{|x|^2}{1+\mathbf{e}_{max}|x|^2} - \frac{\mathbf{e}_{max} \operatorname{tr} Q_0}{\mathbf{e}_{min}|x|^2} \geq \frac{1}{2\mathbf{e}_{max}} - \frac{\mathbf{e}^2_{max} \operatorname{tr} Q_0}{\mathbf{e}_{min} N_*}.$$

Now we choose $N_* > 1$ sufficiently large to obtain a positive term on the right side of the last inequality. So we obtain the drift inequality (3.188) for the Lyapunov function defined in (4.44) with any coefficient $\upsilon^* \geq 1$. The function $g(\theta,y,x)$ can be calculated for any x,y from $\mathbb{R}^p$ and $\theta \in \Theta$ as

$$\begin{aligned} g(\theta,y,x) &= \frac{|G^{-1/2}(x)(y-A_0x)|^2 - |G^{-1/2}(x)(y-\theta x)|^2}{2} \\ &= y^\top G^{-1}(x)(\theta - A_0)x + \frac{x^\top A_0^\top G^{-1}(x)A_0x - x^\top \theta^\top G^{-1}(x)\,\theta\, x}{2}, \end{aligned}$$

where

$$G(x) = \mathsf{E}\,[B_1 x x^\top B_1^\top] + Q_0 = Q_1 \operatorname{Vect}(xx^\top) + Q_0.$$

Taking into account that the matrices Q_0 and Q_1 are positive definite, we obtain that there exists some constant $\mathbf{c}_* > 0$ for which

$$\sup_{x\in\mathbb{R}^p} |G^{-1}(x)| \leq \frac{\mathbf{c}_*}{1+|x|^2}. \tag{4.47}$$

From this we obtain that condition $\mathbf{B}_2$ holds with $\gamma = 1$ and

$$h(y,x) = \mathbf{c}_*(2\theta_{max} + |y|) \quad \text{and} \quad \theta_{max} = \max_{\theta\in\Theta} |\theta|. \tag{4.48}$$

Moreover, note that in this case

$$\widetilde{g}(\theta,x) = \frac{1}{2}|G^{-1/2}(x)(\theta - A_0)x| = \frac{1}{2}x^\top(\theta - A_0)^\top G^{-1}(x)(\theta - A_0)x.$$

The bound (4.47) implies that $g^* = \sup_{x\in\mathbb{R}^p} \sup_{\theta\in\Theta_c} \widetilde{g}(\theta,x) < \infty$.

Now, choosing $V(x) = \upsilon^*(1+x^\top T x^\delta)$ with $\upsilon^* = 1+g^*$ and any fixed $0 < \delta \leq 1$ yields condition $\mathbf{A}_1$. Moreover, for any $r > 0$ and $\delta r \leq 2$

$$\sup_{x\in\mathbb{R}} \sup_{\theta\in\Theta_c} \frac{\sup_{j\geq 1} \mathsf{E}_x^\theta |X_j|^{\delta r}}{1+|x|^{\delta r}} < \infty \quad \text{and} \quad \sup_{j\geq 1} \check{\mathsf{E}}\,|X_j|^{\delta r} < \infty, \tag{4.49}$$

where $\check{\mathsf{E}}$ denotes the expectation with respect to the distribution $\check{\mathsf{P}}$ when $\nu = \infty$. This implies $\mathbf{A}_2$. Therefore, taking into account that δ can be very close to zero and using Theorem 3.12 we get that for any $r > 0$ and any compact set $\Theta_c \subset \Theta \setminus \{A_0\}$ condition $\mathbf{C}_4(r)$ holds with $I_\theta = \mathsf{E}^\theta[\widetilde{g}(\theta, \varsigma_\theta)]$.

Example 4.2 (*Change in the correlation coefficients of the AR(p) model*)**.** Consider the problem of detecting the change of the correlation coefficient in the pth order AR process, assuming that for $n \geq 1$

$$X_n = \mathrm{a}_{1,n} X_{n-1} + \ldots + \mathrm{a}_{p,n} X_{n-p} + w_n, \tag{4.50}$$

where $a_{i,n} = a_i \mathbb{1}_{\{n \le \nu\}} + \theta_i \mathbb{1}_{\{n > \nu\}}$ and $(w_n)_{n \ge 1}$ are i.i.d. Gaussian random variables with $\mathsf{E}[w_1] = 0$, $\mathsf{E}[w_1^2] = 1$. In the sequel, we use the notation $\mathbf{a} = (a_1, \dots, a_p)^\top$ and $\theta = (\theta_1, \dots, \theta_p)^\top$. The process (4.50) is not Markov, but the p-dimensional process

$$\Phi_n = (X_n, \dots, X_{n-p+1})^\top \in \mathbb{R}^p \tag{4.51}$$

is Markov. Note that for $n > \nu$

$$\Phi_n = A\Phi_{n-1} + \widetilde{w}_n, \tag{4.52}$$

where

$$A = A(\boldsymbol{\theta}) = \begin{pmatrix} \theta_1 & \theta_2 & \dots & \theta_p \\ 1 & 0 & \dots & 0 \\ \vdots & \vdots & \ddots & \vdots \\ 0 & 0 & \dots 1 & 0 \end{pmatrix}, \qquad \widetilde{w}_n = (w_n, 0, \dots, 0) \in \mathbb{R}^p.$$

It is clear that

$$\mathsf{E}[\widetilde{w}_n \widetilde{w}_n^\top] = B = \begin{pmatrix} 1 & \dots & 0 \\ \vdots & \ddots & \vdots \\ 0 & \dots & 0 \end{pmatrix}.$$

Assume that the vectors $\mathbf{a}$ and θ belong to the set

$$\Theta = \{\theta \in \mathbb{R}^p : \max_{1 \le j \le p} |\mathbf{e}_j(A(\theta))| < 1\}, \tag{4.53}$$

where $\mathbf{e}_j(A)$ denotes the jth eigenvalue for the matrix A. Note that, in this case, for any θ from some compact set $\Theta_c \subset \Theta$ and any $y = (y_1, \dots, y_p)^\top \in \mathbb{R}^p$ and $x = (x_1, \dots, x_p)^\top \in \mathbb{R}^p$ the function

$$g(\theta, y, x) = y_1 (\theta - \mathbf{a})^\top x + \frac{(\mathbf{a}^\top x)^2 - (\theta^\top x)^2}{2}. \tag{4.54}$$

Obviously, it follows that condition $\mathbf{B}_2$ holds with $\gamma = 1$ and

$$h(y, x) = y_1^2 + (1 + 2\theta_{max})|x|^2,$$

where θ_{max} is defined in (4.48)

Obviously, for any $\theta \in \Theta$ the process $(\Phi_n)_{n > \nu + p}$ is ergodic with the stationary normal distribution

$$\varsigma_\theta = \sum_{i=1}^{\infty} A^{i-1} \widetilde{w}_i \sim \mathcal{N}(0, \mathrm{F}), \quad \text{where} \quad \mathrm{F} = \mathrm{F}(\theta) = \sum_{i=0}^{\infty} A^i B (A^\top)^i. \tag{4.55}$$

Obviously, condition (B1.1) does not hold for the process (4.51). To fulfill this condition we replace this process by the imbedded homogeneous Markov process $\widetilde{\Phi}_{\iota,n} = \Phi_{np+\iota}$ for some $0 \le \iota \le p-1$. This process can be represented as

$$\widetilde{\Phi}_{\iota,n} = A^p \widetilde{\Phi}_{\iota,n-1} + \zeta_{\iota,n} \quad \text{and} \quad \zeta_{\iota,n} = \sum_{j=0}^{p-1} A^j \widetilde{w}_{np+\iota-j}. \tag{4.56}$$

Clearly, $\zeta_{\iota,n}$ is Gaussian with the parameters $(0, Q)$, where

$$Q = Q(\theta) = \sum_{j=0}^{p-1} A^j B (A^\top)^j.$$

One can check directly that this matrix is positive definite. Moreover, it can be easily checked that for any $\theta \in \Theta$ and for any $0 \le \iota \le p-1$ the process (4.56) is ergodic with the same ergodic distribution given in (4.55).

Now, for any fixed $0 < \delta \le 1$ we define the $\mathbb{R}^p \to \mathbb{R}$ function $\mathbf{V}^\theta$ as

$$\mathbf{V}^\theta(x) = \check{\mathbf{c}}(1 + (x^\top T x)^\delta) \quad \text{and} \quad T = T(\theta) = \sum_{l=0}^{\infty} (A^\top)^{pl} A^{pl}, \tag{4.57}$$

where $\check{\mathbf{c}} \ge 1$ will be specified later. Let for any fixed compact set $\Theta_c \subset \Theta \setminus \{\mathbf{a}\}$

$$t_{\max} = \max_{\theta \in \Theta_c} |T(\theta)| \quad \text{and} \quad q_{max} = \max_{\theta \in \Theta_c} |Q(\theta)|.$$

Obviously, $t_{\max} > 1$. Note that, by the Jensen inequality, for any $0 \le \iota < p$

$$\begin{aligned}\mathsf{E}^\theta \left(\mathbf{V}^\theta(\widetilde{\Phi}_{\iota,1}) \,|\, \widetilde{\Phi}_{\iota,0} = x \right) &\le \check{\mathbf{c}} + \check{\mathbf{c}} \left(x^\top (A^p)^\top T A^p x + \operatorname{tr} TQ \right)^\delta \\ &\le \check{\mathbf{c}} + \check{\mathbf{c}} \left(x^\top (A^p)^\top T A^p x + t_{\max} q_{\max} \right)^\delta .\end{aligned}$$

Note that

$$x^\top T x \ge |x|^2 \quad \text{and} \quad \frac{x^\top (A^p)^\top T A^p x}{x^\top T x} = 1 - \frac{|x|^2}{x^\top T x} \le 1 - \frac{1}{t_{max}} = t_* < 1.$$

So, taking into account that $(|a| + |b|)^\delta \le |a|^\delta + |b|^\delta$ for $0 < \delta \le 1$, we obtain

$$\mathsf{E}^\theta \left[\mathbf{V}^\theta(\widetilde{\Phi}_{\iota,1}) \,|\, \widetilde{\Phi}_{\iota,0} = x \right] \le \check{\mathbf{c}} + \check{\mathbf{c}} \left(t_*^\delta (x^\top T x)^\delta + (t_{\max} q_{\max})^\delta \right).$$

Putting

$$N_* = \left(\frac{2(1 + t_{\max}^\delta q_{\max}^\delta)}{1 - t_*^\delta} \right)^{1/2\delta} \quad \text{and} \quad \rho = (1 - t_*^\delta)/2,$$

yields that, for $|x| \ge N_*$,

$$\mathsf{E}^\theta \left[\mathbf{V}^\theta(\widetilde{\Phi}_{\iota,1}) \,|\, \widetilde{\Phi}_{\iota,0} = x \right] \le (1 - \rho) V^\theta(x).$$

Hence, the Markov process (4.56) satisfies the drift inequality (3.191) with

$$C = \{x \in \mathbb{R}^p : |x| \le N_*\} \quad \text{and} \quad D = \check{\mathbf{c}}(1 + t_{max}^\delta N_*^{2\delta} + t_{max}^\delta q_{max}^\delta).$$

Next we need the minorizing measure in condition $\mathbf{B}_1'$ on the Borel σ-field in $\mathbb{R}^p$. To this end, we define $\check{\nu}(\Gamma) = \mathbf{mes}(\Gamma \cap C)/\mathbf{mes}(C)$ for any Borel set Γ in $\mathbb{R}^p$, where $\mathbf{mes}(\cdot)$ is the Lebesgue measure in $\mathbb{R}^p$. Moreover, note that

$$\widetilde{h}(\theta, x) = 1 + (\theta^\top x)^2 + (1 + 2\theta_{max})|x|^2 \le 1 + (1 + 2\theta_{max})|x|^2$$

and

$$\widetilde{g}(\theta, x) = \frac{1}{2} \left((\theta - \mathbf{a}_0)^\top x \right)^2 \le \theta_{max}^2 |x|^2.$$

Therefore, choosing in (4.57) $\check{\mathbf{c}} = 1 + 2\theta_{max} + \theta_{max}^2$, we obtain condition $\mathbf{B}'1.2$. Condition $\mathbf{B}_2$ can be checked in the same way as in Example 3.1 for any $r > 0$ for which $0 < \delta r \le 2$. Therefore, taking into account that δ can be very close to zero, Theorem 3.13 implies that for any $r > 0$ and any compact set $\Theta_c \subset \Theta \setminus \{\mathbf{a}\}$ condition $\mathbf{C}_4(r)$ holds with $I_\theta = \mathsf{E}^\theta [\widetilde{g}(\theta, \varsigma_\theta)]$.

4.4 Monte Carlo Simulations

In this section, we provide Monte Carlo (MC) simulations for the AR(1) model, which is a particular case of Example 3.1 of Subsection 3.11 with Gaussian noise $w_n \sim \mathcal{N}(0,1)$ (see also Example 4.2 of Section 4.3 for $p=1$). Specifically, let the pre-change value $\theta_0 = 0$ and the post-change value $\theta \in \Theta = \{\theta_1,\dots,\theta_N\}$, $-1 < \theta_1 < \theta_2 < \cdots < \theta_N < 1$, $\theta_i \neq 0$, and write

$$L_n^\theta(X_n, X_{n-1}) = \exp\left\{\theta X_n X_{n-1} - \frac{\theta^2 X_{n-1}^2}{2}\right\}, \quad n \ge 1.$$

The MSR stopping time is written as

$$\widetilde{T}_A^W = \inf\left\{n \ge 1 : \sum_{j=1}^N W(\theta_j) R_n(\theta_j) \ge A\right\},$$

where the SR statistic $R_\theta(n)$ tuned to θ satisfies the recursion

$$R_\theta(n) = [1 + R_\theta(n-1)] L_n^\theta(X_n, X_{n-1}), \quad n \ge 1, \quad R_\theta(0) = 0.$$

Thus, the MSR rule can be easily implemented.

The information number $I_\theta = \mathcal{K}_\theta = \theta^2/[2(1-\theta^2)]$, so the first-order approximation (4.40) yields the following approximate formula for the average delay to detection $\mathsf{ADD}_{\nu,\theta}(T_a) = \mathsf{E}_{\nu,\theta}(\widetilde{T}_A^W - \nu \mid \widetilde{T}_A^W > \nu)$:

$$\mathsf{ADD}_{\nu,\theta}(\widetilde{T}_A^W) \approx \mathsf{ADD}_{\nu,\theta}^{app}(\widetilde{T}_A^W) = \frac{2(1-\theta^2)\log A}{\theta^2}. \tag{4.58}$$

In the MC simulations, we set $\theta_{j+1} = \theta_j + 0.1$, $j = 1,\dots,17$ with $\theta_1 = -0.1$, $\theta_{18} = 0.9$ and uniform prior $W(\theta_j) = 1/18$, $j = 1,\dots,18$. The results are presented in Table 4.1 for the upper bound $\beta = 0.01$ on the maximal local conditional probability of false alarm

$$\mathsf{LCPFA}(T) = \sup_{1 \le k \le \ell} \mathsf{P}_\infty(T < k+m \mid T \ge k)$$

and the number of MC runs 10^5. In the table, we compare operating characteristics of the MSR rule $\widetilde{T}_A^W$ with that of the SR rule

$$T_B^* = \inf\{n \ge 1 : R_\theta(n) \ge B\}$$

tuned to the true values of the post-change parameter θ, which are shown in the table (i.e., assuming that the post-change parameter is known and equals one of the values shown in the table). The thresholds A and B (shown in the table) were selected in such a way that the maximal probabilities of false alarm of both rules ($\mathsf{LCPFA}(\widetilde{T}_A^W)$ and $\mathsf{LCPFA}(T_B^*)$) were practically the same. It is seen that for relatively large values of the post-change parameter, $\theta \ge 0.6$, the SR rule only slightly outperforms the MSR rule, but for small parameter values (i.e., for close hypotheses) the difference becomes quite substantial. The worst change point is $\nu = 0$, as expected. Also, the first-order approximation (4.58) is not too accurate, especially for small and large parameter values.

TABLE 4.1
Operating Characteristics of the MSR and SR Detection Procedures.

$\beta = 0.01, \nu = 0$							
θ	A	$\mathsf{ADD}_{\nu,\theta}(\widetilde{T}_A^W)$	$\mathsf{LCPFA}(\widetilde{T}_A^W)$	B	$\mathsf{ADD}_{\nu,\theta}(T_B^*)$	$\mathsf{LCPFA}(T_B^*)$	$\mathsf{ADD}_{\nu,\theta}^{app}(\widetilde{T}_A^W)$
0.9	395	11.74	0.0080	791	11.08	0.0079	2.81
0.8	420	14.72	0.0073	791	13.72	0.0073	6.80
0.7	440	18.97	0.0070	791	17.52	0.0071	12.67
0.6	470	25.32	0.0065	791	23.15	0.0065	21.88
0.5	595	36.35	0.0049	791	31.84	0.0049	38.33
0.4	1040	59.57	0.0024	791	45.88	0.0025	72.94
$\beta = 0.01, \nu = 10$							
θ	A	$\mathsf{ADD}_{\nu,\theta}(\widetilde{T}_A^W)$	$\mathsf{LCPFA}(\widetilde{T}_A^W)$	B	$\mathsf{ADD}_{\nu,\theta}(T_B^*)$	$\mathsf{LCPFA}(T_B^*)$	$\mathsf{ADD}_{\nu,\theta}^{app}(\widetilde{T}_A^W)$
0.9	395	10.05	0.0080	791	9.62	0.0079	2.81
0.8	420	12.72	0.0073	791	11.98	0.0073	6.80
0.7	440	16.59	0.0070	791	15.30	0.0071	12.67
0.6	470	22.55	0.0065	791	20.34	0.0065	21.88
0.5	595	32.96	0.0049	791	28.01	0.0049	38.33
0.4	1040	55.34	0.0024	791	40.83	0.0025	72.94

5

Change Detection Rules Optimal for the Maximal Detection Probability Criterion

5.1 Introduction

In the previous chapters, we considered change detection procedures that accumulate data as long as a detection statistic exceeds a threshold. These procedures are optimal or asymptotically optimal in the sense of minimizing the expected number of observations until detection. Another reasonable problem setting, discussed in Chapter 2 (see Subsection 2.3.7), is to maximize the probability of detection of a change in a prescribed time interval $M \geq 1$. For example, in the Bayesian setting, it is required to find an optimal change detection rule T_{opt} that satisfies the optimality criterion

$$\mathsf{P}_\pi(\nu < T_{\text{opt}} \leq \nu + M | T_{\text{opt}} > \nu) = \sup_{T \in \mathbb{C}_\pi(\alpha)} \mathsf{P}_\pi(\nu < T \leq \nu + M | T > \nu)$$

(see (2.42) in Subsection 2.3.7). However, finding an optimal procedure when $M > 1$ is a very difficult problem even in the i.i.d. case. Reasonable candidates are moving window rules with the window size M. If M is large, then it is perhaps possible to find asymptotically optimal rules as $\alpha \to 0$, as we briefly discuss in Section 5.4.

In Section 5.2, we consider the particular case of $M = 1$, i.e., when one wants to maximize the (instantaneous) probability of detecting a change right at the moment of its onset. Then it is possible to find optimal detection rules in several problem settings – Bayesian and maximin.

In the 1930s, Shewhart [127], motivated by industrial quality control problems, introduced the control chart that raises a warning when a current observation becomes larger than a specific tolerance level.[1] This chart, which is referred to as the Shewhart control chart, was very popular in the past. The usual folklore in the SPC community is that this simplest and even naive detection rule performs well only for detecting large changes, which is intuitively expected. In this chapter, we show that the Shewhart-type rules have certain optimality properties for detecting changes of arbitrarily magnitude. It turns out that the likelihood ratio based Shewhart detection rule, which consists in comparing the running likelihood ratio to a threshold, is optimal with respect to Bayesian as well as maximin criteria, maximizing the instantaneous probability of detection. Furthermore, in certain cases, e.g., for exponential families of distributions, the optimality property holds whether or not the putative value of the post-change parameter agrees with the actual value in contrast to the optimality of CUSUM and Shiryaev's procedures.

In Section 5.3, these results are extended to the case of $M > 1$. In particular, it is shown that a modified Shiryaev-type rule and a modified CUSUM-type rule are optimal when the size of the window M is random having a geometric distribution.

[1] In engineering applications, this kind of rules is called snap-shot algorithms.

5.2 The Shewhart Rule and Its Optimality Properties

5.2.1 Optimality with Respect to the Expected Detection Delay

We begin with considering the simplest moving window detection rule with the size of the window $M = 1$, which is nothing but the LR-based Shewhart detection rule, and recall its optimality property that follows from the optimality of the CUSUM procedure established by Moustakides [93].

Suppose observations $\{X_n\}_{n\in\mathbb{Z}_+}$ are i.i.d. with a density $g(x)$ before the change and i.i.d. with a density $f(x)$ after the change and denote by $\Lambda_n = f(X_n)/g(X_n)$ the likelihood ratio for the nth observation. The Shewhart rule is defined as the first time n such that the LR Λ_n exceeds a positive threshold A:

$$T_A = \inf\{n \ge 1 : \Lambda_n \ge A\}. \tag{5.1}$$

In order to avoid the necessity of randomization on the stopping threshold throughout this chapter we suppose that the LR Λ_n is continuous under both pre-change P_∞ and post-change P_0 measures, i.e., that the distributions $\mathsf{P}_\infty(\Lambda_n \le y)$ and $\mathsf{P}_0(\Lambda_n \le y)$ are continuous as well as strictly increasing.

Let $T_{\rm CS}(A)$ denote the CUSUM stopping time, i.e.,

$$T_{\rm CS} = \inf\{n \ge 1 :\ V_n \ge A\}, \quad V_n = \max\{V_{n-1}, 1\}\Lambda_n, \ V_0 = 1. \tag{5.2}$$

In Section 2.3.5, we already discussed the optimality property of CUSUM:
For any tolerable level $\gamma \ge 1$ of the ARL to false alarm $\mathsf{ARL2FA}(T) = \mathsf{E}_\infty[T]$, *it minimizes Lorden's worst-case delay to detection* $\mathsf{ESADD}(T) = \sup_{0\le\nu<\infty} \operatorname{ess\,sup} \mathsf{E}_\nu[(T-\nu)^+ | \mathscr{F}_\nu^X]$ *in class* $\mathbb{C}_\gamma = \{T : \mathsf{ARL2FA}(T) \ge \gamma\}$, *where* $\nu \in \mathbb{Z}_+$ *is the change point.*

Obviously, for $A \le 1$, the CUSUM stopping time coincides with the Shewhart stopping time T_A (since in this case $V_n = \Lambda_n$). So if the ARL to a false alarm $\mathsf{ARL2FA}(T_A) = \gamma$ is such that $A = A_\gamma \le 1$, then Shewhart's rule is strictly optimal, minimizing Lorden's $\mathsf{ESADD}(T)$ in class $\mathbb{C}_\gamma$. Unfortunately, the range of γ-values when this is true is usually quite restricted, especially for small changes.

To see this, notice that for $i = 0, \infty$ the distributions of the stopping time T_A are geometric

$$\mathsf{P}_i(T_A = n) = \mathsf{P}_i(\Lambda_1 < A, \dots, \Lambda_{n-1} < A, \Lambda_n \ge A) = \mathsf{P}_i(\Lambda_1 \ge A)[\mathsf{P}_i(\Lambda_1 < A)]^{n-1}, \tag{5.3}$$

so

$$\mathsf{E}_i[T_A] = \frac{1}{\mathsf{P}_i(\Lambda_1 \ge A)}, \quad i = 0, \infty. \tag{5.4}$$

Thus, CUSUM is reduced to Shewhart's rule for

$$1 \le \gamma \le \frac{1}{\mathsf{P}_\infty(\Lambda_1 \ge A)}, \quad A \le 1$$

and the maximal value of the ARL to false alarm for which this is true is

$$\gamma_{\max} = \frac{1}{\mathsf{P}_\infty(\Lambda_1 \ge 1)}.$$

For example, if one detects a change from the standard normal distribution $\mathscr{N}(0,1)$ to the normal $\mathscr{N}(\theta,1)$ with mean $\theta \ne 0$, then $\gamma_{\max} = 1/\Phi(-\theta/2)$. For $\theta = 1$ (moderate change), we obtain $\gamma_{\max} \approx 3.25$, which is a really small number for practical purposes. For a large change such as $\theta = 5$, we obtain $\gamma_{\max} \approx 161$. The average delay $\mathsf{ESADD}(T_{A=1}) = 1/[1-\Phi(-\theta/2)]$ equals 1.45 when $\theta = 1$ and 1.006 when $\theta = 5$.

Consider another example of detecting a change in the parameter of the exponential distribution with density $f_\theta(x) = (1+\theta)^{-1} e^{-x/(1+\theta)}, x \ge 0$, from $\theta = 0$ to $\theta = \vartheta > 0$. In this case, $\gamma_{\max} = (1+\vartheta)^{1+1/\vartheta}$ and $\mathsf{ESADD}(T_{A=1}) = (1+\vartheta)^{1/\vartheta}$. For $\vartheta = 1$, we obtain $\gamma_{\max} = 4$ and $\mathsf{ESADD}(T_{A=1}) = 2$. For $\vartheta = 5$, we have $\gamma_{\max} \approx 8.59$ and $\mathsf{ESADD}(T_{A=1}) = 1.43$.

5.2.2 Maximal Average Probability of Detection: the Bayesian Approach

Consider the Bayesian problem with the zero-modified Geometric(q,ρ) prior distribution $\pi_k = \mathsf{P}(\nu = k)$ of the change point ν, i.e,

$$\Pr(\nu < 0) = \pi_{-1} = q \text{ and } \pi_k = \mathsf{P}(\nu = k) = (1-q)\rho(1-\rho)^k \quad \text{for } k \in \mathbb{Z}_+,$$

where $q \in [0,1)$ and $\rho \in (0,1)$. Setting the size of the window $M = 1$ in the general definition of the probability of detection given in (2.41), we obtain that for any stopping time T the probability of detection is

$$\mathsf{PD}_\pi(T) = \mathsf{P}_\pi(T = \nu + 1 | T > \nu) = \frac{\sum_{k=-1}^{\infty} \pi_k \mathsf{P}_k(T = k+1)}{1 - \mathsf{PFA}_\pi(T)}, \tag{5.5}$$

where $\mathsf{P}_\pi(\mathscr{A} \times \mathscr{K}) = \sum_{k\in\mathscr{K}} \pi_k \mathsf{P}_k(\mathscr{A})$. Notice that, as before, for technical purposes it is convenient to assume without loss of generality that for negative values of ν the prior is concentrated at $\nu = -1$ with probability q, which is the probability that the change occurred before the observations became available. Therefore, we can ignore the values of ν smaller than -1 in the definition of $\mathsf{PD}_\pi(T)$. Recall that $\mathbb{C}_\pi(\alpha) = \{T\colon \mathsf{PFA}_\pi(T) \le \alpha\}$ denotes the class of rules for which the probability of false alarm $\mathsf{PFA}_\pi(T) = \sum_{k=0}^{\infty} \pi_k \mathsf{P}_\infty(T \le k)$ does not exceed the prescribed level α.

In the Bayesian setting, the goal is to find $T_{\mathrm{opt}} \in \mathbb{C}_\pi(\alpha)$ such that

$$\mathsf{PD}_\pi(T_{\mathrm{opt}}) = \sup_{T\in\mathbb{C}_\pi(\alpha)} \mathsf{PD}_\pi(T) \quad \text{for every } 0 < \alpha < 1 \tag{5.6}$$

(cf. (2.42) in Chapter 2).

More generally, assume that the post-change distribution belongs to a parametric family $\{F_\theta\}_{\theta\in\Theta}$, $\Theta \subset \mathbb{R}^\ell$, so if $\nu = k$, then the vectors $X_{k+1}, X_{k+2}, \dots$ have density $f_\theta(x)$ (with respect to a σ-finite measure). For a fixed change point $\nu = k$, the post-change hypothesis becomes composite, $\theta \in \Theta_1$. Suppose we are given a prior distribution $W(\theta)$, $\theta \in \Theta_1 \subset \Theta$, on the post-change parameter θ. Write $\Lambda_n^\theta = f_\theta(X_n)/g(X_n)$ for the LR tuned to the parameter θ and define the mixture LR

$$\Lambda_n^W = \int_{\Theta_1} \Lambda_n^\theta \,\mathrm{d}W(\theta) \equiv \frac{\bar{f}_W(X_n)}{g(X_n)},$$

where $\bar{f}_W(x) = \int_{\Theta_1} f_\theta(x)\,\mathrm{d}W(\theta)$. Let $\mathsf{P}_{k,\theta}$ denote probability when the change takes place at $\nu = k$ and the post-change parameter is θ and P_k^W when the change takes place at $\nu = k$ and $\theta \sim W$. Let $\mathsf{P}_\pi^W(\mathscr{A} \times \mathscr{K}) = \sum_{k\in\mathscr{K}} \pi_k \mathsf{P}_k^W(\mathscr{A})$. The Bayesian optimization problem (5.6) is now modified as follows: find $T_{\mathrm{opt}} \in \mathbb{C}_\pi(\alpha)$ such that

$$\mathsf{PD}_\pi^W(T_{\mathrm{opt}}) = \sup_{T\in\mathbb{C}_\pi(\alpha)} \mathsf{PD}_\pi^W(T) \quad \text{for every } 0 < \alpha < 1, \tag{5.7}$$

where

$$\mathsf{PD}_\pi^W(T) = \mathsf{P}_\pi^W(T = \nu + 1 | T > \nu). \tag{5.8}$$

Similarly to (5.1) introduce the detection rule

$$T_A^W = \inf\{n \ge 1 : \Lambda_n^W \ge A\}, \quad A > 0, \tag{5.9}$$

which we will refer to as the *mixture Shewhart rule*.

Obviously, if W is concentrated in a single point θ_1, then the optimization problem (5.7) reduces to (5.6) and the mixture Shewhart rule (5.9) to the Shewhart rule (5.1). Hence, all the optimality results established in Theorem 5.1 below hold for the Shewhart rule $T_A = T_A^{\theta_1}$.

We now provide useful expressions for the probabilities $\mathsf{P}_\pi^W(T=\nu+1)$ and $\mathsf{P}_\pi^W(T>\nu)$, which will be used for proving the optimality of the Shewhart rule. Write $p=\mathsf{P}_\pi^W(T=0)$ for the probability of stopping at 0 without making observations. Since $\mathsf{P}_\pi^W(T\geq 0)=1$ and $\{T>t\}\in\mathscr{F}_t$, changing the measure $\mathsf{P}_k^W\to\mathsf{P}_\infty$ and using Wald's likelihood ratio identity $\mathsf{P}_k^W(T=k+1)=\mathsf{E}_\infty[\Lambda_{k+1}^W\mathbb{1}_{\{T=k+1\}}]$ we obtain

$$
\begin{aligned}
\mathsf{P}_\pi^W(T=\nu+1) &= \mathsf{P}(\nu<0)\mathsf{P}_\pi^W(T=0)+\sum_{k=0}^\infty \pi_k\mathsf{P}_k^W(T=k+1)\\
&= qp+(1-q)\rho\sum_{k=0}^\infty(1-\rho)^k\mathsf{P}_k^W(T=k+1)\\
&= qp+(1-q)\rho\sum_{k=0}^\infty(1-\rho)^k\mathsf{E}_\infty[\Lambda_{k+1}^W\mathbb{1}_{\{T=k+1\}}]\\
&= qp+\frac{(1-q)\rho}{(1-\rho)}\mathsf{E}_\infty[(1-\rho)^T\Lambda_T^W\mathbb{1}_{\{T>0\}}]\\
&= qp+\frac{(1-q)\rho}{(1-\rho)}\mathsf{E}_\infty[(1-\rho)^T\Lambda_T^W|T>0]\mathsf{P}_\pi^W(T>0)\\
&= qp+\frac{(1-q)\rho}{(1-\rho)}\mathsf{E}_\infty[(1-\rho)^T\Lambda_T^W|T>0](1-p)
\end{aligned}
\tag{5.10}
$$

and

$$
\begin{aligned}
\mathsf{P}_\pi^W(T>\nu) &= \mathsf{P}(\nu<0)\mathsf{P}_\pi^W(T\geq 0)+\sum_{k=0}^\infty \pi_k\mathsf{P}_\infty(T>k)\\
&= q+(1-q)\rho\sum_{k=0}^\infty(1-\rho)^k\mathsf{P}_\infty(T>k)\\
&= q+(1-q)\rho\mathsf{E}_\infty\left[\sum_{k=0}^{T-1}(1-\rho)^k\right]\\
&= q+(1-q)\mathsf{E}_\infty[1-(1-\rho)^T]\\
&= q+(1-q)\mathsf{E}_\infty\left[(1-(1-\rho)^T)\mathbb{1}_{\{T>0\}}\right]\\
&= q+(1-q)\mathsf{E}_\infty[1-(1-\rho)^T|T>0]\mathsf{P}_\pi^W(T>0)\\
&= q+(1-q)\{1-\mathsf{E}_\infty[(1-\rho)^T|T>0]\}(1-p).
\end{aligned}
\tag{5.11}
$$

Combining (5.10) and (5.11), we obtain that for any stopping time T

$$
\mathsf{PD}_\pi^W(T)=\frac{\mathsf{P}_\pi^W(T=\nu+1)}{\mathsf{P}_\pi^W(T>\nu)}=\frac{qp+\frac{(1-q)\rho}{1-\rho}\mathsf{E}_\infty[(1-\rho)^T\Lambda_T^W|T>0](1-p)}{q+(1-q)\{1-\mathsf{E}_\infty[(1-\rho)^T|T>0]\}(1-p)}.
\tag{5.12}
$$

Notice that if $\alpha\geq 1-q$, then we should stop without making observations at 0 w.p. 1. Indeed, substitution of $p=\mathsf{P}_\pi^W(T=0)=1$ in (5.12) and (5.11) yields $\mathsf{PD}_\pi^W(T)=1$ and $\mathsf{P}_\pi^W(T>\nu)=1-\mathsf{PFA}_\pi(T)=q$, so that the probability of detection takes a maximal possible value, while $\mathsf{PFA}_\pi(T)=1-q\leq\alpha$, i.e., satisfies the constraint. Thus, it suffices to consider the values of α smaller than $1-q$.

The next theorem shows that a randomized version of the mixture Shewhart rule

$$
\widetilde{T}_A^p=\begin{cases}0 & \text{w.p. } p,\\ T_A^W & \text{w.p. } 1-p\end{cases},
\tag{5.13}
$$

is optimal in the Bayesian problem (5.7) for suitable values of A and p. In other words, the optimal rule is to use the mixture Shewhart rule T_A^W with probability $1-p$ and to stop at 0 with probability p for certain A and p.

Define

$$A^* = \frac{q}{1-q}\frac{1-\rho}{\rho}. \tag{5.14}$$

and

$$p_\alpha = \frac{1}{\rho}\left\{\frac{\alpha}{(1-q)}[\rho \mathsf{P}_\infty(\Lambda_1^W < A^*) + \mathsf{P}_\infty(\Lambda_1^W \geq A^*)] - (1-\rho)\mathsf{P}_\infty(\Lambda_1^W \geq A^*)\right\}. \tag{5.15}$$

Theorem 5.1. *Let the prior distribution be* Geometric(q,ρ). *Let the detection rule* $\widetilde{T}_A^p$ *be defined by* (5.13).

(i) *If* $(1-q)\frac{(1-\rho)\mathsf{P}_\infty(\Lambda_1^W \geq A^*)}{1-(1-\rho)\mathsf{P}_\infty(\Lambda_1^W < A^*)} \leq \alpha < 1-q$, *then the optimal Bayesian detection rule in the optimization problem* (5.7) *is the rule* $\widetilde{T}_A^p$ *with* $A = A^*$ *and* $p = p_\alpha$, *where* A^* *and* p_α *are given by* (5.14) *and* (5.15), *respectively.*

(ii) *If* $\alpha < (1-q)\frac{(1-\rho)\mathsf{P}_\infty(\Lambda_1^W \geq A^*)}{1-(1-\rho)\mathsf{P}_\infty(\Lambda_1^W < A^*)}$, *then the optimal rule is* $\widetilde{T}_{A_\alpha}^{p=0}$, *i.e., it is the mixture Shewhart rule* T_A^W *with threshold* $A = A_\alpha$ *computed from the equation*

$$\mathsf{P}_\infty(\Lambda_1^W \geq A) = \frac{\alpha\rho}{(1-\rho-\alpha)(1-\rho)}. \tag{5.16}$$

(iii) *In both cases, i.e., when* $\alpha < 1-q$, *the probability of detection of the optimal rule is*

$$\mathsf{PD}_\pi^W(T_{\text{opt}}) = \mathsf{P}_0^W(\Lambda_1^W \geq A). \tag{5.17}$$

Proof. Proof of part (i). As we mentioned above, the values $\alpha \geq 1-q$ can be excluded from consideration since in this case the optimal rule is degenerate (stop at 0 with probability 1). Let $\alpha < 1-q$. Since $1-\mathsf{PFA}_\pi(T) \geq 1-\alpha$ for any $T \in \mathbb{C}_\pi(\alpha)$ the denominator in (5.12) is no smaller than $1-\alpha$. It can be shown that it suffices to limit ourselves to stopping times that satisfy the false alarm constraint with equality, in which case the denominator equals $1-\alpha$ (see the proof of Theorem 2.1 in Moustakides [95]). Therefore, the optimization problem reduces to maximizing the numerator in (5.12) subject to the constraint that the denominator is equal to $1-\alpha$, which is performed by the conventional Lagrange multiplier technique, i.e., to maximize the "reward' functional

$$\begin{aligned} G(T) &= qp + \frac{(1-q)\rho}{1-\rho}\mathsf{E}_\infty[(1-\rho)^T\Lambda_T^W | T>0](1-p) \\ &\quad + \lambda\left\{qp + \left[1-(1-q)\mathsf{E}_\infty[(1-\rho)^T | T>0]\right](1-p)\right\} \\ &= q(1+\lambda)p + \left[\lambda + (1-q)\widetilde{G}(T)\right](1-p) \end{aligned} \tag{5.18}$$

over all Markov times $T \geq 0$, where λ is the Lagrange multiplier and

$$\widetilde{G}(T) = \mathsf{E}_\infty\left[(1-\rho)^T\left(\frac{\rho}{1-\rho}\Lambda_T^W - \lambda\right) | T>0\right].$$

The maximization is performed in two steps: first, we maximize $G(T)$ over positive Markov times $T>0$ for a fixed randomizing probability p; second, we maximize over the randomizing probability.

Fix p. Then the problem reduces to maximizing over $T>0$ the functional $\widetilde{G}(T)$. For $A \geq 0$, consider the following specific Lagrange multiplier

$$\lambda = \lambda_A = \frac{A}{1-\rho}\left[1-(1-\rho)\mathsf{P}_\infty(\Lambda_1^W < A)\right] - \mathsf{P}_0^W(\Lambda_1^W \geq A).$$

Using the optimal stopping theory it is straightforward to show that the optimal solution is the mixture Shewhart stopping time T_A^W defined in (5.9) and that the maximal reward is

$$\begin{aligned}\widetilde{G}(T_A^W) &= \sum_{k=1}^{\infty}(1-\rho)^k[\mathsf{P}_\infty(\Lambda_1^W < A)]^{k-1}\left\{\frac{\rho}{1-\rho}\mathsf{P}_0^W(\Lambda_1^W \ge A) - \lambda_A \mathsf{P}_\infty(\Lambda_1^W \ge A)\right\} \\ &= \frac{\rho \mathsf{P}_0^W(\Lambda_1^W \ge A) - \lambda_A(1-\rho)\mathsf{P}_\infty(\Lambda_1^W \ge A)}{1-(1-\rho)\mathsf{P}_\infty(\Lambda_1^W < A)} = \frac{\rho}{1-\rho}A - \lambda_A. \end{aligned} \tag{5.19}$$

Substitution of (5.19) in (5.18) yields the following inequality

$$G(T) \le q(1+\lambda_A)p + \left[q(1+\lambda_A) + (1-q)\frac{\rho}{1-\rho}A - q\right](1-p) := \overline{G}(p,A), \tag{5.20}$$

which holds for any stopping time T. Obviously, it suffices to restrict attention to stopping times $\widetilde{T}_A^p = T_A^W$ with probability $1-p$ and $\widetilde{T}_A^p = 0$ with probability p since the optimal rule $T_{\text{opt}} = \widetilde{T}_A^p$ for some p.

It follows from (5.11) that

$$\mathsf{PFA}_\pi(T) = (1-q)\left\{p + (1-p)\{1 - \mathsf{E}_\infty[(1-\rho)^T | T > 0]\}\right\}.$$

Owing to the fact that $\widetilde{T}_A^p = T_A^W$ on $\{\widetilde{T}_A^p > 0\}$ and that T_A^W has the geometric distribution with the parameter $\mathsf{P}_\infty(\Lambda_1^W \ge A)$ under P_∞ (see (5.3)), we obtain

$$\mathsf{E}_\infty[(1-\rho)^{\widetilde{T}_A^p} | \widetilde{T}_A^p > 0] = \mathsf{E}_\infty[(1-\rho)^{T_A^W}] = \frac{(1-\rho)\mathsf{P}_\infty(\Lambda_1^W \ge A)}{1-(1-\rho)\mathsf{P}_\infty(\Lambda_1^W < A)},$$

so that for a fixed p

$$\mathsf{PFA}_\pi(\widetilde{T}_A^p) = (1-q)\left\{p + \frac{(1-\rho)\mathsf{P}_\infty(\Lambda_1^W \ge A)}{\rho \mathsf{P}_\infty(\Lambda_1^W < A) + \mathsf{P}_\infty(\Lambda_1^W \ge A)}(1-p)\right\},$$

which should be set to α, i.e.,

$$(1-q)\left\{p + \frac{(1-\rho)\mathsf{P}_\infty(\Lambda_1^W \ge A)}{\rho \mathsf{P}_\infty(\Lambda_1^W < A) + \mathsf{P}_\infty(\Lambda_1^W \ge A)}(1-p)\right\} = \alpha. \tag{5.21}$$

Solving this equation yields

$$p_A = \frac{1}{\rho}\left\{\frac{\alpha}{(1-q)}[\rho \mathsf{P}_\infty(\Lambda_1^W < A) + \mathsf{P}_\infty(\Lambda_1^W \ge A)] - (1-\rho)\mathsf{P}_\infty(\Lambda_1^W \ge A)\right\}. \tag{5.22}$$

Finding an optimal value of p reduces to maximizing the upper bound $\overline{G}(p,A)$ in (5.20) over p (under constraint (5.21)), which is a convex combination of $q(1+\lambda_A)$, the reward obtained when stopping at 0, and $q(1+\lambda_A) + (1-q)\frac{\rho}{1-\rho}A - q$, the reward corresponding to the mixture Shewhart stopping time T_A^W. Obviously, we need to put all the probability mass on the largest reward. Let A^* be the threshold when both rewards are the same, i.e.,

$$q(1+\lambda_{A^*}) = q(1+\lambda_{A^*}) + (1-q)\frac{\rho}{1-\rho}A^* - q.$$

Solving this equation, we obtain that A^* is given by (5.14). When $A \le A^*$, the reward of stopping at 0 is larger or equal to the reward of using the mixture Shewhart rule T_A^W, with equality for $A = A^*$.

Therefore, the optimal value of p is given by p_{A^*} defined in (5.22), i.e., it is equal to p_α defined in (5.15). It is easily checked that $0 \le p_\alpha < 1$ whenever

$$(1-q)\frac{(1-\rho)\mathsf{P}_\infty(\Lambda_1^W \ge A^*)}{1-(1-\rho)\mathsf{P}_\infty(\Lambda_1^W < A^*)} \le \alpha < 1-q.$$

This completes the proof of part (i).

Proof of part (ii). Let $A > A^*$. Then the reward corresponding to T_A^W exceeds the reward associated with stopping at 0, and therefore, the optimal value of $p = 0$. Since threshold A must be such that the mixture Shewhart rule satisfies the PFA constraint with equality, i.e., equation (5.21) should hold with $p = 0$, we obtain that the constraint is satisfied when A is computed as in equation (5.16). This equation produces A that exceeds A^* as long as $\alpha < (1-q)\frac{(1-\rho)\mathsf{P}_\infty(\Lambda_1^W \ge A^*)}{1-(1-\rho)\mathsf{P}_\infty(\Lambda_1^W < A^*)}$. The proof of (ii) is complete.

Proof of part (iii). Since T_A^W has the geometric distributions with the parameters $\mathsf{P}_\infty(\Lambda_1^W \ge A)$ and $\mathsf{P}_0^W(\Lambda_1^W \ge A)$ under P_∞ and P_0^W, respectively, we have

$$\mathsf{E}_\infty[(1-\rho)^{T_A^W}] = \frac{(1-\rho)\mathsf{P}_\infty(\Lambda_1^W \ge A)}{1-(1-\rho)\mathsf{P}_\infty(\Lambda_1^W < A)},$$

$$\mathsf{E}_\infty[(1-\rho)^{T_A^W}\Lambda_{T_A}^W] = \mathsf{E}_0^W[(1-\rho)^{T_A^W}] = \frac{(1-\rho)\mathsf{P}_0^W(\Lambda_1^W \ge A)}{1-(1-\rho)\mathsf{P}_\infty(\Lambda_1^W < A)},$$

and using the formula (5.12) we obtain that for any admissible values of $q \in [0,1)$ and $\rho \in (0,1)$ and any $p \in [0,1)$

$$\mathsf{PD}_\pi^W(\widetilde{T}_A^p) = \frac{qp + \frac{(1-q)\rho}{1-\rho}\mathsf{E}_\infty[(1-\rho)^{T_A^W}\Lambda_{T_A^W}^W](1-p)}{q+(1-q)\{1-\mathsf{E}_\infty[(1-\rho)^{T_A^W}]\}(1-p)} = \mathsf{P}_0^W(\Lambda_1^W \ge A). \tag{5.23}$$

Since $\widetilde{T}_A^p = T_{\mathrm{opt}}$ for $p = p_\alpha$ with $A = A^*$ in (i) and for $p = 0$ and $A = A_\alpha$ in (ii) the assertion (iii) follows. □

Remark 5.1. If $q = 0$, then $A^* = 0$ and $\mathsf{P}_\infty(\Lambda_1^W \ge 0) = 1$, so the assertion (ii) holds for $\alpha \ge 1-\rho$. However, this case is in a sense degenerate since the probability of detection $\mathsf{PD}_\pi(T_{\mathrm{opt}}) = \mathsf{P}_0^W(\Lambda_1^W \ge 0) = 1$. In other words, we stop either at 0 w.p. $p_\alpha = [(\alpha-(1-\rho)]/\rho$ or make a single observation w.p. $1-p_\alpha$ and stop no matter what this observation is, and this rule guarantees the perfect score. Fortunately, in most applications $\alpha < 1-\rho$.

As we already mentioned, if the weight $W(\theta)$ is concentrated in a single point ϑ, then $\mathsf{P}_k^W = \mathsf{P}_{k,\vartheta}$ and a direct consequence of Theorem 5.1 is that the randomized version of the classical Shewhart rule T_A^ϑ based on the LR Λ_n^ϑ (defined in (5.1)) is optimal in the Bayesian setting, minimizing the probability of detection $\mathsf{PD}_\pi^\vartheta(T)$ at the point ϑ. More importantly, if the LR Λ_n^θ is monotone with respect to a statistic $S(X_n)$, then the Shewhart rule is uniformly optimal for all post-change parameter values. Specifically, the following corollary follows directly from Theorem 5.1.

Corollary 5.1. *Suppose that the likelihood ratio $\Lambda_1^\theta(X_1)$ ($\theta \in \Theta$) is monotone non-decreasing with respect to a statistic $S(X_1)$ and that for all post-change parameter values $\theta \in \Theta_1$*

$$\alpha < (1-q)\frac{(1-\rho)\mathsf{P}_\infty(\Lambda_1^\theta \ge A^*)}{1-(1-\rho)\mathsf{P}_\infty(\Lambda_1^\theta < A^*)}. \tag{5.24}$$

Then the Shewhart rule

$$T_C = \inf\{n \ge 1 : S(X_n) \ge C\}$$

with threshold $C = C_\alpha$, satisfying the equation

$$\mathsf{P}_\infty(S(X_n) \ge C) = \frac{\alpha\rho}{(1-\rho-\alpha)(1-\rho)}, \tag{5.25}$$

is uniformly optimal in $\theta \in \Theta_1$ in the sense of maximizing the probability of detection $\mathsf{PD}_\pi^\theta(T) = \sum_{k=0}^\infty \pi_k \mathsf{P}_{k,\theta}(T=k+1)/[1-\mathsf{PFA}_\pi(T)]$ over all rules with $\mathsf{PFA}_\pi(T) \le \alpha$ for all $\theta \in \Theta_1$. The probability of detection of this rule is

$$\mathsf{PD}_\pi^\theta(T_C) = \mathsf{P}_{0,\theta}(S(X_1) \ge C), \quad \theta \in \Theta_1. \tag{5.26}$$

Proof. According to Theorem 5.1(ii) the Shewhart rule

$$T_A^\theta = \inf\left\{n \ge 1 : \Lambda_n^\theta \ge A\right\},$$

tuned to some $\theta \in \Theta_1$, is optimal when $\alpha < (1-q)\frac{(1-\rho)\mathsf{P}_\infty(\Lambda_1^\theta \ge A^*)}{1-(1-\rho)\mathsf{P}_\infty(\Lambda_1^\theta < A^*)}$ and threshold A_α is selected from equation (5.16) (with $\Lambda_1^W = \Lambda_1^\theta$). If the LR $\Lambda_1^\theta(X_1)$ is monotone non-decreasing in $S(X_1)$, then the stopping time T_A^θ can be written as

$$T_C = \inf\{n \ge 1 : S(X_n) \ge C\},$$

so that if $C = C_\alpha$ is selected from equation (5.25) both rules T_{C_α} and $T_{A_\alpha}^\theta$ are equivalent. But T_{C_α} does not depend on the post-change parameter θ, which implies that it is uniformly optimal for all $\theta \in \Theta_1$ whenever the optimal rule is non-randomized, i.e., when $\alpha < (1-q)\frac{(1-\rho)\mathsf{P}_\infty(\Lambda_1^\theta \ge A^*)}{1-(1-\rho)\mathsf{P}_\infty(\Lambda_1^\theta < A^*)}$ for all $\theta \in \Theta_1$.

Finally, (5.26) follows from assertion (iii) in Theorem 5.1 and the fact that $\mathsf{P}_{0,\theta}(\Lambda_1^\theta \ge A) = \mathsf{P}_{0,\theta}(S(X_1) \ge C)$. □

We iterate that if $q = 0$, then $A^* = 0$ and $\mathsf{P}_\infty(\Lambda_1^\theta \ge 0) = 1$ for all θ, so that in this case inequality (5.24) turns into $\alpha < 1-\rho$, which typically holds in applications.

For example, the LR is monotone in $S(x)$ if $f_\theta(x) = h(x)\exp\{\theta S(x) - b(\theta)\}$ belongs to the exponential family, where $b(0) = b'(0) = 0$ and $b'(\theta) > 0$ for $\theta > 0$, and if $\theta = 0$ pre-change and $\theta > 0$ post-change.

In practice, the parameters (q,ρ) of the geometric distribution are typically unknown. Then a possible way is finding the worst-case scenario, i.e., finding the values that minimize the probability of detection. A reasonable candidate is an improper prior, i.e., allowing $\rho \to 0$. However, as $\rho \to 0$, the PFA constraint should be replaced by the ARL to false alarm constraint since

$$\frac{\mathsf{P}_\pi(T > \nu)}{\rho} \xrightarrow[\rho\to 0]{} r + (1-q)\,\mathsf{E}_\infty[T], \quad r = \lim_{\rho\to 0}(q/\rho).$$

Thus, consider the following maximin criterion

$$\mathsf{PD}_\pi^W(T_{\text{opt}}) = \sup_{T\in\mathbb{C}_\gamma} \inf_{q,\rho} \mathsf{PD}_{q,\rho}^W(T) \quad \text{for all } \gamma \ge 1, \tag{5.27}$$

where $\mathbb{C}_\gamma = \{T : \mathsf{E}_\infty[T] \ge \gamma\}$.

For $0 \le p < 1$, consider the randomized Shewhart rule $\widetilde{T}_A^p$ defined in (5.13). It follows from (5.23) that for any pair (q,ρ)

$$\mathsf{PD}_{q,\rho}^W(\widetilde{T}_A^p) = \mathsf{P}_0^W(\Lambda_1^W \ge A), \tag{5.28}$$

so that the probability of detection of the rule $\widetilde{T}_A^p$ does not depend on the parameters q and ρ as long as threshold A does not depend on (q,p), which is the case since threshold A should be selected based on the ARL to false alarm constraint γ. In other words, this rule is an equalizer in (q,p). Hence, we may expect that it is a maximin rule when $p = p_\gamma$ and $A = A_\gamma$ are suitably selected.

Since for any p

$$\mathsf{E}_\infty[\widetilde{T}_A^p] = \frac{1-p}{\mathsf{P}_\infty(\Lambda_1^W \geq A)} \tag{5.29}$$

threshold A_γ has to be selected from the equation

$$\frac{1-p_\gamma}{\mathsf{P}_\infty(\Lambda_1^W \geq A)} = \gamma$$

for some $p_\gamma \in [0,1)$. Recall that we assume that the distribution $\mathsf{P}_\infty(\Lambda_1^W \leq y)$ is continuous and monotone. Thus, if there exists a strictly positive randomizing probability p_γ such that this equation has a solution, then the randomized rule (at 0) will perform better than the non-randomized mixture Shewhart rule T_A^W with $p = 0$. Now, if we set $p = p_A$ in (5.29) as a maximal possible value, then it should be optimal and a maximal value is (intuitively) the value of the probability of detection in (5.28), i.e., $p_A = \mathsf{P}_0^W(\Lambda_1^W \geq A)$. As we conjectured, the least-favorable prior is improper uniform.

The following theorem formalizes this semi-heuristic argument. The rigorous proof, which is quite technical, can be found in Moustakides [95].

Theorem 5.2. *The optimal change detection rule in the maximin optimization problem* (5.27) *is the rule $\widetilde{T}_A^p$ defined in* (5.13), *i.e., to use the mixture Shewhart rule T_A^W with threshold $A = A_\gamma$ with probability $1 - p_\gamma$ and to stop at 0 with probability $p_\gamma = \mathsf{P}_0^W(\Lambda_1^W \geq A_\gamma)$, where threshold A_γ is the solution of the equation*

$$\frac{\mathsf{P}_0^W(\Lambda_1^W < A)}{\mathsf{P}_\infty(\Lambda_1^W \geq A)} = \gamma. \tag{5.30}$$

The detection rule $\widetilde{T}_A^p$ is an equalizer over all parameter pairs (q,ρ) and the least-favorable prior is the improper uniform distribution. The probability of detection of this rule is given by (5.28).

Example 5.1 (Change detection in exponential distribution). Let X_n have the exponential distribution

$$f_\theta(x) = \frac{1}{1+\theta}\exp\left\{-\frac{x}{1+\theta}\right\}\mathbb{1}_{\{x\geq 0\}}.$$

The pre-change parameter $\theta = 0$ and the post-change parameter $\theta > 0$. The LR

$$\Lambda_n(X_n) = \frac{1}{1+\theta}\exp\left\{\frac{\theta}{1+\theta}X_n\right\}$$

is a monotone increasing function of X_n, so $T_C = \inf\{n : X_n \geq C\}$. Since $\mathsf{P}_\infty(X_n \geq C) = e^{-C}$ we have

$$C_\alpha = \log\left[\frac{(1-\rho-\alpha)(1-\rho)}{\alpha\rho}\right].$$

If $q = 0$ and $\alpha < 1-\rho$, then by Corollary 5.1 the Shewhart rule T_{C_α} is uniformly optimal for all $\theta > 0$ and the probability of detection is

$$\mathsf{PD}_\pi^\theta(T_{C_\alpha}) = \exp\left\{-\frac{C_\alpha}{1+\theta}\right\} = \left[\frac{\alpha\rho}{(1-\rho-\alpha)(1-\rho)}\right]^{\frac{1}{1+\theta}}.$$

If, for example, $\theta = 4$, $\rho = 0.1$, and $\alpha = 0.1$, then $\mathsf{PD}_\pi(T_{C_\alpha}) \approx 0.674$.

In the maximin Bayesian problem (5.27) the rule $\widetilde{T}_A^p$ is not uniformly optimal, so set a specific post-change parameter $\theta > 0$. Then it follows from Theorem 5.2 that probability $p_\gamma = [(1+\theta)A_\gamma]^{-1/\theta}$, that threshold A_γ is found from the equation

$$\left[1 - [(1+\theta)A]^{-1/\theta}\right][(1+\theta)A]^{1+1/\theta} = \gamma,$$

and that the probability of detection is $\mathsf{PD}_\pi^\theta(\widetilde{T}_{A_\gamma}^p) = [(1+\theta)A_\gamma]^{-1/\theta}$.

Let us now compare the randomized optimal rule with the non-randomized Shewhart rule $T_{\hat{A}}$. The threshold of this rule is computed from the equation $[(1+\theta)\hat{A}]^{1+1/\theta} = \gamma$ and the probability of detection is $\mathsf{PD}_\pi^\theta(T_{\hat{A}_\gamma}) = [(1+\theta)\hat{A}_\gamma]^{-1/\theta}$. Let $\theta = 1$. Then $A_\gamma = (1+\sqrt{1+\gamma})/4$, $\mathsf{PD}_\pi^\theta(\widetilde{T}_{A_\gamma}^p) = (2A_\gamma)^{-1}$ for $\widetilde{T}_{A_\gamma}^p$ and $\hat{A}_\gamma = \sqrt{\gamma/2}$, $\mathsf{PD}_\pi^\theta(T_{\hat{A}_\gamma}) = (2\hat{A}_\gamma)^{-1}$ for $T_{\hat{A}_\gamma}$. For a large $\gamma \gg 1$, we obtain that the gain of $\widetilde{T}_{A_\gamma}^p$ over $T_{\hat{A}_\gamma}$ in the probability of detection is $\mathsf{PD}_\pi^\theta(\widetilde{T}_{A_\gamma}^p)/\mathsf{PD}_\pi^\theta(T_{\hat{A}_\gamma}) \approx 2/\sqrt{2}$, i.e., about 42%.

It follows from the next section that $\mathsf{PD}_\pi^\theta(T_{\hat{A}_\gamma}) = \inf_{k\in\mathbb{Z}_+} \mathsf{P}_k^\theta(T_{\hat{A}_\gamma} = k+1 | T_{\hat{A}_\gamma} > k)$ and that the Shewhart rule is optimal in terms of minimizing the minimal detection probability.

5.2.3 Maximin Frequentist Criteria

Define

$$\mathsf{PD}_{\rm einf}^W(T) = \inf_{\nu\ge 0} \operatorname{ess\,inf} \mathsf{P}_\nu^W(T = \nu+1 | \mathscr{F}_\nu, T > \nu), \quad \mathsf{PD}_{\rm inf}^W(T) = \inf_{\nu\ge 0} \mathsf{P}_\nu^W(T = \nu+1 | T > \nu)$$

and consider the following maximin optimality criteria: find $T_{\rm opt} \in \mathbb{C}_\gamma$ such that

$$\mathsf{PD}_{\rm einf}^W(T_{\rm opt}) = \sup_{T\in\mathbb{C}_\gamma} \mathsf{PD}_{\rm einf}^W(T) \quad \text{for every } \gamma \ge 1 \tag{5.31}$$

and find $T_{\rm opt} \in \mathbb{C}_\gamma$ such that

$$\mathsf{PD}_{\rm inf}^W(T_{\rm opt}) = \sup_{T\in\mathbb{C}_\gamma} \mathsf{PD}_{\rm inf}^W(T) \quad \text{for every } \gamma \ge 1 \tag{5.32}$$

(cf. (2.45)–(2.46) in Chapter 2). In other words, we would like to find optimal detection rules that maximize detection probabilities in the worst-case with respect to the change point $\nu \in \mathbb{Z}_+$ in the class of procedures $\mathbb{C}_\gamma = \{T : \mathsf{E}_\infty[T] \ge \gamma\}$ with the given ARL to false alarm.

Recall that the distributions of the Shewhart stopping time T_A^W are geometric for all $A \ge 0$:

$$\mathsf{P}_k^W(T_A^W = k+1) = \mathsf{P}_0^W(\Lambda_1^W \ge A)[1 - \mathsf{P}_0^W(\Lambda_1^W \ge A)]^k, \quad k \in \mathbb{Z}_+,$$
$$\mathsf{P}_\infty(T_A^W = k+1) = \mathsf{P}_\infty^W(\Lambda_1^W \ge A)[1 - \mathsf{P}_\infty^W(\Lambda_1^W \ge A)]^k, \quad k \in \mathbb{Z}_+.$$

Hence, the probability of detection of this rule is

$$\mathsf{P}_\nu^W(T_A^W = \nu+1 | T_A^W > \nu) = \frac{\mathsf{P}_\nu^W(T_A^W = \nu+1)}{\mathsf{P}_\nu^W(T_A^W > \nu)} = \mathsf{P}_0^W(\Lambda_1^W \ge A) \quad \text{for all } \nu \in \mathbb{Z}_+ \tag{5.33}$$

and the ARL to false alarm is

$$\mathsf{E}_\infty[T_A^W] = \frac{1}{\mathsf{P}_\infty^W(\Lambda_1^W \ge A)}. \tag{5.34}$$

That is, if threshold $A = A_\gamma$ is found from the equation

$$\frac{1}{\mathsf{P}_\infty^W(\Lambda_1^W \ge A)} = \gamma, \tag{5.35}$$

then $\mathsf{E}_\infty[T_{A_\gamma}^W] = \gamma$ for all $\gamma \geq 1$ and the probability of detection does not depend on the change point ν, meaning that the rule $T_{A_\gamma}^W$ is an equalizer. According to the decision theory the maximin rule must be an equalizer, so the mixture Shewhart rule $T_{A_\gamma}^W$ may be a maximin rule.

The following theorem shows that the mixture Shewhart rule is indeed optimal in problems (5.31) and (5.32).

Theorem 5.3. *The optimal change detection rule in the maximin problems* (5.31) *and* (5.32) *is the mixture Shewhart rule T_A^W with threshold $A = A_\gamma$ computed from the equation* (5.35) *and the maximal probability of detection is*

$$\mathsf{PD}_{\mathrm{einf}}^W(T_{A_\gamma}^W) = \mathsf{PD}_{\mathrm{inf}}^W(T_{A_\gamma}^W) = \mathsf{P}_0^W(\Lambda_1^W \geq A_\gamma).$$

Proof. We provide a proof only for the maximin criterion (5.32). The proof for the optimization problem (5.31) is essentially similar. In order to prove the assertion of the theorem we have to show that

$$\sup_{T \in \mathbb{C}_\gamma} \mathsf{PD}_{\mathrm{inf}}^W(T) \leq \mathsf{PD}_{\mathrm{inf}}^W(T_{A_\gamma}^W) = \mathsf{P}_0^W(\Lambda_1^W \geq A_\gamma), \tag{5.36}$$

where A_γ satisfies the equation (5.35).

It suffices to consider stopping times with a finite ARL to false alarm, $\gamma \leq \mathsf{E}_\infty[T] < \infty$.

Obviously, for any stopping time $T \geq 0$ and all $\nu \in \mathbb{Z}_+$

$$\mathsf{PD}_{\mathrm{inf}}^W(T) \leq \mathsf{P}_\nu^W(T = \nu+1 | T > \nu) = \frac{\mathsf{P}_\nu^W(T = \nu+1)}{\mathsf{P}_\nu^W(T \geq \nu+1)} = \frac{\mathsf{P}_\nu^W(T = \nu+1)}{\mathsf{P}_\infty^W(T > \nu)},$$

so changing the measure we obtain

$$\mathsf{PD}_{\mathrm{inf}}^W(T)\mathsf{P}_\infty(T > \nu) \leq \mathsf{P}_\nu^W(T = \nu+1) = \mathsf{E}_\infty[\Lambda_{\nu+1}^W \mathbb{1}_{\{T=\nu+1\}}].$$

Consequently,

$$\mathsf{PD}_{\mathrm{inf}}^W(T) \sum_{\nu=0}^{\infty} \mathsf{P}_\infty(T > \nu) = \mathsf{PD}_{\mathrm{inf}}^W(T)\mathsf{E}_\infty[T] \leq \sum_{\nu=0}^{\infty} \mathsf{E}_\infty[\Lambda_{\nu+1}^W \mathbb{1}_{\{T=\nu+1\}}] = \mathsf{E}_\infty[\Lambda_T^W],$$

where we set $\Lambda_0^W = 0$ since the left-hand side equals 0 for $T = 0$. Hence, for any stopping time $T \geq 0$ such that $T = 0$ w.p. p and $T > 0$ w.p. $1-p$ ($p \in [0,1)$) we have the inequality

$$\mathsf{PD}_{\mathrm{inf}}^W((T) \leq \frac{\mathsf{E}_\infty[\Lambda_T^W]}{\mathsf{E}_\infty[T]} = \frac{(1-p)\mathsf{E}_\infty[\Lambda_T^W | T > 0]}{(1-p)\mathsf{E}_\infty[T | T > 0]} = \frac{\mathsf{E}_\infty[\Lambda_T^W | T > 0]}{\mathsf{E}_\infty[T | T > 0]}.$$

Note that $\mathsf{E}_\infty[\Lambda_T^W]/\mathsf{E}_\infty[T]$ does not depend on p. So if $\mathsf{E}_\infty[T] = (1-p)\mathsf{E}_\infty[T|T>0] > \gamma$ we can replace p by $p^* > p$ such that $(1-p^*)\mathsf{E}_\infty[T|T>0] = \gamma$ and the value of the ratio remains the same. Hence, it suffices to consider stopping times that satisfy the constraint with equality, i.e., class $\mathbb{C}_\gamma^* = \{T : \mathsf{E}_\infty[T] = \gamma\}$ and we obtain

$$\sup_{T \in \mathbb{C}_\gamma} \mathsf{PD}_{\mathrm{inf}}^W((T) \leq \sup_{T \in \mathbb{C}_\gamma^*} \frac{\mathsf{E}_\infty[\Lambda_T^W]}{\mathsf{E}_\infty[T]} = \gamma^{-1} \sup_{T \in \mathbb{C}_\gamma^*} \mathsf{E}_\infty[\Lambda_T^W]. \tag{5.37}$$

As usual, maximization of $\mathsf{E}_\infty[\Lambda_T^W]$ over all stopping times in class $\mathbb{C}_\gamma^*$ is reduced to an unconstraint optimization problem using the Lagrange multiplier technique. Specifically, define the reward functional

$$G_p(T) = \mathsf{E}_\infty[\Lambda_T^W - \lambda T] = (1-p)\mathsf{E}_\infty[\Lambda_T^W - \lambda T | T > 0]$$

that has to be maximized over all nonnegative Markov times T. Let the multiplier λ be equal to

$$\lambda_A = \mathsf{P}_0^W(\Lambda_1^W \geq A) - A\mathsf{P}_\infty(\Lambda_1^W \geq A),$$

where $A = A_\gamma$ is the solution of the equation $\gamma \mathsf{P}_\infty(\Lambda_1^W \geq A) = 1$.

Fix p and consider the optimization over positive stopping times, $T > 0$. Using optimal stopping theory it can be shown that for $T > 0$ the reward $G_p(T)$ attains maximum by the Shewhart rule T_A^W. Since $\mathsf{E}_\infty[T_A^W] = 1/\mathsf{P}_\infty(\Lambda_1^W \geq A) = \gamma$ and

$$\mathsf{E}_\infty\left[\Lambda_{T_A^W}^W\right] = \mathsf{P}_0^W(\Lambda_1^W \geq A)/\mathsf{P}_\infty(\Lambda_1^W \geq A) = \gamma \mathsf{P}_0^W(\Lambda_1^W \geq A),$$

we obtain

$$G_p(T) \leq G_p(T_A^W) = (1-p)\mathsf{E}_\infty[\Lambda_{T_A^W}^W - \lambda_A T_A^W] = (1-p)A \leq A,$$

which implies that the optimal value of p is $p = 0$. Thus, for any $T \in \mathbb{C}_\gamma^*$

$$G_0(T) = \mathsf{E}_\infty[\Lambda_T^W] - \lambda_A\gamma \leq G_0(T_A^W) = \mathsf{E}_\infty[\Lambda_{T_A^W}^W] - \lambda_A\gamma,$$

which implies $\mathsf{E}_\infty[\Lambda_T^W] \leq \mathsf{E}_\infty[\Lambda_{T_A^W}^W] = \gamma \mathsf{P}_0^W(\Lambda_1^W \geq A)$. By (5.33), $\mathsf{PD}_{\inf}^W(T_A^W) = \mathsf{P}_0^W(\Lambda_1^W \geq A)$ and using (5.37), we obtain

$$\sup_{T \in \mathbb{C}_\gamma} \mathsf{PD}_{\inf}^W(T) \leq \gamma^{-1} \sup_{T \in \mathbb{C}_\gamma^*} \mathsf{E}_\infty[\Lambda_T^W] \leq \gamma^{-1}\{\gamma \mathsf{P}_0^W(\Lambda_1^W \geq A_\gamma)\} = \mathsf{P}_0^W(\Lambda_1^W \geq A_\gamma) = \mathsf{PD}_{\inf}^W(T_{A_\gamma}^W),$$

so that the inequality (5.36) holds and the proof is complete. □

Analogously to Corollary 5.1 we have the following corollary regarding uniform maximin optimality of the Shewhart rule when the LR $\Lambda_n(X_n) = \Lambda(S(X_n))$ is a monotone non-decreasing function of a statistic $S(X_n)$.

Corollary 5.2. *Suppose that the likelihood ratio $\Lambda_1^\theta(X_1)$ ($\theta \in \Theta$) is the monotone non-decreasing function of a statistic $S(X_1)$. Then the Shewhart rule*

$$T_C = \inf\{n \geq 1 : S(X_n) \geq C\} \tag{5.38}$$

with threshold $C = C_\gamma$, satisfying the equation

$$\mathsf{P}_\infty(S(X_n) \geq C) = \gamma, \tag{5.39}$$

is uniformly maximin optimal in $\theta \in \Theta_1$ in the sense of maximizing the probabilities of detection $\mathsf{PD}_{\mathrm{einf}}^\theta(T)$ and $\mathsf{PD}_{\inf}^\theta(T)$ over all rules with $\mathsf{E}_\infty[T] \geq \gamma$ for all $\theta \in \Theta_1$. The probability of detection of this rule is

$$\mathsf{PD}_{\mathrm{einf}}^\theta(T_{C_\gamma}) = \mathsf{PD}_{\inf}^\theta(T_{C_\gamma}) = \mathsf{P}_0^\theta(S(X_1) \geq C_\gamma), \quad \theta \in \Theta_1. \tag{5.40}$$

Yet another interesting (perhaps even more interesting) optimality criterion is to find a rule that maximizes the minimal detection probabilities $\mathsf{PD}_{\mathrm{einf}}^W(T)$ and $\mathsf{PD}_{\inf}^W(T)$ in the class of rules

$$\mathbb{C}(m,\beta) = \left\{T : \sup_{\ell \geq 0} \mathsf{P}_\infty(T \leq \ell + m | T > \ell) \leq \beta\right\} \tag{5.41}$$

for which the maximal conditional probability of false alarm in the time window of size m does not exceed a predefined number $\beta \in (0,1)$ (see (2.51)). Specifically, we are now concerned with finding a rule $T_{\mathrm{opt}} \in \mathbb{C}_\gamma$ such that

$$\mathsf{PD}_{\mathrm{einf}}^W(T_{\mathrm{opt}}) = \sup_{T \in \mathbb{C}(m,\beta)} \mathsf{PD}_{\mathrm{einf}}^W(T) \quad \text{for every } \beta \in (0,1),\ m \geq 1 \tag{5.42}$$

and

$$\mathsf{PD}^W_{\mathrm{inf}}(T_{\mathrm{opt}}) = \sup_{T\in\mathbb{C}_\gamma} \mathsf{PD}^W_{\mathrm{inf}}(T) \quad \text{for every } \beta \in (0,1),\ m \geq 1 \tag{5.43}$$

An appeal to Lemmas 2.1 and 2.2 gives the following result.

Theorem 5.4. *Let distributions $\mathsf{P}_\infty(\Lambda_1(X_1) \leq y)$ and $\mathsf{P}_0(\Lambda_1(X_1) \leq y)$ be continuous. Then the following assertions hold.*

(i) *The optimal change detection rule in the maximin problems (5.42) and (5.43) is the mixture Shewhart rule T^W_A with threshold $A = A_{m,\beta}$ computed from the equation*

$$\mathsf{P}_\infty(\Lambda^W_1 \geq A) = (1-\beta)^{1/m},$$

and the maximal probability of detection is

$$\mathsf{PD}^W_{\mathrm{einf}}(T^W_{A_{m,\beta}}) = \mathsf{PD}^W_{\mathrm{inf}}(T^W_{A_{m,\beta}}) = \mathsf{P}^W_0(\Lambda^W_1 \geq A_{m,\beta}).$$

(ii) *Moreover, if the likelihood ratio $\Lambda^\theta_1(X_1)$ ($\theta \in \Theta$) is the monotone non-decreasing function of a statistic $S(X_1)$, then the Shewhart rule T_C defined in (5.38) with threshold $C = C_{m,\beta}$, satisfying the equation*

$$\mathsf{P}_\infty(S(X_n) \geq C) = (1-\beta)^{1/m}, \tag{5.44}$$

is uniformly maximin optimal in $\theta \in \Theta_1$ in the sense of maximizing the probabilities of detection $\mathsf{PD}^\theta_{\mathrm{einf}}(T)$ and $\mathsf{PD}^\theta_{\mathrm{inf}}(T)$ over all rules with $\sup_{\ell\geq 0}\mathsf{P}_\infty(T \leq \ell+m|T > \ell) \leq \beta$ for all $\theta \in \Theta_1$. The probability of detection of this rule is

$$\mathsf{PD}^\theta_{\mathrm{einf}}(T_{C_{m,\beta}}) = \mathsf{PD}^\theta_{\mathrm{inf}}(T_{C_{m,\beta}}) = \mathsf{P}^\theta_0(S(X_1) \geq C_{m,\beta}), \quad \theta \in \Theta_1. \tag{5.45}$$

Proof. Owing to the fact that the P_∞-distribution of the Shewhart stopping time is geometric with the parameter $\mathsf{P}_\infty(\Lambda^W_1 \geq A)$, we obtain

$$\mathsf{P}_\infty(T^W_A \leq \ell+m|T^W_A > \ell) = 1 - [\mathsf{P}_\infty(\Lambda^W_1 \geq A)]^m \quad \text{for all } \ell \in \mathbb{Z}_+.$$

Consequently, if threshold $A = A_{m,\beta}$ is found from the equation

$$\mathsf{P}_\infty(\Lambda^W_1 \geq A) = (1-\beta)^{1/m},$$

then $\sup_{\ell\in\mathbb{Z}_+}\mathsf{P}_\infty(T^W_A \leq \ell+m|T^W_A > \ell) = \beta$.

Next, it follows from Lemmas 2.1 and 2.2 that the constraint in (5.41) on the maximal conditional probability of false alarm is more stringent than the constraint on the ARL to false alarm, so that we have inclusion $\mathbb{C}(m,\beta) \subset \mathbb{C}_\gamma$ for some $\gamma = \gamma(m,\beta)$. This implies that

$$\sup_{T\in\mathbb{C}(m,\beta)} \mathsf{PD}^W_{\mathrm{einf}}(T) \geq \sup_{T\in\mathbb{C}_{\gamma(m,\beta)}} \mathsf{PD}^W_{\mathrm{einf}}(T),$$

$$\sup_{T\in\mathbb{C}(m,\beta)} \mathsf{PD}^W_{\mathrm{inf}}(T) \geq \sup_{T\in\mathbb{C}_{\gamma(m,\beta)}} \mathsf{PD}^W_{\mathrm{inf}}(T).$$

But by Theorem 5.3 the lower bounds are attained for the Shewhart rule if $\gamma(m,\beta) = (1-\beta)^{-m}$ due to the fact that the distribution of T^W_A is geometric. This completes the proof of (i).

Part (ii) follows immediately from (i). □

5.3 Bayesian and Maximin Sequential Detection in Windows with Arbitrarily Size

For practical purposes, it is important to generalize previous results to time-windows of the size $M > 1$. In particular, let $\mathbb{C}(m,\beta)$ be the class of detection rules defined in (5.41) for which the maximal conditional probability of false alarm in the time window of size m does not exceed a predefined number $\beta \in (0,1)$ and consider the maximin criterion

$$\inf_{\nu\geq 0} \mathsf{P}_\nu(T_{\rm opt} \leq \nu+M | T_{\rm opt} > \nu) = \sup_{T\in\mathbb{C}(m,\beta)} \inf_{\nu\geq 0} \mathsf{P}_\nu(\nu < T \leq \nu+M | T>\nu), \tag{5.46}$$

which was introduced in Section 2.3.7 (see (2.51)). While finding a strictly optimal solution to this optimization problem is very difficult, we believe that window-limited detection rules, in particular, the sliding window LLR change detection rule[2]

$$T_a^M = \inf\left\{ n \geq M : \sum_{t=n-M+1}^{n} \log\left[\frac{f(X_t)}{g(X_t)}\right] \geq a \right\} \tag{5.47}$$

are asymptotically optimal when $M = M_\beta \to \infty$ and $m = m_\beta \to \infty$ approach infinity with certain rates as $\beta \to 0$ and $a = a_\beta \to \infty$. This rule was optimized asymptotically in [52, 53] in a different problem of detecting finite-length transient changes, but its asymptotic optimality was not established.

A similar sliding window rule is expected to be asymptotically optimal when $a = a_\alpha \to \infty$ and $M = M_\alpha \to \infty$ in the Bayesian problem

$$\mathsf{P}_\pi(T_{\rm opt} \leq \nu+M | T_{\rm opt} > \nu) = \sup_{T\in\mathbb{C}_\pi(\alpha)} \mathsf{P}_\pi(T \leq \nu+M | T>\nu). \tag{5.48}$$

While for a fixed window size $M > 1$ the solutions of optimization problems (5.46) and (5.48) are intractable and the structure of the optimal rule may be found only in asymptotic settings as the false alarm probability goes to 0, the optimal detection rule can be found in the case where M is random having geometric distribution. This assumption makes perfect sense at least when M is an unknown (and nuisance) parameter characterizing duration of a change (i.e., the moment of "signal" disappearance), as discussed in Subsection 2.3.7 where we introduced the optimality criteria (2.43), (2.47), and (2.48). Also, this assumption may be interpreted as considering the exponentially decaying window. Having said this, let M be Geometric(ρ_0), $\pi_i^M = \rho_0(1-\rho_0)^{i-1}$, $i = 1,2,\ldots$, $\rho_0 \in (0,1)$.

5.3.1 Bayes Optimal Change Detection Rule

In the case where the size of the window M is random, it is natural to exploit the optimization criterion (2.43); that is, to introduce the average probability of detection

$$\mathsf{PD}_\pi(T) := \mathsf{P}_\pi(T \leq \nu+M | T>\nu) = \frac{\sum_{k=-1}^{\infty} \pi_k \sum_{i=1}^{\infty} \pi_i^M \mathsf{P}_k(k < T \leq k+i)}{1-\mathsf{PFA}_\pi(T)} \tag{5.49}$$

and to consider the Bayesian optimization problem of finding a rule $T_{\rm opt}$ that maximizes this probability in the class of rules with the upper bound on the $\mathsf{PFA}_\pi(T)$

$$\mathsf{PD}_\pi(T_{\rm opt}) = \sup_{T\in\mathbb{C}_\pi(\alpha)} \mathsf{PD}_\pi(T). \tag{5.50}$$

[2]This procedure is often referred to as the finite moving average test.

Note that when M is Geometric(ρ_0) we have

$$\sum_{i=1}^{\infty} \pi_i^M \mathsf{P}_k(k < T \le k+i) = \mathsf{E}_k\left[\sum_{i=1}^{\infty} \rho_0(1-\rho_0)^{i-1} \mathbb{1}_{\{0<T-k\le i\}}\right]$$
$$= \mathsf{E}_k\left[\sum_{i=T-k}^{\infty} \rho_0(1-\rho_0)^{i-1} \mathbb{1}_{\{T>k\}}\right] = (1-\rho_0)^{-1}\mathsf{E}_k\left[(1-\rho_0)^{T-k}\mathbb{1}_{\{T>k\}}\right]. \tag{5.51}$$

Hence, the detection probability of the rule T can be equivalently represented as

$$\mathsf{PD}_\pi(T) = \frac{1}{[1-\mathsf{PFA}_\pi(T)](1-\rho_0)} \sum_{k=-1}^{\infty} \pi_k \mathsf{E}_k\left[(1-\rho_0)^{T-k}\mathbb{1}_{\{T>k\}}\right]$$
$$= \frac{1}{[1-\mathsf{PFA}_\pi(T)](1-\rho_0)} \mathsf{E}_\pi\left[(1-\rho_0)^{T-\nu}\mathbb{1}_{\{T>\nu\}}\right], \tag{5.52}$$

and therefore, the optimization problem (5.50) reduces to finding a rule $T_{\rm opt}$ that solves

$$\mathsf{PD}_\pi(T_{\rm opt}) = \sup_{T\in\mathbb{C}_\pi(\alpha)} \frac{1}{[1-\mathsf{PFA}_\pi(T)](1-\rho_0)} \mathsf{E}_\pi\left[(1-\rho_0)^{T-\nu}\mathbb{1}_{\{T>\nu\}}\right]. \tag{5.53}$$

The following theorem establishes the structure of the optimal detection rule, which turns out to be a modified Shiryaev-type rule.

Theorem 5.5. *Let observations $\{X_n\}_{n\in\mathbb{Z}_+}$ be i.i.d. with a density $g(x)$ pre-change and with a density $f(x)$ post-change. Assume that the P_∞-distribution of the LR $\Lambda_n = f(X_n)/g(X_n)$ is continuous.*

(i) *If $1 \ge \alpha \ge 1-q$, then for any prior distribution of the change point $\{\pi_k\}_{k\ge -1}$ ($\pi_{-1} = q$, $0 < q < 1$) and any distribution of M the optimal Bayesian rule prescribes stopping immediately without making observations, i.e., $T_{\rm opt} = 0$ w.p. 1.*

(ii) *If the prior distribution of the change point ν is zero-modified* Geometric(q,ρ),

$$\pi_{-1} = q, \quad \pi_k = (1-q)\rho(1-\rho)^k, \quad k \in \mathbb{Z}_+,$$

and the distribution of the window size M is Geometric(ρ_0), *then the change detection rule*

$$T_{A_\alpha} = \inf\left\{n \ge 1 : R_{\rho,\rho_0}(n) \ge A_\alpha\right\}, \tag{5.54}$$

where the statistic $R_{\rho,\rho_0}(n)$ is given by the recursion

$$R_{\rho,\rho_0}(n+1) = [1+R_{\rho,\rho_0}(n)]\Lambda_{n+1}\frac{(1-\rho_0)}{1-\rho}, \quad n\in\mathbb{Z}_+, \quad R_{\rho,\rho_0}(0) = \frac{(1-\rho_0)q}{\rho(1-q)} \tag{5.55}$$

and threshold A_α satisfies $\mathsf{PFA}_\pi(T_{A_\alpha}) = \alpha$, is Bayes optimal in the problem (5.50) *for all $0 < \alpha < 1-q$.*

Proof. The proof of part (i). Let $1 \ge \alpha \ge 1-q$. Obviously,

$$\mathsf{PFA}_\pi(T=0) = \sum_{k=0}^{\infty} \pi_k = 1-q,$$

so the rule $T = 0$ belongs to class $\mathbb{C}_\pi(\alpha)$ for $\alpha \ge 1-q$. Also, using (5.49), we obtain

$$\mathsf{PD}_\pi(T=0) = \frac{\sum_{k=-1}^{\infty}\pi_k \sum_{i=1}^{\infty}\Pr(M=i))}{1-\mathsf{PFA}_\pi(T=0)} = \frac{q}{q} = 1,$$

which is the best one can do.

The proof of part (ii). Let $0 < \alpha < 1 - q$. Obviously, the optimization problem (5.53) is equivalent to minimizing the following functional over all stopping times in class $\mathbb{C}_\pi(\alpha)$:

$$\inf_{T \in \mathbb{C}_\pi(\alpha)} \frac{1}{1 - \mathsf{PFA}_\pi(T)} \mathsf{E}_\pi \left[\frac{1 - (1-\rho_0)^{T-\nu} \mathbb{1}_{\{T>\nu\}}}{\rho_0} \right].$$

It is easily seen that we can restrict attention to stopping times that achieve the constraint on the PFA with equality, $\mathsf{PFA}_\pi(T) = \alpha$. Denote such a class as $\widetilde{\mathbb{C}}_\pi(\alpha)$. Indeed, if $\mathsf{PFA}(T) < \alpha$, then we can produce a stopping time that achieves the constraint with equality without decreasing the average probability of detection PD_π, simply by randomizing between T and the stopping time that is identically zero. Therefore, it suffices to consider the unnormalized optimization problem

$$\inf_{T \in \widetilde{\mathbb{C}}_\pi(\alpha)} \mathsf{E}_\pi \left[\frac{1 - (1-\rho_0)^{T-\nu} \mathbb{1}_{\{T>\nu\}}}{\rho_0} \right].$$

This latter problem is reduced to the following (unconditional) purely Bayesian problem of minimizing the integrated risk

$$\mathscr{R}_c(T) = \mathsf{PFA}(T) + c\,\mathsf{E}_\pi \left[\frac{1 - (1-\rho_0)^{T-\nu} \mathbb{1}_{\{T>\nu\}}}{\rho_0} \right]$$

over all positive Markov times $T \in \mathscr{M}$, using the standard Lagrange multiplier method, i.e., $\inf_{T>0, T \in \mathscr{M}} \mathscr{R}_c(T)$, where $c = c_\alpha$ is a multiplier that depends on the given PFA constraint α. Theorem 4.1 of Poor [117] implies that the optimal solution of this purely Bayes problem is delivered by the stopping time T_A defined in (5.54) with an appropriate positive threshold $A = A(c_\alpha)$, which in our constrained problem must be so selected that $\mathsf{PFA}_\pi(T_A) = \alpha$. By the assumption the distribution $\mathsf{P}_\infty(\Lambda_1 \le y)$ is continuous, so there exists $A = A_\alpha$ that satisfies this latter equation for any $\alpha < 1 - q$. This completes the proof. □

Remark 5.2. The condition of continuity of the P_∞-distribution of the LR Λ_1 is imposed to guarantee the existence of threshold $A = A_\alpha$ that satisfies $\mathsf{PFA}_\pi(T_A) = \alpha$ for any $\alpha < 1 - q$, since if the LR is not continuous a complication arises due to the fact that the solution of this equation may not exist for some values of α. If the distribution $\mathsf{P}_\infty(\Lambda_1 \le y)$ is not continuous the assertion of Theorem 5.5(ii) holds for a randomized procedure with a randomization on the boundary A.

Note also that the optimal rule (5.54) is nothing but a modified Shiryaev change detection rule with an additional factor $1 - \rho_0$ in the recursion for the statistic $R_{\rho,\rho_0}(n)$. It is interesting that this modified rule is still optimal in a very different problem setting of minimizing the probability of miss detection of a change in place of minimizing the exponential expected delay [117].

It is interesting to compare operating characteristics of the obtained optimal rule with the Shiryaev rule T_B^* which is based on thresholding the statistic $R_\rho(n)$ given by the recursion

$$R_\rho(n+1) = [1 + R_\rho(n)] \frac{\Lambda_{n+1}}{1-\rho}, \quad n \in \mathbb{Z}_+, \quad R_\rho(0) = \frac{q}{\rho(1-q)}.$$

To this end, we performed MC simulations for the Gaussian model $X_n = \mathbb{1}_{\{\nu<n\}}\theta + \xi_n$, where $\{\xi_n\}_{n\ge1}$ is the i.i.d. Gaussian sequence with mean zero and unit variance. Simulations show that typically the optimal rule only slightly outperforms the Shiryaev rule for the probabilities of detection $\mathsf{PD}_\pi = 0.8$ and higher, which are of interest for applications. The difference becomes visible for relatively small values of the PD_π such as 0.5 and smaller. For example, if $\rho = \rho_0 = 0.05$, $\theta = 1.8$, and $\alpha = 10^{-3}$, then $\mathsf{PD}_\pi(T_A) = 0.905$ and $\mathsf{PD}_\pi(T_B^*) = 0.902$; if $\rho = \rho_0 = 0.3$, $\theta = 3$, and $\alpha = 10^{-4}$, then $\mathsf{PD}_\pi(T_A) = 0.464$ and $\mathsf{PD}_\pi(T_B^*) = 0.402$.

5.3.2 Maximin Optimal Change Detection Rule

Let us now focus on the maximin criterion (2.47) introduced in Subsection 2.3.7. Specifically, introduce the minimal average over the distribution of M probability of detection

$$\mathsf{PD}_{\mathrm{einf}}(T) = \sum_{i=1}^{\infty} \pi_i^M \inf_{\nu \in \mathbb{Z}_+} \operatorname{ess\,inf} \mathsf{P}_\nu(T \le \nu + i | T > \nu, \mathscr{F}_\nu, M = i) \tag{5.56}$$

Consider the maximin problem of maximizing the minimal probability of detection (5.56) subject to the constraint imposed on the ARL to false alarm $\mathsf{E}_\infty[T] \ge \gamma$, i.e., the optimization problem is to find a rule T_{opt} that solves

$$\mathsf{PD}_{\mathrm{einf}}(T_{\mathrm{opt}}) = \sup_{T \in \mathbb{C}_\gamma} \mathsf{PD}_{\mathrm{einf}}(T), \tag{5.57}$$

where $\mathbb{C}_\gamma = \{T : \mathsf{E}_\infty[T] \ge \gamma\}$ $(\gamma \ge 1)$.

Consider also the maximin problem of maximizing the minimal probability of detection (5.56) subject to the constraint on the maximal LCPFA $\sup_{\ell \in \mathbb{Z}_+} \mathsf{P}_\infty(T \le \ell + m | T > \ell) \le \beta$, i.e., the optimization problem is to find a rule T_{opt} that solves

$$\mathsf{PD}_{\mathrm{einf}}(T_{\mathrm{opt}}) = \sup_{T \in \mathbb{C}(m,\beta)} \mathsf{PD}_{\mathrm{einf}}(T) \quad \text{for every } \beta \in (0,1),\ m \ge 1 \tag{5.58}$$

where class $\mathbb{C}(m,\beta)$ is defined in (5.41).

The following theorem establishes the structure of the optimal detection rule, which turns out to be a modified CUSUM-type rule.

Theorem 5.6. *Let observations $\{X_n\}_{n \in \mathbb{Z}_+}$ be i.i.d. with a density $g(x)$ pre-change and with a density $f(x)$ post-change. Suppose the distribution of the window size M is* Geometric(ρ_0). *Further, assume that the P_∞-distribution of the LR $\Lambda_1 = f(X_1)/g(X_1)$ is continuous.*

(i) *Then the change detection rule*

$$T^*_{B_\beta} = \inf\{n \ge 1 : V_{\rho_0}(n) \ge B_\gamma\} \tag{5.59}$$

with the statistic $V_{\rho_0}(n)$ given by the recursion

$$V_{\rho_0}(n+1) = \max\{1, V_{\rho_0}(n)\} \Lambda_{n+1}(1-\rho_0), \quad n \in \mathbb{Z}_+, \quad V_{\rho_0}(0) = 1 \tag{5.60}$$

and with threshold B_γ that satisfies $\mathsf{E}_\infty[T^*_{B_\gamma}] = \gamma$ is maximin optimal in the problem (5.57) *for all $\gamma > 1$.*

(ii) *If threshold $B_{m,\beta}$ satisfies $\sup_{\ell \in \mathbb{Z}_+} \mathsf{P}_\infty(T^*_B \le \ell + m | T^*_B > \ell) = \beta$, then the rule* (5.59) *is maximin optimal in the problem* (5.58) *for all $0 < \beta < 1$.*

Proof. Proof of (i). We have

$$\sum_{i=1}^{\infty} \pi_i^M \mathsf{P}_k(k < T \le k+i | T > k, \mathscr{F}_k, M = i) = \mathsf{E}_k\left[\sum_{i=1}^{\infty} \rho_0(1-\rho_0)^{i-1} \mathbb{1}_{\{0 < T-k \le i\}} | T > k, \mathscr{F}_k, M = i\right]$$

$$= \mathsf{E}_k\left[\sum_{i=T-k}^{\infty} \rho_0(1-\rho_0)^{i-1} | T > k, \mathscr{F}_k\right] = (1-\rho_0)^{-1} \mathsf{E}_k\left[(1-\rho_0)^{T-k} | T > k, \mathscr{F}_k\right].$$

As a result, the original maximin problem (5.57) reduces to

$$\inf_T \inf_{\nu \in \mathbb{Z}_+} \operatorname{ess\,sup} \frac{1}{\rho_0} \mathsf{E}_\nu\left[1 - (1-\rho_0)^{T-\nu} | T > \nu, \mathscr{F}_\nu\right] \quad \text{subject to } \mathsf{E}_\infty[T] \ge \gamma.$$

By Theorem 2.1 of Poor [117] the optimal solution of this problem is delivered by the stopping time T_B^* defined in (5.59) when threshold $B = B_\gamma$ satisfies $\mathsf{E}_\infty[T_{B_\gamma}^*] = \gamma$.

Proof of (ii). An appeal to Lemmas 2.1 and 2.2 proves assertion (ii) since Lemmas 2.1 and 2.2 show that $\mathbb{C}(m,\beta) \subset \mathbb{C}_\gamma$ for some $\gamma = \gamma(m,\beta)$ (the constraint on the maximal LCPFA is more stringent than the constraint on the ARL to false alarm). This implies that

$$\sup_{T\in\mathbb{C}(m,\beta)} \mathsf{PD}_{\mathrm{einf}}(T) \geq \sup_{T\in\mathbb{C}_{\gamma(m,\beta)}} \mathsf{PD}_{\mathrm{einf}}(T),$$

where by part (i) the lower bound is attained for the modified CUSUM rule T_B^* if threshold $B = B_{m,\beta}$ satisfies $\mathsf{E}_\infty[T_{B_{m,\beta}}^*] = \gamma(m,\beta)$. □

Note that the optimal rule (5.59) is nothing but a modified CUSUM rule with an additional factor $1-\rho_0$ in the recursion for the statistic $V_{\rho_0}(n)$. It is interesting that this modified rule is still optimal in a very different problem setting of minimizing the probability of miss detection of a change in place of minimizing the exponential expected delay in the worst-case scenario [117]. If the distribution $\mathsf{P}_\infty(\Lambda_1 \leq y)$ is not continuous the assertion of Theorem 5.6 holds for a randomized procedure with a randomization on the boundary B.

5.4 Concluding Remarks

1. Pollak and Krieger [110] established optimality of the nonrandomized mixture Shewhart rule T_A^W in a Bayesian problem with the loss function $L(T,\nu+1) = \mathbb{1}_{\{T\neq\nu+1\}}$ and geometric prior Geometric$(0,\rho)$, which is substantially different from the one considered in Theorem 5.1.

2. It would be useful to generalize the results obtained in previous sections to non-i.i.d. models (2.1). In this case, a procedure that is based on thresholding the LR with a threshold $A_n(\mathbf{X}^{n-1})$ that depends on the previous $n-1$ observations $\mathbf{X}^{n-1}$ may be optimal. In the case of independent but non-identically distributed observations with pre-change densities $g_n(X_n)$ $(n \leq \nu)$ and post-change densities $f_{\theta,n}(X_n)$ $(n \geq \nu+1)$, Moustakides [95] established optimality of the Shewhart rule with a time-varying threshold A_n. Specifically, define the Shewhart-type stopping time $T_{A(\beta)}^W = \inf\{n \geq 1 : \Lambda_n^W \geq A_n(\beta)\}$, where $\Lambda_n^W = \int f_{\theta,n}(X_n)\,\mathrm{d}W(\theta)/g_n(X_n)$ is the LR for the nth observation and the sequence of thresholds $A(\beta) = \{A_n(\beta)\}_{n\geq 1}$ is found from the equations $\mathsf{P}_{0,n}^W\{\Lambda_n \geq A_n(\beta)\} = \beta$, $n \geq 1$. Then $T_{A(\beta)}^W$ is an equalizer with constant probability of detection β and if $\beta = \beta_\gamma$ satisfies the equality

$$\mathsf{E}_\infty[T_{A(\beta)}^W] = 1 + \sum_{n=1}^{\infty}\prod_{t=1}^{n} \mathsf{P}_{\infty,t}(\Lambda_n^W < A_n(\beta)) = \gamma,$$

$T_{A(\beta)}^W$ is the optimal detection rule in maximin problems (5.31) and (5.32). For a generalization to Markov models, see Moustakides [96]. In the general non-i.i.d. case, this problem is still open.

6

Quickest Change Detection in Multiple Streams

6.1 Introduction

The problem of changepoint detection in multiple data streams (sensors, populations or in multi-channel systems) arises in numerous applications similar to those discussed in Chapter 1 where we considered the problem of hypothesis testing in multiple streams. These applications include the medical sphere (detection of an epidemic present in only a fraction of hospitals [23, 43, 145, 172]); environmental monitoring (detection of the presence of hazardous materials or intruders [42, 125]); military defense (detection of an unknown number of targets by multichannel sensor systems [4, 162]); cyber security (detection of attacks in computer networks [148, 159, 168, 169]); detection of malicious activity in social networks [119, 120], to name a few.

In most surveillance applications with unknown points of change, including the classical Statistical Process Control sphere, the baseline (pre-change, in-control) distribution of observed data is known, but the post-change out-of-control distribution is not completely known. As discussed in Chapter 3 there are three conventional approaches in this case: (i) to select a representative value of the post-change parameter and apply efficient detection rules tuned to this value such as the Shiryaev rule, the Shiryaev–Roberts rule or CUSUM, (ii) to select a mixing measure over the parameter space and apply mixture-type rules, (iii) to estimate the parameter and apply adaptive schemes. In Chapters 3 and 4, a single stream case was considered. In this chapter, we consider a more general case where the change occurs in multiple data streams and more general multi-stream double-mixture-type change detection rules, assuming that the number and location of affected data streams are also unknown.

To be more specific, suppose there are N data streams $\{X_n(i)\}_{n\geq 1}$, $i=1,\dots,N$, observed sequentially in time subject to a change at an unknown time $\nu \in \{0,1,2,\dots\}$, so that $X_1(i),\dots,X_\nu(i)$ are generated by one stochastic model and $X_{\nu+1}(i), X_{\nu+2}(i),\dots$ by another model when the change occurs in the ith stream. The change in distributions happens at a subset of streams $\mathscr{B} \subseteq \{1,\dots,N\}$ of a size (cardinality) $1 \leq |\mathscr{B}| \leq K \leq N$, where K is an assumed maximal number of streams that can be affected, which can be and often is substantially smaller than N. A sequential detection rule is a stopping time T with respect to an observed sequence $\{\mathbf{X}_n\}_{n\geq 1}$, $\mathbf{X}_n = (X_n(1),\dots,X_n(N))$, i.e., T is an integer-valued random variable, such that the event $\{T=n\}$ belongs to the sigma-algebra $\mathscr{F}_n = \sigma(\mathbf{X}^n)$ generated by observations $\mathbf{X}_1,\dots,\mathbf{X}_n$. A false alarm is raised when the detection is declared before the change occurs. We want to detect the change with as small a delay as possible while controlling for a risk of false alarms.

Assume that the observations have a fairly general stochastic structure. So if we let $\mathbf{X}^n(i) = (X_1(i),\dots,X_n(i))$ denote the sample of size n in the ith stream and if $\{f_{\theta_i,n}(X_n(i)|\mathbf{X}^{n-1}(i))\}_{n\geq 1}$, $\theta_i \in \Theta_i$ is a parametric family of conditional densities of $X_n(i)$ given $\mathbf{X}^{n-1}(i)$, then when $\nu = \infty$ (there is no change) the parameter θ_i is equal to the known value $\theta_{i,0}$, i.e., $f_{\theta_i,n}(X_n(i)|\mathbf{X}^{n-1}(i)) = f_{\theta_{i,0},n}(X_n(i)|\mathbf{X}^{n-1}(i))$ for all $n \geq 1$ and when $\nu = k < \infty$, then $\theta_i = \theta_{i,1} \neq \theta_{i,0}$, i.e., $f_{\theta_i,n}(X_n(i)|\mathbf{X}^{n-1}(i)) = f_{\theta_{i,0},n}(X_n(i)|\mathbf{X}^{n-1}(i))$ for $n \leq k$ and $f_{\theta_i,n}(X_n(i)|\mathbf{X}^{n-1}(i)) = f_{\theta_{i,1},n}(X_n(i)|\mathbf{X}^{n-1}(i))$ for $n > k$. Not only the point of change ν, but also the subset $\mathscr{B}$, its size $|\mathscr{B}|$, and the post-change parameters $\theta_{i,1}$ are unknown.

In the case where $f_{\theta_i,n}(X_n(i)|\mathbf{X}^{n-1}(i)) = f_{\theta_i}(X_n(i))$, i.e., when the observations in the ith stream are i.i.d. according to a distribution with density $f_{\theta_{i,0}}(X_n(i))$ in the pre-change mode and with density $f_{\theta_{1,i}}(X_n(i))$ in the post-change mode this problem was considered in [39, 89, 155, 164, 168, 192]. Specifically, in the case of a known post-change parameter and $K = 1$ (i.e., when only one stream can be affected but it is unknown which one), Tartakovsky [155] proposed to use a multi-chart CUSUM procedure that raises an alarm when one of the partial CUSUM statistics exceeds a threshold. This procedure is very simple, but it is not optimal and performs poorly when many data streams are affected. To avoid this drawback, Mei [89] suggested a SUM-CUSUM rule based on the sum of CUSUM statistics in streams and evaluated its first-order performance, which shows that this detection scheme is first-order asymptotically minimax minimizing the maximal expected delay to detection when the ARL to false alarm approaches infinity. Fellouris and Sokolov [39] suggested more efficient generalized and mixture-based CUSUM rules that are second-order minimax. Xie and Siegmund [192] considered a particular Gaussian model with an unknown post-change mean. They suggested a rule that combines mixture likelihood ratios that incorporate an assumption about the proportion of affected data streams with the generalized CUSUM statistics in streams and then add up the resulting local statistics. They also performed a detailed asymptotic analysis of the proposed detection rule in terms of the ARL to false alarm and the expected delay as well as MC simulations. Chan [22] studied a version of the mixture likelihood ratio rule for detecting a change in the mean of the normal population assuming independence of data streams and established its asymptotic optimality in a minimax setting as well as dependence of operating characteristics on the fraction of affected streams.

In the present chapter, we consider a Bayesian problem with a general prior distribution of the change point and we generalize the results of Chapter 3 for a single data stream and a general stochastic model to multiple data streams with an unknown pattern, i.e., when the size and location of the affected streams are unknown. It is assumed that the observations can be dependent and non-identically distributed in data streams and even across the streams. We introduce two double-mixture detection rules – the first one mixes the Shiryaev-type statistic over the distributions of the unknown pattern and unknown post-change parameter; the second one is the double-mixture Shiryaev–Roberts statistic. The resulting statistics are then compared to appropriate thresholds. In this chapter, we present a general theory for very general stochastic models, providing sufficient conditions under which the suggested detection rules are first-order asymptotically optimal.

6.2 A Multistream Model and Change Detection Rules Based on Mixtures

6.2.1 The General Multistream Model

Consider the multistream scenario where the observations $\mathbf{X} = (X(1),\dots,X(N))$ are sequentially acquired in N streams, i.e., in the ith stream one observes a sequence $X(i) = \{X_n(i)\}_{n\ge 1}$, where $i \in \mathcal{N} := \{1,\dots,N\}$. Let P_∞ denote the probability measure corresponding to the sequence of observations $\{\mathbf{X}_n\}_{n\ge 1}$ from all N streams when there is never a change ($\nu = \infty$) in any of the components and, for $k = 0, 1, \dots$ and $\mathscr{B} \subset \mathcal{N}$, let $\mathsf{P}_{k,\mathscr{B}}$ denote the measure corresponding to the sequence $\{\mathbf{X}_n\}_{n\ge 1}$ when $\nu = k < \infty$ and the change occurs in a subset $\mathscr{B}$ of the set $\mathscr{P}$ (i.e., $X_{\nu+1}(i), i \in \mathscr{B}$ is the first post-change observation). By $\mathsf{H}_\infty : \nu = \infty$ we denote the hypothesis that the change never occurs, and by $\mathsf{H}_{k,\mathscr{B}}$ – the hypothesis that the change occurs at time $0 \le k < \infty$ in the subset of streams $\mathscr{B} \subset \mathscr{P}$. The set $\mathscr{P}$ is a class of subsets of $\mathcal{N}$ that incorporates available prior information regarding the subset $\mathscr{B}$ where the change may occur. For example, when it is known that *exactly K* streams can be affected after the change occurs, then $\mathscr{P} = \widetilde{\mathscr{P}}_K$, and when it is known that *at most*

K streams can be affected, then $\mathscr{P} = \mathscr{P}_K$, where

$$\widetilde{\mathscr{P}}_K = \{\mathscr{B} \subset \mathscr{N} : |\mathscr{B}| = K\}, \quad \mathscr{P}_K = \{\mathscr{B} \subset \mathscr{N} : 1 \le |\mathscr{B}| \le K\}. \tag{6.1}$$

Hereafter we denote by $|\mathscr{B}|$ the size of a subset $\mathscr{B}$, i.e., the number of affected streams under $\mathsf{H}_{k,\mathscr{B}}$ and $|\mathscr{P}|$ denotes the size of class $\mathscr{P}$, i.e., the number of possible alternatives in $\mathscr{P}$. Note that $|\mathscr{P}|$ takes maximum value when there is no prior information regarding the subset of affected streams, i.e., when $\mathscr{P} = \mathscr{P}_N$, in which case $|\mathscr{P}| = 2^N - 1$.

The problem is to detect the change as soon as possible after it occurs regardless of the subset $\mathscr{B}$, i.e., we are interested in detecting the event $\cup_{\mathscr{B}\in\mathscr{P}} \mathsf{H}_{k,\mathscr{B}}$ that the change has occurred in some subset but not in identifying the subset of streams where it occurs.

We will write $\mathbf{X}^n(i) = (X_0(i), X_1(i), \dots, X_n(i))$ for the concatenation of the first n observations from the ith data stream with an added initial condition $X_i(0)$ and $\mathbf{X}^n = (\mathbf{X}_0, \mathbf{X}_1, \dots, \mathbf{X}_n)$ for the concatenation of the first n observations from all N data streams with added initial values $\mathbf{X}_0$ (which typically are not observed). Let $\{g(\mathbf{X}_n|\mathbf{X}^{n-1})\}_{n\in\mathbb{Z}_+}$ and $\{f_{\mathscr{B}}(\mathbf{X}_n|\mathbf{X}^{n-1})\}_{n\in\mathbb{Z}_+}$ be sequences of conditional densities of $\mathbf{X}_n$ given $\mathbf{X}^{n-1}$, which may depend on n, i.e., $g = g_n$ and $f_{\mathscr{B}} = f_{\mathscr{B},n}$. For the general non-i.i.d. changepoint model, which we are interested in, the joint density $p(\mathbf{X}^n|\mathsf{H}_{k,\mathscr{B}})$ under hypothesis $\mathsf{H}_{k,\mathscr{B}}$ can be written as follows

$$p(\mathbf{X}^n|\mathsf{H}_{k,\mathscr{B}}) = f_\infty(\mathbf{X}^n) = \prod_{t=0}^{n} g(\mathbf{X}_t|\mathbf{X}^{t-1}) \quad \text{for } \nu = k \ge n, \tag{6.2}$$

$$p(\mathbf{X}^n|\mathsf{H}_{k,\mathscr{B}}) = \prod_{t=0}^{k} g(\mathbf{X}_t|\mathbf{X}^{t-1}) \times \prod_{t=k+1}^{n} f_{\mathscr{B}}(\mathbf{X}_t|\mathbf{X}^{t-1}) \quad \text{for } \nu = k < n, \tag{6.3}$$

where $\mathscr{B} \subset \mathscr{P}$. Therefore, $g(\mathbf{X}_n|\mathbf{X}^{n-1})$ is the pre-change conditional density and $f_{\mathscr{B}}(\mathbf{X}_n|\mathbf{X}^{n-1})$ is the post-change conditional density given that the change occurs in the subset $\mathscr{B}$.

In most practical applications, the post-change distribution is not completely known – it depends on an unknown (generally multidimensional) parameter $\theta \in \Theta$, so that the model (6.3) may be treated only as a benchmark for a more practical case where the post-change densities $f_{\mathscr{B}}(\mathbf{X}_t|\mathbf{X}^{t-1})$ are replaced by $f_{\mathscr{B},\theta}(\mathbf{X}_t|\mathbf{X}^{t-1})$, i.e.,

$$p(\mathbf{X}^n|H_{k,\mathscr{B}}, \theta) = \prod_{t=0}^{k} g(\mathbf{X}_t|\mathbf{X}^{t-1}) \times \prod_{t=k+1}^{n} f_{\mathscr{B},\theta}(\mathbf{X}_t|\mathbf{X}^{t-1}) \quad \text{for } \nu = k < n. \tag{6.4}$$

As in Chapter 3, we assume that the change point ν is a random variable independent of the observations with prior distribution $\pi_k = \mathsf{P}(\nu = k)$, $k = 0, 1, 2, \dots$ with $\pi_k > 0$ for $k \in \{0, 1, 2, \dots\} = \mathbb{Z}_+$ and that a change point may take negative values, but the detailed structure of the distribution $\mathsf{P}(\nu = k)$ for $k = -1, -2, \dots$ is not important. Only the total probability $q = \mathsf{P}(\nu \le -1)$ of the change being in effect before the observations become available matters.

6.2.2 Double-Mixture Change Detection Rules

We begin by considering the most general scenario where the observations across streams are dependent and then go on tackling the scenario where the streams are mutually independent.

6.2.2.1 The General Case

Let $\mathscr{L}_{\mathscr{B},\theta}(n) = f_{\mathscr{B},\theta}(\mathbf{X}_n|\mathbf{X}^{n-1})/g(\mathbf{X}_n|\mathbf{X}^{n-1})$. Note that in the general non-i.i.d. case the statistic $\mathscr{L}_{\mathscr{B},\theta}(n) = \mathscr{L}^{(k)}_{\mathscr{B},\theta}(n)$ may depend on the change point $\nu = k$ since the post-change density $f_{\mathscr{B},\theta}(\mathbf{X}_n|\mathbf{X}^{n-1})$ may depend on k. The likelihood ratio (LR) of the hypothesis $\mathsf{H}_{k,\mathscr{B}}$ that the change occurs at $\nu = k$ in the subset of streams $\mathscr{B}$ against the no-change hypothesis H_∞ based on the sample $\mathbf{X}^n = (\mathbf{X}_0, \mathbf{X}_1, \dots, \mathbf{X}_n)$ is given by the product

$$LR_{\mathscr{B},\theta}(k,n) = \prod_{t=k+1}^{n} \mathscr{L}_{\mathscr{B},\theta}(t), \quad n > k$$

and we set $LR_{\mathscr{B},\theta}(k,n) = 1$ for $n \le k$. For $\mathscr{B} \in \mathscr{P}$ and $\theta \in \Theta$, where $\mathscr{P}$ is an arbitrarily class of subsets of $\mathscr{N}$, define the statistic

$$S^{\pi}_{\mathscr{B},\theta}(n) = \frac{1}{\mathsf{P}(\nu \ge n)} \left[qLR_{\mathscr{B},\theta}(0,n) + \sum_{k=0}^{n-1} \pi_k LR_{\mathscr{B},\theta}(k,n) \right], \quad n \ge 1 \tag{6.5}$$

(with the initial condition $S^{\pi}_{\mathscr{B},\theta}(0) = q/(1-q)$) which is the Shiryaev-type statistic for detection of a change when it happens in a subset of streams $\mathscr{B}$ and the post-change parameter is θ.

Next, let

$$\mathbf{p} = \{p_{\mathscr{B}}, \mathscr{B} \in \mathscr{P}\}, \quad p_{\mathscr{B}} > 0 \,\forall\, \mathscr{B} \in \mathscr{P}, \quad \sum_{\mathscr{B} \in \mathscr{P}} p_{\mathscr{B}} = 1$$

be the probability mass function on $\mathscr{N}$ (mixing measure), and define the mixture statistic

$$S^{\pi}_{\mathbf{p},\theta}(n) = \sum_{\mathscr{B} \in \mathscr{P}} p_{\mathscr{B}} S^{\pi}_{\mathscr{B},\theta}(n), \quad S^{\pi}_{\mathbf{p},\theta}(0) = q/(1-q). \tag{6.6}$$

This statistic can be also represented as

$$S^{\pi}_{\mathbf{p},\theta}(n) = \frac{1}{\mathsf{P}(\nu \ge n)} \left[q\Lambda_{\mathbf{p},\theta}(0,n) + \sum_{k=0}^{n-1} \pi_k \Lambda_{\mathbf{p},\theta}(k,n) \right], \tag{6.7}$$

where

$$\Lambda_{\mathbf{p},\theta}(k,n) = \sum_{\mathscr{B} \in \mathscr{P}} p_{\mathscr{B}} LR_{\mathscr{B},\theta}(k,n)$$

is the mixture LR.

When the parameter θ is unknown there are two conventional approaches – either to maximize or average (mix) over θ. Introduce a mixing measure $W(\theta)$, $\int_\Theta \mathrm{d}W(\theta) = 1$, which can be interpreted as a prior distribution and define the double LR-mixture (average LR)

$$\Lambda_{\mathbf{p},W}(k,n) = \int_\Theta \sum_{\mathscr{B} \in \mathscr{P}} p_{\mathscr{B}} LR_{\mathscr{B},\theta}(k,n) \,\mathrm{d}W(\theta) = \int_\Theta \Lambda_{\mathbf{p},\theta}(k,n) \,\mathrm{d}W(\theta), \quad k < n \tag{6.8}$$

and the double-mixture Shiryaev-type statistic

$$S^{\pi}_{\mathbf{p},W}(n) = \int_\Theta \sum_{\mathscr{B} \in \mathscr{P}} p_{\mathscr{B}} S^{\pi}_{\mathscr{B},\theta}(n) \,\mathrm{d}W(\theta) = \frac{1}{\mathsf{P}(\nu \ge n)} \left[q\Lambda_{\mathbf{p},W}(0,n) + \sum_{k=0}^{n-1} \pi_k \Lambda_{\mathbf{p},W}(k,n) \right]. \tag{6.9}$$

The corresponding double-mixture LR-based detection rule is given by the stopping rule which is the first time $n \ge 1$ such that the statistic $S^{\pi}_{\mathbf{p},W}(n)$ hits the level $A > 0$, i.e.,

$$T^{\mathbf{p},W}_A = \inf\left\{n \ge 1 : S^{\pi}_{\mathbf{p},W}(n) \ge A\right\}. \tag{6.10}$$

Introduce the generalized SR statistic

$$R_{\mathscr{B},\theta}(n) = \omega LR_{\mathscr{B},\theta}(0,n) + \sum_{k=0}^{n-1} LR_{\mathscr{B},\theta}(k,n), \quad n \ge 1, \ R_{\mathscr{B},\theta}(0) = \omega \tag{6.11}$$

with a non-negative head-start $\omega \geq 0$ and, for a fixed value of θ, introduce the mixture statistic

$$R_{\mathbf{p},\theta}(n) = \sum_{\mathscr{B}\in\mathscr{P}} p_{\mathscr{B}} R_{\mathscr{B},\theta}(n) = \omega\Lambda_{\mathbf{p},\theta}(0,n) + \sum_{k=0}^{n-1} \Lambda_{\mathbf{p},\theta}(k,n) \quad n \geq 1, \;\; R_{\mathbf{p},\theta}(0) = \omega, \tag{6.12}$$

and the generalized double-mixture SR statistic

$$R_{\mathbf{p},W}(n) = \int_{\Theta} \sum_{\mathscr{B}\in\mathscr{P}} p_{\mathscr{B}} R_{\mathscr{B},\theta}(n) \, \mathrm{d}W(\theta) = \omega\Lambda_{\mathbf{p},W}(0,n) + \sum_{k=0}^{n-1} \Lambda_{\mathbf{p},W}(k,n), \quad n \geq 1, \;\; R_{\mathbf{p},W}(0) = \omega \tag{6.13}$$

(with a non-negative head-start ω) as well as the corresponding stopping rule

$$\widetilde{T}_A^{\mathbf{p},W} = \inf\{n \geq 1 : R_{\mathbf{p},W}(n) \geq A\}, \quad A > 0. \tag{6.14}$$

Note that we consider a very general stochastic model where not only the observations in streams may be dependent and non-identically distributed, but also the streams may be mutually dependent. In this very general case, computing statistics $S^{\pi}_{\mathbf{p},W}(n)$ and $R_{\mathbf{p},W}(n)$ is problematic even when the statistics in data streams $S^{\pi}_i(n)$ and $R_i(n)$, $i = 1,\dots,N$, can be computed. The computational problem becomes manageable when the data between data streams are independent, as discussed in the next subsection.

6.2.2.2 Independent Streams

Consider now a special scenario where the data across streams are independent. Let $\theta_i \in \Theta_i$ denote the unknown post-change parameter (generally multidimensional) in the ith stream ($i = 1,\dots,N$). Thus, we have

$$\begin{aligned}
p(\mathbf{X}^n | H_{k,\mathscr{B}}, \boldsymbol{\theta}_{\mathscr{B}}) &= \prod_{t=0}^{n}\prod_{i=1}^{N} g_i(X_t(i)|\mathbf{X}^{t-1}(i)) \quad \text{for } \nu = k \geq n, \\
p(\mathbf{X}^n | H_{k,\mathscr{B}}, \boldsymbol{\theta}_{\mathscr{B}}) &= \prod_{t=0}^{k}\prod_{i=1}^{N} g_i(X_t(i)|\mathbf{X}^{t-1}(i)) \times \\
& \prod_{t=k+1}^{n}\prod_{i\in\mathscr{B}} f_{i,\theta_i}(X_t(i)|\mathbf{X}^{t-1}(i)) \prod_{i\notin\mathscr{B}} g_i(X_t(i)|\mathbf{X}^{t-1}(i)) \quad \text{for } \nu = k < n,
\end{aligned} \tag{6.15}$$

where $g_i(X_t(i)|\mathbf{X}^{t-1}(i))$ and $f_{i,\theta_i}(X_t(i)|\mathbf{X}^{t-1}(i))$ are conditional pre- and post-change densities in the ith data stream, respectively, and $\boldsymbol{\theta}_{\mathscr{B}} = (\theta_i, i \in \mathscr{B})$. So the LRs are

$$LR_{\mathscr{B},\boldsymbol{\theta}_{\mathscr{B}}}(k,n) = \prod_{i\in\mathscr{B}} LR_{i,\theta_i}(k,n), \quad LR_{i,\theta_i}(k,n) = \prod_{t=k+1}^{n} \mathscr{L}_{i,\theta_i}(t), \quad n > k, \tag{6.16}$$

where $\mathscr{L}_{i,\theta_i}(t) = f_{i,\theta_i}(X_t(i)|\mathbf{X}^{t-1}(i))/g_i(X_t(i)|\mathbf{X}^{t-1}(i))$.

Assume in addition that the mixing measure is such that

$$p_{\mathscr{B}} = C(\mathscr{P}_K) \prod_{i\in\mathscr{B}} p_i, \quad C(\mathscr{P}_K) = \left(\sum_{\mathscr{B}\in\mathscr{P}_K} \prod_{i\in\mathscr{B}} p_i\right)^{-1}.$$

Then the mixture LR is

$$\Lambda_{\mathbf{p},\boldsymbol{\theta}}(k,n) = C(\mathscr{P}_K) \sum_{i=1}^{K} \sum_{\mathscr{B}\in\widetilde{\mathscr{P}}_i} \prod_{j\in\mathscr{B}} p_j LR_{j,\theta_i}(k,n),$$

and its computational complexity is polynomial in the number of data streams. Moreover, in the special, perhaps most interesting and difficult case of $K = N$ and $p_j = p$, we obtain

$$\Lambda_{\mathbf{p},\boldsymbol{\theta}}(k,n) = C(\mathscr{P}_N)\left[\prod_{i=1}^{N}\left(1+pLR_{i,\theta_i}(k,n)\right)-1\right], \tag{6.17}$$

so its computational complexity is only $O(N)$. The representation (6.17) corresponds to the case when each stream is affected independently with probability $p/(1+p)$, the assumption that was made in [192].

6.3 Asymptotic Optimality Problems and Assumptions

Let $\mathsf{E}_{k,\mathscr{B},\theta}$ and E_∞ denote expectations under $\mathsf{P}_{k,\mathscr{B},\theta}$ and P_∞, respectively, where $\mathsf{P}_{k,\mathscr{B},\theta}$ corresponds to model (6.4) with an unknown parameter $\theta \in \Theta$. Define the probability measure $\mathsf{P}^\pi_{\mathscr{B},\theta}(\mathscr{A}\times\mathscr{K}) = \sum_{k\in\mathscr{K}} \pi_k \mathsf{P}_{k,\mathscr{B},\theta}(\mathscr{A})$ under which the change point ν has distribution $\pi = \{\pi_k\}$ and the model for the observations is of the form (6.2),(6.4), i.e., $\mathbf{X}(t)$ has conditional density $g(\mathbf{X}(t)|\mathbf{X}^{t-1})$ if $\nu \le k$ and conditional density $f_{\mathscr{B},\theta}(\mathbf{X}(t)|\mathbf{X}^{t-1})$ if $\nu > k$ and the change occurs in the subset $\mathscr{B}$ with the parameter θ. Let $\mathsf{E}^\pi_{\mathscr{B},\theta}$ denote the corresponding expectation.

For $r \ge 1$, $\nu = k \in \mathbb{Z}_+$, $\mathscr{B} \in \mathscr{P}$, and $\theta \in \Theta$ introduce the risk associated with the conditional rth moment of the detection delay $\mathscr{R}^r_{k,\mathscr{B},\theta}(T) = \mathsf{E}_{k,\mathscr{B},\theta}\left[(T-k)^r \,|\, T>k\right]$. In a Bayesian setting, the risk associated with the moments of delay to detection is

$$\overline{\mathscr{R}}^r_{\mathscr{B},\theta}(T) := \mathsf{E}^\pi_{\mathscr{B},\theta}[(T-\nu)^r|T>\nu] = \frac{\sum_{k=0}^\infty \pi_k \mathscr{R}^r_{k,\mathscr{B},\theta}(T)\mathsf{P}_\infty(T>k)}{1-\mathsf{PFA}_\pi(T)}, \tag{6.18}$$

where

$$\mathsf{PFA}_\pi(T) = \mathsf{P}^\pi_{\mathscr{B},\theta}(T\le\nu) = \sum_{k=0}^{\infty}\pi_k \mathsf{P}_\infty(T\le k) \tag{6.19}$$

is the weighted probability of false alarm that corresponds to the risk associated with a false alarm. In (6.18), we set $\pi_0 = \Pr(\nu \le 0)$ but not $\pi_0 = \Pr(\nu = 0)$. This allows us to avoid summation over $k \le -1$ which requires setting $\mathscr{R}^r_{k,\mathscr{B},\theta}(T) = \mathsf{E}_0[T]$ in place of $\mathsf{E}_k[T-k]$. However, in (6.19), $\pi_0 = \Pr(\nu = 0)$. Note that in (6.18) and (6.19) we used the fact that $\mathsf{P}_{k,\mathscr{B},\theta}(T\le k) = \mathsf{P}_\infty(T\le k)$ since the event $\{T\le k\}$ depends on the observations $\mathbf{X}_1,\dots\mathbf{X}_k$ generated by the pre-change probability measure P_∞ since by our convention $\mathbf{X}_k$ is the last pre-change observation if $\nu = k$.

In this chapter, we are interested in the Bayesian (constrained) optimization problem

$$\inf_{\{T:\mathsf{PFA}_\pi(T)\le\alpha\}} \overline{\mathscr{R}}^r_{\mathscr{B},\theta}(T) \quad \text{for all } \mathscr{B}\in\mathscr{P},\ \theta\in\Theta. \tag{6.20}$$

However, in general this problem is intractable for every value of the PFA $\alpha \in (0,1)$. So we will focus on the asymptotic problem assuming that the PFA α approaches zero. Specifically, we will be interested in proving that the double-mixture detection rule $T_A^{\mathbf{p},W}$ is first-order uniformly asymptotically optimal for all possible subsets $\mathscr{B}\in\mathscr{P}$ where the change may occur and all parameter values $\theta \in \Theta$, i.e.,

$$\lim_{\alpha\to 0}\frac{\inf_{T\in\mathbb{C}_\pi(\alpha)} \overline{\mathscr{R}}^r_{\mathscr{B},\theta}(T)}{\overline{\mathscr{R}}^r_{\mathscr{B},\theta}(T_A^{\mathbf{p},W})} = 1 \quad \text{for all } \mathscr{B}\in\mathscr{P},\ \theta\in\Theta \tag{6.21}$$

and

$$\lim_{\alpha\to 0}\frac{\inf_{T\in\mathbb{C}_\pi(\alpha)} \mathscr{R}^r_{k,\mathscr{B},\theta}(T)}{\mathscr{R}^r_{k,\mathscr{B},\theta}(T_A^{\mathbf{p},W})} = 1 \quad \text{for all } \mathscr{B}\in\mathscr{P},\ \theta\in\Theta \text{ and all } \nu=k\in\mathbb{Z}_+, \tag{6.22}$$

where $\mathbb{C}_\pi(\alpha) = \{T : \mathsf{PFA}_\pi(T) \le \alpha\}$ is the class of detection rules for which the PFA does not exceed a prescribed number $\alpha \in (0,1)$ and $A = A_\alpha$ is suitably selected.

First-order asymptotic optimality properties of the double-mixture SR-type detection rule $\widetilde{T}_A^{\mathbf{p},W}$ under certain conditions will be also established.

Instead of the constrained optimization problem (6.20) one may be also interested in the unconstrained Bayes problem with the average (integrated) risk function

$$\begin{aligned}\rho_{\pi,\mathbf{p},W}^{c,r}(T) &= \mathsf{E}\left[\mathbb{1}_{\{T\le\nu\}} + c(T-\nu)^r \mathbb{1}_{\{T>\nu\}}\right] \\ &= \mathsf{PFA}_\pi(T) + c\sum_{\mathscr{B}\in\mathscr{P}} p_{\mathscr{B}} \int_\Theta \mathsf{E}_{\mathscr{B},\theta}^\pi[(T-\nu)^+]^r \,\mathrm{d}W(\theta),\end{aligned} \tag{6.23}$$

where $c > 0$ is the cost of delay per unit of time and $r \ge 1$. An unknown post-change parameter θ and an unknown location of the change pattern $\mathscr{B}$ are now assumed random and the weight functions $W(\theta)$ and $p_{\mathscr{B}}$ are interpreted as the prior distributions of θ and $\mathscr{B}$, respectively. The first-order asymptotic problem is

$$\lim_{c\to 0} \frac{\inf_{T\in\mathscr{M}} \rho_{\pi,\mathbf{p},W}^{c,r}(T)}{\rho_{\pi,\mathbf{p},W}^{c,r}(T_A^{\mathbf{p},W})} = 1, \tag{6.24}$$

where $\mathscr{M}$ is the class of all Markov stopping times and threshold $A = A_{c,r}$ that depends on the cost c should be suitably selected.

While we consider a general prior and a very general stochastic model for the observations in streams and between streams, to study asymptotic optimality properties we still need to impose certain constraints on the prior distribution $\pi = \{\pi_k\}$ and on the general stochastic model (6.2)–(6.3) that guarantee asymptotic stability of the detection statistics as the sample size increases.

Regarding the prior distribution we will assume that the same conditions $\mathbf{CP}_1$ and $\mathbf{CP}_2$ as in Chapter 3 are satisfied (see (3.20)–(3.21)). More generally, if $\pi = \pi^\alpha$ depends on the PFA α, then we will assume that conditions $\mathbf{CP}_1^{(\alpha)}$, $\mathbf{CP}_2^{(\alpha)}$, and $\mathbf{CP}_3^{(\alpha)}$ in Chapter 3 hold (see (3.65)–(3.67)).

For $\mathscr{B} \in \mathscr{P}$ and $\theta \in \Theta$, introduce the log-likelihood ratio (LLR) process $\{\lambda_{\mathscr{B},\theta}(k,n)\}_{n\ge k+1}$ between the hypotheses "$\mathsf{H}_{k,\mathscr{B}},\theta$" ($k = 0,1,\dots$) and H_∞:

$$\lambda_{\mathscr{B},\theta}(k,n) = \sum_{t=k+1}^n \log \frac{f_{\mathscr{B},\theta}(\mathbf{X}_t|\mathbf{X}^{t-1})}{g(\mathbf{X}_t|\mathbf{X}^{t-1})}, \quad n > k$$

($\lambda_{\mathscr{B},\theta}(k,n) = 0$ for $n \le k$).

Define

$$p_{M,k}(\varepsilon,\mathscr{B},\theta) = \mathsf{P}_{k,\mathscr{B},\theta}\left\{\frac{1}{M}\max_{1\le n\le M} \lambda_{\mathscr{B},\theta}(k,k+n) \ge (1+\varepsilon)I_{\mathscr{B},\theta}\right\}$$

and for $\delta > 0$ define $\Gamma_{\delta,\theta} = \{\vartheta \in \Theta : |\vartheta - \theta| < \delta\}$ and

$$\Upsilon_r(\varepsilon,\mathscr{B},\theta) = \lim_{\delta\to 0} \sum_{n=1}^\infty n^{r-1} \sup_{k\in\mathbb{Z}_+} \mathsf{P}_{k,\mathscr{B},\theta}\left\{\frac{1}{n}\inf_{\vartheta\in\Gamma_{\delta,\theta}} \lambda_{\mathscr{B},\vartheta}(k,k+n) < I_{\mathscr{B},\theta} - \varepsilon\right\}.$$

Regarding the general model for the observations (6.2), (6.4) we assume that the following two conditions are satisfied:

$\mathbf{C}_1$. *There exist positive and finite numbers $I_{\mathscr{B},\theta}$ ($\mathscr{B} \in \mathscr{P}$, $\theta \in \Theta$) such that $n^{-1}\lambda_{\mathscr{B},\theta}(k,k+n) \to I_{\mathscr{B},\theta}$ in $\mathsf{P}_{k,\mathscr{B},\theta}$-probability and for any $\varepsilon > 0$*

$$\lim_{M\to\infty} p_{M,k}(\varepsilon,\mathscr{B},\theta) = 0 \quad \text{for all } k \in \mathbb{Z}_+, \mathscr{B} \in \mathscr{P}, \theta \in \Theta; \tag{6.25}$$

$\mathbf{C}_2$. *For any $\varepsilon > 0$ and some $r \ge 1$*

$$\Upsilon_r(\varepsilon,\mathscr{B},\theta) < \infty \quad \text{for all } \mathscr{B} \in \mathscr{P}, \theta \in \Theta. \tag{6.26}$$

Hereafter we ignore parameter values with W-measure null, i.e., we assume in the rest of this chapter that $W(\Gamma_{\delta,\theta}) > 0$ for all $\theta \in \Theta$ and $\delta > 0$.

6.4 Asymptotic Lower Bounds for Moments of the Detection Delay and Average Risk Function

In order to establish asymptotic optimality of detection rules we first obtain, under condition $\mathbf{C}_1$, asymptotic (as $\alpha \to 0$) lower bounds for moments of the detection delay $\overline{\mathscr{R}}^r_{\mathscr{B},\theta}(T) = \mathsf{E}^\pi_{\mathscr{B},\theta}[(T-\nu)^r | T > \nu]$ and $\mathscr{R}^r_{k,\mathscr{B},\theta} = \mathsf{E}_{k,\mathscr{B},\theta}[(T-k)^r | T > k]$ of any detection rule T from class $\mathbb{C}_\pi(\alpha)$. In the following sections, we show that under condition $\mathbf{C}_2$ these bounds are attained for the double-mixture rule T_A^W uniformly for all $\mathscr{B} \in \mathscr{P}$ and $\theta \in \Theta$ and that the same is true for the double-mixture rules $\widetilde{T}_A^{\mathbf{p},W}$ when the prior distribution is either heavy-tailed or has an exponential tail with a small parameter μ. We also establish the asymptotic lower bound for the integrated risk $\rho^{c,r}_{\pi,\mathbf{p},W}(T)$ as $c \to 0$ in the class of all Markov times and show that it is attained by the double-mixture rules T_A^W and $\widetilde{T}_A^{\mathbf{p},W}$.

Define

$$\overline{\mathscr{R}}^r_{\mathbf{p},W}(T) = \sum_{\mathscr{B}\in\mathscr{P}} p_{\mathscr{B}} \int_\Theta \overline{\mathscr{R}}^r_{\mathscr{B},\theta}(T)\,\mathrm{d}W(\theta) \tag{6.27}$$

and

$$D_{\mu,r} = \sum_{\mathscr{B}\in\mathscr{P}} p_{\mathscr{B}} \int_\Theta \left(\frac{1}{I_{\mathscr{B},\theta}+\mu}\right)^r \mathrm{d}W(\theta) \tag{6.28}$$

Asymptotic lower bounds for all positive moments of the detection delay and the integrated risk are specified in the following lemma. Recall that we say that $\pi \in \mathbf{C}(\mu)$ if condition $\mathbf{CP}_1$ is satisfied (see (3.20)) and that $\pi^\alpha \in \mathbf{C}^{(\alpha)}(\mu)$ if condition $\mathbf{CP}_1^{(\alpha)}$ is satisfied (see (3.65)).

Theorem 6.1. *Let, for some $\mu \ge 0$, the prior distribution belong to class $\mathbf{C}(\mu)$. Assume that for some positive and finite numbers $I_{\mathscr{B},\theta}$ ($\mathscr{B} \in \mathscr{P}$, $\theta \in \Theta$) condition $\mathbf{C}_1$ holds. Then for all $r > 0$ and all $\mathscr{B} \in \mathscr{P}$, $\theta \in \Theta$*

$$\liminf_{\alpha\to 0} \frac{\inf_{T\in\mathbb{C}_\pi(\alpha)} \mathscr{R}^r_{k,\mathscr{B},\theta}(T)}{|\log\alpha|^r} \ge \frac{1}{(I_{\mathscr{B},\theta}+\mu)^r}, \quad k \in \mathbb{Z}_+, \tag{6.29}$$

$$\liminf_{\alpha\to 0} \frac{\inf_{T\in\mathbb{C}_\pi(\alpha)} \overline{\mathscr{R}}^r_{\mathscr{B},\theta}(T)}{|\log\alpha|^r} \ge \frac{1}{(I_{\mathscr{B},\theta}+\mu)^r}, \tag{6.30}$$

and for all $r > 0$

$$\liminf_{c\to 0} \frac{\inf_{T\ge 0} \rho^{c,r}_{\pi,\mathbf{p},W}(T)}{c|\log c|^r} \ge D_{\mu,r}. \tag{6.31}$$

Proof. The methodology of the proof is essentially analogous to that used in the proofs of the lower bounds in Theorem 3.1, Lemma 3.3 and Lemma 3.5 for a single stream change detection problem by replacing θ with $\tilde{\theta} = (\mathscr{B}, \theta)$. Indeed, now the combined vector $(\mathscr{B}, \theta) = \tilde{\theta}$ is an unknown parameter, so extending the parameter θ to $\tilde{\theta}$ in the proofs of Theorem 3.1 and Lemma 3.3 in Chapter 3 yields the lower bounds (6.29) and (6.30). Also, by an argument similar to the proof of Lemma 3.5, we deduce the lower bound (6.31) if $\mu_\alpha = \mu$ for all α (i.e., when π_k^α does not depend on α) and if $\mu_\alpha \to \mu = 0$. For all technical details the reader is referred to the paper by Tartakovsky [161]. □

6.5 Asymptotic Optimality of Double-Mixture Detection Rules in Class $\mathbb{C}_\pi(\alpha)$

We now proceed with establishing asymptotic optimality properties of the double-mixture detection rules $T_A^{\mathbf{p},W}$ and $\widetilde{T}_A^{\mathbf{p},W}$ in class $\mathbb{C}_\pi(\alpha)$ as $\alpha \to 0$.

6.5.1 Asymptotic Optimality of the Double-Mixture Rule $T_A^{\mathbf{p},W}$

The following lemma provides the upper bound for the PFA of the double-mixture rule $T_A^{\mathbf{p},W}$ defined in (6.10).

Lemma 6.1. *For all $A > q/(1-q)$ and any prior distribution of the change point, the PFA of the rule T_A^W satisfies the inequality*

$$\mathsf{PFA}_\pi(T_A^{\mathbf{p},W}) \le 1/(1+A), \tag{6.32}$$

so that if $A = A_\alpha = (1-\alpha)/\alpha$, then $\mathsf{PFA}_\pi(T_{A_\alpha}^{\mathbf{p},W}) \le \alpha$ for $0 < \alpha < 1-q$, i.e., $T_{A_\alpha}^{\mathbf{p},W} \in \mathbb{C}_\pi(\alpha)$.

Proof. Using the Bayes rule and the fact that $\Lambda_{\mathbf{p},W}(k,n) = 1$ for $k \ge n$, we obtain

$$\begin{aligned}
&\mathsf{P}(\nu = k|\mathscr{F}_n) = \\
&\frac{\sum_{\mathscr{B}\in\mathscr{P}} p_{\mathscr{B}} \pi_k \prod_{t=0}^k g(\mathbf{X}_t|\mathbf{X}^{t-1}) \int_\Theta \prod_{t=k+1}^n f_{\mathscr{B},\theta}(\mathbf{X}_t|\mathbf{X}^{t-1}) \mathrm{d}W(\theta)}{\sum_{j=-1}^\infty \sum_{\mathscr{B}\in\mathscr{P}} p_{\mathscr{B}} \pi_j \prod_{t=0}^j g(\mathbf{X}_t|\mathbf{X}^{t-1}) \int_\Theta \prod_{t=j+1}^n f_{\mathscr{B},\theta}(\mathbf{X}_t|\mathbf{X}^{t-1}) \mathrm{d}W(\theta)} \\
&= \frac{\pi_k \Lambda_{\mathbf{p},W}(k,n)}{q\Lambda_{\mathbf{p},W}(0,n) + \sum_{j=0}^{n-1} \pi_j \Lambda_{\mathbf{p},W}(j,n) + \mathsf{P}(\nu \ge n)}.
\end{aligned}$$

It follows that

$$\begin{aligned}
\mathsf{P}(\nu \ge n|\mathscr{F}_n) &= \frac{\sum_{k=n}^\infty \pi_k \Lambda_{\mathbf{p},W}(k,n)}{q\Lambda_{\mathbf{p},W}(0,n) + \sum_{j=0}^{n-1} \pi_j \Lambda_{\mathbf{p},W}(j,n) + \mathsf{P}(\nu \ge n)} \\
&= \frac{\mathsf{P}(\nu \ge n)}{q\Lambda_{\mathbf{p},W}(0,n) + \sum_{j=0}^{n-1} \pi_j \Lambda_{\mathbf{p},W}(j,n) + \mathsf{P}(\nu \ge n)} \\
&= \frac{1}{S_{\mathbf{p},W}^\pi(n) + 1}.
\end{aligned} \tag{6.33}$$

By the definition of the stopping time $T_A^{\mathbf{p},W}$, $S_{\mathbf{p},W}^\pi(T_A^{\mathbf{p},W}) \ge A$ on $\{T_A^{\mathbf{p},W} < \infty\}$ and $\mathsf{PFA}_\pi(T_A^{\mathbf{p},W}) = \mathsf{E}_{\mathscr{B},\theta}^\pi[\mathsf{P}(T_A^{\mathbf{p},W} \le \nu|\mathscr{F}_{T_A^{\mathbf{p},W}}); T_A^{\mathbf{p},W} < \infty]$, so that

$$\mathsf{PFA}_\pi(T_A^{\mathbf{p},W}) = \mathsf{E}_{\mathscr{B},\theta}^\pi\left[(1 + S_{\mathbf{p},W}^\pi(T_A^{\mathbf{p},W}))^{-1}; T_A^{\mathbf{p},W} < \infty\right] \le 1/(1+A),$$

which completes the proof of inequality (6.32). □

The following proposition provides asymptotic operating characteristics of the double-mixture rule $T_A^{\mathbf{p},W}$ for large values of threshold A. It requires conditions $\mathbf{CP}_1^{(\alpha)}$, $\mathbf{CP}_2^{(\alpha)}$, and $\mathbf{CP}_3^{(\alpha)}$ with $\alpha = 1/A$ for the prior distribution $\pi^A = \{\pi_k^A\}$ that depends on threshold A, specifically

$\mathbf{CP}_1^{(A)}$. *For some $0 \le \mu_A < \infty$ and $0 \le \mu < \infty$,*

$$\lim_{n\to\infty} \frac{1}{n} \left|\log \Pi_n^A\right| = \mu_A \quad \text{and} \quad \lim_{A\to\infty} \mu_A = \mu, \tag{6.34}$$

$\mathbf{CP}_2^{(A)}$. *If $\mu_A > 0$ for all A and $\mu = 0$, then μ_A approaches zero at such rate that for some $r \geq 1$*

$$\lim_{A\to\infty} \frac{\sum_{k=0}^{\infty} \pi_k^A |\log \pi_k^A|^r}{|\log A|^r} = 0. \tag{6.35}$$

$\mathbf{CP}_3^{(A)}$. *For all $k \in \mathbb{Z}_+$*

$$\lim_{A\to 0} \frac{|\log \pi_k^A|}{|\log A|} = 0. \tag{6.36}$$

In the case when condition (6.34) is satisfied we say that π^A belongs to class $\mathbf{C}^{(A)}(\mu)$.

Proposition 6.1. *Let the prior distribution of the change point π^A belong to class $\mathbf{C}^{(A)}(\mu)$. Let $r \geq 1$ and assume that for some $0 < I_{\mathscr{B},\theta} < \infty$ ($\mathscr{B} \in \mathscr{P}$, $\theta \in \Theta$), right-tail and left-tail conditions $\mathbf{C}_1$ and $\mathbf{C}_2$ are satisfied.*
(i) *If condition $\mathbf{CP}_3^{(A)}$ holds, then, for all $0 < m \leq r$ and all $\mathscr{B} \in \mathscr{P}$, $\theta \in \Theta$ as $A \to \infty$*

$$\lim_{A\to\infty} \frac{\mathscr{R}^m_{k,\mathscr{B},\theta}(T_A^{\mathbf{p},W})}{|\log A|^m} = \frac{1}{(I_{\mathscr{B},\theta}+\mu)^m} \quad \text{for all } k \in \mathbb{Z}_+. \tag{6.37}$$

(ii) *If condition $\mathbf{CP}_2^{(A)}$ holds, then, for all $0 < m \leq r$ and all $\mathscr{B} \in \mathscr{P}$, $\theta \in \Theta$ as $A \to \infty$*

$$\lim_{A\to\infty} \frac{\overline{\mathscr{R}}^m_{\mathscr{B},\theta}(T_A^{\mathbf{p},W})}{|\log A|^m} = \frac{1}{(I_{\mathscr{B},\theta}+\mu)^m}. \tag{6.38}$$

Proof. To prove asymptotic approximations (6.37) and (6.38) note first that by (6.32) the detection rule $T_A^{\mathbf{p},W}$ belongs to class $\mathbb{C}(1/(A+1))$, so replacing α by $1/(A+1)$ in the asymptotic lower bounds (6.29) and (6.30) in Theorem 6.1, we obtain that under the right-tail condition $\mathbf{C}_1$ the following asymptotic lower bounds hold for all $r > 0$, $\mathscr{B} \in \mathscr{P}$ and $\theta \in \Theta$:

$$\liminf_{A\to\infty} \frac{\mathscr{R}^r_{k,\mathscr{B},\theta}(T_A^{\mathbf{p},W})}{(\log A)^r} \geq \frac{1}{(I_{\mathscr{B},\theta}+\mu)^r}, \quad k \in \mathbb{Z}_+, \tag{6.39}$$

$$\liminf_{A\to\infty} \frac{\overline{\mathscr{R}}^r_{\mathscr{B},\theta}(T_A^{\mathbf{p},W})}{(\log A)^r} \geq \frac{1}{(I_{\mathscr{B},\theta}+\mu)^r}. \tag{6.40}$$

Therefore, to prove the assertions of the proposition it suffices to show that, under the left-tail condition $\mathbf{C}_2$, for all $0 < m \leq r$, $\mathscr{B} \in \mathscr{P}$ and $\theta \in \Theta$

$$\limsup_{A\to\infty} \frac{\mathscr{R}^m_{k,\mathscr{B},\theta}(T_A^{\mathbf{p},W})}{(\log A)^m} \leq \frac{1}{(I_{\mathscr{B},\theta}+\mu)^m}, \quad k \in \mathbb{Z}_+, \tag{6.41}$$

$$\limsup_{A\to\infty} \frac{\overline{\mathscr{R}}^m_{\mathscr{B},\theta}(T_A^{\mathbf{p},W})}{(\log A)^m} \leq \frac{1}{(I_{\mathscr{B},\theta}+\mu)^m}. \tag{6.42}$$

The proof of part (i). For $\varepsilon \in (0, I_{\mathscr{B},\theta}+\mu)$, define

$$N_A = N_A(\varepsilon, \mathscr{B}, \theta) = 1 + \left\lfloor \frac{\log(A/\pi_k^A)}{I_{\mathscr{B},\theta}+\mu-\varepsilon} \right\rfloor.$$

Obviously, for any $n \geq 1$,

$$\begin{aligned}\log S^{\pi}_{\mathbf{p},W}(k+n) &\geq \log \Lambda_{\mathbf{p},W}(k,k+n) + \log \pi_k^A - \log \Pi^A_{k-1+n} \\ &\geq \inf_{\vartheta\in\Gamma_{\delta,\theta}} \lambda_{\mathscr{B},\vartheta}(k,k+n) + \log W(\Gamma_{\delta,\theta}) + \log p_{\mathscr{B}} + \log \pi_k^A - \log \Pi^A_{k-1+n},\end{aligned}$$

where $\Gamma_{\delta,\theta} = \{\vartheta \in \Theta : |\vartheta - \theta| < \delta\}$, so that for any $\mathscr{B} \in \mathscr{P}$, $\theta \in \Theta$, $k \in \mathbb{Z}_+$

$$\begin{aligned}\mathsf{P}_{k,\mathscr{B},\theta}\left(T_A^{\mathbf{p},W} - k > n\right) &\le \mathsf{P}_{k,\mathscr{B},\theta}\left\{\frac{1}{n}\log S_{\mathbf{p},W}^{\pi}(k+n) < \frac{1}{n}\log A\right\}\\ &\le \mathsf{P}_{k,\mathscr{B},\theta}\left\{\frac{1}{n}\inf_{\vartheta\in\Gamma_{\delta,\theta}} \lambda_{\mathscr{B},\vartheta}(k,k+n) < \frac{1}{n}\log\left(\frac{A}{\pi_k^A}\right)\right.\\ &\qquad\left. + \frac{1}{n}\left|\log\left[p_{\mathscr{B}} W(\Gamma_{\delta,\theta})\right]\right| - \frac{1}{n}\left|\log \Pi_{k-1+n}^A\right|\right\}.\end{aligned}$$

It is easy to see that for $n \ge N_A$ the last probability does not exceed the probability

$$\mathsf{P}_{k,\mathscr{B},\theta}\left\{\frac{1}{n}\inf_{\vartheta\in\Gamma_{\delta,\theta}} \lambda_{\mathscr{B},\vartheta}(k,k+n) < I_{\mathscr{B},\theta} + \mu - \varepsilon \right.$$
$$\left. + \frac{1}{N_A}\left|\left[\log p_{\mathscr{B}} W(\Gamma_{\delta,\theta})\right]\right| - \frac{1}{n}\left|\log \Pi_{k-1+n}^A\right|\right\}.$$

Since, by condition $\mathbf{CP}_1^{(A)}$, $N_A^{-1}|\log \Pi_{k-1+N_A}^A| \to \mu$ as $A \to \infty$, for a sufficiently large value of A there exists a small κ such that

$$\left|\mu - \frac{|\log \Pi_{k-1+N_A}^A|}{N_A}\right| < \kappa. \tag{6.43}$$

Hence, for $\varepsilon > 0$, $n \ge N_A$ and all sufficiently large A such that $\kappa + |\log[p_{\mathscr{B}} W(\Gamma_{\delta,\theta})]|/N_A < \varepsilon/2$ we have

$$\mathsf{P}_{k,\mathscr{B},\theta}\left(T_A^{\mathbf{p},W} - k > n\right) \le \mathsf{P}_{k,\mathscr{B},\theta}\left\{\frac{1}{n}\inf_{\vartheta\in\Gamma_{\delta,\theta}} \lambda_{\mathscr{B},\vartheta}(k,k+n) < I_{\mathscr{B},\theta} - \varepsilon/2\right\}. \tag{6.44}$$

By Lemma A.1, for any $k \in \mathbb{Z}_+$, $\mathscr{B} \in \mathscr{P}$ and $\theta \in \Theta$ we have the following inequality

$$\mathsf{E}_{k,\mathscr{B},\theta}\left[(T_A^{\mathbf{p},W} - k)^+\right]^r \le N_A^r + r2^{r-1}\sum_{n=N_A}^{\infty} n^{r-1}\mathsf{P}_{k,\mathscr{B},\theta}\left(T_A^{\mathbf{p},W} - k > n\right), \tag{6.45}$$

which along with (6.44) yields

$$\mathsf{E}_{k,\mathscr{B},\theta}\left[\left(T_A^{\mathbf{p},W} - k\right)^+\right]^r \le \left(1 + \left\lfloor \frac{\log(A/\pi_k^A)}{I_{\mathscr{B},\theta} + \mu - \varepsilon}\right\rfloor\right)^r + r2^{r-1}\Upsilon_r(\varepsilon/2, \mathscr{B}, \theta). \tag{6.46}$$

Now, note that

$$\mathsf{PFA}_\pi(T_A^{\mathbf{p},W}) \ge \sum_{i=k}^{\infty} \pi_i^A \mathsf{P}_\infty(T_A^{\mathbf{p},W} \le i) \ge \mathsf{P}_\infty(T_A^{\mathbf{p},W} \le k)\sum_{i=k}^{\infty}\pi_i^A = \mathsf{P}_\infty(T_A^{\mathbf{p},W} \le k)\Pi_{k-1}^A,$$

and hence,

$$\mathsf{P}_\infty(T_A^{\mathbf{p},W} > k) \ge 1 - \mathsf{PFA}_\pi(T_A^{\mathbf{p},W})/\Pi_{k-1}^A \ge 1 - [(A+1)\Pi_{k-1}^A]^{-1}, \quad k \in \mathbb{Z}_+. \tag{6.47}$$

It follows from (6.46) and (6.47) that

$$\mathscr{R}_{k,\mathscr{B},\theta}^r(T_A^{\mathbf{p},W}) = \frac{\mathsf{E}_{k,\mathscr{B},\theta}\left[\left(T_A^{\mathbf{p},W} - k\right)^+\right]^r}{\mathsf{P}_\infty(T_A^{\mathbf{p},W} > k)} \le \frac{\left(1 + \left\lfloor \frac{\log(A/\pi_k^A)}{I_{\mathscr{B},\theta}+\mu-\varepsilon}\right\rfloor\right)^r + r2^{r-1}\Upsilon_r(\varepsilon/2, \mathscr{B}, \theta)}{1 - 1/(A\Pi_{k-1}^A)}. \tag{6.48}$$

Since by condition $\mathbf{C}_2$, $\Upsilon_r(\varepsilon,\mathscr{B},\theta)<\infty$ for all $\mathscr{B}\in\mathscr{P}$, $\theta\in\Theta$ and $\varepsilon>0$ and, by condition $\mathbf{CP}_3^{(A)}$, $(A\Pi_{k-1}^A)^{-1}\to 0$, $|\log\pi_k^A|/\log A\to 0$ as $A\to\infty$, inequality (6.48) implies the asymptotic inequality

$$\overline{\mathscr{R}}^r_{k,\mathscr{B},\theta}(T_A^{\mathbf{p},W})\le\left(\frac{\log A}{I_{\mathscr{B},\theta}+\mu-\varepsilon}\right)^r(1+o(1)),\quad A\to\infty.$$

Since ε can be arbitrarily small, this implies the asymptotic upper bound (6.41) (for all $0<m\le r$, $\mathscr{B}\in\mathscr{P}$ and $\theta\in\Theta$). This upper bound and the lower bound (6.39) prove the asymptotic relation (6.37). The proof of (i) is complete.

The proof of part (ii). Using the inequalities (6.48) and $1-\mathsf{PFA}_\pi(T_A^{\mathbf{p},W})\ge A/(1+A)$, we obtain that for any $0<\varepsilon<I_{\mathscr{B},\theta}+\mu$

$$\overline{\mathscr{R}}^r_{\mathscr{B},\theta}(T_A^{\mathbf{p},W})=\frac{\sum_{k=0}^\infty\pi_k^A\mathsf{E}_{k,\mathscr{B},\theta}\left[(T_A^{\mathbf{p},W}-k)^+\right]^r}{1-\mathsf{PFA}_\pi(T_A^{\mathbf{p},W})}\le\frac{\sum_{k=0}^{\infty}\pi_k^A\left(1+\left\lfloor\frac{\log(A/\pi_k^A)}{I_{\mathscr{B},\theta}+\mu-\varepsilon}\right\rfloor\right)^r+r2^{r-1}\Upsilon_r(\varepsilon/2,\mathscr{B},\theta)}{A/(1+A)}. \tag{6.49}$$

By condition $\mathbf{C}_2$, $\Upsilon_r(\varepsilon,\mathscr{B},\theta)<\infty$ for any $\varepsilon>0$, $\mathscr{B}\in\mathscr{P}$, and $\theta\in\Theta$ and, by condition $\mathbf{CP}_2^{(A)}$, $\sum_{k=0}^\infty\pi_k^A|\log\pi_k^A|^r=o(|\log A|^r)$ as $A\to\infty$, which implies that for all $\mathscr{B}\in\mathscr{P}$ and $\theta\in\Theta$

$$\overline{\mathscr{R}}^r_{\mathscr{B},\theta}(T_A^{\mathbf{p},W})\le\left(\frac{\log A}{I_{\mathscr{B},\theta}+\mu-\varepsilon}\right)^r(1+o(1)),\quad A\to\infty.$$

Since ε can be arbitrarily small, the asymptotic upper bound (6.42) follows and the proof of the asymptotic approximation (6.38) is complete. □

The following theorem shows that the double-mixture detection rule $T_A^{\mathbf{p},W}$ attains the asymptotic lower bounds (6.29)–(6.30) in Theorem 6.1 for the moments of the detection delay under conditions postulated in Proposition 6.1, being therefore first-order asymptotically optimal in class $\mathbb{C}_\pi(\alpha)$ as $\alpha\to 0$ in the general non-i.i.d. case.

Theorem 6.2. *Let the prior distribution of the change point belong to class $\mathbf{C}^{(\alpha)}(\mu)$. Let $r\ge 1$ and assume that for some $0<I_{\mathscr{B},\theta}<\infty$, $\mathscr{B}\in\mathscr{P}$, $\theta\in\Theta$, right-tail and left-tail conditions $\mathbf{C}_1$ and $\mathbf{C}_2$ are satisfied. Assume that $A=A_\alpha$ is so selected that $\mathsf{PFA}_\pi(T_{A_\alpha}^{\mathbf{p},W})\le\alpha$ and $\log A_\alpha\sim|\log\alpha|$ as $\alpha\to 0$, in particular $A_\alpha=(1-\alpha)/\alpha$.*
(i) *If condition $\mathbf{CP}_3^{(\alpha)}$ holds, then $T_{A_\alpha}^{\mathbf{p},W}$ is first-order asymptotically optimal as $\alpha\to 0$ in class $\mathbb{C}_\pi(\alpha)$, minimizing conditional moments of the detection delay $\mathscr{R}^m_{k,\mathscr{B},\theta}(T)$ up to order r, i.e., for all $0<m\le r$ and all $\mathscr{B}\in\mathscr{P}$, $\theta\in\Theta$ as $\alpha\to 0$*

$$\inf_{T\in\mathbb{C}_\pi(\alpha)}\mathscr{R}^m_{k,\mathscr{B},\theta}(T)\sim\left(\frac{|\log\alpha|}{I_{\mathscr{B},\theta}+\mu}\right)^m\sim\mathscr{R}^m_{k,\mathscr{B},\theta}(T_{A_\alpha}^{\mathbf{p},W})\quad\text{for all }k\in\mathbb{Z}_+. \tag{6.50}$$

(ii) *If condition $\mathbf{CP}_2^{(\alpha)}$ holds, then $T_{A_\alpha}^{\mathbf{p},W}$ is first-order asymptotically optimal as $\alpha\to 0$ in class $\mathbb{C}_\pi(\alpha)$, minimizing moments of the detection delay $\overline{\mathscr{R}}^m_{\mathscr{B},\theta}(T)$ up to order r, i.e., for all $0<m\le r$ and all $\mathscr{B}\in\mathscr{P}$, $\theta\in\Theta$ as $\alpha\to 0$*

$$\inf_{T\in\mathbb{C}_\pi(\alpha)}\overline{\mathscr{R}}^m_{\mathscr{B},\theta}(T)\sim\left(\frac{|\log\alpha|}{I_{\mathscr{B},\theta}+\mu}\right)^m\sim\overline{\mathscr{R}}^m_{\mathscr{B},\theta}(T_{A_\alpha}^{\mathbf{p},W}). \tag{6.51}$$

Proof. The proof of part (i). If threshold A_α is so selected that $\log A_\alpha\sim|\log\alpha|$ as $\alpha\to 0$, it follows from Proposition 6.1(i) that for $0<m\le r$ and all $\mathscr{B}\in\mathscr{P}$, $\theta\in\Theta$ as $\alpha\to 0$

$$\mathscr{R}^m_{k,\mathscr{B},\theta}(T_{A_\alpha}^{\mathbf{p},W})\sim\left(\frac{|\log\alpha|}{I_{\mathscr{B},\theta}+\mu}\right)^m\quad\text{for all }k\in\mathbb{Z}_+.$$

This asymptotic approximation shows that the asymptotic lower bound (6.29) in Theorem 6.1 is attained by $T^W_{A_\alpha}$, proving the assertion (i) of the theorem.

The proof of part (ii). If threshold A_α is so selected that $\log A_\alpha \sim |\log \alpha|$ as $\alpha \to 0$, it follows from Proposition 6.1(ii) that for $0 < m \le r$ and all $\mathscr{B} \in \mathscr{P}$, $\theta \in \Theta$ as $\alpha \to 0$

$$\overline{\mathscr{R}}^m_{\mathscr{B},\theta}(T_{A_\alpha}) \sim \left(\frac{|\log \alpha|}{I_{\mathscr{B},\theta} + \mu}\right)^m.$$

This asymptotic approximation along with the asymptotic lower bound (6.30) in Theorem 6.1 proves the assertion (ii) of the theorem. □

Remark 6.1. If the prior distribution π of the change point does not depend on α, then condition $\mathbf{CP}_3^{(\alpha)}$ is satisfied automatically, so the assertion (i) of Theorem 6.2 holds under condition $\mathbf{CP}_1$ (see (3.20) in Chapter 3) and the assertion (ii) holds under condition $\mathbf{CP}_2$ (see (3.21) in Chapter 3).

6.5.2 Asymptotic Optimality of the Double-Mixture Rule $\widetilde{T}_A^{\mathbf{p},W}$

Consider now the double-mixture detection rule $\widetilde{T}_A^{\mathbf{p},W}$ defined in (6.13) and (6.14).

Note that

$$\mathsf{E}_\infty[R_{\mathbf{p},W}(n)|\mathscr{F}_{n-1}] = \sum_{\mathscr{B}\in\mathscr{P}} p_{\mathscr{B}} \int_\Theta \mathrm{d}W(\theta) + \sum_{\mathscr{B}\in\mathscr{P}} p_{\mathscr{B}} \int_\Theta R_{\mathscr{B},\theta}(n-1)\,\mathrm{d}W(\theta) = 1 + R_{\mathbf{p},W}(n-1),$$

and hence, $\{R_{\mathbf{p},W}(n)\}_{n\ge1}$ is a $(\mathsf{P}_\infty, \mathscr{F}_n)$–submartingale with mean $\mathsf{E}_\infty[R_{\mathbf{p},W}(n) = \omega + n$. Therefore, by Doob's submartingale inequality,

$$\mathsf{P}_\infty(\widetilde{T}_A^{\mathbf{p},W} \le k) \le (\omega + k)/A, \quad k = 1, 2, \dots, \tag{6.52}$$

which implies the following lemma that establishes an upper bound for the PFA of the rule $\widetilde{T}_A^{\mathbf{p},W}$.

Lemma 6.2. *For all $A > 0$ and any prior distribution of ν with finite mean $\bar{\nu} = \sum_{k=1}^\infty k\pi_k$, the PFA of the rule $\widetilde{T}_A^{\mathbf{p},W}$ satisfies the inequality*

$$\mathsf{PFA}_\pi(\widetilde{T}_A^{\mathbf{p},W}) \le \frac{\omega b + \bar{\nu}}{A}, \tag{6.53}$$

where $b = \sum_{k=1}^\infty \pi_k$, so that if $A = A_\alpha = (\omega b + \bar{\nu})/\alpha$, then $\mathsf{PFA}_\pi(\widetilde{T}_{A_\alpha}^{\mathbf{p},W}) \le \alpha$, i.e., $\widetilde{T}_{A_\alpha}^{\mathbf{p},W} \in \mathbb{C}_\pi(\alpha)$.

As before, the prior distribution may depend on the PFA α, so the mean $\bar{\nu}_\alpha = \sum_{k=0}^\infty k\pi_k^\alpha$ depends on α. We also suppose that in general the head-start $\omega = \omega_\alpha$ depends on α and may go to infinity as $\alpha \to 0$. Throughout this subsection we assume that $\omega_\alpha \to \infty$ and $\bar{\nu}_\alpha \to \infty$ with such a rate that the following condition holds:

$$\lim_{\alpha\to0} \frac{\log(\omega_\alpha + \bar{\nu}_\alpha)}{|\log \alpha|} = 0. \tag{6.54}$$

However, in the following proposition, which establishes asymptotic operating characteristics of the rule $\widetilde{T}_A^{\mathbf{p},W}$ for large threshold A regardless of the PFA constraint, we suppose that the mean of the prior distribution $\bar{\nu}_A$ and the head-start ω_A may depend on A and approach infinity as $A \to \infty$ in such a way that

$$\lim_{A\to\infty} \frac{\log(\omega_A + \bar{\nu}_A)}{\log A} = 0. \tag{6.55}$$

Proposition 6.2. *Suppose that condition (6.55) holds and there exist positive and finite numbers $I_{\mathscr{B},\theta}$, $\mathscr{B} \in \mathscr{P}$, $\theta \in \Theta$, such that right-tail and left-tail conditions $\mathbf{C}_1$ and $\mathbf{C}_2$ are satisfied. Then, for all $0 < m \le r$, $\mathscr{B} \in \mathscr{P}$, and $\theta \in \Theta$*

$$\lim_{A\to\infty} \frac{\mathscr{R}^m_{k,\mathscr{B},\theta}(\widetilde{T}^{\mathbf{p},W}_A)}{(\log A)^m} = \frac{1}{I^m_{\mathscr{B},\theta}} \quad \text{for all } k \in \mathbb{Z}_+ \tag{6.56}$$

and

$$\lim_{A\to\infty} \frac{\overline{\mathscr{R}}^m_{\mathscr{B},\theta}(\widetilde{T}^{\mathbf{p},W}_A)}{(\log A)^m} = \frac{1}{I^m_{\mathscr{B},\theta}}. \tag{6.57}$$

Proof. For $\varepsilon \in (0,1)$, let

$$M_A = M_A(\varepsilon, \mathscr{B}, \theta) = (1-\varepsilon)\frac{\log A}{I_{\mathscr{B},\theta}}.$$

Recall that

$$\mathsf{P}_{k,\mathscr{B},\theta}(\widetilde{T}^{\mathbf{p},W}_A > k) = \mathsf{P}_\infty(\widetilde{T}^{\mathbf{p},W}_A > k) \ge 1 - \frac{k+\omega_A}{A}, \quad k \in \mathbb{Z}_+$$

(see (6.52)), so using Chebyshev's inequality, we obtain

$$\begin{aligned}
\mathscr{R}^r_{k,\mathscr{B},\theta}(\widetilde{T}^{\mathbf{p},W}_A) &\ge M_A^r \mathsf{P}_{k,\mathscr{B},\theta}(\widetilde{T}^{\mathbf{p},W}_A - k > M_A) \\
&\ge M_A^r \left[\mathsf{P}_{k,\mathscr{B},\theta}(\widetilde{T}^{\mathbf{p},W}_A > k) - \mathsf{P}_{k,\mathscr{B},\theta}(k < \widetilde{T}^{\mathbf{p},W}_A < k + M_A)\right] \\
&\ge M_A^r \left[1 - \frac{\omega_A + k}{A} - \mathsf{P}_{k,\mathscr{B},\theta}(k < \widetilde{T}^{\mathbf{p},W}_A < k + M_A)\right].
\end{aligned} \tag{6.58}$$

Analogously to (3.38),

$$\mathsf{P}_{k,\mathscr{B},\theta}(k < T < k + M_A) \le U_{M_A,k}(T) + p_{M_A,k}(\varepsilon, \mathscr{B}, \theta). \tag{6.59}$$

Since

$$\mathsf{P}_\infty\left(0 < \widetilde{T}^{\mathbf{p},W}_A - k < M_A\right) \le \mathsf{P}_\infty\left(\widetilde{T}^{\mathbf{p},W}_A < k + M_A\right) \le (k + \omega_A + M_A)/A,$$

we have

$$U_{M_A,k}(\widetilde{T}^{\mathbf{p},W}_A) \le \frac{k + \omega_A + (1-\varepsilon)I^{-1}_{\mathscr{B},\theta}\log A}{A^{\varepsilon^2}}. \tag{6.60}$$

Condition (6.55) implies that $\omega_A = o(A^\gamma)$ as $A \to \infty$ for any $\gamma > 0$. Therefore, $U_{M_A,k}(\widetilde{T}^{\mathbf{p},W}_A) \to 0$ as $A \to \infty$ for any fixed k. Also, $p_{M_A,k}(\varepsilon, \mathscr{B}, \theta) \to 0$ by condition $\mathbf{C}_1$, so that $\mathsf{P}_{k,\mathscr{B},\theta}\left(0 < \widetilde{T}^{\mathbf{p},W}_A - k < M_A\right) \to 0$ for any fixed k. It follows from (6.58) that for an arbitrarily $\varepsilon \in (0,1)$ as $A \to \infty$

$$\mathscr{R}^r_{k,\mathscr{B},\theta}(\widetilde{T}^{\mathbf{p},W}_A) \ge \left(\frac{(1-\varepsilon)\log A}{I_{\mathscr{B},\theta}}\right)^r (1 + o(1)),$$

which yields the asymptotic lower bound (for any fixed $k \in \mathbb{Z}_+$, $\mathscr{B} \in \mathscr{P}$ and $\theta \in \Theta$)

$$\liminf_{A\to\infty} \frac{\mathscr{R}^r_{k,\mathscr{B},\theta}(\widetilde{T}^{\mathbf{p},W}_A)}{(\log A)^r} \ge \frac{1}{I^r_{\mathscr{B},\theta}}. \tag{6.61}$$

To prove (6.56) it suffices to show that this bound is attained by $\widetilde{T}^{\mathbf{p},W}_A$, i.e.,

$$\limsup_{A\to\infty} \frac{\mathscr{R}^r_{k,\mathscr{B},\theta}(\widetilde{T}^{\mathbf{p},W}_A)}{(\log A)^r} \le \frac{1}{I^r_{\mathscr{B},\theta}}. \tag{6.62}$$

Define

$$\widetilde{M}_A = \widetilde{M}_A(\varepsilon, \mathscr{B}, \theta) = 1 + \left\lfloor \frac{\log A}{I_{\mathscr{B},\theta} - \varepsilon} \right\rfloor.$$

By Lemma A.1, for any $k \in \mathbb{Z}_+$, $\mathscr{B} \in \mathscr{P}$, and $\theta \in \Theta$,

$$\mathsf{E}_{k,\mathscr{B},\theta}\left[(\widetilde{T}_A^{\mathbf{p},W} - k)^+\right]^r \le \widetilde{M}_A^r + r2^{r-1} \sum_{n=\widetilde{M}_A}^{\infty} n^{r-1} \mathsf{P}_{k,\mathscr{B},\theta}\left(\widetilde{T}_A^{\mathbf{p},W} > n\right), \tag{6.63}$$

and since for any $n \ge 1$,

$$\log R_{\mathbf{p},W}^{\pi}(k+n) \ge \log \Lambda_{\mathbf{p},W}(k,k+n) \ge \inf_{\vartheta \in \Gamma_{\delta,\theta}} \lambda_{\mathscr{B},\vartheta}(k,k+n) + \log W(\Gamma_{\delta,\theta}) + \log p_{\mathscr{B}},$$

in just the same way as in the proof of Proposition 6.1 (setting $\pi_k^A = 1$) we obtain that for all $n \ge \widetilde{M}_A$

$$\mathsf{P}_{k,\mathscr{B},\theta}\left(\widetilde{T}_A^{\mathbf{p},W} > n\right) \le \mathsf{P}_{k,\mathscr{B},\theta}\left\{\frac{1}{n} \inf_{\vartheta \in \Gamma_{\delta,\theta}} \lambda_{\mathscr{B},\vartheta}(k,k+n) < I_{\mathscr{B},\theta} - \varepsilon + \frac{1}{\widetilde{M}_A} |\log[p_{\mathscr{B}} W(\Gamma_{\delta,\theta})]|\right\}.$$

Hence, for all $n \ge \widetilde{M}_A$ and all sufficiently large A such that $|\log[p_{\mathscr{B}} W(\Gamma_{\delta,\theta})]|/\widetilde{M}_A < \varepsilon/2$,

$$\mathsf{P}_{k,\mathscr{B},\theta}\left(\widetilde{T}_A^{\mathbf{p},W} - k > n\right) \le \mathsf{P}_{k,\mathscr{B},\theta}\left(\frac{1}{n} \inf_{\vartheta \in \Gamma_{\delta,\theta}} \lambda_{\mathscr{B},\vartheta}(k,k+n) < I_{\mathscr{B},\theta} - \varepsilon/2\right). \tag{6.64}$$

Using (6.63) and (6.64), we obtain

$$\mathsf{E}_{k,\mathscr{B},\theta}\left[\left(\widetilde{T}_A^{\mathbf{p},W} - k\right)^+\right]^r \le \left(1 + \left\lfloor \frac{\log A}{I_{\mathscr{B},\theta} - \varepsilon} \right\rfloor\right)^r + r2^{r-1} \Upsilon_r(\varepsilon/2, \mathscr{B}, \theta), \tag{6.65}$$

which along with the inequality $\mathsf{P}_\infty(\widetilde{T}_A^{\mathbf{p},W} > k) > 1 - (\omega_A + k)/A$ (see (6.52)) implies the inequality

$$\begin{aligned} \mathscr{R}_{k,\mathscr{B},\theta}^r(\widetilde{T}_A^{\mathbf{p},W}) &= \frac{\mathsf{E}_{k,\mathscr{B},\theta}\left[\left(\widetilde{T}_A^{\mathbf{p},W} - k\right)^+\right]^r}{\mathsf{P}_\infty(\widetilde{T}_A^{\mathbf{p},W} > k)} \\ &\le \frac{\left(1 + \left\lfloor \frac{\log A}{I_{\mathscr{B},\theta} - \varepsilon} \right\rfloor\right)^r + r2^{r-1} \Upsilon_r(\varepsilon/2, \mathscr{B}, \theta)}{1 - (\omega_A + k)/A}. \end{aligned} \tag{6.66}$$

Since due to (6.55) $\omega_A/A \to 0$ and, by condition $\mathbf{C}_2$, $\Upsilon_r(\varepsilon, \mathscr{B}, \theta) < \infty$ for all $\varepsilon > 0$, $\mathscr{B} \in \mathscr{P}$, $\theta \in \Theta$, inequality (6.66) implies the asymptotic inequality

$$\mathscr{R}_{k,\mathscr{B},\theta}^r(\widetilde{T}_A^{\mathbf{p},W}) \le \left(\frac{\log A}{I_{\mathscr{B},\theta} - \varepsilon}\right)^r (1 + o(1)), \quad A \to \infty.$$

Since ε can be arbitrarily small the asymptotic upper bound (6.62) follows and the proof of the asymptotic approximation (6.56) is complete.

In order to prove (6.57) note first that, using (6.58) yields the lower bound

$$\overline{\mathscr{R}}_{\mathscr{B},\theta}^r(\widetilde{T}_A) \ge M_A^r \left[1 - \frac{\bar{\nu}_A + \omega_A}{A} - \mathsf{P}_{\mathscr{B},\theta}^{\pi}\left(0 < \widetilde{T}_A - \nu < M_A\right)\right]. \tag{6.67}$$

Let K_A be an integer number that approaches infinity as $A \to \infty$ with rate $O(A^\gamma)$, $\gamma > 0$. Now, using (6.59) and (6.60), we obtain

$$\mathsf{P}_{\mathscr{B},\theta}^{\pi}(0 < \widetilde{T}_A^{\mathbf{p},W} - \nu < M_A) = \sum_{k=0}^{\infty} \pi_k^A \mathsf{P}_{k,\mathscr{B},\theta}\left(0 < \widetilde{T}_A^{\mathbf{p},W} - k < M_A\right)$$

$$\le \mathsf{P}(\nu > K_A) + \sum_{k=0}^{\infty} \pi_k^A U_{M_A,k}(\widetilde{T}_A^{\mathbf{p},W}) + \sum_{k=0}^{K_A} \pi_k^A p_{M_A,k}(\varepsilon, \mathscr{B}, \theta)$$

$$\le \mathsf{P}(\nu > K_A) + \frac{\bar{\nu}_A + \omega_A + (1-\varepsilon) I_{\mathscr{B},\theta}^{-1} \log A}{A^{\varepsilon^2}} + \sum_{k=0}^{K_A} \pi_k^A p_{M_A,k}(\varepsilon, \mathscr{B}, \theta). \tag{6.68}$$

Note that due to (6.55) $(\omega_A + \bar{\nu}_A)/A^\gamma \to 0$ as $A \to \infty$ for any $\gamma > 0$. As a result, the first two terms in (6.68) go to zero as $A \to \infty$ (by Markov's inequality $\mathsf{P}(\nu > K_A) \le \bar{\nu}_A / K_A = \bar{\nu}_A / O(A^\gamma) \to 0$) and the last term also goes to zero by condition $\mathbf{C}_1$ and Lebesgue's dominated convergence theorem. Thus, for all $0 < \varepsilon < 1$, $\mathsf{P}_{\mathscr{B}}^{\pi}(0 < \widetilde{T}_A^{\mathbf{p},W} - \nu < M_A)$ approaches 0 as $A \to \infty$. Using inequality (6.67), we obtain that for any $0 < \varepsilon < 1$ as $A \to \infty$

$$\overline{\mathscr{R}}^r_{\mathscr{B},\theta}(\widetilde{T}_A^{\mathbf{p},W}) \ge (1-\varepsilon)^r \left(\frac{\log A}{I_{\mathscr{B},\theta}}\right)^r (1+o(1)),$$

which yields the asymptotic lower bound (for any $r > 0$, $\mathscr{B} \in \mathscr{P}$, and $\theta \in \Theta$)

$$\liminf_{A \to \infty} \frac{\overline{\mathscr{R}}^r_{\pi,\mathscr{B},\theta}(\widetilde{T}_A^{\mathbf{p},W})}{(\log A)^r} \ge \frac{1}{I^r_{\mathscr{B},\theta}}. \tag{6.69}$$

To obtain the upper bound it suffices to use inequality (6.65), which along with the fact that $\mathsf{PFA}_\pi(\widetilde{T}_A^{\mathbf{p},W}) \le (\bar{\nu}_A + \omega_A)/A$ yields (for every $0 < \varepsilon < I_{\mathscr{B},\theta}$)

$$\overline{\mathscr{R}}^r_{\mathscr{B},\theta}(\widetilde{T}_A^{\mathbf{p},W}) = \frac{\sum_{k=0}^{\infty} \pi_k^A \mathsf{E}_{k,\mathscr{B},\theta}[(\widetilde{T}_A^{\mathbf{p},W} - k)^+]^r}{1 - \mathsf{PFA}_\pi(\widetilde{T}_A^{\mathbf{p},W})} \le \frac{\left(1 + \frac{\log A}{I_{\mathscr{B},\theta} - \varepsilon}\right)^r + r 2^{r-1} \Upsilon_r(\varepsilon/2, \mathscr{B}, \theta)}{1 - (\omega_A + \bar{\nu}_A)/A}.$$

Since $(\omega_A + \bar{\nu}_A)/A \to 0$ and, by condition $\mathbf{C}_2$, $\Upsilon_r(\varepsilon, \mathscr{B}, \theta) < \infty$ for any $\varepsilon > 0$, $\mathscr{B} \in \mathscr{P}$, and $\theta \in \Theta$ we obtain that, for every $0 < \varepsilon < I_{\mathscr{B},\theta}$ as $A \to \infty$,

$$\overline{\mathscr{R}}^r_{\mathscr{B},\theta}(\widetilde{T}_A^{\mathbf{p},W}) \le \left(\frac{\log A}{I_{\mathscr{B},\theta} - \varepsilon}\right)^r (1+o(1)),$$

which implies

$$\limsup_{A \to \infty} \frac{\overline{\mathscr{R}}^r_{\mathscr{B},\theta}(\widetilde{T}_A^{\mathbf{p},W})}{(\log A)^r} \le \frac{1}{I^r_{\mathscr{B},\theta}} \tag{6.70}$$

since ε can be arbitrarily small.

Applying the bounds (6.69) and (6.70) yields (6.57). □

Using asymptotic approximations (6.56)–(6.57) in Proposition 6.2, we now can easily prove that the double-mixture rule $\widetilde{T}_A^{\mathbf{p},W}$ attains the asymptotic lower bounds (6.29)–(6.30) in Theorem 6.1 for moments of the detection delay when $\mu = 0$. This means that the rule $\widetilde{T}_{A_\alpha}^{\mathbf{p},W}$ is first-order asymptotically optimal as $\alpha \to 0$ if the prior distribution of the change point belongs to class $\mathbf{C}^{(\alpha)}(\mu = 0)$.

Theorem 6.3. *Assume that the head-start $\omega = \omega_\alpha$ of the statistic $R_{\mathbf{p},W}(n)$ and the mean value $\bar{\nu} = \bar{\nu}_\alpha$ of the prior distribution $\{\pi_k^\alpha\}$ approach infinity as $\alpha \to 0$ with such a rate that condition (6.54) is satisfied. Suppose further that for some $0 < I_{\mathscr{B},\theta} < \infty$ and $r \ge 1$ conditions $\mathbf{C}_1$ and $\mathbf{C}_2$ are satisfied. If threshold A_α is so selected that $\mathsf{PFA}_\pi(\widetilde{T}_{A_\alpha}^{\mathbf{p},W}) \le \alpha$ and $\log A_\alpha \sim |\log \alpha|$ as $\alpha \to 0$, in particular $A_\alpha = (\omega_\alpha b_\alpha + \bar{\nu}_\alpha)/\alpha$, then for all $0 < m \le r$, $\mathscr{B} \in \mathscr{P}$, and $\theta \in \Theta$ as $\alpha \to 0$*

$$\begin{aligned} \overline{\mathscr{R}}^m_{\mathscr{B},\theta}(\widetilde{T}_{A_\alpha}^{\mathbf{p},W}) &\sim \left(\frac{|\log \alpha|}{I_{\mathscr{B},\theta}}\right)^m, \\ \mathscr{R}^m_{k,\mathscr{B},\theta}(\widetilde{T}_{A_\alpha}^{\mathbf{p},W}) &\sim \left(\frac{|\log \alpha|}{I_{\mathscr{B},\theta}}\right)^m \quad \text{for all } k \in \mathbb{Z}_+. \end{aligned} \tag{6.71}$$

If the prior distribution belongs to class $\mathbf{C}^{(\alpha)}(\mu = 0)$*, then for all* $0 < m \le r$*,* $\mathscr{B} \in \mathscr{P}$*, and* $\theta \in \Theta$ *as* $\alpha \to 0$

$$\begin{aligned} &\inf_{T \in \mathbb{C}_\pi(\alpha)} \overline{\mathscr{R}}^m_{\mathscr{B},\theta}(T) \sim \overline{\mathscr{R}}^m_{\mathscr{B},\theta}(\widetilde{T}^{\mathbf{p},W}_{A_\alpha}), \\ &\inf_{T \in \mathbb{C}_\pi(\alpha)} \mathscr{R}^m_{k,\mathscr{B},\theta}(T) \sim \mathscr{R}^m_{k,\mathscr{B},\theta}(\widetilde{T}^{\mathbf{p},W}_{A_\alpha}) \quad \text{for all } k \in \mathbb{Z}_+. \end{aligned} \tag{6.72}$$

Therefore, the rule $\widetilde{T}^{\mathbf{p},W}_{A_\alpha}$ *is asymptotically optimal as* $\alpha \to 0$ *in class* $\mathbb{C}_\pi(\alpha)$*, minimizing moments of the detection delay up to order r, if the prior distribution of the change point belongs to class* $\mathbf{C}^{(\alpha)}(\mu)$ *with* $\mu = 0$*.*

Proof. If A_α is so selected that $\log A_\alpha \sim |\log \alpha|$ as $\alpha \to 0$, then asymptotic approximations (6.71) follow immediately from asymptotic approximations (6.56)–(6.57) in Proposition 6.2. Since these approximations are the same as the asymptotic lower bounds (6.29)–(6.30) in Theorem 6.1 for $\mu = 0$, this shows that these bounds are attained by the detection rule $\widetilde{T}^{\mathbf{p},W}_{A_\alpha}$ whenever the prior distribution belongs to class $\mathbf{C}^{(\alpha)}(\mu = 0)$, which completes the proof of assertions (6.72). □

Obviously, the assertions of Theorem 6.3 hold when $\bar{\nu}$ and ω do not depend on α. Thus, the double MSR rule $\widetilde{T}^{\mathbf{p},W}_{A_\alpha}$ is asymptotically optimal for heavy-tailed priors (when $\mu = 0$ in (3.20)). Also, comparing (6.71) with the assertion of Theorem 6.2 (see (6.50) and (6.51)) we can see that the rule $\widetilde{T}^{\mathbf{p},W}_{A_\alpha}$ is not asymptotically optimal when $\mu > 0$, i.e., for the priors with asymptotic exponential tails. But it is asymptotically optimal for priors with exponential tails when the parameter μ is small. This can be expected since the statistic $R_{\mathbf{p},W}(n)$ uses the uniform prior distribution of the change point.

6.6 Asymptotic Optimality with Respect to the Average Risk

In this section, instead of the constrained optimization problem (6.20) we are interested in the unconstrained Bayes problem (6.24) with the average (integrated) risk function $\rho^{c,r}_{\pi,\mathbf{p},W}(T)$ defined in (6.23), where $c > 0$ is the cost of delay per time unit and $r \ge 1$. Below we show that the double-mixture rule $T^{\mathbf{p},W}_A$ with a certain threshold $A = A_{c,r}$ that depends on the cost c is asymptotically optimal, minimizing the average risk $\rho^{c,r}_{\pi,\mathbf{p},W}(T)$ to first order over all stopping times as the cost vanishes, $c \to 0$.

Let $\mathbf{C}^{(c)}(\mu)$ denote the class of prior distributions $\pi^c = \{\pi^c_k\}$ that satisfy

$$\lim_{c \to 0} \frac{1}{n} |\log \Pi^c_n| = \mu_c, \quad \lim_{c \to 0} \mu_c = \mu \quad (\mu_c, \mu \ge 0).$$

Define

$$G_{c,r}(A) = \frac{1}{A} + D_{\mu,r} c (\log A)^r \tag{6.73}$$

and recall that $\overline{\mathscr{R}}^r_{\mathbf{p},W}(T)$, $D_{\mu,r}$ are defined in (6.27), (6.28), respectively. Since $\mathsf{PFA}_\pi(T^{\mathbf{p},W}_A) \approx 1/(1+A)$ (ignoring an excess over the boundary), using the asymptotic formula (6.38) we obtain that for a large A

$$\overline{\mathscr{R}}^r_{\mathbf{p},W}(T^{\mathbf{p},W}_A) \approx \sum_{\mathscr{B} \in \mathscr{P}} p_{\mathscr{B}} \int_\Theta \left(\frac{\log A}{I_{\mathscr{B},\theta} + \mu} \right)^r \mathrm{d}W(\theta) = (\log A)^r D_{\mu,r}.$$

So for large A the average risk of the rule $T_A^{\mathbf{p},W}$ is approximately equal to

$$\rho_{\pi,\mathbf{p},W}^{c,r}(T_A^{\mathbf{p},W}) = \mathsf{PFA}_\pi(T_A^{\mathbf{p},W}) + c\,[1-\mathsf{PFA}_\pi(T_A^{\mathbf{p},W})]\overline{\mathscr{R}}_{\mathbf{p},W}^{r}(T_A^{\mathbf{p},W}) \approx G_{c,r}(A).$$

The rule $T_{A_{c,r}}^{\mathbf{p},W}$ with the threshold value $A = A_{c,r}$ that minimizes $G_{c,r}(A)$, $A > 0$, which is a solution of the equation (6.75) (see below), is a reasonable candidate for being asymptotically optimal in the Bayesian sense as $c \to 0$, i.e., in the asymptotic problem (6.24). The next theorem shows that this is true under conditions $\mathbf{C}_1$ and $\mathbf{C}_2$ when the set Θ is compact and that the same is true for the rule $\widetilde{T}_{A_{c,r}}^{\mathbf{p},W}$ with certain threshold $A_{c,r}$ in class of priors $\mathbf{C}^{(c)}(\mu)$ with $\mu = 0$.

Theorem 6.4. *Assume that for some $0 < I_{\mathscr{B},\theta} < \infty$, $\mathscr{B} \in \mathscr{P}$, $\theta \in \Theta$, right-tail and left-tail conditions $\mathbf{C}_1$ and $\mathbf{C}_2$ are satisfied and that Θ is a compact set.*

(i) *If the prior distribution of the change point $\pi^c = \{\pi_k^c\}$ belongs to class $\mathbf{C}^{(c)}(\mu)$ and*

$$\lim_{c\to 0} \frac{\sum_{k=0}^{\infty} \pi_k^c |\log \pi_k^c|^r}{|\log c|^r} = 0, \tag{6.74}$$

and if threshold $A = A_{c,r}$ of the rule $T_A^{\mathbf{p},W}$ is the solution of the equation

$$r D_{\mu,r} A (\log A)^{r-1} = 1/c, \tag{6.75}$$

then

$$\inf_{T\in\mathscr{M}} \rho_{\pi,\mathbf{p},W}^{c,r}(T) \sim D_{\mu,r}\, c\, |\log c|^r \sim \rho_{\pi,\mathbf{p},W}^{c,r}(T_{A_{c,r}}^{\mathbf{p},W}) \quad \text{as } c \to 0, \tag{6.76}$$

i.e., $T_{A_{c,r}}^{\mathbf{p},W}$ is first-order asymptotically Bayes as $c \to 0$.

(ii) *If the head-start ω_c and the mean of the prior distribution $\bar{\nu}_c$ approach infinity at such rate that*

$$\lim_{c\to 0} \frac{\log(\omega_c + \bar{\nu}_c)}{|\log c|} = 0 \tag{6.77}$$

and if threshold $A = A_{c,r}$ of the rule $\widetilde{T}_A^{\mathbf{p},W}$ is the solution of the equation

$$r D_{0,r} A (\log A)^{r-1} = (\omega_c b_c + \bar{\nu}_c)/c, \tag{6.78}$$

then

$$\rho_{\pi,\mathbf{p},W}^{c,r}(\widetilde{T}_{A_{c,r}}^{\mathbf{p},W}) \sim D_{0,r}\, c\, |\log c|^r \quad \text{as } c \to 0, \tag{6.79}$$

i.e., $\widetilde{T}_{A_{c,r}}^{\mathbf{p},W}$ is first-order asymptotically Bayes as $c \to 0$ in the class of priors $\mathbf{C}(\mu = 0)$.

Proof. The proof of part (i). Since Θ is compact it follows from Proposition 6.1(ii) (cf. the asymptotic approximation (6.38)) that under conditions $\mathbf{C}_1$, $\mathbf{C}_2$, and (6.74) as $A \to \infty$

$$\overline{\mathscr{R}}_{\mathbf{p},W}^{r}(T_A^{\mathbf{p},W}) \sim \sum_{\mathscr{B}\in\mathscr{P}} p_{\mathscr{B}} \int_{\Theta} \left(\frac{\log A}{I_{\mathscr{B},\theta} + \mu}\right)^r \mathrm{d}W(\theta) = D_{\mu,r}(\log A)^r.$$

By Lemma 6.1, $\mathsf{PFA}_\pi(T_A^{\mathbf{p},W}) \le 1/(A+1)$. Obviously, if threshold $A_{c,r}$ satisfies equation (6.75), then $\log A_{c,r} \sim |\log c|$ and $\mathsf{PFA}_\pi(T_{A_{c,r}}^{W}) = o(c|\log c|^r)$ as $c \to 0$ (for any $r \ge 1$ and $\mu \ge 0$). As a result,

$$\rho_{\pi,\mathbf{p},W}^{c,r}(T_{A_{c,r}}^{\mathbf{p},W}) \sim D_{\mu,r}\, c\, |\log c|^r \quad \text{as } c \to 0,$$

which along with the lower bound (6.31) in Theorem 6.1 completes the proof of assertion (i).

The proof of part (ii). Since Θ is compact it follows from the asymptotic approximation (6.57) in Proposition 6.2 that under conditions $\mathbf{C}_1$, $\mathbf{C}_2$, and (6.77) as $A \to \infty$

$$\overline{\mathscr{R}}^r_{\mathbf{p},W}(\widetilde{T}_A^{\mathbf{p},W}) \sim \sum_{\mathscr{B}\in\mathscr{P}} p_{\mathscr{B}} \int_\Theta \left(\frac{\log A}{I_{\mathscr{B},\theta}}\right)^r \mathrm{d}W(\theta) = D_{0,r}(\log A)^r.$$

Define

$$\widetilde{G}_{c,r}(A) = (\omega_c b_c + \bar{\nu}_c)/A + c D_{0,r}(\log A)^r.$$

By Lemma 6.2, $\mathsf{PFA}_\pi(\widetilde{T}_A^{\mathbf{p},W}) \le (\omega_c b_c + \bar{\nu}_c)/A$, so that for a sufficiently large A,

$$\rho^{c,r}_{\pi,\mathbf{p},W}(\widetilde{T}_A^{\mathbf{p},W}) \approx \widetilde{G}_{c,r}(A).$$

Threshold $A_{c,r}$, which satisfies equation (6.78), minimizes $\widetilde{G}_{c,r}(A)$. By assumption (6.77), $\omega_c b_c + \bar{\nu}_c = o(|\log c|)$ as $c \to 0$, so that $\log A_{c,r} \sim |\log c|$ and $\mathsf{PFA}_\pi(\widetilde{T}_{A_{c,r}}^{\mathbf{p},W}) \le (\omega_c b_c + \bar{\nu}_c)/A_{c,r} = o(c|\log c|^r)$ as $c \to 0$. Hence, it follows that

$$\rho^{c,r}_{\pi,\mathbf{p},W}(\widetilde{T}_{A_{c,r}}^{\mathbf{p},W}) \sim \widetilde{G}_{c,r}(A_{c,r}) \sim D_{0,r}\, c\,|\log c|^r \quad \text{as } c \to 0.$$

This implies the asymptotic approximation (6.79). Asymptotic optimality of $\widetilde{T}_{A_{c,r}}^{\mathbf{p},W}$ in the class of priors $\mathbf{C}(\mu = 0)$ follows from (6.76) and (6.79). □

6.7 Asymptotic Optimality for a Putative Value of the Post-Change Parameter

If the value of the post-change parameter $\theta = \vartheta$ is known or its putative value ϑ is of special interest, representing a nominal change, then it is reasonable to turn the double-mixture rules $T_A^{\mathbf{p},W}$ and $\widetilde{T}_A^{\mathbf{p},W}$ in single-mixture rules $T_A^{\mathbf{p},\vartheta}$ and $\widetilde{T}_A^{\mathbf{p},\vartheta}$ by taking the degenerate weight function W concentrated at ϑ. These rules are of the form

$$T_A^{\mathbf{p},\vartheta} = \inf\left\{n \ge 1 : S^\pi_{\mathbf{p},\vartheta}(n) \ge A\right\}, \quad \widetilde{T}_A^{\mathbf{p},\vartheta} = \inf\left\{n \ge 1 : R_{\mathbf{p},\vartheta}(n) \ge A\right\},$$

and they have first-order asymptotic optimality properties at the point $\theta = \vartheta$ (and only at this point) with respect to $\mathscr{R}_{k,\mathscr{B},\vartheta}(T)$ and $\overline{\mathscr{R}}_{\mathscr{B},\vartheta}(T)$ when the right-tail condition $\mathbf{C}_1$ is satisfied for $\theta = \vartheta$ and the following left-tail condition holds:

$\widetilde{\mathbf{C}}_2$. *For every* $\mathscr{B} \in \mathscr{P}$, $\varepsilon > 0$, *and for some* $r \ge 1$

$$\sum_{n=1}^{\infty} n^{r-1} \sup_{k\in\mathbb{Z}_+} \mathsf{P}_{k,\mathscr{B},\vartheta}\left(\frac{1}{n}\lambda_{\mathscr{B},\vartheta}(k,k+n) < I_{\mathscr{B},\vartheta} - \varepsilon\right) < \infty. \tag{6.80}$$

The assertions of Theorem 6.4 also hold under conditions $\mathbf{C}_1$ and $\widetilde{\mathbf{C}}_2$ for the average risk

$$\rho^{c,r}_{\pi,\mathbf{p},\vartheta}(T) = \mathsf{PFA}_\pi(T) + c \sum_{\mathscr{B}\in\mathscr{P}} p_{\mathscr{B}} \mathsf{E}^\pi_{\mathscr{B},\vartheta}[(T-\nu)^+]^r$$

with

$$D_{\mu,r} = \sum_{\mathscr{B}\in\mathscr{P}} p_{\mathscr{B}} \left(\frac{1}{I_{\mathscr{B},\vartheta} + \mu}\right)^r.$$

6.8 Asymptotic Optimality in the Case of Independent Streams

A particular, still very general scenario where the data streams are mutually independent (but have a general statistical structure) is of special interest for many applications. In this case, the model is given by (6.15) and, as discussed in Subsection 6.2.2.2, the implementation of detection rules may be feasible since the LR process $\Lambda_{\mathbf{p},\theta}(k,n)$ can be easily computed (see (6.17)). Recall that we write θ_i for a post-change parameter in the ith stream and bold $\boldsymbol{\theta}_{\mathscr{B}} = (\theta_i, i \in \mathscr{B}) \in \boldsymbol{\Theta}_{\mathscr{B}}$ for the vector of post-change parameters in the subset of streams $\mathscr{B}$.

Since the data are independent across streams, for an assumed value of the change point $\nu = k$, stream $i \in \mathscr{N}$, and the post-change parameter value in the ith stream θ_i, the LLR of observations accumulated by time $k+n$ is given by

$$\lambda_{i,\theta_i}(k,k+n) = \sum_{t=k+1}^{k+n} \log \frac{f_{i,\theta_i}(X_t(i)|\mathbf{X}^{t-1}(i))}{g_i(X_t(i))|\mathbf{X}^{t-1}(i))}, \quad n \geq 1.$$

Let

$$p_{M,k}(\varepsilon,i,\theta_i) = \mathsf{P}_{k,i,\theta_i}\left\{\frac{1}{M}\max_{1\leq n\leq M} \lambda_{i,\theta_i}(k,k+n) \geq (1+\varepsilon)I_{i,\theta_i}\right\},$$

$$\Upsilon_r(\varepsilon,i,\theta_i) = \lim_{\delta\to 0}\sum_{n=1}^{\infty} n^{r-1} \sup_{k\in\mathbb{Z}_+} \mathsf{P}_{k,i,\theta_i}\left\{\frac{1}{n}\inf_{\vartheta\in\Gamma_{\delta,\theta_i}} \lambda_{i,\vartheta}(k,k+n) < I_{i,\theta_i} - \varepsilon\right\},$$

where $\Gamma_{\delta,\theta_i} = \{\vartheta \in \Theta_i : |\vartheta - \theta_i| < \delta\}$.

Assume that the following conditions are satisfied for local statistics in data streams:

$\mathbf{C}_1^{(i)}$. *There exist positive and finite numbers I_{i,θ_i}, $\theta_i \in \Theta_i$, $i \in \mathscr{N}$, such that for any $\varepsilon > 0$*

$$\lim_{M\to\infty} p_{M,k}(\varepsilon,i,\theta_i) = 0 \quad \text{for all } k \in \mathbb{Z}_+, \theta_i \in \Theta_i, i \in \mathscr{N}; \tag{6.81}$$

$\mathbf{C}_2^{(i)}$. *For any $\varepsilon > 0$ and some $r \geq 1$*

$$\Upsilon_r(\varepsilon,i,\theta_i) < \infty \quad \text{for all } \theta_i \in \Theta_i, i \in \mathscr{N}. \tag{6.82}$$

As before, we assume that $W(\Gamma_{\delta,\theta_i}) > 0$ for all $\delta > 0$ and $i \in \mathscr{N}$.

Let $I_{\mathscr{B},\boldsymbol{\theta}_{\mathscr{B}}} = \sum_{i\in\mathscr{B}} I_{i,\theta_i}$. Since the LLR process $\lambda_{\mathscr{B},\boldsymbol{\theta}_{\mathscr{B}}}(k,k+n)$ is the sum of independent local LLRs, $\lambda_{\mathscr{B},\boldsymbol{\theta}_{\mathscr{B}}}(k,k+n) = \sum_{i\in\mathscr{B}} \lambda_{i,\theta_i}(k,k+n)$ (see (6.16)), it is easy to show that

$$p_{M,k}(\varepsilon,\mathscr{B},\boldsymbol{\theta}_{\mathscr{B}}) \leq \sum_{i\in\mathscr{B}} p_{M,k}(\varepsilon,i,\theta_i),$$

so that local conditions $\mathbf{C}_1^{(i)}$ imply global right-tail condition $\mathbf{C}_1$. This is true, in particular, if the normalized local LLRs $n^{-1}\lambda_{i,\theta_i}(k,k+n)$ converge $\mathsf{P}_{k,i,\theta_i}$-a.s. to I_{i,θ_i}, $i = 1,\dots,N$, in which case the SLLN for the global LLR (6.89) holds with $I_{\mathscr{B},\theta_{\mathscr{B}}} = \sum_{i\in\mathscr{B}} I_{i,\theta_i}$. Also,

$$\Upsilon_r(\varepsilon,\mathscr{B},\boldsymbol{\theta}_{\mathscr{B}}) \leq \sum_{i\in\mathscr{B}} \Upsilon_r(\varepsilon,\mathscr{B},\theta_i),$$

which shows that local left-tail conditions $\mathbf{C}_2^{(i)}$ imply global left-tail condition $\mathbf{C}_2$.

Thus, Theorem 6.2 and Theorem 6.3 imply the following results on asymptotic properties of the double-mixture rules $T_A^{\mathbf{p},W}$ and $\widetilde{T}_A^{\mathbf{p},W}$.

Corollary 6.1. *Let $r \geq 1$ and assume that for some positive and finite numbers I_{i,θ_i}, $\theta_i \in \Theta_i$, $i = 1\dots,N$, right-tail and left-tail conditions $\mathbf{C}_1^{(i)}$ and $\mathbf{C}_2^{(i)}$ for local data streams are satisfied.*

(i) *Let the prior distribution of the change point belong to class $\mathbf{C}^{(\alpha)}(\mu)$. If $A = A_\alpha$ is so selected that $\mathsf{PFA}_\pi(T_{A_\alpha}^{\mathbf{p},W}) \leq \alpha$ and $\log A_\alpha \sim |\log\alpha|$ as $\alpha \to 0$, in particular $A_\alpha = (1-\alpha)/\alpha$, and if conditions $\mathbf{CP}_2^{(\alpha)}$ and $\mathbf{CP}_3^{(\alpha)}$ are satisfied, then asymptotic formulas (6.50) and (6.51) hold with $I_{\mathscr{B},\theta} = I_{\mathscr{B},\boldsymbol{\theta}_\mathscr{B}} = \sum_{i\in\mathscr{B}} I_{i,\theta_i}$, and therefore, $T_{A_\alpha}^{\mathbf{p},W}$ is first-order asymptotically optimal as $\alpha \to 0$ in class $\mathbb{C}_\pi(\alpha)$, minimizing moments of the detection delay up to order r uniformly for all $\mathscr{B} \in \mathscr{P}$ and $\boldsymbol{\theta}_\mathscr{B} \in \boldsymbol{\Theta}_\mathscr{B}$.*

(ii) *If threshold A_α is so selected that $\mathsf{PFA}_\pi(\widetilde{T}_{A_\alpha}^{\mathbf{p},W}) \leq \alpha$ and $\log A_\alpha \sim |\log\alpha|$ as $\alpha \to 0$, in particular $A_\alpha = (\omega_\alpha b_\alpha + \bar{\nu}_\alpha)/\alpha$, and if condition (6.54) is satisfied, then asymptotic formulas (6.71) and (6.72) hold with $I_{\mathscr{B},\theta} = I_{\mathscr{B},\boldsymbol{\theta}_\mathscr{B}} = \sum_{i\in\mathscr{B}} I_{i,\theta_i}$, and therefore, $\widetilde{T}_{A_\alpha}^{\mathbf{p},W}$ is first-order asymptotically optimal as $\alpha \to 0$ in class $\mathbb{C}_\pi(\alpha)$, minimizing moments of the detection delay up to order r uniformly for all $\mathscr{B} \in \mathscr{P}$ and $\boldsymbol{\theta}_\mathscr{B} \in \boldsymbol{\Theta}_\mathscr{B}$ if the prior distribution of the change point belongs to class $\mathbf{C}^{(\alpha)}(\mu)$ with $\mu = 0$.*

Remark 6.2. The assertions of Corollary 6.1 also hold for different distribution functions W_i, $i \in \mathscr{N}$ in streams if we assume that $W_i(\Gamma_{\delta,\theta_i}) > 0$ for all $\delta > 0$ and $i \in \mathscr{N}$. A modification in the proof is trivial.

Remark 6.3. Obviously, the following condition implies condition $\mathbf{C}_2^{(i)}$:
$\mathbf{C}_3^{(i)}$. *Let the $\Theta_i \to \mathbb{R}_+$ functions $I_i(\theta_i) = I_{i,\theta_i}$ be continuous and assume that for every compact set $\Theta_{c,i} \subseteq \Theta_i$, every $\varepsilon > 0$, and some $r \geq 1$*

$$\Upsilon_r^*(\varepsilon, i, \Theta_{c,i}) := \sup_{\theta\in\Theta_{c,i}} \Upsilon_r(\varepsilon, i, \theta_i) < \infty \quad \text{for all } i \in \mathscr{N}. \tag{6.83}$$

Define

$$\Upsilon_r^{**}(\varepsilon, \Theta_{c,i}) := \sum_{n=1}^{\infty} n^{r-1} \sup_{k\in\mathbb{Z}_+} \sup_{\theta_i\in\Theta_{c,i}} \mathsf{P}_{k,i,\theta_i}\left(\sup_{\vartheta_i\in\Theta_{c,i}} \left|\frac{1}{n}\lambda_{i,\vartheta_i}(k,k+n) - I_i(\vartheta_i,\theta_i)\right| > \varepsilon\right).$$

Note also that if there exists continuous $\Theta_i \times \Theta_i \to \mathbb{R}_+$ functions $I_i(\vartheta_i, \theta_i)$ such that for any $\varepsilon > 0$, any compact $\Theta_{c,i} \subseteq \Theta_i$ and some $r \geq 1$

$$\Upsilon_r^{**}(\varepsilon, \Theta_{c,i}) < \infty \quad \text{for all } i \in \mathscr{N}, \tag{6.84}$$

then conditions $\mathbf{C}_3^{(i)}$, and hence, conditions $\mathbf{C}_2^{(i)}$ are satisfied with $I_{\mathscr{B},\boldsymbol{\theta}_\mathscr{B}} = \sum_{i\in\mathscr{B}} I_i(\vartheta_i, \theta_i)$ since

$$\mathsf{P}_{k,i,\theta_i}\left(\frac{1}{n}\inf_{|\vartheta_i-\theta_i|<\delta} \lambda_{i,\vartheta_i}(k,k+n) < I_{i,\theta_i} - \varepsilon\right) \leq \mathsf{P}_{k,i,\theta_i}\left(\sup_{\vartheta_i\in\Theta_{c,i}} \left|\frac{1}{n}\lambda_{i,\vartheta_i}(k,k+n) - I_i(\vartheta_i,\theta_i)\right| > \varepsilon\right).$$

Conditions $\mathbf{C}_3^{(i)}$ and (6.84) are useful for establishing asymptotic optimality of proposed detection rules in particular examples.

6.9 Examples

Example 6.1 (*Detection of signals with unknown amplitudes in a multichannel system*)**.** In this subsection, we consider the N-channel quickest detection problem, which is an interesting real-world example, arising in multichannel radar systems and electro-optic imaging systems where it is required to detect an unknown number of randomly appearing signals from objects in clutter and noise (cf., e.g., [4, 162, 164]).

Specifically, we are interested in the quickest detection of deterministic signals $\theta_i S_{i,n}$ with unknown amplitudes $\theta_i > 0$ that appear at an unknown time ν in additive noises $\xi_{i,n}$ in an N-channel system, i.e., observations in the ith channel have the form

$$X_n(i) = \theta_i S_{i,n} \mathbb{1}_{\{n>\nu\}} + \xi_{i,n}, \quad n \geq 1,\ i = 1,\dots,N.$$

Assume that mutually independent noise processes $\{\xi_{i,n}\}_{n\in\mathbb{Z}_+}$ are pth order Gaussian autoregressive processes AR(p) that obey recursions

$$\xi_{i,n} = \sum_{j=1}^{p} \beta_{i,j} \xi_{i,n-j} + w_{i,n}, \quad n \geq 1, \tag{6.85}$$

where $\{w_{i,n}\}_{n\geq 1}$, $i = 1,\dots,N$, are mutually independent i.i.d. normal $\mathcal{N}(0,\sigma_i^2)$ sequences ($\sigma_i > 0$), so the observations in channels $X_{1,n},\dots,X_{N,n}$ are independent of each other. The initial values $\xi_{i,1-p},\xi_{i,2-p},\dots,\xi_{i,0}$ are arbitrarily random or deterministic numbers; in particular we may set zero initial conditions $\xi_{i,1-p} = \xi_{i,2-p} = \cdots = \xi_{i,0} = 0$. The coefficients $\beta_{i,1},\dots,\beta_{i,p}$ and variances σ_i^2 are known and all roots of the equation $z^p - \beta_{i,1} z^{p-1} - \cdots - \beta_{i,p} = 0$ are in the interior of the unit circle, so that the AR(p) processes are stable.

Define the p_n-th order residual

$$\widetilde{Y}_{i,n} = Y_{i,n} - \sum_{j=1}^{p_n} \beta_{i,j} Y_{i,n-j}, \quad n \geq 1,$$

where $p_n = p$ if $n > p$ and $p_n = n$ if $n \leq p$. It is easy to see that the conditional pre-change and post-change densities in the ith channel are

$$g_i(X_n(i)|\mathbf{X}^{n-1}(i)) = f_{0,i}(X_n(i)|\mathbf{X}^{n-1}(i)) = \frac{1}{\sqrt{2\pi\sigma_i^2}} \exp\left\{-\frac{\widetilde{X}_n(i)^2}{2\sigma_i^2}\right\},$$

$$f_{\theta_i}(X_n(i)|\mathbf{X}^{n-1}(i)) = \frac{1}{\sqrt{2\pi\sigma_i^2}} \exp\left\{-\frac{(\widetilde{X}_n(i) - \theta_i \widetilde{S}_{i,n})^2}{2\sigma_i^2}\right\}, \quad \theta_i \in \Theta = (0,\infty),$$

and that for all $k \in \mathbb{Z}_+$ and $n \geq 1$ the LLR in the ith channel has the form

$$\lambda_{i,\theta_i}(k,k+n) = \frac{\theta_i}{\sigma_i^2} \sum_{j=k+1}^{k+n} \widetilde{S}_{i,j} \widetilde{X}_{i,j} - \frac{\theta_i^2 \sum_{j=k+1}^{k+n} \widetilde{S}_{i,j}^2}{2\sigma_i^2}.$$

Since under measure $\mathsf{P}_{k,i,\vartheta_i}$ the random variables $\{\widetilde{X}_{i,n}\}_{n\geq k+1}$ are independent Gaussian random variables $\mathcal{N}(\vartheta_i \widetilde{S}_{i,n},\sigma_i^2)$, under $\mathsf{P}_{k,i,\vartheta_i}$ the LLR $\lambda_{i,\theta_i}(k,k+n)$ is a Gaussian process (with independent non-identically distributed increments) with mean and variance

$$\mathsf{E}_{k,i,\vartheta_i}[\lambda_{i,\theta_i}(k,k+n)] = \frac{2\theta_i\vartheta_i - \theta_i^2}{2\sigma_i^2} \sum_{j=k+1}^{k+n} \widetilde{S}_{i,j}^2, \quad \mathsf{Var}_{k,i,\vartheta_i}[\lambda_{i,\theta_i}(k,k+n)] = \frac{\theta_i^2}{\sigma_i^2} \sum_{j=k+1}^{k+n} \widetilde{S}_{i,j}^2. \tag{6.86}$$

Assume that

$$\lim_{n\to\infty} \frac{1}{n} \sup_{k\in\mathbb{Z}_+} \sum_{j=k+1}^{k+n} \widetilde{S}_{i,j}^2 = Q_i,$$

where $0 < Q_i < \infty$. This is typically the case in most signal processing applications, e.g., in radar applications where the signals $\theta_i S_{i,n}$ are the sequences of harmonic pulses. Then for all $k \in \mathbb{Z}_+$ and $\theta_i \in (0, \infty)$

$$\frac{1}{n}\lambda_{i,\theta_i}(k, k+n) \xrightarrow[n\to\infty]{\mathsf{P}_{k,i,\theta_i}-\text{a.s.}} \frac{\theta_i^2 Q_i}{2\sigma_i^2} = I_{i,\theta_i},$$

so that condition $\mathbf{C}_1^{(i)}$ holds (cf. Lemma B.1 in Appendix B). Furthermore, since all moments of the LLR are finite it can be shown, as in Example 3.5 in Chapter 3, that condition $\widetilde{\mathbf{C}}_2^{(i)}$ (and hence, condition $\mathbf{C}_2^{(i)}$) holds for all $r \geq 1$. Thus, by Corollary 6.1, the double-mixture rule $T_A^{\mathbf{p},W}$ minimizes as $\alpha \to 0$ all positive moments of the detection delay and asymptotic formulas (6.50) and (6.51) hold with

$$I_{\mathscr{B},\boldsymbol{\theta}_\mathscr{B}} = \sum_{i\in\mathscr{B}} \frac{\theta_i^2 Q_i}{2\sigma_i^2}.$$

Example 6.2 (*Detection of non-additive changes in mixtures*). Assume that the observations across streams are independent. Let $p_{1,i}(X_n(i))$, $p_{2,i}(X_n(i))$ and $f_{\theta_i}(X_n(i))$ be distinct densities, $i = 1, \dots, N$. Consider an example with non-additive changes where the observations in the ith stream in the normal mode follow the pre-change joint density

$$g_i(\mathbf{X}^n(i)) = \beta_i \prod_{j=1}^n p_{1,i}(X_j(i)) + (1-\beta_i) \prod_{j=1}^n p_{2,i}(X_j(i)),$$

which is the mixture density with a mixing probability $0 < \beta_i < 1$, and in the abnormal mode the observations follow the post-change joint density

$$f_{\theta_i}(\mathbf{X}^n(i)) = \prod_{j=1}^n f_{\theta_i}(X_j(i)), \quad \theta_i \in \Theta_i.$$

Therefore, the observations $\{X_n(i)\}_{n\geq 1}$ in the ith stream are dependent with the conditional probability density

$$g_i(X_n(i)) \mid \mathbf{X}^{n-1}(i)) = \frac{\beta_i \prod_{j=1}^n p_{1,i}(X_j(i)) + (1-\beta_i)\prod_{j=1}^n p_{2,i}(X_j(i))}{\beta_i \prod_{j=1}^{n-1} p_{1,i}(X_j(i)) + (1-\beta_i)\prod_{j=1}^{n-1} p_{2,i}(X_j(i))}, \quad \nu > n$$

before the change occurs and i.i.d. with density $f_{\theta_i}(X_n(i))$ after the change occurs ($n \geq \nu$). Note that in contrast to the previous example, pre-change densities g_i do not belong to the same parametric family as post-change densities f_{θ_i}.

Define $\mathscr{L}_{i,n}^{(s)}(\theta_i) = \log[f_{\theta_i}(X_n(i))/p_{s,i}(X_n(i))]$; $I_{\theta_i}^{(s)} = \mathsf{E}_{0,\theta_i}[\mathscr{L}_{i,1}^{(s)}(\theta_i)]$, $s = 1,2$; $\Delta G_{i,n} = p_{1,i}(X_n(i))/p_{2,i}(X_n(i))$; $G_{i,n} = \prod_{j=1}^n \Delta G_{i,j}$; and $\mathrm{v}_i = \beta_i/(1-\beta_i)$. It is easily seen that

$$\frac{f_{\theta_i}(X_n(i))}{g_i(X_n(i) \mid \mathbf{X}^{n-1}(i))} = \exp\left\{\mathscr{L}_{i,n}^{(2)}(\theta_i)\right\} \frac{1-\beta_i+\beta_i G_{i,n-1}}{1-\beta_i+\beta_i G_{i,n}}.$$

Observing that

$$\prod_{j=k+1}^{k+n} \frac{1-\beta_i+\beta_i G_{i,j-1}}{1-\beta_i+\beta_i G_{i,j}} = \frac{1+\mathrm{v}_i G_{i,k}}{1+\mathrm{v}_i G_{i,k+n}},$$

we obtain

$$\prod_{j=k+1}^{k+n} \frac{f_{\theta_i}(X_j(i))}{g_i(X_j(i) \mid \mathbf{X}^{j-1}(i))} = \exp\left\{\sum_{j=k+1}^{k+n} \mathscr{L}_{i,j}^{(2)}(\theta_i)\right\} \frac{1+\mathrm{v}_i G_{i,k}}{1+\mathrm{v}_i G_{i,k+n}},$$

and therefore,

$$\lambda_{i,\theta_i}(k, k+n) = \sum_{j=k+1}^{k+n} \mathscr{L}_{i,j}^{(2)}(\theta_i) + \log \frac{1+\mathrm{v}_i G_{i,k}}{1+\mathrm{v}_i G_{i,k+n}}. \tag{6.87}$$

Assume that $I_{\theta_i}^{(1)} > I_{\theta_i}^{(2)}$. Then $\mathsf{E}_{k,i,\theta_i}[\log \Delta G_{in}] = I_{\theta_i}^{(2)} - I_{\theta_i}^{(1)} < 0$ for $k \le n$, and hence, for all $k \in \mathbb{Z}_+$

$$G_{i,k+n} = G_{i,k} \prod_{j=k+1}^{k+n} \Delta G_{i,j} \xrightarrow[n\to\infty]{\mathsf{P}_{k,i,\theta_i}\text{-a.s.}} 0$$

and

$$\frac{1}{n} \log \frac{1 + \mathrm{v}_i G_{i,k}}{1 + \mathrm{v}_i G_{i,k+n}} \xrightarrow[n\to\infty]{\mathsf{P}_{k,i,\theta_i}\text{-a.s.}} 0.$$

Since under $\mathsf{P}_{k,i,\theta_i}$ the random variables $\mathscr{L}_{i,n}^{(2)}(\theta_i)$, $n = k+1, k+2, \dots$ are i.i.d. with mean $I_{\theta_i}^{(2)}$, we have

$$\frac{1}{n} \lambda_{i,\theta_i}(k, k+n) \xrightarrow[n\to\infty]{\mathsf{P}_{k,i,\theta_i}-\text{a.s.}} I_{\theta_i}^{(2)},$$

and hence, condition $\mathbf{C}_1^{(i)}$ holds with $I_{i,\theta_i} = I_{\theta_i}^{(2)}$.

Now, under $\mathsf{P}_{k,i,\theta_i}$ the LLR $\lambda_{i,\vartheta_i}(k, k+n)$ can be written as

$$\lambda_{i,\vartheta_i}(k, k+n) = \sum_{j=k+1}^{k+n} \mathscr{L}_{i,j}^{(2)}(\vartheta_i, \theta_i) + \psi_i(k,n),$$

where $\mathscr{L}_{i,j}^{(2)}(\vartheta_i, \theta_i)$ is the statistic $\mathscr{L}_{i,j}^{(2)}(\vartheta_i)$ under $\mathsf{P}_{k,i,\theta_i}$ and

$$\psi_i(k,n) = \log \frac{1 + \mathrm{v}_i G_{i,k}}{1 + \mathrm{v}_i G_{i,k+n}} \le \log(1 + \mathrm{v}_i G_{i,k}) = \psi_{i,k}^{\star}$$

for any $n \ge 1$. Since $\psi_{i,k}^{\star} \ge 0$ and $\{\mathscr{L}_{i,j}^{(2)}(\vartheta_i)\}_{j>k}$ are i.i.d. under $\mathsf{P}_{k,i,\theta_i}$, we have

$$\begin{aligned}
&\mathsf{P}_{k,i,\theta_i}\left(\frac{1}{n} \inf_{\vartheta_i \in \Gamma_{\delta,\theta_i}} \lambda_{\vartheta_i}(k, k+n) < I_{i,\theta_i} - \varepsilon \right) \\
&\le \mathsf{P}_{k,i,\theta_i}\left(\frac{1}{n} \inf_{\vartheta_i \in \Gamma_{\delta,\theta_i}} \sum_{j=k+1}^{k+n} \mathscr{L}_{i,j}^{(2)}(\vartheta_i) < I_{i,\theta_i} - \varepsilon - \frac{1}{n}\psi_{i,k}^{\star} \right) \\
&\le \mathsf{P}_{k,i,\theta_i}\left(\frac{1}{n} \inf_{\vartheta_i \in \Gamma_{\delta,\theta_i}} \sum_{j=k+1}^{k+n} \mathscr{L}_{i,j}^{(2)}(\vartheta_i) < I_{i,\theta_i} - \varepsilon \right) \\
&= \mathsf{P}_{0,i,\theta_i}\left(\frac{1}{n} \inf_{\vartheta_i \in \Gamma_{\delta,\theta_i}} \sum_{j=1}^{n} \mathscr{L}_{i,j}^{(2)}(\vartheta_i) < I_{i,\theta_i} - \varepsilon \right)
\end{aligned}$$

and, consequently, conditions $\mathbf{C}_2^{(i)}$ are satisfied as long as

$$\sum_{n=1}^{\infty} n^{r-1} \sup_{\theta_i \in \Theta_{i,c}} \mathsf{P}_{0,i,\theta_i}\left(\frac{1}{n} \inf_{\vartheta_i \in \Gamma_{\delta,\theta_i}} \sum_{j=1}^{n} \mathscr{L}_{i,j}^{(2)}(\vartheta_i) < I_{i,\theta_i} - \varepsilon \right) < \infty. \tag{6.88}$$

Typically condition (6.88) holds if the $(r+1)$th absolute moment of $\mathscr{L}_{i,1}^{(2)}(\vartheta_i)$ is finite, $\mathsf{E}_{0,i,\theta_i}|\mathscr{L}_{i,1}^{(2)}(\vartheta_i)|^{r+1} < \infty$.

For example, let us consider the following Gaussian model:

$$f_{i,\theta_i}(x) = \frac{1}{\sqrt{2\pi\sigma_i^2}} \exp\left\{ \frac{(x-\theta_i)^2}{2\sigma_i^2} \right\}, \quad p_{s,i}(x) = \frac{1}{\sqrt{2\pi\sigma_i^2}} \exp\left\{ \frac{(x-\mu_{i,s})^2}{2\sigma_i^2} \right\}, \quad s = 1,2,$$

where $\theta_i > 0$, $\mu_{i,1} > \mu_{i,2} = 0$. Then

$$\mathscr{L}_{i,n}^{(s)}(\theta_i) = \frac{\theta_i - \mu_{i,s}}{\sigma_i^2} X_n(i) - \frac{(\theta_i - \mu_{i,s})^2}{2\sigma_i^2},$$

$I_{i,\theta_i}^{(s)} = (\theta_i - \mu_{i,s})^2/2\sigma_i^2$, $I_{i,\theta_i}^{(2)} > I_{i,\theta_i}^{(1)}$ and

$$\frac{1}{n}\sum_{j=1}^{n} \mathscr{L}_{i,j}^{(2)}(\vartheta_i, \theta_i) = \frac{\vartheta_i\theta_i - \vartheta_i^2/2}{\sigma_i^2} + \frac{\vartheta_i}{\sigma_i n}\sum_{j=1}^{n} \eta_{i,j},$$

where $\eta_{i,j} \sim \mathscr{N}(0,1)$, $j = 1,2,\dots$ are i.i.d. standard normal random variables. Since all moments of $\eta_{i,j}$ are finite, by the same argument as in the previous example, condition (6.88) holds for all $r \geq 1$, and hence, the detection rule $T_{A_\alpha}^{\mathbf{p},W}$ is asymptotically optimal as $\alpha \to 0$, minimizing all positive moments of the detection delay.

Example 6.3 (*Change detection in a multistream exponential model*). Let

$$f_\theta(x) = \frac{1}{1+\theta} \exp\left(-\frac{x}{1+\theta}\right) \mathbb{1}_{\{[0,\infty)\}}(x)$$

be the density of the exponential distribution with the parameter $1+\theta$, $\theta \geq 0$. Assume that the observations are independent across streams and also independent in streams having the exponential distributions with pre-change and post-change densities $g_i(X_n(i)|\mathbf{X}^{n-1}(i)) = f_0(X_n(i))$ and $f_{i,\theta_i}(X_n(i)|\mathbf{X}^{n-1}(i)) = f_{\theta_i}(X_n(i))$, $\theta_i > 0$. That is, the pre-change parameter is $\theta = 0$ and the post-change parameter in the ith stream is $\theta = \theta_i > 0$. This model is of interest in many applications, including detecting objects that appear at unknown points in time in multichannel radar systems (range and doppler channels) based on preprocessed data (at the output of square-law detectors after matched filtering) [4]. In this case, θ_i is the signal-to-noise ratio in the ith stream.

The LLR in the ith stream is

$$\lambda_{i,\theta_i}(k,k+n) = \sum_{t=k+1}^{k+n} \Delta\lambda_i(t), \quad \Delta\lambda_i(t) = \frac{\theta_i}{1+\theta_i} X_t(i) - \log(1+\theta_i).$$

Define the Kullback-Leibler numbers

$$\mathscr{K}_{i,\theta_i} = \mathsf{E}_{0,i,\theta_i}[\Delta\lambda_i(1)] = \int_0^\infty \log\left[\frac{f_{\theta_i}(x)}{g_i(x)}\right] f_{\theta_i}(x)\mathrm{d}x = \theta_i - \log(1+\theta_i), \quad i = 1,\dots,N.$$

Conditions $\mathbf{C}_1^{(i)}$ hold with $I_{i,\theta_i} = \mathscr{K}_{i,\theta_i}$ since $n^{-1}\lambda_{i,\theta_i}(k,k+n) \to \mathscr{K}_{i,\theta_i}$ as $n \to \infty$ a.s. under $\mathsf{P}_{k,i,\theta}$ by the SLLN. Conditions $\mathbf{C}_2^{(i)}$ also hold for all $r \geq 1$ since all positive moments of the LLR $\mathsf{E}_{0,i}|\Delta\lambda_i(1)|^r$ are finite. Thus, by Corollary 6.1, the double-mixture rule $T_A^{\mathbf{p},W}$ minimizes as $\alpha \to 0$ all positive moments of the detection delay and asymptotic formulas (6.50) and (6.51) hold with

$$I_{\mathscr{B},\boldsymbol{\theta}_{\mathscr{B}}} = \sum_{i\in\mathscr{B}} [\theta_i - \log(1+\theta_i)].$$

6.10 Discussion and Remarks

1. Note again that condition $\mathbf{C}_1$ holds whenever $\lambda_{\mathscr{B},\theta}(k,k+n)/n$ converges almost surely to $I_{\mathscr{B},\theta}$ under $\mathsf{P}_{k,\mathscr{B},\theta}$,

$$\frac{1}{n}\lambda_{\mathscr{B},\theta}(k,k+n) \xrightarrow[n\to\infty]{\mathsf{P}_{k,\mathscr{B},\theta}-\text{a.s.}} I_{\mathscr{B},\theta} \tag{6.89}$$

(cf. Lemma B.1 in Appendix B). However, the a.s. convergence is not sufficient for asymptotic optimality of the detection rules with respect to moments of the detection delay. In fact, the average detection delay may even be infinite under the a.s. convergence (6.89). The left-tail condition $\mathbf{C}_2$ guarantees finiteness of first r moments of the detection delay and asymptotic optimality of the detection rules in Theorem 6.2, Theorem 6.3, and Theorem 6.4. Note also that the uniform r-complete convergence conditions for $n^{-1}\lambda_{\mathscr{B},\theta}(k,k+n)$ and $n^{-1}\log\Lambda_{\mathbf{p},W}(k,k+n)$ to $I_{\mathscr{B},\theta}$ under $\mathsf{P}_{k,\mathscr{B},\theta}$, i.e., when for all $\varepsilon>0$, $\mathscr{B}\in\mathscr{P}$ and $\theta\in\Theta$

$$\sum_{n=1}^{\infty} n^{r-1} \sup_{k\in\mathbb{Z}_+} \mathsf{P}_{k,\mathscr{B},\theta}\left(\left|\frac{1}{n}\lambda_{\mathscr{B},\theta}(k,k+n)-I_{\mathscr{B},\theta}\right|>\varepsilon\right)<\infty,$$

$$\sum_{n=1}^{\infty} n^{r-1} \sup_{k\in\mathbb{Z}_+} \mathsf{P}_{k,\mathscr{B},\theta}\left(\left|\frac{1}{n}\log\Lambda_{\mathbf{p},W}(k,k+n)-I_{\mathscr{B},\theta}\right|>\varepsilon\right)<\infty,$$

are sufficient for asymptotic optimality results presented in Theorems 6.2–6.4. However, on the one hand these conditions are stronger than conditions $\mathbf{C}_1$ and $\mathbf{C}_2$, and on the other hand, verification of the r-complete convergence conditions is more difficult than checking conditions $\mathbf{C}_1$ and $\mathbf{C}_2$ for the local values of the LLR in the vicinity of the true parameter value, which is especially true for the weighted LLR $\log\Lambda_{\mathbf{p},W}(k,k+n)$. Still the r-complete convergence conditions are intuitively appealing since they define the rate of convergence in the strong law of large numbers (6.89).

2. As already discussed in Section 3.10 for a single stream case, even for independent streams and general non-i.i.d. models in streams the computational complexity and memory requirements of the double mixture Shiryaev and the double mixture Shiryaev–Roberts rules $T_A^{\mathbf{p},W}$ and $\widetilde{T}_A^{\mathbf{p},W}$ can be quite high, so it is reasonable to use window-limited versions of double-mixture detection rules where the summation over potential change points k is restricted to the sliding window of size ℓ. In the window-limited versions of $T_A^{\mathbf{p},W}$ and $\widetilde{T}_A^{\mathbf{p},W}$, defined in (6.10) and (6.14), the statistics $S_{\mathbf{p},W}^{\pi}(n)$ and $R_{\mathbf{p},W}(n)$ are replaced by the window-limited statistics

$$\widehat{S}_{\mathbf{p},W}^{\pi}(n)=S_{\mathbf{p},W}^{\pi}(n)\quad\text{for } n\le\ell;\quad \widehat{S}_{\mathbf{p},W}^{\pi}(n)=\frac{1}{\mathsf{P}(\nu\ge n)}\sum_{k=n-(\ell+1)}^{n-1}\pi_k\Lambda_{\mathbf{p},W}(k,n)\quad\text{for } n>\ell$$

and

$$\widehat{R}_{\mathbf{p},W}(n)=R_{\mathbf{p},W}(n)\quad\text{for } n\le\ell;\quad \widehat{R}_{\mathbf{p},W}(n)=\sum_{k=n-(\ell+1)}^{n-1}\Lambda_{\mathbf{p},W}(k,n)\quad\text{for } n>\ell.$$

Following guidelines of Lai [75] and the techniques developed in Section 3.10 for the single stream scenario, it can be shown that these window-limited versions also have first-order asymptotic optimality properties as long as the size of the window $\ell(A)$ approaches infinity as $A\to\infty$ with $\ell(A)/\log A\to\infty$. Since thresholds $A=A_\alpha$ in detection rules should be selected in such a way that $\log A_\alpha\sim|\log\alpha|$ as $\alpha\to 0$, the value of the window size $\ell(\alpha)$ should satisfy $\lim_{\alpha\to 0}\frac{\ell(\alpha)}{|\log\alpha|}=\infty$.

3. It is expected that first-order approximations for the moments of the detection delay are inaccurate in most cases, so higher-order approximations are in order. However, it is not feasible to obtain such approximations in the general non-i.i.d. case considered in this chapter. Higher order approximations for the expected delay to detection and the probability of false alarm in the "i.i.d." scenario, assuming that the observations in streams are independent and also independent across streams, can be derived based on the renewal theory and nonlinear renewal theory.

7

Joint Changepoint Detection and Identification

7.1 Introduction

Often, in many applications, one needs not only to detect an abrupt change as quickly as possible but also provide a detailed diagnosis of the occurred change – to determine which type of change is in effect. For example, the problem of detection and diagnosis is important for rapid detection and isolation of intrusions in large-scale distributed computer networks, target detection with radar, sonar and optical sensors in a cluttered environment, detecting terrorists' malicious activity, fault detection and isolation in dynamic systems and networks, and integrity monitoring of navigation systems, to name a few (see [164, Ch 10] for an overview and references). In other words, there are several kinds of changes that can be associated with several different post-change distributions and the goal is to detect the change and to identify which distribution corresponds to the change. As a result, formally, the problem of changepoint detection and diagnosis is a generalization of the quickest change detection problem considered in Chapter 6 to the case of $N \geq 2$ post-change hypotheses, and it can be formulated as a joint change detection and identification problem. In the literature, this problem is usually called *change detection and isolation*. The detection–isolation problem has been considered in both Bayesian and minimax settings. In 1995, Nikiforov [99] suggested a minimax approach to the change detection–isolation problem and showed that the multihypothesis version of the CUSUM rule is asymptotically optimal when the mean time to false alarm and the mean time to false isolation become large. Several versions of the multihypothesis CUSUM-type and SR-type procedures, which have minimax optimality properties in the classes of rules with constraints imposed on the ARL to false alarm and conditional probabilities of false isolation, are proposed by Nikiforov [100, 101] and Tartakovsky [156]. These rules asymptotically minimize maximal expected delays to detection and isolation as the ARL to false alarm is large and the probabilities of wrong isolations are small. Dayanik et al. [28] proposed an asymptotically optimal Bayesian detection–isolation rule assuming that the prior distribution of the change point is geometric. In all these papers, the optimality results were restricted to the case of independent observations and simple post-change hypotheses. A generalization for the non-i.i.d. case and composite hypotheses was done by Lai [76]. See Chapter 10 in Tartakovsky et al. [164] for a detailed overview.

One of the most challenging and important versions of the change detection–isolation problem is the multidecision and multistream detection problem which generalizes the multistream changepoint detection problem considered in Chapter 6 in that it is also required to identify the streams where the change happens with given probabilities of misidentification. Specifically, there are N data streams (populations) and the change occurs in some of them at an unknown point in time. It is necessary to detect the change in distribution as soon as possible and indicate which streams are "corrupted." Both the rates of false alarms and misidentification should be controlled by given (usually low) levels. In the following we will refer to this problem as the *Multistream/Multisample Sequential Change Detection–Identification* problem.

In this chapter, we address a simplified multistream detection–identification scenario where the change can occur only in a single stream and we need to determine in which stream. We focus on a semi-Bayesian setting assuming that the change point ν is random possessing the distribution

$\pi_k = \Pr(\nu = k)$, $k = -1, 0, 1, \dots$ However, we do not suppose that there is a prior distribution on post-change hypotheses. Specifically, we show that under certain very general conditions the proposed multihypothesis detection–identification rule asymptotically minimizes the trade-off between positive moments of the detection delay and the false alarm/misclassification rates expressed via the weighted probabilities of false alarm and false identification.

7.2 The Model and the Detection–Identification Rule

Consider a scenario with independent streams given by (6.15). Furthermore, we will assume throughout this chapter that only one stream can be affected, i.e., $\mathscr{B} = i \in \mathscr{N} = \{1, \dots, N\}$, so we are interested in a "multisample slippage" changepoint model (given ν, $\mathscr{B} = i$ and θ_i) of the form

$$
\begin{aligned}
p(\mathbf{X}^n | \mathsf{H}_{\nu,i}, \theta_i) &= p(\mathbf{X}^n | \mathsf{H}_\infty) = \prod_{t=0}^{n} \prod_{i=1}^{N} g_i(X_t(i) | \mathbf{X}^{t-1}(i)) \quad \text{for } \nu \geq n, \\
p(\mathbf{X}^n | \mathsf{H}_{\nu,i}, \theta_i) &= \prod_{t=0}^{\nu} g_i(X_t(i) | \mathbf{X}^{t-1}(i)) \times \prod_{t=\nu+1}^{n} f_{i,\theta_i}(X_t(i) | \mathbf{X}^{t-1}(i)) \times \\
&\quad \prod_{j \in \mathscr{N} \setminus \{i\}} \prod_{t=0}^{n} g_j(X_t(j) | \mathbf{X}^{t-1}(j)) \quad \text{for } \nu < n,
\end{aligned}
\tag{7.1}
$$

where $g_i(X_t(i) | \mathbf{X}^{t-1}(i))$ and $f_{i,\theta_i}(X_t(i) | \mathbf{X}^{t-1}(i))$ are conditional pre- and post-change densities in the ith data stream, respectively. In other words, all components $X_t(i)$, $i = 1, \dots, N$, have conditional densities $g_i(X_t(i) | \mathbf{X}^{t-1}(i))$ before the change occurs and $X_t(i)$ has conditional density $f_{i,\theta_i}(X_t(i) | \mathbf{X}^{t-1}(i))$ after the change occurs in the ith stream and the rest of the components $X_t(j)$, $j \in \mathscr{N} \setminus \{i\}$ have conditional densities $g_j(X_t(j) | \mathbf{X}^{t-1}(j))$. The parameters $\theta_i \in \Theta_i$, $i = 1, \dots, N$ of the post-change distributions are unknown.

It follows from (7.1) that for an assumed value of the change point $\nu = k$, stream $i \in \mathscr{N}$, and the post-change parameter value in the ith stream θ_i, the likelihood ratio $LR_{i,\theta_i}(k,n) = p(\mathbf{X}^n | \mathsf{H}_{k,i}, \theta_i) / p(\mathbf{X}^n | \mathsf{H}_\infty)$ between the hypotheses $\mathsf{H}_{k,i}$ and H_∞ for observations accumulated by the time n has the form

$$
LR_{i,\theta_i}(k,n) = \prod_{t=k+1}^{n} \mathscr{L}_{i,\theta_i}(t), \quad i \in \mathscr{N}, \; n > k, \; k = -1, 0, 1, \dots, \tag{7.2}
$$

where $\mathscr{L}_{i,\theta_i}(t) = f_{i,\theta_i}(X_t(i) | \mathbf{X}^{t-1}(i)) / g_i(X_t(i) | \mathbf{X}^{t-1}(i))$. It is supposed that $\mathscr{L}_{i,\theta_i}(0) = 1$, so that $LR_{i,\theta_i}(-1,n) = LR_{i,\theta_i}(0,n)$. Define the average (over the prior π_k) LR statistics

$$
\Lambda^\pi_{i,\theta_i}(n) = \sum_{k=-1}^{n-1} \pi_k LR_{i,\theta_i}(k,n), \quad i = 1, \dots, N. \tag{7.3}
$$

Let $W_i(\theta_i)$, $\int_{\Theta_i} \mathrm{d}W_i(\theta_i) = 1$, $i = 1, \dots, N$ be mixing measures and define the LR-mixtures

$$
LR_{i,W}(k,n) = \int_{\Theta_i} LR_{i,\theta_i}(k,n) \, \mathrm{d}W_i(\theta_i), \quad k < n, \; i = 1, \dots, N \tag{7.4}
$$

and the statistics

$$
\Lambda^\pi_{i,W}(n) = \begin{cases} \sum_{k=-1}^{n-1} \pi_k LR_{i,W}(k,n), & i \in \mathscr{N} \\ \mathsf{P}(\nu \geq n) & i = 0 \end{cases}, \tag{7.5}
$$

$$S_{ij}^{\pi}(n) = \frac{\Lambda_{i,W}^{\pi}(n)}{\sum_{k=-1}^{n-1} \pi_k \sup_{\theta_j \in \Theta_j} LR_{j,\theta_j}(k,n)}, \quad i,j = 1,\dots,N, \ i \neq j, \ n \geq 1;$$
$$S_{i0}^{\pi}(n) = \frac{\Lambda_{i,W}^{\pi}(n)}{\mathsf{P}(\nu \geq n)}, \quad i = 1,\dots,N, \ n \geq 1. \tag{7.6}$$

Write $\mathscr{N}_0 = \{0,1,\dots,N\}$. For the set of positive thresholds $A = (A_{ij})$, $j \in \mathscr{N}_0 \setminus \{i\}$, $i \in \mathscr{N}$, the change detection–identification rule $\delta_A = (d_A, T_A)$ is defined as follows:

$$T_A^{(i)} = \inf\left\{n \geq 1 : S_{ij}^{\pi}(n) \geq A_{ij} \ \text{ for all } \ j \in \mathscr{N}_0 \setminus \{i\}\right\}, \quad i \in \mathscr{N}; \tag{7.7}$$

$$T_A = \min\left\{T_A^{(1)}, \dots, T_A^{(N)}\right\}, \quad d_A = i \ \text{ if } \ T_A = T_A^{(i)}. \tag{7.8}$$

If $T_A = T_A^{(i)}$ for several values of i then any of them can be taken.

7.3 The Optimization Problem and Assumptions

Let $\mathsf{E}_{k,i,\theta_i}$ and E_∞ denote expectations under $\mathsf{P}_{k,i,\theta_i}$ and P_∞, respectively, where $\mathsf{P}_{k,i,\theta_i}$ corresponds to model (7.1) with an assumed value of the parameter $\theta_i \in \Theta_i$. Define the probability measure $\mathsf{P}_{i,\theta_i}^{\pi}(\mathscr{A} \times \mathscr{K}) = \sum_{k \in \mathscr{K}} \pi_k \mathsf{P}_{k,i,\theta_i}(\mathscr{A})$ under which the change point ν has distribution $\pi = \{\pi_k\}$ and the model for the observations is of the form (7.1) and let $\mathsf{E}_{i,\theta_i}^{\pi}$ denote the corresponding expectation.

For $r \geq 1$, $\nu = k \in \mathbb{Z}_+$, $\theta_i \in \Theta_i$ and $i \in \mathscr{N}$ introduce the risk associated with the conditional rth moment of the detection delay $\mathscr{R}_{k,i,\theta_i}^r(\delta) = \mathsf{E}_{k,i,\theta_i}[(T-k)^r; d = i \,|\, T > k]$, where for $k = -1$ we set $T - k = T$, but not $T+1$, as well as the integrated (over prior π) risk associated with the moments of delay to detection

$$\overline{\mathscr{R}}_{i,\theta_i}^r(\delta) := \mathsf{E}_{i,\theta_i}^{\pi}[(T-\nu)^r; d = i | T > \nu] = \frac{\sum_{k=-1}^{\infty} \pi_k \mathscr{R}_{k,i,\theta_i}^r(\delta) \mathsf{P}_\infty(T > k; d = i)}{1 - \mathsf{PFA}_i^{\pi}(\delta)}, \tag{7.9}$$

where

$$\mathsf{PFA}_i^{\pi}(\delta) = \mathsf{P}_{i,\theta_i}^{\pi}(T \leq \nu; d = i) = \sum_{k=0}^{\infty} \pi_k \mathsf{P}_\infty(T \leq k; d = i) \tag{7.10}$$

is the weighted probability of false alarm on the event $\{d = i\}$, i.e., the probability of raising the alarm with the decision $d = i$ that there is a change in the ith stream when there is no change at all. The loss associated with wrong identification is reasonable to measure by the maximal probabilities of wrong decisions (misidentification)

$$\mathsf{PMI}_{ij}^{\pi}(\delta) = \sup_{\theta_i \in \Theta_i} \mathsf{P}_{i,\theta_i}^{\pi}(d = j; T < \infty | T > \nu), \quad i,j = 1,\dots,N, \ i \neq j.$$

Note that

$$\mathsf{P}_{i,\theta_i}^{\pi}(d = j; T < \infty | T > \nu) = \frac{\sum_{k=-1}^{\infty} \pi_k \mathsf{P}_{k,i,\theta_i}(d = j; k < T < \infty)}{1 - \mathsf{PFA}_i^{\pi}(\delta)}.$$

Define the class of change detection–identification rules δ with constraints on the probabilities of false alarm $\mathsf{PFA}_i^{\pi}(\delta)$ and the probabilities of misidentification $\mathsf{PMI}_{ij}^{\pi}(\delta)$:

$$\mathbb{C}_\pi(\boldsymbol{\alpha}, \boldsymbol{\beta}) = \left\{\delta : \mathsf{PFA}_i^{\pi}(\delta) \leq \alpha_i, \ i \in \mathscr{N}, \ \ \mathsf{PMI}_{ij}^{\pi}(\delta) \leq \beta_{ij}, \ i,j \in \mathscr{N}, i \neq j\right\}, \tag{7.11}$$

where $\boldsymbol{\alpha} = (\alpha_1, \dots, \alpha_N)$ and $\boldsymbol{\beta} = ||\beta_{ij}||$, $i,j \in \mathscr{N}, i \neq j$ are the sets of prescribed probabilities $\alpha_i \in (0,1)$ and $\beta_{ij} \in (0,1)$.

Ideally, we would be interested in finding an optimal rule $\delta_{\rm opt} = (d_{\rm opt}, T_{\rm opt})$ that solves the optimization problem

$$\overline{\mathscr{R}}^r_{i,\theta_i}(\delta_{\rm opt}) = \inf_{\delta \in \mathbb{C}_\pi(\boldsymbol{\alpha},\boldsymbol{\beta})} \overline{\mathscr{R}}^r_{i,\theta_i}(\delta) \quad \text{for all } \theta_i \in \Theta_i,\ i \in \mathscr{N}. \tag{7.12}$$

However, this problem is intractable for arbitrarily values of $\alpha_i \in (0,1)$ and $\beta_{ij} \in (0,1)$. For this reason, we will focus on the asymptotic problem assuming that the given probabilities α_i and β_{ij} approach zero. To be more specific, we will be interested in proving that the proposed detection–identification rule $\delta_A = (d_A, T_A)$ defined in (7.7)–(7.8) is first-order uniformly asymptotically optimal in the following sense

$$\lim_{\alpha_{\max}\to 0, \beta_{\max}\to 0} \frac{\inf_{\delta \in \mathbb{C}_\pi(\boldsymbol{\alpha},\boldsymbol{\beta})} \overline{\mathscr{R}}^r_{i,\theta_i}(\delta)}{\overline{\mathscr{R}}^r_{i,\theta_i}(\delta_A)} = 1 \quad \text{for all } \theta_i \in \Theta_i \text{ and } i \in \mathscr{N} \tag{7.13}$$

where $A = A(\boldsymbol{\alpha},\boldsymbol{\beta})$ is the set of suitably selected thresholds such that $\delta_A \in \mathbb{C}(\boldsymbol{\alpha},\boldsymbol{\beta})$. Hereafter $\alpha_{\max} = \max_{i \in \mathscr{N}} \alpha_i$, $\beta_{\max} = \max_{i,j \in \mathscr{N}, i \neq j} \beta_{ij}$.

It is also of interest to consider the class of detection–identification rules

$$\mathbb{C}^\star_\pi(\alpha, \bar{\boldsymbol{\beta}}) = \{\delta : \mathsf{PFA}^\pi(\delta) \le \alpha \ \text{ and } \ \mathsf{PMI}^\pi_i(\delta) \le \bar{\beta}_i,\ i = 1, \dots, N\} \tag{7.14}$$

($\bar{\boldsymbol{\beta}}_i = (\bar{\beta}_1, \dots, \bar{\beta}_N)$) with constraints on the total probability of false alarm

$$\mathsf{PFA}^\pi(\delta) = \mathsf{P}^\pi(T \le \nu) = \sum_{k=0}^\infty \pi_k \mathsf{P}_\infty(T \le k)$$

regardless of the decision $d = i$ which is made under hypothesis H_∞ and on the misidentification probabilities

$$\mathsf{PMI}^\pi_i(\delta) = \sup_{\theta_i \in \Theta_i} \mathsf{P}^\pi_{i,\theta_i}(d \neq i; T < \infty | T > \nu), \quad i = 1, \dots, N.$$

Obviously, $\mathsf{PFA}^\pi(\delta) = \sum_{i=1}^N \mathsf{PFA}^\pi_i(\delta)$ and $\mathsf{PMI}^\pi_i(\delta) = \sum_{j \neq i} \mathsf{PMI}^\pi_{ij}(\delta)$.

As before in (3.22), we assume that mixing measures W_i, $i = 1, \dots, N$, satisfy the condition:

$$W_i\{\vartheta \in \Theta_i : |\vartheta - \theta_i| < \varkappa\} > 0 \quad \text{for any } \varkappa > 0 \text{ and any } \theta_i \in \Theta_i. \tag{7.15}$$

By (7.2), for the assumed values of $\nu = k$, $i \in \mathscr{N}$ and $\theta_i \in \Theta_i$, the LLR $\lambda_{i,\theta_i}(k, k+n) = \log \mathscr{L}_{i,\theta_i}(k, k+n)$ of observations accumulated by the time $k+n$ is

$$\lambda_{i,\theta_i}(k, k+n) = \sum_{t=k+1}^{k+n} \log \frac{f_{i,\theta_i}(X_t(i) | \mathbf{X}^{t-1}(i))}{g_i(X_t(i)) | \mathbf{X}^{t-1}(i))}, \quad n \ge 1$$

and the LLR between the hypotheses $\mathsf{H}_{k,i}$ and $\mathsf{H}_{k,j}$ of observations accumulated by the time $k+n$ is

$$\lambda_{i,\theta_i;j,\theta_j}(k, k+n) = \log \frac{p(\mathbf{X}^{k+n} | \mathsf{H}_{k,i})}{p(\mathbf{X}^{k+n} | \mathsf{H}_{k,j})} \equiv \lambda_{i,\theta_i}(k, k+n) - \lambda_{j,\theta_j}(k, k+n), \quad n \ge 1.$$

For $j = 0$, we set $\lambda_{0,\theta_0}(k, k+n) = 0$, so that $\lambda_{i,\theta_i;0,\theta_0}(k, k+n) = \lambda_{i,\theta_i}(k, k+n)$.

For $\varkappa > 0$, let $\Gamma_{\varkappa,\theta_i} = \{\vartheta \in \Theta_i : |\vartheta - \theta_i| < \varkappa\}$ and for $0 < I_{ij}(\theta_i, \theta_j) < \infty$, $j \in \mathscr{N}_0 \setminus \{i\}$, $i \in \mathscr{N}$, define

$$p_{M,k}(\varepsilon; i, \theta_i; j, \theta_j) = \mathsf{P}_{k,i,\theta_i}\left\{\frac{1}{M} \max_{1 \le n \le M} \lambda_{i,\theta_i;j,\theta_j}(k, k+n) \ge (1+\varepsilon) I_{ij}(\theta_i, \theta_j)\right\},$$

$$\Upsilon_r(\varepsilon;i,\theta_i) = \lim_{\varkappa\to 0}\sum_{n=1}^{\infty} n^{r-1} \sup_{k\in\mathbb{Z}_+} \mathsf{P}_{k,i,\theta_i}\left\{\frac{1}{n}\inf_{\vartheta\in\Gamma_{\varkappa,\theta_i}} \lambda_{i,\vartheta}(k,k+n) < I_i(\theta_i)-\varepsilon\right\}, \tag{7.16}$$

where $I_{i0}(\theta_i,\theta_0) = I_i(\theta_i)$, so that

$$p_{M,k}(\varepsilon;i,\theta_i;0,\theta_0) = p_{M,k}(\varepsilon;i,\theta_i) = \mathsf{P}_{k,i,\theta_i}\left\{\frac{1}{M}\max_{1\le n\le M}\lambda_{i,\theta_i}(k,k+n) \ge (1+\varepsilon)I_i(\theta_i)\right\}.$$

Regarding the model for the observations (7.1), we assume that the following conditions are satisfied (for local LLRs in data streams):

$\mathbf{C}_1$. *There exist positive and finite numbers $I_i(\theta_i)$, $\theta_i\in\Theta_i$, $i\in\mathscr{N}$ and $I_{ij}(\theta_i,\theta_j)$, $\theta_j\in\Theta_j$, $j\in\mathscr{N}\setminus\{i\}$, $\theta_i\in\Theta_i$, $i\in\mathscr{N}$, such that for any $\varepsilon>0$*

$$\lim_{M\to\infty} p_{M,k}(\varepsilon;i,\theta_i;j,\theta_j) = 0 \quad \text{for all } k\in\mathbb{Z}_+,\ \theta_i\in\Theta_i,\ \theta_j\in\Theta_j,\ j\in\mathscr{N}_0\setminus\{i\},\ i\in\mathscr{N}. \tag{7.17}$$

$\mathbf{C}_2$. *For any $\varepsilon>0$ and some $r\ge 1$*

$$\Upsilon_r(\varepsilon;i,\theta_i) < \infty \quad \text{for all } \theta_i\in\Theta_i,\ i\in\mathscr{N}. \tag{7.18}$$

Regarding the prior distribution $\pi_k = \Pr(\nu=k)$ we assume that it is fully supported (i.e., $\pi_k>0$ for all $k\in\mathbb{Z}_+$ and $\pi_\infty=0$) and the same conditions $\mathbf{CP}_1$ and $\mathbf{CP}_2$ as in Chapter 3 are satisfied (see (3.20) and (3.21)). Recall that in this case we say that the prior distribution $\pi=\{\pi_k\}$ belongs to class $\mathbf{C}(\mu)$. Recall also that we allow the change point ν to take negative values but the detailed distribution for all $k\le -1$ is not important, so without loss of generality we suppose that it is concentrated at $k=-1$ and $\pi_{-1}=q\in[0,1)$, where q is the total probability of the event $\{\nu\le -1\}$ that the change occurs before the observations become available. More generally, if $\pi=\pi^\alpha$ depends on the PFA α, then we will assume that conditions $\mathbf{CP}_1^{(\alpha)}$, $\mathbf{CP}_2^{(\alpha)}$, and $\mathbf{CP}_3^{(\alpha)}$ in Chapter 3 hold (see (3.65)–(3.67)).

In order to obtain lower bounds for moments of the detection delay we need only right-tail conditions (7.17). However, to establish the asymptotic optimality property of the rule δ_A both right-tail and left-tail conditions (7.17) and (7.18) are needed.

7.4 Upper Bounds on Probabilities of False Alarm and Misidentification of the Detection–Identification Rule δ_A

Let $\widetilde{\mathsf{P}}^{\pi}_{i,\theta_i}(\mathscr{A}) = \mathsf{P}^{\pi}_{i,\theta_i}(\mathscr{A},\nu<n)$ denote the measure $\mathsf{P}^{\pi}_{i,\theta_i}$ on the event $\{\nu<n\}$ and $\widehat{\mathsf{P}}^{\pi}_{i}(\mathscr{A}) = \mathsf{P}^{\pi}_{i}(\mathscr{A},\nu\ge n)$. The distribution $\widetilde{\mathsf{P}}^{\pi}_{i,\theta_i}(\mathbf{X}^n\in\mathscr{X}^n)$ has density

$$f^{\pi}_{i,\theta_i}(\mathbf{X}^n) = \sum_{k=-1}^{n-1}\pi_k\left[\prod_{t=0}^{k} g_i(X_t(i)|\mathbf{X}^{t-1}(i)) \prod_{t=k+1}^{n} f_{i,\theta_i}(X_t(i)|\mathbf{X}^{t-1}(i))\right]\times$$
$$\mathsf{P}(\nu<n)\prod_{j\in\mathscr{N}\setminus\{i\}}\prod_{t=0}^{n} g_j(X_t(j)|\mathbf{X}^{t-1}(j)),$$

where $\prod_{t=0}^{-1} g_i(X_t(i)|\mathbf{X}^{t-1}(i)) = 1$, and $\widehat{\mathsf{P}}^{\pi}_{i}(\mathbf{X}^n\in\mathscr{X}^n)$ has density (see (7.1))

$$\hat{f}(\mathbf{X}^n) = \mathsf{P}(\nu\ge n)\prod_{t=0}^{n}\prod_{i=1}^{N} g_i(X_t(i)|\mathbf{X}^{t-1}(i)).$$

Write

$$f^{\pi}_{i,W}(\mathbf{X}^n) = \int_{\Theta_i} f^{\pi}_{i,\theta_i}(\mathbf{X}^n)\, \mathrm{d}W_i(\theta_i).$$

Next, define the statistic $\widetilde{S}^{\pi}_{i,j,\theta_j}(n) = \Lambda^{\pi}_{i,W}(n)/\Lambda^{\pi}_{j,\theta_j}(n)$ and the measure

$$\widetilde{\mathsf{P}}^{\pi}_{\ell,W}(\mathscr{A}) = \int_{\Theta_\ell} \widetilde{\mathsf{P}}^{\pi}_{\ell,\theta_\ell}(\mathscr{A})\mathrm{d}W_\ell(\theta_\ell).$$

Obviously,

$$\widetilde{S}^{\pi}_{i,j,\theta_j}(n) = \frac{\mathrm{d}\widetilde{\mathsf{P}}^{\pi}_{i,W}}{\mathrm{d}\widetilde{\mathsf{P}}^{\pi}_{j,\theta_j}}\Bigg|_{\mathscr{F}_n}, \quad i \neq j,$$

and hence, the statistic $\widetilde{S}^{\pi}_{i,j,\theta_j}(n)$ is a $(\widetilde{\mathsf{P}}^{\pi}_{j,\theta_j}, \mathscr{F}_n)$-martingale with unit expectation for all $\theta_j \in \Theta_j$. Therefore, by the Wald–Doob identity, for any stopping time T and all $\theta_j \in \Theta_j$,

$$\widetilde{\mathsf{E}}^{\pi}_{i,\theta_i}\left[\widetilde{S}^{\pi}_{j,i,\theta_i}(T)\mathbb{1}_{\{\mathscr{A},T<\infty\}}\right] = \widetilde{\mathsf{E}}^{\pi}_{j,W}\left[\mathbb{1}_{\{\mathscr{A},T<\infty\}}\right] = \widetilde{\mathsf{P}}^{\pi}_{j,W}(\mathscr{A}\cap\{T<\infty\}), \tag{7.19}$$

where $\widetilde{\mathsf{E}}^{\pi}_{j,W}$ and $\widetilde{\mathsf{E}}^{\pi}_{j,\theta_j}$ stand for the operators of expectation under $\widetilde{\mathsf{P}}^{\pi}_{j,W}$ and $\widetilde{\mathsf{P}}^{\pi}_{j,\theta_j}$, respectively.

The following theorem establishes upper bounds for the PFA and PMI of the proposed detection–identification rule δ_A. Note that these bounds are valid in the most general case – neither of the conditions on the model $\mathbf{C}_1$, $\mathbf{C}_2$ or on the prior distribution $\mathbf{CP}_1$ are required.

Theorem 7.1. *Let δ_A be the changepoint detection–identification rule defined in (7.7)–(7.8). The following upper bounds for the PFA and PMI of rule δ_A hold*

$$\mathsf{PFA}^{\pi}_i(\delta_A) \le (1+A_{i0})^{-1}, \quad i=1,\dots,N, \tag{7.20}$$

$$\mathsf{PFA}^{\pi}(\delta_A) \le \sum_{i=1}^{N}(1+A_{i0})^{-1} \tag{7.21}$$

and

$$\mathsf{PMI}^{\pi}_{ij}(\delta_A) \le \frac{1+A_{i0}}{A_{i0}A_{ji}}, \quad i,j=1,\dots,N,\ i\neq j, \tag{7.22}$$

$$\mathsf{PMI}^{\pi}_{i}(\delta_A) \le \frac{1+A_{i0}}{A_{i0}} \sum_{j\in\mathscr{N}\setminus\{i\}} \frac{1}{A_{ji}}, \quad i=1,\dots,N. \tag{7.23}$$

Thus, if $\alpha_{\max} < 1-\pi_{-1}$, then

$$A_{i0} = \frac{1-\alpha_i}{\alpha_i} \ \text{ and } \ A_{ij} = \frac{1}{(1-\alpha_j)\beta_{ji}} \quad \text{imply } \ \delta_A \in \mathbb{C}_{\pi}(\boldsymbol{\alpha},\boldsymbol{\beta}) \tag{7.24}$$

and if $A_{i0} = A_0$ for $i \in \mathscr{N}$ and $A_{ij} = A_j$ for $j \in \mathscr{N}\setminus\{i\}$, then

$$A_0 = \frac{N}{\alpha}(1-\alpha/N) \ \text{ and } \ A_j = \frac{N-1}{(1-\alpha/N)\bar{\beta}_j} \quad \text{imply } \ \delta_A \in \mathbb{C}^{\star}_{\pi}(\alpha,\bar{\boldsymbol{\beta}}). \tag{7.25}$$

Proof. Clearly,

$$\mathsf{PFA}_i(\delta_A) = \mathsf{P}^{\pi}_{i,\theta_i}(T^{(i)}_A \le \nu, T_A = T^{(i)}_A) \le \mathsf{P}^{\pi}_{i,\theta_i}(T^{(i)}_A \le \nu).$$

In just the same way as in the proof of Lemma 3.1 we have

$$\mathsf{P}(\nu \ge n|\mathscr{F}_n) = \frac{\mathsf{P}(\nu\ge n)}{\Lambda_{i,W}(n)+\mathsf{P}(\nu\ge n)} = \frac{1}{S^{\pi}_{i0}(n)+1}.$$

Therefore, taking into account that $\mathsf{P}^{\pi}_{i,\theta_i}(T_A^{(i)} \le \nu) = \mathsf{E}^{\pi}_{i,\theta_i}[\mathsf{P}(T_A^{(i)} \le \nu | \mathscr{F}_{T_A^{(i)}})]$ and that $S_{i0}(T_A^{(i)}) \ge A_{i0}$ on $\{T_A^{(i)} < \infty\}$, we obtain

$$\mathsf{PFA}_i(\delta_A) \le \mathsf{E}^{\pi}_{i,\theta_i}[(1+S_{i0}(T_A^{(i)}))^{-1}; T_A^{(i)} < \infty] \le 1/(1+A_{i0})$$

and inequalities (7.20) follow. Inequality (7.21) follows immediately from the fact that $\mathsf{PFA}^{\pi}(\delta) = \sum_{i=1}^{N} \mathsf{PFA}^{\pi}_i(\delta)$.

To prove the upper bound (7.22) note that $\widetilde{S}^{\pi}_{j,i,\theta_i}(n) \ge S^{\pi}_{ji}(n)$ for all $n \ge 1$ and $\theta_i \in \Theta_i$ and that $S^{\pi}_{ji}(T_A^{(j)}) \ge A_{ji}$ on $\{T_A^{(j)} < \infty\}$ and we have

$$\begin{aligned}
\mathsf{P}^{\pi}_{i,\theta_i}(d_A = j, \nu < T_A < \infty) &= \mathsf{P}^{\pi}_{i,\theta_i}(T_A = T_A^{(j)}, \nu < T_A^{(j)} < \infty) = \widetilde{\mathsf{P}}^{\pi}_{i,\theta_i}(T_A = T_A^{(j)}, T_A^{(j)} < \infty) \\
&= \widetilde{\mathsf{E}}^{\pi}_{i,\theta_i}\left[\frac{S^{\pi}_{ji}(T_A^{(j)})}{S^{\pi}_{ji}(T_A^{(j)})} \mathbb{1}_{\{T_A = T_A^{(j)}, T_A^{(j)} < \infty\}}\right] \le \frac{1}{A_{ji}} \widetilde{\mathsf{E}}^{\pi}_{i,\theta_i}\left[S^{\pi}_{ji}(T_A^{(j)}) \mathbb{1}_{\{T_A = T_A^{(j)}, T_A^{(j)} < \infty\}}\right] \\
&\le \frac{1}{A_{ji}} \widetilde{\mathsf{E}}^{\pi}_{i,\theta_i}\left[\widetilde{S}^{\pi}_{j,i,\theta_i}(T_A^{(j)}) \mathbb{1}_{\{T_A = T_A^{(j)}, T_A^{(j)} < \infty\}}\right] \quad \text{for all } \theta_i \in \Theta_i,
\end{aligned}$$

where, by equality (7.19), the last term is equal to

$$\frac{1}{A_{ji}} \widetilde{\mathsf{P}}^{\pi}_{j,W}(T_A = T_A^{(j)} \cap \{T_A^{(j)} < \infty\}).$$

This yields

$$\mathsf{P}^{\pi}_{i,\theta_i}(d_A = j, \nu < T_A < \infty) \le \frac{1}{A_{ji}} \widetilde{\mathsf{P}}^{\pi}_{j,W}(T_A = T_A^{(j)} \cap \{T_A^{(j)} < \infty\}) \le \frac{1}{A_{ji}} \quad \text{for all } \theta_i \in \Theta_i.$$

Since $\mathsf{P}^{\pi}_{i,\theta_i}(d_A = j | T_A > \nu) = \mathsf{P}^{\pi}_{i,\theta_i}(d_A = j, \nu < T_A < \infty)/\mathsf{P}^{\pi}_{i,\theta_i}(T_A > \nu)$ and, by (7.20), $\mathsf{P}^{\pi}_{i,\theta_i}(T_A > \nu) \ge A_i/(1+A_i)$, the upper bound (7.22) follows. The upper bound (7.23) follows from (7.22) and the fact that $\mathsf{PMI}^{\pi}_i(\delta) = \sum_{j \ne i} \mathsf{PMI}^{\pi}_{ij}(\delta)$.

Implications (7.24) and (7.25) are obvious. □

7.5 Lower Bounds on the Moments of the Detection Delay

Define

$$\Psi_i(\boldsymbol{\alpha}, \boldsymbol{\beta}) = \max\left\{\frac{|\log \alpha_i|}{I_i(\theta_i) + \mu}, \max_{j \in \mathscr{N} \setminus \{i\}} \frac{|\log \beta_{ji}|}{\inf_{\theta_j \in \Theta_j} I_{ij}(\theta_i, \theta_j)}\right\}, \quad i \in \mathscr{N} \tag{7.26}$$

and

$$\Psi^{\star}_i(\alpha, \bar{\boldsymbol{\beta}}) = \max\left\{\frac{|\log \alpha|}{I_i(\theta_i) + \mu}, \max_{j \in \mathscr{N} \setminus \{i\}} \frac{|\log \bar{\beta}_j|}{\inf_{\theta_j \in \Theta_j} I_{ij}(\theta_i, \theta_j)}\right\}, \quad i \in \mathscr{N}. \tag{7.27}$$

The following theorem establishes asymptotic lower bounds on moments of the detection delay $\overline{\mathscr{R}}^{r}_{i,\theta_i}(\delta) = \mathsf{E}^{\pi}_{i,\theta_i}[(T-\nu)^r; d = i | T > \nu]$ in the classes of detection–identification rules $\mathbb{C}_{\pi}(\boldsymbol{\alpha}, \boldsymbol{\beta})$ and $\mathbb{C}^{\star}_{\pi}(\alpha, \bar{\boldsymbol{\beta}})$ defined in (7.11) and (7.14), respectively. These bounds will be used in the next section for proving asymptotic optimality of the detection–identification rule δ_A with suitable thresholds.

Theorem 7.2. *Let, for some $\mu \ge 0$, the prior distribution belong to class $\mathbf{C}(\mu)$. Assume that for some positive and finite numbers $I_i(\theta_i)$ ($\theta_i \in \Theta_i$, $i \in \mathscr{N}$) and $I_{ij}(\theta_i, \theta_j)$ ($\theta_i \in \Theta_i$, $\theta_j \in \Theta_j$, $i \in \mathscr{N}$, $j \in \mathscr{N} \setminus \{i\}$) condition $\mathbf{C}_1$ holds. If $\inf_{\theta_j \in \Theta_j} I_{ij}(\theta_i, \theta_j) > 0$ for all $j \ne i$, then for all $r > 0$ and all $\theta_i \in \Theta_i$ and $i \in \mathscr{N}$,*

$$\liminf_{\alpha_{\max}, \beta_{\max} \to 0} \frac{\inf_{\delta \in \mathbb{C}_\pi(\boldsymbol{\alpha}, \boldsymbol{\beta})} \overline{\mathscr{R}}^r_{i,\theta_i}(\delta)}{[\Psi_i(\boldsymbol{\alpha}, \boldsymbol{\beta})]^r} \ge 1 \tag{7.28}$$

and

$$\liminf_{\alpha_{\max}, \beta_{\max} \to 0} \frac{\inf_{\delta \in \mathbb{C}^\star_\pi(\alpha, \bar{\boldsymbol{\beta}})} \overline{\mathscr{R}}^r_{i,\theta_i}(\delta)}{[\Psi^\star_i(\alpha, \bar{\boldsymbol{\beta}})]^r} \ge 1, \tag{7.29}$$

where $\Psi_i(\boldsymbol{\alpha}, \boldsymbol{\beta})$ and $\Psi^\star_i(\alpha, \bar{\boldsymbol{\beta}})$ are defined in (7.26) and (7.27), respectively.

Proof. We only provide the proof of the asymptotic lower bound (7.28). The proof of (7.29) is essentially similar.

Notice that the proof can be split in two parts since if we show that, on one hand, for any rule $\delta \in \mathbb{C}_\pi(\boldsymbol{\alpha}, \boldsymbol{\beta})$

$$\overline{\mathscr{R}}^r_{i,\theta_i}(\delta) \ge \left(\frac{|\log \alpha_i|}{I_i(\theta_i) + \mu}\right)^r (1 + o(1)) \text{ as } \alpha_{\max} \to 0, \beta_{\max} \to 0, \tag{7.30}$$

and on the other hand

$$\overline{\mathscr{R}}^r_{i,\theta_i}(\delta) \ge \max_{j \in \mathscr{N} \setminus \{i\}} \left[\frac{|\log \beta_{ji}|}{\inf_{\theta_j \in \Theta_j} I_{ij}(\theta_i, \theta_j)}\right]^r (1 + o(1)) \text{ as } \alpha_{\max} \to 0, \beta_{\max} \to 0 \tag{7.31}$$

then, obviously, combining inequalities (7.30) and (7.31) yields (7.28).

Part 1: Proof of asymptotic inequalities (7.30).

Changing the measure $\mathsf{P}_\infty \to \mathsf{P}_{k,i,\theta_i}$ and using an argument similar to the proof of Lemma 3.1 with N_α replaced by

$$M^*_{\alpha_i} = \frac{(1-\varepsilon)|\log \alpha_i|}{I_i(\theta_i) + \mu + \varepsilon_1}$$

(see (3.37)) along with the fact that for any $\delta \in \mathbb{C}_\pi(\boldsymbol{\alpha}, \boldsymbol{\beta})$

$$\mathsf{P}_\infty(T \le k, d = i) \le \alpha_i / \Pi_{k-1}, \quad k \ge 1,$$

we obtain

$$\begin{aligned} \mathsf{P}_{k,i,\theta_i}\left\{0 < T - k \le M^*_{\alpha_i}, d = i\right\} &\le e^{(1+\varepsilon)I_i(\theta_i)M^*_{\alpha_i}} \mathsf{P}_\infty\left\{0 < T - k \le M^*_{\alpha_i}, d = i\right\} \\ &+ \mathsf{P}_{k,i,\theta_i}\left\{\frac{1}{M^*_{\alpha_i}} \max_{1 \le n \le M^*_{\alpha_i}} \lambda_{i,\theta_i}(k, k+n) \ge (1+\varepsilon) I_i(\theta_i)\right\}, \end{aligned} \tag{7.32}$$

where

$$\begin{aligned} e^{(1+\varepsilon)I_i(\theta_i)M^*_{\alpha_i}} \mathsf{P}_\infty\left\{0 < T - k \le M^*_{\alpha_i}, d = i\right\} &\le e^{(1+\varepsilon)I_i(\theta_i)M^*_{\alpha_i}} \alpha_i / \Pi_{k-1+M^*_{\alpha_i}} \\ &\le \alpha_i \exp\left\{(1+\varepsilon)I_i(\theta_i)M^*_{\alpha_i} + (k - 1 + M^*_{\alpha_i})\frac{|\log \Pi_{k-1+M^*_{\alpha_i}}|}{k - 1 + M^*_{\alpha_i}}\right\}. \end{aligned}$$

Due to condition (3.20), for all sufficiently large $M^*_{\alpha_i}$ (small α_i), there exists a (small) ε_1 such that $|\log \Pi_{k-1+M^*_{\alpha_i}}|/(k - 1 + M^*_{\alpha_i}) \le \mu + \varepsilon_1$, so for a sufficiently small α_i

$$e^{(1+\varepsilon)I_i(\theta_i)M^*_{\alpha_i}} \mathsf{P}_\infty\left\{0 < T - k \le M^*_{\alpha_i}, d = i\right\} \le \alpha_i \exp\left\{(1+\varepsilon)I_i(\theta_i)M^*_{\alpha_i} + (k - 1 + M^*_{\alpha_i})(\mu + \varepsilon_1)\right\}$$

$$\le \exp\left\{-\varepsilon^2|\log\alpha_i| + |\log\alpha_i| - |\log\alpha_i| + (\mu+\varepsilon_1)(k-1)\right\}$$
$$= \exp\left\{-\varepsilon^2|\log\alpha_i| + (\mu+\varepsilon_1)(k-1)\right\} := \overline{U}_{\alpha_i,k}(\varepsilon,\varepsilon_1) \tag{7.33}$$

for all $\varepsilon \in (0,1)$. Note that $\overline{U}_{\alpha_i,k}(\varepsilon,\varepsilon_1)$ does not depend on δ.

Write

$$K_{\alpha_i} = K_{\alpha_i}(\varepsilon,\mu,\varepsilon_1) = \left\lfloor \frac{\varepsilon^3|\log\alpha_i|}{\mu+\varepsilon_1} \right\rfloor.$$

Using inequalities (7.32) and (7.33), we obtain

$$\begin{aligned}
\mathsf{P}^{\pi}_{i,\theta_i}(0 < T-\nu \le M^*_{\alpha_i}, d=i) &= \sum_{k=0}^{\infty} \pi_k \mathsf{P}_{k,i,\theta_i}\left(0 < T-k \le M^*_{\alpha_i}, d=i\right) \\
&= \sum_{k=0}^{K_{\alpha_i}} \pi_k \mathsf{P}_{k,i,\theta_i}\left(0 < T-k \le M^*_{\alpha_i}, d=i\right) + \sum_{k=K_{\alpha_i}+1}^{\infty} \pi_k \mathsf{P}_{k,i,\theta_i}\left(0 < T-k \le M^*_{\alpha_i}, d=i\right) \\
&\le \sum_{k=0}^{K_{\alpha_i}} \pi_k \overline{U}_{\alpha_i,k}(\varepsilon,\varepsilon_1) + \sum_{k=0}^{K_{\alpha_i}} \pi_k p_{M^*_{\alpha_i},k}(\varepsilon;i,\theta_i) + \sum_{k=K_{\alpha_i}+1}^{\infty} \pi_k \\
&\le \Pi_{K_{\alpha_i}} + \max_{0\le k\le K_{\alpha_i}} \overline{U}_{\alpha_i,k}(\varepsilon,\varepsilon_1) + \sum_{k=0}^{K_{\alpha_i}} \pi_k p_{M^*_{\alpha_i},k}(\varepsilon;i,\theta_i) \\
&= \Pi_{K_{\alpha_i}} + \overline{U}_{\alpha_i,K_{\alpha_i}}(\varepsilon,\varepsilon_1) + \sum_{k=0}^{K_{\alpha_i}} \pi_k p_{M^*_{\alpha_i},k}(\varepsilon;i,\theta_i).
\end{aligned}$$

If $\mu > 0$, by condition (3.20), $\log \Pi_{K_{\alpha_i}} \sim -\mu K_{\alpha_i}$ as $\alpha_{\max} \to 0$, so $\Pi_{K_{\alpha_i}} \to 0$. If $\mu = 0$, this probability goes to 0 as $\alpha_{\max} \to 0$ as well since, by condition (3.21),

$$\Pi_{K_{\alpha_i}} < \sum_{k=K_{\alpha_i}}^{\infty} \pi_k |\log \pi_k| \xrightarrow[\alpha_{\max}\to 0]{} 0.$$

Obviously, the second term $\overline{U}_{\alpha_i,K_{\alpha_i}}(\varepsilon,\varepsilon_1) \to 0$ as $\alpha_{\max} \to 0$. By condition $\mathbf{C}_1$ and Lebesgue's dominated convergence theorem, the third term goes to 0, and therefore, all three terms go to zero as $\alpha_{\max}, \beta_{\max} \to 0$ for all $\varepsilon, \varepsilon_1 > 0$, so that

$$\mathsf{P}^{\pi}_{i,\theta_i}(0 < T-\nu \le M^*_{\alpha_i}, d=i) \to 0 \quad \text{as } \alpha_{\max}, \beta_{\max} \to 0.$$

Since

$$\mathsf{P}^{\pi}_{i,\theta_i}(T-\nu > M^*_{\alpha_i}, d=i) = \mathsf{P}^{\pi}_{i,\theta_i}(T > \nu, d=i) - \mathsf{P}^{\pi}_{i,\theta_i}(0 < T-\nu \le M^*_{\alpha_i}, d=i)$$

and $\mathsf{P}^{\pi}_{i,\theta_i}(T > \nu, d=i) \ge 1-\alpha_i \to 1$ as $\alpha_{\max}, \beta_{\max} \to 0$ for any $\delta \in \mathbb{C}_\pi(\boldsymbol{\alpha},\boldsymbol{\beta})$, it follows that

$$\mathsf{P}^{\pi}_{i,\theta_i}(T-\nu > M^*_{\alpha_i}, d=i) \to 1 \quad \text{as } \alpha_{\max}, \beta_{\max} \to 0.$$

Finally, by the Chebyshev inequality,

$$\overline{\mathscr{R}}^r_{i,\theta_i}(\delta) \ge \mathsf{E}^{\pi}_{i,\theta_i}[(T-\nu)^r, d=i, T>\nu] \ge (M^*_{\alpha_i})^r \mathsf{P}^{\pi}_{i,\theta_i}(T-\nu > M^*_{\alpha_i}, d=i),$$

which implies that for any $\delta \in \mathbb{C}_\pi(\boldsymbol{\alpha},\boldsymbol{\beta})$

$$\overline{\mathscr{R}}^r_{i,\theta_i}(\delta) \ge \left[\frac{(1-\varepsilon)|\log\alpha_i|}{I_i(\theta_i)+\mu+\varepsilon_1}\right]^r (1+o(1)) \quad \text{as } \alpha_{\max}, \beta_{\max} \to 0.$$

Owing to the fact that ε and ε_1 can be arbitrarily small the inequality (7.30) follows.

Part 2: Proof of asymptotic inequalities (7.31).

To prove (7.31) define $M_{\beta_{ji}} = M_{\beta_{ji}}(\varepsilon, \theta_i, \theta_j) = (1-\varepsilon)|\log \beta_{ji}|/I_{ij}(\theta_i, \theta_j)$ and note first that by the Chebyshev inequality, for every $\varepsilon \in (0,1)$ and $r > 0$

$$\mathsf{E}^{\pi}_{i,\theta_i}\left[(T-\nu)^r; d=i; T>\nu\right] \ge M^r_{\beta_{ji}} \mathsf{P}^{\pi}_{i,\theta_i}\left\{T-\nu > M_{\beta_{ji}}, d=i\right\}.$$

Therefore,

$$\mathsf{E}^{\pi}_{i,\theta_i}\left[(T-\nu)^r; d=i; T>\nu\right] \ge \left[\frac{(1-\varepsilon)|\log \beta_{ji}|}{I_{ij}(\theta_i,\theta_j)}\right]^r (1+o(1)) \text{ for all } \theta_j \in \Theta_j \text{ and } j \in \mathcal{N}_0 \setminus \{i\}$$

whenever for all $\delta \in \mathbb{C}_{\pi}(\boldsymbol{\alpha}, \boldsymbol{\beta})$ and all $\varepsilon \in (0,1)$

$$\lim_{\alpha_{\max}, \beta_{\max} \to 0} \mathsf{P}^{\pi}_{i,\theta_i}\left\{T-\nu > M_{\beta_{ji}}, d=i\right\} = 1 \text{ for all } \theta_j \in \Theta_j \text{ and } j \in \mathcal{N}_0 \setminus \{i\} \tag{7.34}$$

and inequality (7.31) follows since ε can be arbitrarily small and

$$\overline{\mathscr{R}}^r_{i,\theta_i}(\delta) \ge \mathsf{E}^{\pi}_{i,\theta_i}\left[(T-\nu)^r; d=i; T>\nu\right].$$

Hence, we now focus on proving (7.34).

Let $\mathscr{A}_{k,\beta} = \{k < T \le k + M_{\beta_{ji}}\}$. For the sake of brevity, we will write $\lambda_{i,j}(k,k+n)$ for the LLR $\lambda_{i,\theta_i;j,\theta_j}(k,k+n)$. Changing the measure $\mathsf{P}_{k,j,\theta_j} \to \mathsf{P}_{k,i,\theta_i}$, for any $C>0$ we obtain

$$\begin{aligned}
\mathsf{P}_{k,j,\theta_j}(d=i, T<\infty) &= \mathsf{E}_{k,j,\theta_j}\left[\mathbb{1}_{\{d=i,T<\infty\}}\right] = \mathsf{E}_{k,i,\theta_i}\left[\mathbb{1}_{\{d=i,T<\infty\}} e^{-\lambda_{i,j}(k,T)}\right] \\
&\ge \mathsf{E}_{k,i,\theta_i}\left[\mathbb{1}_{\{d=i,\mathscr{A}_{k,\beta},\lambda_{i,j}(k,T)<C\}} e^{-\lambda_{i,j}(k,T)}\right] \ge e^{-C} \mathsf{E}_{k,i,\theta_i}\left[\mathbb{1}_{\{d=i,\mathscr{A}_{k,\beta},\lambda_{i,j}(k,T)<C\}}\right] \\
&= e^{-C} \mathsf{P}_{k,i,\theta_i}\left\{\{d=i, \mathscr{A}_{k,\beta}\} \bigcap \{\max_{k<n\le k+M_{\beta_{ji}}} \lambda_{i,j}(k,n) < C\}\right\} \\
&\ge e^{-C}\left[\mathsf{P}_{k,i,\theta_i}(d=i, \mathscr{A}_{k,\beta}) - \mathsf{P}_{k,i,\theta_i}\left\{\max_{1\le n\le M_{\beta_{ji}}} \lambda_{i,j}(k,k+n) \ge C\right\}\right],
\end{aligned}$$

where the last inequality follows from the trivial inequality $\mathsf{P}(A \cap B) \ge \mathsf{P}(A) - \mathsf{P}(B^c)$. It follows that

$$\mathsf{P}_{k,i,\theta_i}(\mathscr{A}_{k,\beta}, d=i) \le \mathsf{P}_{k,j,\theta_j}(d=i, T<\infty) e^C + \mathsf{P}_{k,i,\theta_i}\left\{\max_{1\le n\le M_{\beta_{ji}}} \lambda_{i,j}(k,k+n) \ge C\right\}.$$

Setting $C = (1+\varepsilon) I_{ij}(\theta_i,\theta_j) M_{\beta_{ji}} = (1-\varepsilon^2)|\log \beta_{ji}|$, multiplying both sides by π_k and summing over $k \ge 0$, we obtain

$$\mathsf{P}^{\pi}_{i,\theta_i}\left\{0 < T-\nu < M_{\beta_{ji}}, d=i\right\} \le \mathsf{P}^{\pi}_{j,\theta_j}(d=i, T<\infty) e^{(1-\varepsilon^2)|\log \beta_{ji}|} + \sum_{k=0}^{\infty} \pi_k p_{M_{\beta_{ji}},k}(\varepsilon; i, \theta_i; j, \theta_j).$$

Since $\sup_{\theta_j \in \Theta_j} \mathsf{P}^{\pi}_{j,\theta_j}(d=i, T<\infty) \le \beta_{ji}$, it follows that

$$\mathsf{P}^{\pi}_{i,\theta_i}\left\{0 < T-\nu < M_{\beta_{ji}}, d=i\right\} \le \beta_{ji}^{\varepsilon^2} + \mathsf{P}(\nu > K_\beta) + \sum_{k=0}^{K_\beta} \pi_k p_{M_{\beta_{ji}},k}(\varepsilon; i, \theta_i; j, \theta_j),$$

where K_β is an arbitrarily integer which goes to infinity as $\beta_{\max} \to 0$. Obviously, the first term goes to 0 as $\beta_{\max} \to 0$. The second term $\mathsf{P}(\nu > K_\beta) \to 0$ by conditions (3.20) and (3.21). The third term also goes to 0 due to condition $\mathbf{C}_1$ and Lebesgue's dominated convergence theorem. Hence, for any $\delta \in \mathbb{C}(\boldsymbol{\alpha},\boldsymbol{\beta})$,

$$\mathsf{P}^{\pi}_{i,\theta_i}\left\{0 < T - \nu < M_{\beta_{ji}}, d = i\right\} \to 0 \quad \text{as } \alpha_{\max}, \beta_{\max} \to 0.$$

Since

$$\mathsf{P}^{\pi}_{i,\theta_i}\left\{T - \nu > M_{\beta_{ji}}, d = i\right\} = \mathsf{P}^{\pi}_{i,\theta_i}(T > \nu, d = i) - \mathsf{P}^{\pi}_{i,\theta_i}\left\{0 < T - \nu < M_{\beta_{ji}}, d = i\right\},$$

where $\mathsf{P}^{\pi}_{i,\theta_i}(T > \nu, d = i) \ge 1 - \alpha_i \to 1$ as $\alpha_{\max} \to 0$, this yields (7.34), and therefore, inequalities (7.31) . □

7.6 Asymptotic Optimality of the Detection–Identification Rule δ_A

The following proposition establishes first-order asymptotic approximations for the moments of the detection delay of the detection–identification rule δ_A when thresholds A_{ij} go to infinity regardless of the PFA and PMI constraints. Write $A_{\min} = \min_{i\in\mathcal{N}, j\in\mathcal{N}_0\setminus\{i\}} A_{ij}$.

Proposition 7.1. *Let $r \ge 1$ and let the prior distribution of the change point belong to class $\mathbf{C}(\mu)$. Assume that for some $0 < I_i(\theta_i) < \infty$, $\theta_i \in \Theta_i$, $i \in \mathcal{N}$ and $0 < I_{ij}(\theta_i, \theta_j) < \infty$, $\theta_i \in \Theta_i$, $\theta_j \in \Theta_j$, $i \in \mathcal{N}$, $j \in \mathcal{N} \setminus \{i\}$ right-tail and left-tail conditions $\mathbf{C}_1$ and $\mathbf{C}_2$ are satisfied and that $\inf_{\theta_j\in\Theta_j} I_{ij}(\theta_i, \theta_j) > 0$ for all $j \in \mathcal{N} \setminus \{i\}$, $i \in \mathcal{N}$. Then, for all $0 < m \le r$*

$$\overline{\mathscr{R}}^m_{i,\theta_i}(\delta_A) \sim [\Psi_i(A, \theta_i, \mu)]^m \quad \text{as } A_{\min} \to \infty \quad \text{for all } \theta_i \in \Theta_i \text{ and } i \in \mathcal{N}, \tag{7.35}$$

where

$$\Psi_i(A, \theta_i, \mu) = \max\left\{\frac{\log A_{i0}}{I_i(\theta_i) + \mu}, \max_{j\in\mathcal{N}\setminus\{i\}} \frac{\log A_{ij}}{\inf_{\theta_j\in\Theta_j} I_{ij}(\theta_i, \theta_j)}\right\}. \tag{7.36}$$

In order to prove this proposition we need the following lemma which generalizes Lemma 3.2. For $i = 1, \dots, N$, define $\lambda_{i,W}(k, k+n) = \log LR_{i,W}(k, k+n)$, $\lambda^\pi_i(n) = \log[\sum_{k=-1}^{n-1} \pi_k \sup_{\theta_i\in\Theta_i} LR_{i,\theta_i}(k, n)]$,

$$\widetilde{\Psi}_i(A, \pi_k, \theta_i, \mu, \varepsilon) = \max\left\{\frac{\log(A_{i0}/\pi_k)}{I_i(\theta_i) + \mu - \varepsilon}, \max_{j\in\mathcal{N}\setminus\{i\}} \frac{\log(A_{ij}/\pi_k)}{\inf_{\theta_j\in\Theta_j} I_{ij}(\theta_i, \theta_j) - \varepsilon}\right\}, \tag{7.37}$$

$$M_i(A) = M_i(A, \pi_k, \theta_i, \mu, \varepsilon) = 1 + \left\lfloor \widetilde{\Psi}_i(A, \pi_k, \theta_i, \mu, \varepsilon) \right\rfloor.$$

Lemma 7.1. *Let $r \ge 1$ and let the prior distribution of the change point satisfy condition* (3.20)*. Then, for a sufficiently large $A_{\min}$, any $0 < \varepsilon < J_{ij}(\theta_i, \mu)$ and all $k \in \mathbb{Z}_+$,*

$$\begin{aligned} \mathsf{E}_{k,i,\theta_i}[(T_A - k)^+]^r &\le \left[1 + \widetilde{\Psi}_i(A, \pi_k, \theta_i, \mu, \varepsilon)\right]^r \\ &\quad + r2^{r-1} \sum_{n=M_i(A)}^{\infty} n^{r-1} \mathsf{P}_{k,i,\theta_i}\left\{\frac{1}{n} \inf_{\vartheta\in\Gamma_{\varkappa,\theta_i}} \lambda_{i,\vartheta}(k, k+n) < I_i(\theta_i) - \varepsilon\right\}, \end{aligned} \tag{7.38}$$

where $J_{ij}(\theta_i, \mu) = \min\left\{I_i(\theta_i) + \mu, \min_{j\in\mathcal{N}\setminus\{i\}} \inf_{\theta_j\in\Theta_j} I_{ij}(\theta_i, \theta_j)\right\}$.

Proof. For $k \in \mathbb{Z}_+$, define the exit times

$$\tau_i^{(k)}(A) = \inf\left\{n \geq 1 : \lambda_{i,W}(k,k+n) - \lambda_j^{\pi}(k+n) \geq \log(A_{ij}/\pi_k)\ \forall\, j \in \mathscr{N}_0 \setminus \{i\}\right\},\ i \in \mathscr{N},$$

where $\lambda_0^{\pi}(k+n) = \log \mathsf{P}(\nu \geq k+n) = \log \Pi_{k+n-1}$.

Obviously, for any $n > k$ and $k \in \mathbb{Z}_+$,

$$\log S_{ij}^{\pi}(n) \geq \log\left(\frac{\pi_k LR_{i,W}(k,n)}{\sum_{\ell=-1}^{n-1} \pi_\ell \sup_{\theta_j \in \Theta_j} LR_{j,\theta_j}(\ell,n)}\right) = \lambda_{i,W}(k,n) - \lambda_j^{\pi}(n) + \log \pi_k,$$

so for every set A of positive thresholds $A_{ij} > 0$, $(T_A - k)^+ \leq (T_A^{(i)} - k)^+ \leq \tau_i^{(k)}(A)$ and, hence, $\mathsf{E}_{k,i,\theta_i}[(T_A^W - k)^+]^r \leq \mathsf{E}_{k,i,\theta_i}[(\tau_i^{(k)}(A))^r]$.

Setting $\tau = \tau_i^{(k)}(A)$ and $N = M_i(A)$ in inequality (A.1) in Lemma A.1 (see Appendix A) we obtain that the following inequality holds:

$$\mathsf{E}_{k,i,\theta_i}\left[\left(\tau_i^{(k)}(A)\right)^r\right] \leq [M_i(A)]^r + r2^{r-1} \sum_{n=M_i(A)}^{\infty} n^{r-1} \mathsf{P}_{k,\theta}\left(\tau_i^{(k)}(A) > n\right). \tag{7.39}$$

Next, we have

$$\begin{aligned}\mathsf{P}_{k,i,\theta_i}\left(\tau_i^{(k)}(A) > n\right) &= \mathsf{P}_{k,i,\theta_i}\left\{\frac{\lambda_{i,W}(k,k+n) - \lambda_j^{\pi}(k+n)}{n} < \frac{1}{n}\log\left(\frac{A_{ij}}{\pi_k}\right), j \in \mathscr{N}_0 \setminus \{i\}\right\} \\ &\leq \mathsf{P}_{k,i,\theta_i}\left\{\frac{\lambda_{i,W}(k,k+n) - \log \Pi_{k+n-1}}{n} < \frac{1}{n}\log\left(\frac{A_{i0}}{\pi_k}\right)\right\}.\end{aligned}$$

Let

$$\widetilde{M}_i(A_{i0}) = 1 + \left\lfloor \frac{\log(A_{i0}/\pi_k)}{I_i(\theta_i) + \mu - \varepsilon} \right\rfloor.$$

Clearly, for all $n \geq \widetilde{M}_i(A_{i0})$ the last probability does not exceed the probability

$$\mathsf{P}_{k,i,\theta_i}\left\{\frac{\lambda_{i,W}(k,k+n)}{n} < I_i(\theta_i) + \mu - \varepsilon - \frac{|\log \Pi_{k+n-1}|}{n}\right\}$$

and, by condition $\mathbf{CP}_1$, for a sufficiently large value of A_{i0} there exists a small κ such that

$$\left|\mu - \frac{|\log \Pi_{k+\widetilde{M}_i(A_{i0})-1}|}{\widetilde{M}_i(A_{i0})}\right| < \kappa.$$

Therefore, for all sufficiently large n,

$$\mathsf{P}_{k,i,\theta_i}\left(\tau_i^{(k)}(A) > n\right) \leq \mathsf{P}_{k,i,\theta_i}\left(\frac{1}{n}\lambda_{i,W}(k,k+n) < I_i(\theta_i) - \varepsilon + \kappa\right).$$

Also,

$$\lambda_{i,W}(k,k+n) \geq \inf_{\vartheta \in \Gamma_{\varkappa,\theta_i}} \lambda_{i,\vartheta}(k,k+n) + \log W_i(\Gamma_{\varkappa,\theta_i}),$$

where $\Gamma_{\varkappa,\theta_i} = \{\vartheta \in \Theta_i : |\vartheta - \theta_i| < \varkappa\}$. Thus, for all sufficiently large n and $A_{\min}$, for which $\kappa + |\log W(\Gamma_{\varkappa,\theta_i})|/n < \varepsilon/2$, we have

$$\mathsf{P}_{k,i,\theta_i}\left(\tau_i^{(k)}(A) > n\right) \leq \mathsf{P}_{k,i,\theta_i}\left(\frac{1}{n}\inf_{\vartheta \in \Gamma_{\varkappa,\theta_i}} \lambda_{i,\vartheta}(k,k+n) < I_i(\theta_i) - \varepsilon + \kappa + \frac{1}{n}|\log W(\Gamma_{\varkappa,\theta_i})|\right)$$

$$\le \mathsf{P}_{k,i,\theta_i}\left(\frac{1}{n}\inf_{\vartheta\in\Gamma_{\varkappa,\theta_i}} \lambda_{i,\vartheta}(k,k+n) < I_i(\theta_i)-\varepsilon/2\right). \tag{7.40}$$

Using (7.39) and (7.40) yields inequality (7.38) and the proof is complete. □

Proof of Proposition 7.1. By Theorem 7.1, the rule δ_A belongs to class $\mathbb{C}_\pi(\boldsymbol{\alpha},\boldsymbol{\beta})$ when

$$\alpha_i = \frac{1}{1+A_{i0}}, \quad i\in\mathscr{N}; \quad \beta_{ij} = \frac{1+A_{i0}}{A_{i0}A_{ji}}, \quad j\in\mathscr{N}\setminus\{i\}, i\in\mathscr{N},$$

and hence, Theorem 7.2 implies (under condition $\mathbf{C}_1$) the lower bounds

$$\overline{\mathscr{R}}^r_{i,\theta_i}(\delta_A) \ge [\Psi_i(A,\theta_i,\mu)]^r(1+o(1)) \quad \text{as } A_{\min}\to\infty \tag{7.41}$$

which hold for all $r>0$, $\theta_i\in\Theta_i$, and $i\in\mathscr{N}$. Thus, to prove the validity of the asymptotic approximation (7.35) it suffices to show that, under the left-tail condition $\mathbf{C}_2$, for $0<m\le r$ and all $\theta_i\in\Theta_i$ and $i\in\mathscr{N}$

$$\overline{\mathscr{R}}^m_{i,\theta_i}(\delta_A) \le [\Psi_i(A,\theta_i,\mu)]^m(1+o(1)) \quad \text{as } A_{\min}\to\infty. \tag{7.42}$$

It follows from Lemma 7.1 that for any $0<\varepsilon<J_{ij}(\theta_i,\mu)$

$$\begin{aligned}\mathsf{E}^\pi_{i,\theta_i}[(T_A-\nu)^+]^r &= \sum_{k=-1}^{\infty}\pi_k \mathsf{E}_{k,i,\theta_i}\left[(T_A-k)^+\right]^r \\ &\le \sum_{k=-1}^{\infty}\pi_k\left[1+\widetilde{\Psi}_i(A,\pi_k,\theta_i,\mu,\varepsilon)\right]^r + r2^{r-1}\Upsilon_r(\varepsilon;i,\theta_i),\end{aligned} \tag{7.43}$$

where $\Upsilon_r(\varepsilon;i,\theta_i)$ is defined in (7.16). Recall that we set $T_A-k=T_A$ for $k=-1$. Applying inequality (7.43) together with $1-\mathsf{PFA}^\pi_i(T_A)\ge A_{i0}/(1+A_{i0})$ (see (7.20)) yields

$$\begin{aligned}\overline{\mathscr{R}}^r_{i,\theta_i}(\delta_A) &= \frac{\sum_{k=-1}^{\infty}\pi_k\mathsf{E}_{k,i,\theta_i}[(T_A-k)^+]^r}{1-\mathsf{PFA}^\pi_i(T_A)} \\ &\le \frac{\sum_{k=-1}^{\infty}\pi_k\left[1+\widetilde{\Psi}_i(A,\pi_k,\theta_i,\mu,\varepsilon)\right]^r + r2^{r-1}\Upsilon_r(\varepsilon;i,\theta_i)}{A_{i0}/(1+A_{i0})}.\end{aligned} \tag{7.44}$$

By condition $\mathbf{C}_2$, $\Upsilon_r(\varepsilon;i,\theta_i)<\infty$ for any $\varepsilon>0$ and any $\theta_i\in\Theta_i$ and, by condition (3.21), $\sum_{k=0}^{\infty}\pi_k|\log\pi_k|^r<\infty$. This implies that, as $A_{\min}\to\infty$, for all $0<m\le r$, all $\theta_i\in\Theta_i$, and all $i\in\mathscr{N}$

$$\overline{\mathscr{R}}^r_{i,\theta_i}(\delta_A) \le \left[\widetilde{\Psi}_i(A,\pi_k=1,\theta_i,\mu,\varepsilon)\right]^r(1+o(1)).$$

Since ε can be arbitrarily small and $\lim_{\varepsilon\to0}\widetilde{\Psi}_i(A,\pi_k=1,\theta_i,\mu,\varepsilon)=\Psi_i(A,\theta_i,\mu)$, the upper bound (7.42) follows and the proof of the asymptotic approximation (7.35) is complete. □

Theorem 7.1, Theorem 7.2 and Proposition 7.1 allow us to conclude that the detection–identification rule δ_A is asymptotically first-order optimal in classes $\mathbb{C}_\pi(\boldsymbol{\alpha},\boldsymbol{\beta})$ and $\mathbb{C}^\star_\pi(\alpha,\bar{\boldsymbol{\beta}})$ as $\alpha_{\max},\beta_{\max}\to0$.

Theorem 7.3. *Let $r\ge1$ and let the prior distribution of the change point belong to class $\mathbf{C}(\mu)$. Assume that for some $0<I_i(\theta_i)<\infty$, $\theta_i\in\Theta_i$, $i\in\mathscr{N}$ and $0<I_{ij}(\theta_i,\theta_j)<\infty$, $\theta_i\in\Theta_i$, $\theta_j\in\Theta_j$, $i\in\mathscr{N}$, $j\in\mathscr{N}\setminus\{i\}$ right-tail and left-tail conditions $\mathbf{C}_1$ and $\mathbf{C}_2$ are satisfied and that $\inf_{\theta_j\in\Theta_j}I_{ij}(\theta_i,\theta_j)>0$ for all $j\in\mathscr{N}\setminus\{i\}$, $i\in\mathscr{N}$.*

(i) *If thresholds A_{i0}, $i\in\mathscr{N}$ and A_{ij}, $j\in\mathscr{N}\setminus\{i\}$, $i\in\mathscr{N}$ are so selected that $\mathsf{PFA}^\pi_i(\delta_A)\le\alpha_i$, $\mathsf{PMI}_{ij}(\delta_A)\le\beta_{ij}$ and $\log A_{i0}\sim|\log\alpha_i|$, $\log A_{ij}\sim|\log\beta_{ji}|$ as $\alpha_{\max},\beta_{\max}\to0$, in particular as $A_{i0}=$*

$(1-\alpha_i)/\alpha_i$ and $A_{ij} = [(1-\alpha_j)\beta_{ji}]^{-1}$, then δ_A is first-order asymptotically optimal as $\alpha_{\max}, \beta_{\max} \to 0$ in class $\mathbb{C}_\pi(\boldsymbol{\alpha}, \boldsymbol{\beta})$, minimizing moments of the detection delay up to order r: for all $0 < m \le r$,

$$\inf_{\delta \in \mathbb{C}_\pi(\boldsymbol{\alpha},\boldsymbol{\beta})} \overline{\mathscr{R}}^m_{i,\theta_i}(\delta) \sim \max\left\{\frac{|\log \alpha_i|}{I_i(\theta_i)+\mu}, \max_{j\in\mathscr{N}\setminus\{i\}} \frac{|\log \beta_{ji}|}{\inf_{\theta_j\in\Theta_j} I_{ij}(\theta_i,\theta_j)}\right\}^m \tag{7.45}$$
$$\sim \overline{\mathscr{R}}^m_{i,\theta_i}(\delta_A) \quad \text{as } \alpha_{\max}, \beta_{\max} \to 0 \text{ for all } \theta_i \in \Theta_i \text{ and all } i \in \mathscr{N}.$$

(ii) *If thresholds $A_{i0} = A_0$ and $A_{ij} = A_j$, $j \in \mathscr{N} \setminus \{i\}$, $i \in \mathscr{N}$ are so selected that $\mathsf{PFA}^\pi(\delta_A) \le \alpha$, $\mathsf{PMI}_i(\delta_A) \le \bar{\beta}_i$ and $\log A_0 \sim |\log \alpha|$, $\log A_j \sim |\log \bar{\beta}_j|$ as $\alpha, \bar{\beta}_{\max} \to 0$, in particular as $A_0 = N(1-\alpha/N)/\alpha$ and $A_j = (N-1)[(1-\alpha/N)\bar{\beta}_j]^{-1}$, then δ_A is first-order asymptotically optimal as $\alpha, \bar{\beta}_{\max} \to 0$ in class $\mathbb{C}^\star_\pi(\alpha, \bar{\boldsymbol{\beta}})$, minimizing moments of the detection delay up to order r: for all $0 < m \le r$,*

$$\inf_{\delta \in \mathbb{C}^\star_\pi(\alpha,\bar{\boldsymbol{\beta}})} \overline{\mathscr{R}}^m_{i,\theta_i}(\delta) \sim \max\left\{\frac{|\log \alpha|}{I_i(\theta_i)+\mu}, \max_{j\in\mathscr{N}\setminus\{i\}} \frac{|\log \bar{\beta}_j|}{\inf_{\theta_j\in\Theta_j} I_{ij}(\theta_i,\theta_j)}\right\}^m \tag{7.46}$$
$$\sim \overline{\mathscr{R}}^m_{i,\theta_i}(\delta_A) \quad \text{as } \alpha, \bar{\beta}_{\max} \to 0 \text{ for all } \theta_i \in \Theta_i \text{ and all } i \in \mathscr{N}.$$

Proof. Proof of (i). Setting $\log A_{i0} \sim |\log \alpha_i|$ and $\log A_{ij} \sim |\log \beta_{ji}|$ in (7.35) yields as $\alpha_{\max}, \beta_{\max} \to 0$

$$\overline{\mathscr{R}}^m_{i,\theta_i}(\delta_A) \sim \max\left\{\frac{|\log \alpha_i|}{I_i(\theta_i)+\mu}, \max_{j\in\mathscr{N}\setminus\{i\}} \frac{|\log \beta_{ji}|}{\inf_{\theta_j\in\Theta_j} I_{ij}(\theta_i,\theta_j)}\right\}^m, \quad i \in \mathscr{N}. \tag{7.47}$$

In particular, $\log A_{i0} \sim |\log \alpha_i|$ and $\log A_{ij} \sim |\log \beta_{ji}|$ if $A_{i0} = (1-\alpha_i)/\alpha_i$ and $A_{ij} = [(1-\alpha_j)\beta_{ji}]^{-1}$, and by Theorem 7.1, $\mathsf{PFA}^\pi_i(\delta_A) \le \alpha_i$ and $\mathsf{PMI}_{ij}(\delta_A) \le \beta_{ij}$ with this choice of thresholds (see 7.24). Comparing asymptotic approximations (7.47) with the lower bounds (7.28) in Theorem 7.2 completes the proof of (7.45).

Proof of (ii). Setting $\log A_0 \sim |\log \alpha|$ and $\log A_i \sim |\log \bar{\beta}_j|$ in (7.35) yields as $\alpha_{\max}, \bar{\beta}_{\max} \to 0$

$$\overline{\mathscr{R}}^m_{i,\theta_i}(\delta_A) \sim \max\left\{\frac{|\log \alpha|}{I_i(\theta_i)+\mu}, \max_{j\in\mathscr{N}\setminus\{i\}} \frac{|\log \bar{\beta}_j|}{\inf_{\theta_j\in\Theta_j} I_{ij}(\theta_i,\theta_j)}\right\}^m, \quad i \in \mathscr{N}. \tag{7.48}$$

In particular, $\log A_0 \sim |\log \alpha|$ and $\log A_j \sim |\log \bar{\beta}_j|$ if $A_0 = N(1-\alpha/N)/\alpha$ and $A_j = (N-1)[(1-\alpha/N)\bar{\beta}_j]^{-1}$, and by Theorem 7.1, $\mathsf{PFA}^\pi(\delta_A) \le \alpha$ and $\mathsf{PMI}_i(\delta_A) \le \bar{\beta}_i$ with this choice of thresholds (see 7.25). Comparing asymptotic approximations (7.48) with the lower bounds (7.29) in Theorem 7.2 completes the proof of (7.46). □

Remark 7.1. It can be also shown that asymptotics (7.45) and (7.46) hold for the conditional moments of the detection delay $\mathscr{R}^m_{k,i,\theta_i}(\delta) = \mathsf{E}_{k,i,\theta_i}[(T-k)^m | T > k]$ for all fixed $k \in \mathbb{Z}_+$, i.e., that the rule δ_A is asymptotically uniformly optimal to first order for all change points.

Remark 7.2. If the prior distribution $\pi = \pi^\alpha$ depends on the PFA α and conditions $\mathbf{CP}^{(\alpha)}_1$, $\mathbf{CP}^{(\alpha)}_2$, and $\mathbf{CP}^{(\alpha)}_3$ in Chapter 3 hold (see (3.65)–(3.67)) with $\mu_\alpha \to 0$ as $\alpha \to 0$, then a modification of the preceding argument along the lines of the proofs presented in Section 3.4 for the single stream case can be used to show that the assertions of Theorem 7.3 hold with $\mu = 0$.

Note that conditions (7.17) are satisfied if

$$\frac{1}{n}\lambda_{i,\theta_i;j,\theta_j}(k,k+n) \xrightarrow[n\to\infty]{\mathsf{P}_{i,\theta_i}\text{-a.s.}} I_{ij}(\theta_i,\theta_j)$$

(see Lemma B.1 in Appendix B). Assume also that for some positive and finite numbers $I_{0,i}(\theta_i)$, $i \in \mathcal{N}$,

$$-\frac{1}{n}\lambda_{i,\theta_i}(k,k+n) \xrightarrow[n\to\infty]{\mathsf{P}_\infty-\text{a.s.}} I_{0,i}(\theta_i).$$

In particular, in the i.i.d. case, these conditions hold with

$$\begin{aligned} I_{ij}(\theta_i,\theta_j) &\equiv \mathcal{K}_{ij}(\theta_i,\theta_j) = \int \left(\log\frac{f_{i,\theta_i}(x)}{f_{j,\theta_j}(x)}\right) f_{i,\theta_i}(x)\mathrm{d}x, \\ I_{0,i}(\theta_i) &\equiv \mathcal{K}_{0,i}(\theta_i) = \int \left(\log\frac{g_i(x)}{f_{i,\theta_i}(x)}\right) f_{i,\theta_i}(x)\mathrm{d}x \end{aligned} \tag{7.49}$$

being the Kullback–Leibler information numbers. Then, $I_{ij}(\theta_i,\theta_j) = I_i(\theta_i) + I_{0,j}(\theta_j) \ge I_i(\theta_i)$. Therefore, if the prior distribution of the change point is heavy-tailed (i.e., $\mu = 0$) and the PFA is smaller than the PMI, $\alpha_i < \beta_{ji}$, $\alpha < \bar\beta_j$, which is typical in many applications, then asymptotics (7.45) and (7.46) are reduced to

$$\inf_{\delta\in\mathbb{C}_\pi(\boldsymbol{\alpha},\boldsymbol{\beta})} \overline{\mathcal{R}}^m_{i,\theta_i}(\delta) \sim \left(\frac{|\log\alpha_i|}{I_i(\theta_i)}\right)^m \sim \overline{\mathcal{R}}^m_{i,\theta_i}(\delta_A) \quad \text{as } \alpha_{\max},\beta_{\max}\to 0. \tag{7.50}$$

and

$$\inf_{\delta\in\mathbb{C}^\star_\pi(\alpha,\boldsymbol{\beta})} \overline{\mathcal{R}}^m_{i,\theta_i}(\delta) \sim \left(\frac{|\log\alpha|}{I_i(\theta_i)}\right)^m \sim \overline{\mathcal{R}}^m_{i,\theta_i}(\delta_A) \quad \text{as } \alpha,\bar\beta_{\max}\to 0. \tag{7.51}$$

7.7 An Example: Detection–Identification of Signals with Unknown Amplitudes in a Multichannel System

Consider Example 6.1 of Chapter 6 in the context of detection and identification. To be more specific, there is an N-channel sensor system. The observations $X_n(i)$ in the ith channel have the form

$$X_n(i) = \theta_i S_{i,n} \mathbb{1}_{\{n>\nu\}} + \xi_{i,n}, \quad n\ge 1,\ i=1,\dots,N,$$

where $\theta_i S_{i,n}$ is a deterministic signal (from an object) with an unknown intensity $\theta_i > 0$ and $\{\xi_{i,n}\}_{n\in\mathbb{Z}_+}$, $i\in\mathcal{N}$ are mutually independent noises which are AR(p) Gaussian stable processes that obey recursions

$$\xi_{i,n} = \sum_{t=1}^{p} \beta_{i,t}\xi_{i,n-t} + w_{i,n}, \quad n\ge 1. \tag{7.52}$$

Here $\{w_{i,n}\}_{n\ge1}$, $i=1,\dots,N$, are mutually independent i.i.d. Gaussian $\mathcal{N}(0,\sigma^2)$ sequences ($\sigma>0$). The coefficients $\beta_{i,1},\dots,\beta_{i,p}$ and variance σ^2 are known.

It is assumed that there is only one object and the signal from this object may appear only in one channel and should be detected quickly. Also, the number of a channel where the signal appears should be identified along with detection.

Define $\widetilde{Y}_{i,n} = Y_{i,n} - \sum_{t=1}^{p_n}\beta_{i,t}Y_{i,n-t}$, where $p_n = p$ if $n>p$ and $p_n = n$ if $n\le p$. The LLRs have the form

$$\lambda_{i,\theta_i}(k,k+n) = \frac{\theta_i}{\sigma^2}\sum_{t=k+1}^{k+n}\widetilde{S}_{i,t}\widetilde{X}_{i,t} - \frac{\theta_i^2}{2\sigma^2}\sum_{t=k+1}^{k+n}\widetilde{S}^2_{i,t},$$
$$\lambda_{i,\theta_i;j,\theta_j}(k,k+n) = \lambda_{i,\theta_i}(k,k+n) - \lambda_{j,\theta_j}(k,k+n).$$

Under measure $\mathsf{P}_{k,i,\vartheta}$ the LLR $\lambda_{i,\theta_i;j,\theta_j}(k,k+n)$ is a Gaussian process (with independent non-identically distributed increments) with mean and variance

$$\begin{aligned}\mathsf{E}_{k,i,\vartheta}[\lambda_{i,\theta_i;j,\theta_j}(k,k+n)] &= \frac{1}{2\sigma^2}\left[(2\theta_i\vartheta-\theta_i^2)\sum_{t=k+1}^{k+n}\widetilde{S}_{i,t}^2+\theta_j^2\sum_{t=k+1}^{k+n}\widetilde{S}_{j,t}^2\right],\\ \mathsf{Var}_{k,i,\vartheta}[\lambda_{i,\theta_i;j,\theta_j}(k,k+n)] &= \frac{1}{\sigma^2}\left[\theta_i^2\sum_{t=k+1}^{k+n}\widetilde{S}_{i,t}^2+\theta_j^2\sum_{t=k+1}^{k+n}\widetilde{S}_{j,t}^2\right].\end{aligned} \tag{7.53}$$

Let $\Theta_i=(0,\infty)$, $i\in\mathscr{N}$ and assume that

$$\lim_{n\to\infty}\frac{1}{n}\sup_{k\in\mathbb{Z}_+}\sum_{t=k+1}^{k+n}\widetilde{S}_{i,t}^2=Q_i,$$

where $0<Q_i<\infty$. Then for all $k\in\mathbb{Z}_+$ and $\theta_i,\theta_j\in(0,\infty)$

$$\begin{aligned}\frac{1}{n}\lambda_{i,\theta_i;j,\theta_j}(k,k+n) &\xrightarrow[n\to\infty]{\mathsf{P}_{k,i,\theta_i}-\text{a.s.}}\frac{\theta_i^2Q_i+\theta_j^2Q_j}{2\sigma^2}=I_{ij}(\theta_i,\theta_j), \quad j\in\mathscr{N}\setminus\{i\},\ i\in\mathscr{N},\\ \frac{1}{n}\lambda_{i,\theta_i}(k,k+n) &\xrightarrow[n\to\infty]{\mathsf{P}_{k,i,\theta_i}-\text{a.s.}}\frac{\theta_i^2Q_i}{2\sigma^2}=I_i(\theta_i), \quad i\in\mathscr{N},\end{aligned}$$

so that condition $\mathbf{C}_1$ holds by Lemma B.1 in Appendix B. Furthermore, since all moments of the LLR are finite it can be shown, as in Example 3.5 in Chapter 3, that condition $\mathbf{C}_2$ holds for all $r\geq 1$. Obviously, $\inf_{\theta_j\in(0,\infty)}I_{ij}(\theta_i,\theta_j)=\theta_i^2Q_i/(2\sigma^2)=I_i(\theta_i)>0$. Therefore, by Theorem 7.3, the detection–identification rule δ_A is asymptotically first-order optimal with respect to all positive moments of the detection delay and asymptotic formulas (7.45) and (7.46) hold with

$$\inf_{\theta_j\in(0,\infty)}I_{ij}(\theta_i,\theta_j)=I_i(\theta_i)=\frac{\theta_i^2Q_i}{2\sigma^2}.$$

If $\max_{j\neq i}\beta_{ji}\geq\alpha_i$, $\max_{j\neq i}\beta_j\geq\alpha$, and $\mu=0$, then asymptotic formulas (7.50) and (7.51) hold.

7.8 Concluding Remarks

1. While we focused on the multistream detection–identification problem (7.1), it should be noted that similar results also hold in the "scalar" detection–isolation problem when the observations $\{X_n\}_{n\geq 1}$ represent either a scalar process or a vector process but all components of this process change at time ν. Let $\{f_\theta(X_t|\mathbf{X}^{t-1}),\theta\in\Theta\}$ be a parametric family of densities and for $i=1,\dots,N$ and $\Theta_i\in\Theta$ consider the model

$$\begin{aligned}p(\mathbf{X}^n|\mathsf{H}_{\nu,i},\theta) &= p(\mathbf{X}^n|\mathsf{H}_\infty)=\prod_{t=0}^{n}g(X_t|\mathbf{X}^{t-1}) \quad \text{for } \nu\geq n,\\ p(\mathbf{X}^n|\mathsf{H}_{\nu,i},\theta) &= \prod_{t=0}^{\nu}g(X_t|\mathbf{X}^{t-1})\times\prod_{t=\nu+1}^{n}f_\theta(X_t|\mathbf{X}^{t-1}) \quad \text{for } \nu<n,\ \theta\in\Theta_i,\end{aligned}$$

where $g(X_t|\mathbf{X}^{t-1})$ and $f_\theta(X_t|\mathbf{X}^{t-1})$ are conditional pre- and post-change densities, $\mathbf{X}^n=(X_0,X_1,\dots,X_n)$. In other words, there are N types of change and for the ith type of change the value of the post-change parameter θ belongs to a subset Θ_i of the parameter space Θ. It is necessary to detect and isolate a change as rapidly as possible, i.e., to identify what type of change has occurred. The change detection–identification rule $\delta_A = (d_A, T_A)$ is defined as in (7.8) where the statistics $S_{ij}^{\pi}(n)$ get modified as follows

$$S_{ij}^{\pi}(n) = \frac{\sum_{k=-1}^{n-1} \pi_k \int_{\Theta_i} LR_{\theta}(k,n)\, \mathrm{d}W_i(\theta)}{\sum_{k=-1}^{n-1} \pi_k \sup_{\theta \in \Theta_j} LR_{\theta}(k,n)}, \quad i,j = 1,\dots,N,\ i \neq j;$$

$$S_{i0}^{\pi}(n) = \frac{\sum_{k=-1}^{n-1} \pi_k \int_{\Theta_i} LR_{\theta}(k,n)\, \mathrm{d}W_i(\theta)}{\mathsf{P}(\nu \geq n)}, \quad i = 1,\dots,N$$

with the likelihood ratio

$$LR_{\theta}(k,n) = \prod_{t=k+1}^{n} \frac{f_{\theta}(X_t|\mathbf{X}^{t-1})}{g(X_t|\mathbf{X}^{t-1})}.$$

Write $\lambda_{\theta}(k,k+n) = \log LR_{\theta}(k,n)$ and $\lambda_{\theta,\theta^*}(k,k+n) - \lambda_{\theta^*}(k,k+n)$, where $\lambda_{\theta^*}(k,k+n) = 0$ for $\theta^* = \theta_0$, i.e., when there is no change. Conditions $\mathbf{C}_1$ and $\mathbf{C}_2$ also get modified as

$\mathbf{C}_1$. *There exist positive and finite numbers* $I(\theta,\theta_0) = I(\theta)$, $\theta \in \Theta_i$, $i \in \mathscr{N}$ *and* $I(\theta,\theta^*)$, $\theta^* \in \Theta_j$, $j \in \mathscr{N} \setminus \{i\}$, $\theta \in \Theta_i$, $i \in \mathscr{N}$, *such that for any* $\varepsilon > 0$

$$\lim_{M\to\infty} p_{M,k}(\varepsilon;\theta,\theta^*) = 0 \quad \text{for all } k \in \mathbb{Z}_+,\ \theta \in \Theta_i,\ \theta^* \in \Theta_j\ j \in \mathscr{N}_0 \setminus \{i\},\ i \in \mathscr{N}.$$

$\mathbf{C}_2$. *For any* $\varepsilon > 0$ *and some* $r \geq 1$

$$\Upsilon_r(\varepsilon;\theta) < \infty \quad \text{for all } \theta \in \Theta_i,\ i \in \mathscr{N},$$

where

$$p_{M,k}(\varepsilon;\theta;\theta^*) = \mathsf{P}_{k,\theta}\left\{\frac{1}{M}\max_{1\leq n\leq M} \lambda_{\theta,\theta^*}(k,k+n) \geq (1+\varepsilon) I(\theta,\theta^*)\right\},$$

$$\Upsilon_r(\varepsilon;\theta) = \lim_{\varkappa\to 0} \sum_{n=1}^{\infty} n^{r-1} \sup_{k\in\mathbb{Z}_+} \mathsf{P}_{k,\theta}\left\{\frac{1}{n} \inf_{\{\vartheta\in\Theta:|\vartheta-\theta|<\varkappa\}} \lambda_{\vartheta}(k,k+n) < I(\theta) - \varepsilon\right\}.$$

Essentially the same argument shows that all previous results hold in this case too. In particular, the assertions of Theorem 7.3 are correct:

$$\inf_{\delta\in\mathbb{C}_\pi(\boldsymbol{\alpha},\boldsymbol{\beta})} \overline{\mathscr{R}}_{\theta}^{m}(\delta) \sim \max\left\{\frac{|\log\alpha_i|}{I(\theta)+\mu}, \max_{j\in\mathscr{N}\setminus\{i\}} \frac{|\log\beta_{ji}|}{\inf_{\theta^*\in\Theta_j} I(\theta,\theta^*)}\right\}^m$$
$$\sim \overline{\mathscr{R}}_{\theta}^{m}(\delta_A) \quad \text{as } \alpha_{\max},\beta_{\max} \to 0 \text{ for all } \theta \in \Theta_i \text{ and all } i \in \mathscr{N},$$

i.e., the detection–identification rule δ_A is asymptotically optimal to first order.

Note also that, in general, these asymptotics are not reduced to (7.50) even when $\alpha_i = \beta_{ji}$. Everything depends on the configuration of hypotheses. Indeed, consider the example of Section 7.7 with a single stream and two simple post-change hypotheses $\theta = \theta_1$ or $\theta = \theta_2$, $\theta_2 > \theta_1 > 0$, i.e.,

$$X_n = \theta S_n \mathbb{1}_{\{n>\nu\}} + \xi_n,$$

where ξ_n is given by (7.52) with $\beta_{i,t} = \beta_t$. Then $I(\theta) = \theta^2 Q/(2\sigma^2)$ for $\theta = \theta_1, \theta_2$ and $I(\theta_1,\theta_2) = I(\theta_2,\theta_1) = (\theta_1-\theta_2)^2 Q/(2\sigma^2)$, where $Q = \lim_{n\to\infty} n^{-1}\sum_{t=1}^{n} \widetilde{S}_t^2$. Obviously, $I(\theta_1) > I(\theta_1,\theta_2)$ if $\theta_1 > \theta_2/2$ and $I(\theta_2) > I(\theta_2,\theta_1)$ for any $\theta_1,\theta_2 > 0$. Therefore, if $\alpha_i = \beta_{ji} = \alpha$, we have

$$\overline{\mathscr{R}}_{\theta_1}^{m}(\delta_A) \sim \frac{|\log\alpha|}{I(\theta_1,\theta_2)} = \frac{2\sigma^2|\log\alpha|}{(\theta_1-\theta_2)^2}, \quad \overline{\mathscr{R}}_{\theta_2}^{m}(\delta_A) \sim \frac{|\log\alpha|}{I(\theta_2,\theta_1)} = \frac{2\sigma^2|\log\alpha|}{(\theta_1-\theta_2)^2}$$

for any $\mu \geq 0$.

2. In the case of i.i.d. observations in streams when $g_i(X_t(i)|\mathbf{X}^{t-1}(i)) = g_i(X_t(i))$ and $f_{i,\theta_i}(X_t(i)|\mathbf{X}^{t-1}(i)) = f_{i,\theta_i}(X_t(i))$ in (7.1), the assertions of Theorem 7.3 hold with $I_i(\theta_i) = \mathscr{K}_i(\theta_i)$ and $I_{ij}(\theta_i, \theta_j) = \mathscr{K}_{ij}(\theta_i, \theta_j)$, where $\mathscr{K}_i(\theta_i)$, $\mathscr{K}_{ij}(\theta_i, \theta_j)$ are Kullback–Leibler information numbers given by (7.49), assuming that they are positive and finite. Then condition $\mathbf{C}_1$ is satisfied by the SLLN since

$$\frac{1}{n}\lambda_{i,\theta_i;j,\theta_j}(k,k+n) \xrightarrow[n\to\infty]{\mathsf{P}_{k,i,\theta_i}-\text{a.s.}} \mathscr{K}_{ij}(\theta_i,\theta_j), \quad j \in \mathscr{N}_0\setminus\{i\},\ i \in \mathscr{N}$$

(cf. Lemma B.1 in Appendix B). Condition $\mathbf{C}_2$ typically holds under the $(r+1)$th moment condition for the LLR, $\mathsf{E}_{0,i,\theta_i}|\lambda_{i,\theta_i;j,\theta_j}(0,1)|^{r+1} < \infty$. However, in the i.i.d. case, this condition can be relaxed into the first moment condition.

For simple post-change hypotheses, higher order approximations for the average detection delay may be obtained using nonlinear renewal theory, modifying an argument used in Dragalin et al. [35] and Tartakovsky et al. [164, Sec 4.3.3] for multihypothesis sequential tests. Specifically, in the case of simple hypotheses integration over prior W is not performed, so statistics $S_{ij}^{\pi}(n)$ have the form

$$S_{ij}^{\pi}(n) = \frac{\sum_{k=-1}^{n-1}\pi_k LR_{\theta_i}(k,n)}{\sum_{k=-1}^{n-1}\pi_k LR_{\theta_j}(k,n)}, \quad i,j=1,\dots,N,\ i\neq j;$$

$$S_{i0}^{\pi}(n) = \frac{\sum_{k=-1}^{n-1} LR_{\theta_i}(k,n)}{\mathsf{P}(\nu \ge n)}, \quad i=1,\dots,N.$$

Assume that the prior distribution of the change point π_k is zero-modified Geometric(q,ρ), that $A_{i0} = A_{ij} = A$ and that $\mathscr{K}_i(\theta_i) + |\log(1-\rho)| < \min_{j\neq i}\mathscr{K}_{ij}(\theta_i,\theta_j)$. Then, it can be proved that

$$\overline{\mathscr{R}}_{\theta_i}^{1}(T_A) = \overline{\mathscr{R}}_{\theta_i}^{1}(T_A^{(i)}) + o(1) \quad \text{as } A\to\infty$$

and, as in Section 3.9, using the nonlinear renewal theory for perturbed random walks (see Appendix F) one can also derive higher order approximations for the PFA and the expected detection delay. Specifically, applying the Second Nonlinear Renewal Theorem (Theorem F.2) yields

$$\overline{\mathscr{R}}_{i\theta_i}^{1}(T_A^{(i)}) = \frac{1}{\mathscr{K}_i + |\log(1-\rho)|}\left[\log(A/\rho) - C_i(\rho,,\mathscr{K}_i) + \varkappa_i(\rho,,\mathscr{K}_i)\right] + o(1),$$

where all constants are defined similarly to that in Section 3.9 for a single stream case, assuming that we deal only with one stream i.

3. For independent observations as well as for many Markov and certain hidden Markov models the decision statistics $S_{ij}^{\pi}(n)$ can be computed effectively, so implementation of the proposed detection–identification rule is not an issue. Still, in general, as already noticed in Section 3.10 and Section 6.10, the computational complexity and memory requirements of the rule δ_A are high. To avoid this complication, the rule δ_A can be modified into a window-limited version where the summation in the statistics $S_{ij}^{\pi}(n)$ over potential change points k is restricted to the sliding window of size ℓ. More specifically, in the Markov times $T_A^{(i)}$, defined in (7.7), the statistics $S_{ij}^{\pi}(n)$ are replaced by the statistics

$$\widehat{S}_{ij}^{\pi}(n) = S_{ij}^{\pi}(n) \quad i,j=1,\dots,N,\ i\neq j \text{ for } n\le \ell,$$

$$\widehat{S}_{ij}^{\pi}(n) = \frac{\sum_{k=n-\ell-1}^{n-1}\pi_k \int_{\Theta_i} LR_{\theta}(k,n)\,\mathrm{d}W_i(\theta_i)}{\sum_{k=n-\ell+1}^{n-1}\pi_k \sup_{\theta_j\in\Theta_j} LR_{j,\theta_j}(k,n)}, \quad i,j=1,\dots,N,\ i\neq j, \text{ for } n>\ell;$$

$$\widehat{S}_{i0}^{\pi}(n) = \frac{\sum_{k=n-\ell-1}^{n-1}\pi_k \int_{\Theta_i} LR_{\theta}(k,n)\,\mathrm{d}W_i(\theta_i))}{\mathsf{P}(\nu\ge n)}, \quad i=1,\dots,N, \text{ for } n>\ell.$$

Following guidelines of Section 3.10, it can be shown that the window-limited version also has first-order asymptotic optimality properties as long as the size of the window $\ell(A)$ approaches infinity as $A \to \infty$ at such a rate that $\ell(A)/\log A \to \infty$, i.e.,

$$\lim_{\alpha_{\max} \to 0} \frac{\ell(\alpha_{\max})}{|\log \alpha_{\max}|} = \infty,$$

assuming that the ratio $|\log \alpha_{\max}|/|\log \beta_{\max}|$ is bounded away from zero and infinity.

4. If $\pi \in \mathbf{C}(\mu = 0)$ or π^{α} depends on α and $\mu_{\alpha} \to 0$ as $\alpha \to 0$, then an alternative detection–identification rule $\delta_A^* = (d^*, T_A^*)$ defined as in (7.7)–(7.8) where in the definition of $T_A^{(i)}$ the statistics $S_{ij}^{\pi}(n)$ are replaced by the statistics

$$R_{ij}(n) = \frac{\sum_{k=0}^{n-1} \int_{\Theta_i} LR_{i,\theta_i}(k,n)\,\mathrm{d}W_i(\theta_i)}{\sum_{k=0}^{n-1} \sup_{\theta_j \in \Theta_j} LR_{j,\theta_j}(k,n)}, \quad i,j = 1,\dots,N, \ i \neq j;$$

$$R_{i0}(n) = \sum_{k=0}^{n-1} \int_{\Theta_i} LR_{i,\theta_i}(k,n)\,\mathrm{d}W_i(\theta_i), \quad i = 1,\dots,N,$$

is also asymptotically optimal to first order. Specifically, with a suitable selection of thresholds asymptotic approximations (7.50) and (7.51) hold for δ_A^*.

8

Applications

In this chapter, we show that the changepoint detection theory developed in previous chapters has clear practical applications. It allows for the development of efficient algorithms that are both easily implemented and have certain optimality properties.

8.1 Application to Object Track Management in Sonar Systems

In this section, we address the problem that is motivated by certain multisensor target track management applications in active sonar systems [16]. In object tracking by sonar systems due to drastically changing undersea environment, in particular, due to heavy clutter generated by reverberation, signal-to-noise ratio (SNR) frequently undergoes drastic changes (fading). As a result, the target's detection probability in single scans can change suddenly between high and low values. It is of interest to examine the performance of track management (in particular, termination) strategies where target detections are modeled according to a two-state HMM with high and low detection states. Obviously, track termination routines can be treated as the quickest change detection problem – terminate the track as rapidly as possible after target disappearance to avoid penalization of the tracker. Below we study three track termination strategies based on the Shiryaev, the SR and the CUSUM change detection rules and compare their performance. Despite the fact that the asymptotic theory developed in Section 3.12 for HMMs suggests that the Shiryaev track termination rule must perform better than the SR rule as long as the generalized Kullback–Leibler information number $\mathcal{K}$ is comparable with the parameter μ that characterizes the amount of information coming from the prior distribution of the change point, it is shown that the SR rule has nearly the same performance as the Shiryaev rule for a wide range of values of the parameter μ, even when $\mu > \mathcal{K}$. At the same time, the CUSUM test performs very poorly compared to the SR rule. Thus, our study allows us to recommend the SR detection rule for practical implementation for track termination problems.

In the context of termination of tracks from objects with drastically fluctuating signal-to-noise ratios in active sonar systems, this drastic fluctuation is modeled as Markovian switches between low and high intensity signals, which lead to low and high probabilities of detection. To be specific, let $Z_n \in \{1,2\}$ be a hidden two-state Markov chain with transition probabilities $\mathsf{P}_j(Z_n = 1|Z_{n-1} = 2) = p_j(2,1) = p$ and $\mathsf{P}_j(Z_n = 2|Z_{n-1} = 1) = p_j(1,2) = \tilde{p}$, $n \geq 1$ and initial stationary distribution $\mathsf{P}_j(Z_0 = 1) = \omega_j(1) = p/(p+\tilde{p})$ for both $j = \infty$ and $j = 0$. The states $Z_n = 1$ and $Z_n = 2$ correspond to the high and low values of the SNR, respectively. The observations $X_n \in \{0,1\}$ are detections in single scans $n \geq 1$, where $X_n = 1$ corresponds to detection and $X_n = 0$ otherwise. Change point ν is the unknown moment of target track disappearance, which should be detected as soon as possible. Under the pre-change hypothesis $\mathsf{H}_\infty : \nu = \infty$ (track exists), the conditional probability of the observation X_n is

$$p(X_n|\mathbf{X}_0^{n-1}, Z_n = l, \mathsf{H}_\infty) = g_l(X_n) \quad \text{for } l = 1,2,\ n \geq 1,$$

and under the post-change hypothesis $\mathsf{H}_k : \nu = k$ (target track disappeared at $\nu = k$), the observations $X_{k+1}, X_{k+2}, \dots$ are i.i.d. with density $f(x)$, i.e.,

$$p(X_n|\mathbf{X}_0^{n-1}, Z_n = l, \mathsf{H}_k) = f(X_n) \quad \text{for } n > k.$$

As a result, this scenario leads to the Bernoulli model with binary observations $X_n \in \{0,1\}$ with probabilities $P_d^1 = P_d^{\text{High}}$ and $P_d^2 = P_d^{\text{Low}}$ being local probabilities of detection (in single scans) for high and low intensity signals, respectively, and P_{fa} is the probability of a false alarm that satisfies inequalities $P_d^1 > P_d^2 > P_{fa}$, i.e.,

$$\begin{aligned} g_l(X_n) &= (P_d^l)^{X_n}(1-P_d^l)^{1-X_n}, \ l = 1,2; \\ f(X_n) &= (P_{fa})^{X_n}(1-P_{fa})^{1-X_n}. \end{aligned} \tag{8.1}$$

The pre-change conditional density is given by

$$p_\infty(X_i|\mathbf{X}_0^{i-1}) = \sum_{l=1}^{2} g_l(X_i)\mathsf{P}(Z_i = l|\mathbf{X}_0^{i-1}), \quad i \ge 1, \tag{8.2}$$

where the prediction term $\mathsf{P}(Z_i = l|\mathbf{X}_0^{i-1}) = P_{i|i-1}(l)$ used for the update of the posterior probability

$$\mathsf{P}(Z_i = l|\mathbf{X}_0^i) := P_i(l) = \frac{g_l(X_i)P_{i|i-1}(l)}{\sum_{s=1}^2 g_s(X_i)P_{i|i-1}(s)}, \tag{8.3}$$

is computed as follows

$$\begin{aligned} P_{i|i-1}(2) &= P_{i-1}(2)(1-p) + P_{i-1}(1)\tilde{p}, \\ P_{i|i-1}(1) &= P_{i-1}(1)(1-\tilde{p}) + P_{i-1}(2)p. \end{aligned} \tag{8.4}$$

Hence, the likelihood ratio $\mathscr{L}_n = f(X_n)/p_\infty(X_n|\mathbf{X}_0^{n-1})$ can be easily computed using formulas (8.1)–(8.4).

In the rest of this section, we suppose that the prior distribution of the change point is zero-modified Geometric(q,ρ), i.e.,

$$\pi_{-1} = q, \quad \pi_k = (1-q)\rho(1-\rho)^k, \quad k \in \mathbb{Z}_+.$$

Then the Shiryaev statistic $R_\rho(n) = S_\pi(n)/\rho$ and the SR statistic $R(n)$ are computed recursively as

$$R_\rho(n) = [1 + R_\rho(n-1)]\frac{\mathscr{L}_n}{1-\rho}, \quad n \ge 1, \ R_\rho(0) = \frac{q}{\rho(1-q)}$$

and

$$R(n) = [1 + R(n-1)]\mathscr{L}_n, \quad n \ge 1, \ R(0) = \ell = q/\rho.$$

In the Shiryaev track termination rule, the target track is terminated at the first time the statistic $R_\rho(n)$ exceeds threshold A and, in the SR track termination rule, at the first time the statistic $R(n)$ exceeds threshold B. That is, the tracks are terminated at the random stopping times

$$T_A = \inf\{n \ge 1 : R_\rho(n) \ge A\}, \quad \widetilde{T}_B = \inf\{n \ge 1 : R(n) \ge B\}.$$

Therefore, both the Shiryaev and the SR track termination rules can be easily implemented on-line.

Note first that condition $\widehat{\mathbf{C}}_1^h$ obviously holds and that condition (3.243) in $\widehat{\mathbf{C}}_2^h$ also holds since

$$\sum_{x=0}^{1} x^{r+1} f(x) < \infty, \quad \sum_{x=0}^{1} x^{r+1} g_l(x) < \infty \quad \text{for all } r \geq 1.$$

Therefore, by Theorem 3.7(ii), the Shiryaev track termination rule is nearly optimal, minimizing asymptotically all positive moments of the detection delay and the SR track termination rule is also asymptotically optimal as long as $\mu = |\log(1-\rho)| \ll \mathscr{K}$.

In order to examine real operating characteristics of both track termination rules we performed extensive Monte Carlo simulations. In Table 8.1, we present the results of simulations with 10^6 MC runs for $P_{fa} = 0.01$ and the following parameters of the HMM: $p = 1/30$, $\tilde{p} = 0.1$, $P_d^1 = 0.9$, $P_d^2 = 0.1$. The values of $\widehat{\mathsf{ADD}}(T)$ and $\widehat{\mathsf{PFA}}(T)$ correspond to MC estimates and $\mathsf{ADD}(T)$ to the first-order asymptotic approximations, i.e.,

$$\mathsf{ADD}(T_A) = \frac{\log A}{\mathscr{K} + |\log(1-\rho)|}, \quad \mathsf{ADD}(\widetilde{T}_B) = \frac{\log B}{\mathscr{K}}. \tag{8.5}$$

The generalized Kullback-Leibler information number was estimated by Monte Carlo as $\mathscr{K} = 0.1047$. The thresholds A and B were adjusted so that the PFAs of both procedures were almost the same and close to the given values of α. The data in the table allow us to conclude that the SR track termination rule has almost the same performance as the Shiryaev track termination rule not only when the parameter ρ of the prior distribution is small, which can be expected from the theory, but practically for any reasonable values of ρ. Even for the value of ρ as large as 0.5 the difference between $\widehat{\mathsf{ADD}}(T_A)$ and $\widehat{\mathsf{ADD}}(\widetilde{T}_B)$ is relatively small for all tested values of α. This fact can be explained by a conjecture made in Tartakovsky and Moustakides [163] that the upper bound $O(1)/B$ on the $\mathsf{PFA}(\widetilde{T}_B)$ (see (3.74)) is not accurate, but rather the formula $\mathsf{PFA}(\widetilde{T}_B) \approx O(1)/B^{s(\rho)}$ with some $s(\rho) > 1$ is a correct approximation for the PFA of the SR rule. This conjecture was confirmed in [163] by numerical study for the i.i.d. exponential model. Our current MC simulations for the HMM also confirm this fact.

Also, the first-order approximations (8.5) are not accurate in most cases. This is especially true for the approximation for $\mathsf{ADD}(\widetilde{T}_B)$. Higher-order asymptotic approximations presented in Section 3.12.3 are difficult to implement since the renewal-theoretic constants cannot be usually computed. Therefore, the development of two-dimensional integral equations for performance metrics and accurate numerical techniques for their solution is a challenging open problem.

In addition, we examined the performance of the popular CUSUM procedure

$$T_C^* = \inf\{n \geq 1 : V(n) \geq C\}, \quad V(n) = \max[1, V(n-1)]\mathscr{L}_n, \quad n \geq 1, \; V(0) = 1.$$

The results of comparison with the Shirayev and SR track termination rules are presented in Table 8.2.[1] It is seen that CUSUM performs poorly in this problem when ρ is not very small. Its performance is still worse than that of the SR rule even for small values of ρ, such as 10^{-4}, but in this case the difference is not dramatic. Thus, we conclude that the SR track termination rule is very efficient and robust (w.r.t. to the parameter of prior ρ), so that we recommend it for practical implementation.

[1]Thanks to V. Spivak for MC simulations.

TABLE 8.1
Operating characteristics of the Shiryaev and SR track termination rules T_A and $\widetilde{T}_B$. HMM parameters: $p = 1/30$, $\tilde{p} = 0.1$, $P_d^1 = 0.9$, $P_d^2 = 0.1$, $P_{fa} = 0.01$. Number of MC runs 10^6.

Parameters Setting			Simulation Results					
q	ρ	α	$\widehat{\mathsf{ADD}}(T_A)$	$\mathsf{ADD}(T_A)$	$\widehat{\mathsf{ADD}}(\widetilde{T}_B)$	$\mathsf{ADD}(\widetilde{T}_B)$	$\widehat{\mathsf{PFA}}(T_A)$	$\widehat{\mathsf{PFA}}(\widetilde{T}_B)$
0	0.5	0.01	4.176	6.113	4.644	46.59	7.62×10^{-3}	7.62×10^{-3}
0	0.5	0.001	7.052	9.009	7.749	68.18	7.56×10^{-4}	7.96×10^{-4}
0	0.5	0.0001	10.17	12.57	10.37	90.50	7.40×10^{-5}	7.70×10^{-5}
0	0.1	0.01	19.36	22.41	19.58	44.96	9.05×10^{-3}	9.07×10^{-3}
0	0.1	0.001	30.27	33.46	30.44	67.14	8.87×10^{-4}	9.76×10^{-4}
0	0.1	0.0001	41.21	44.2	42.58	88.69	9.30×10^{-5}	9.40×10^{-5}
0	0.01	0.01	50.24	40.61	50.3	44.51	9.48×10^{-3}	9.48×10^{-3}
0	0.01	0.001	70.44	60.93	70.47	66.78	9.21×10^{-4}	9.22×10^{-4}
0	0.01	0.0001	90.47	80.46	90.89	88.19	9.80×10^{-5}	9.90×10^{-5}
0	0.001	0.01	75.59	44.17	75.6	44.59	9.40×10^{-3}	9.40×10^{-3}
0	0.001	0.001	97.52	66.02	97.53	66.65	9.34×10^{-4}	9.35×10^{-4}
0	0.001	0.0001	119.30	88.48	119.40	89.33	8.70×10^{-5}	8.70×10^{-5}
0.5	0.5	0.01	4.153	5.959	4.891	44.53	8.62×10^{-3}	9.46×10^{-3}
0.5	0.5	0.001	7.210	9.046	8.049	68.58	7.34×10^{-4}	7.63×10^{-4}
0.5	0.5	0.0001	10.12	11.68	10.50	88.39	9.00×10^{-5}	9.60×10^{-5}
0.5	0.1	0.01	18.30	22.32	18.34	44.79	9.21×10^{-3}	9.87×10^{-3}
0.5	0.1	0.001	29.26	33.27	29.88	66.76	9.23×10^{-4}	9.82×10^{-4}
0.5	0.1	0.0001	40.29	44.3	41.88	88.9	9.10×10^{-5}	9.10×10^{-5}
0.5	0.01	0.01	43.79	40.82	44.3	44.74	9.25×10^{-3}	9.26×10^{-3}
0.5	0.01	0.001	64.08	60.95	64.33	66.81	9.19×10^{-4}	9.28×10^{-4}
0.5	0.01	0.0001	83.98	81.30	84.30	89.11	8.90×10^{-5}	9.40×10^{-5}
0.5	0.001	0.01	59.32	44.23	59.94	44.65	9.34×10^{-3}	9.36×10^{-3}
0.5	0.001	0.001	81.12	65.68	81.79	66.31	9.68×10^{-4}	9.69×10^{-4}
0.5	0.001	0.0001	103.00	88.48	103.20	89.33	8.70×10^{-5}	9.40×10^{-5}

8.2 Application to Detection of Traces of Space Objects

Processing of streams of images (frames) received from electro-optic/infrared (EO/IR) sensors is based on two main technologies: intraframe and interframe data processing with the accumulation of signals from objects. Throughout this section, we are interested in detection and tracking of near-Earth space objects with ground-based telescopes. The technology of in-frame signal accumulation

TABLE 8.2
Comparison of operating characteristics of the Shiryaev, SR and CUSUM track termination rules T_A, $\widetilde{T}_B$ and T_C^*. HMM parameters: $p = 1/30$, $\tilde{p} = 0.1$, $P_d^1 = 0.9$, $P_d^2 = 0.1$, $P_{fa} = 0.01$. Parameter of prior $q = 0$. Number of MC runs 10^6.

ρ	α	ADD(T_A)	ADD($\widetilde{T}_B$)	ADD(T_C^*)	PFA(T_A)	PFA($\widetilde{T}_B$)	PFA(T_C^*)
0.5	0.1	1.247	1.276	2.317	0.09980	0.09952	0.09253
0.5	0.01	4.177	4.647	8.951	0.00759	0.00769	0.00760
0.5	0.001	7.067	7.744	15.404	0.000766	0.000870	0.000814
0.5	0.0001	10.098	10.374	18.995	0.000074	0.000084	0.000079
0.1	0.1	8.910	8.974	12.504	0.09132	0.09131	0.09214
0.1	0.01	19.357	19.584	23.378	0.00902	0.00914	0.00912
0.1	0.001	30.048	30.435	34.821	0.000943	0.000996	0.000991
0.1	0.0001	41.210	42.561	46.607	0.000094	0.000097	0.000091
0.01	0.1	29.985	30.006	32.442	0.09343	0.09309	0.09141
0.01	0.01	50.284	50.290	52.313	0.00941	0.00951	0.00973
0.01	0.001	70.450	70.550	73.074	0.000960	0.000916	0.000953
0.01	0.0001	90.503	90.870	95.500	0.000086	0.000099	0.000083

allows for the preliminary detection and evaluation of astrometric parameters of faint space objects in each received frame. The accumulation of signals is necessary when reliable detection based on thresholding of single pixels is impossible due to low SNR. It is used to increase the detecting capability and the reliability of information about the detected faint objects.

In this section, we discuss the problem of in-frame detection of streaks of faint space objects with unknown orbits in CCD frames.[2] Input frames are captured with a telescope mounted at the equator. Thus, the traces of space objects will be located almost vertically in a small area at the center of the frame in the form of streaks with unknown points of appearance and disappearance (coordinates of the beginning and end, see Figure 8.1). For simplicity, it is assumed that there is a streak of only one space object. The problem is solved for streaks with a very low pixel signal-to-noise ratio of the order of 1 and smaller. An effective solution of this problem can be obtained on the basis of joint testing of hypotheses and estimation of unknown parameters (beginning and end of the streak) by maximizing the likelihood ratio, i.e., by the generalized likelihood ratio method. But such a method, when applied to large input frames of 4000×2000 pixels and larger, requires a huge amount of computation that can be performed only by a powerful supercomputer.

Observe that this problem is similar to the changepoint detection problem (but in space) since the distribution of observations changes abruptly when the trace appears and disappears (starts and ends), so we suggest an alternative two-stage approach. At the first stage, we determine the direction of a streak and its approximate location (streak localization) in the area shown in Figure 8.1. At the second stage, we estimate the position of the streak, maximizing the likelihood ratio. The solution of this task is based on the search for local maxima of the decision statistics in each of the selected directions. Then the global maximum is found among all directions and the streak is detected and localized in the direction corresponding to the global maximum. Once the streak is localized its position is further estimated more accurately in the localized area. The parameters (intensity and position) that deliver a global maximum of the likelihood ratio will correspond to the most accurate position of the streak. As a result, the maximum likelihood ratio method can be successfully applied in a localized area on a standard laptop.

[2]Thanks to N. Berenkov for implementing the algorithms as well as for simulations and processing of semi-real data.

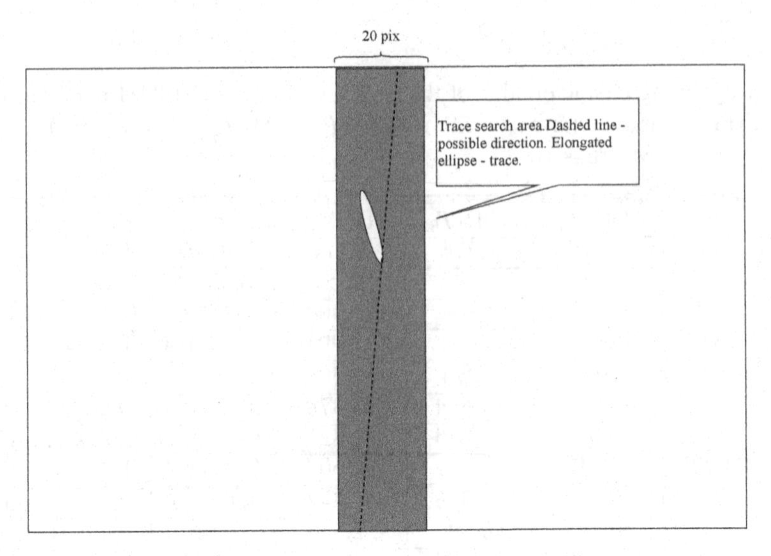

FIGURE 8.1
Search area of the object's trace and directions of search.

The situation is aggravated by the presence of stars in the images that produce very intense discrete clutter. A typical image with clutter from stars and a barely visible trace from the object in the middle of the image is shown in Figure 8.2(a). In this case, the image should be preprocessed. Spacial-only processing is usually not too effective for clutter removal. An effective way is the spatiotamporal technique [162] based on subtraction of a background estimate from the current image, which requires very accurate image registration (alignment). The clutter suppressed frame is shown in Figure 8.2(b). It is seen that the data in the image $\{X_{i,j}\}_{(i,j)\in\Omega}$ were whitened very well.

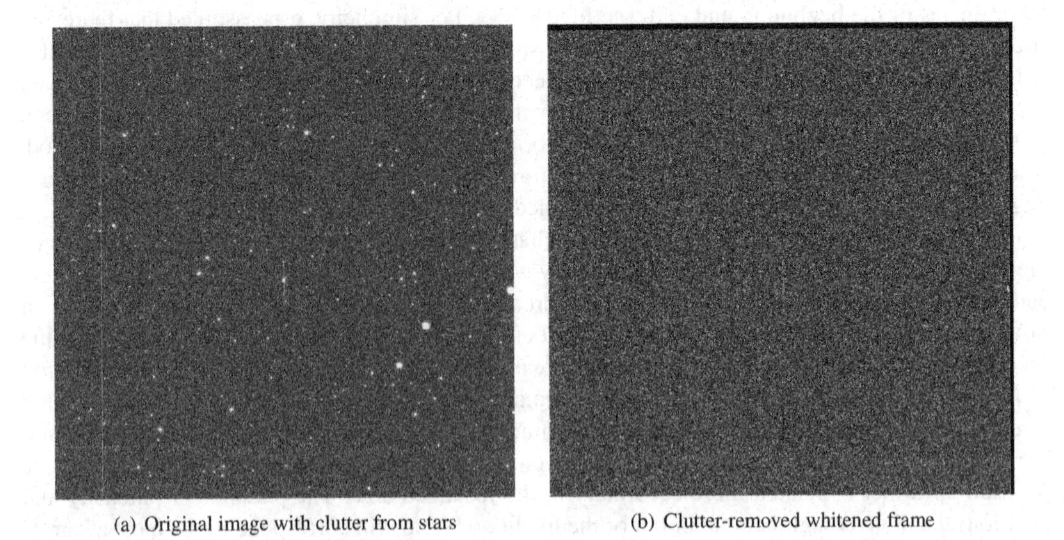

(a) Original image with clutter from stars (b) Clutter-removed whitened frame

FIGURE 8.2
Original raw image with clutter from stars and clutter-suppressed image with a barely visible streak.

The following model for the observation $X_{i,j}$ at the pixel (i,j) of the 2-D $n_x \times n_y$ image is accepted

$$X_{i,j} = A \int_0^T \int_i^{i+1} \int_j^{j+1} F(x - x_\tau, y - y_\tau) \mathrm{d}x \mathrm{d}y \mathrm{d}\tau + \xi_{i,j},$$

where $i = 1, \dots, n_x$, $j = 1, \dots, n_y$, T is the exposition time, A is the object radiation intensity, (x_τ, y_τ) is the location of the object in the frame at the time moment τ; $F(x,y)$ is the point spread function (PSF) of the sensor, $\{\xi_{i,j}\}$ is the residual noise after preprocessing with mean zero and the variance σ^2. It is assumed that noise is Gaussian, $\xi_{i,j} \sim \mathscr{N}(0, \sigma^2)$, and independent from pixel to pixel and that the object moves in the frame uniformly and rectilinearly, i.e., $x_\tau = x_0 + \mathrm{v}_x \tau$, $y_\tau = y_0 + \mathrm{v}_y \tau$. Thus, the object's trace is given by the vector of its beginning and end $Z = (x_0, y_0, x_T, y_T)$.

Denoting

$$S_{i,j}(Z) = A \int_0^T \int_i^{i+1} \int_j^{j+1} F(x - x_0 - \mathrm{v}_x \tau, y - y_0 - \mathrm{v}_y \tau) \mathrm{d}x \mathrm{d}y \mathrm{d}\tau, \tag{8.6}$$

we obtain

$$X_{i,j} = A S_{i,j}(Z) + \xi_{i,j}.$$

Figure 8.3 shows the trace profile $\{S_{i,j}(Z)\}$ that was computed using (8.6) for the Gaussian PSF F.

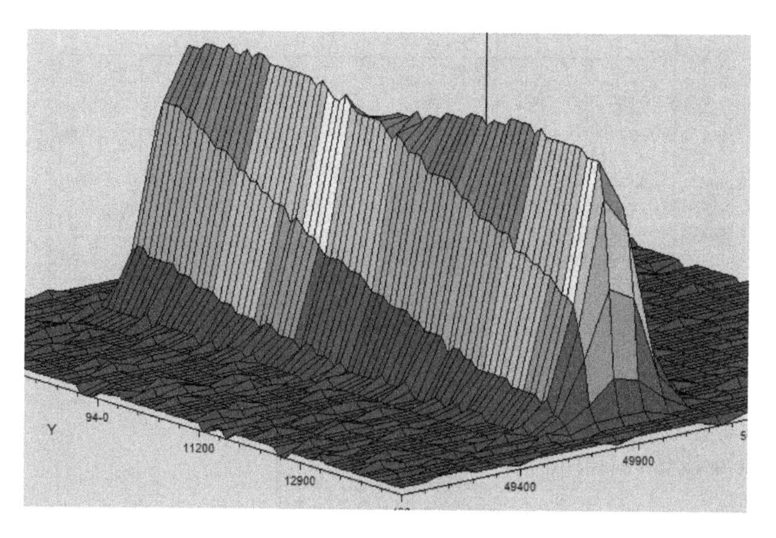

FIGURE 8.3
Trace profile.

Let $\Omega_s \subset \Omega$ denote the given search area of the streak, which in our case is shown in Figure 8.1 (taking into account the described geometry corresponding to the telescope set at the equator). Let H_0 be the hypothesis that there is no object's streak in the frame and $\mathsf{H}_1(A, Z)$ that there is a streak with local intensity A and coordinates Z of the beginning and the end, which are unknown parameters. The problem is to test these two hypotheses, based on the observations $X_{i,j}$, $(i,j) \in \Omega_s$, and if the hypothesis H_1 is accepted also to estimate the coordinates Z.

We solve the problem using the following two-step procedure.

Step 1: Detection and localization of the streak. In the area Ω_s, we localize a streak in some direction using the moving average rule (5.47) discussed in Section 5.3. More specifically, we select a 2-D window and slide this window in various directions. The form of this window is matched with the trace profile, shown in Figure 8.3. Denote the window in direction d by $M_d(t)$, where t is the end of the window. Since for the Gaussian model, the log-likelihood ratio of the hypotheses is a monotone nondecreasing function of the statistic

$$R_{M_d(t)} = \sum_{(i,j) \in M_d(t)} S_{i,j} X_{i,j}$$

we exploit this statistic along the multiple directions in the search area Ω_s. The maximal statistic $\max_d R_{M_d(t)}$ is selected and when it crosses a threshold the decision on the presence of a streak is made in the direction that maximizes this statistic. This is nothing but a spacial changepoint detection algorithm, which allows us to determine the position of the streak typically with an accuracy of about $5-10$ pixels. Figure 8.4 shows the behavior of the moving average detection statistic in the case of a very low SNR $A/\sigma = 1$. The streak was localized at the points $(47, 117)$ while the true values are $Z = (50, 120)$. The window $M_d(t)$ of the size 8×10 pixels was used.

FIGURE 8.4
Moving average detection statistic R.

Step 2: More accurate evaluation of streak's position. After detection of a streak the area $(47, 117)$, shown in Figure 8.4 (that corresponds to the approximate coordinates of the streak position), is extended by N pixels in each direction (typically we used $N = 10$ pixels) and in this extended area, which will be denoted as Π, the maximum likelihood estimates $\hat{A}$ and $\hat{Z}$ of the intensity A and coordinates Z are calculated,

$$(\hat{A}, \hat{Z}) = \arg\min_{(A,Z)} \sum_{(i,j) \in \Pi} [X_{i,j} - A S_{i,j}(Z)]^2.$$

It is easily seen that

$$\hat{A} = \frac{\sum_{(i,j) \in \Pi} X_{i,j} S_{i,j}(Z)}{\sum_{(i,j) \in \Pi} S_{i,j}^2(Z)}$$

and

$$\hat{Z} = \arg\min_{Z} \sum_{(i,j) \in \Pi} \left[X_{i,j} - \frac{\sum_{(i,j) \in \Pi} X_{i,j} S_{i,j}(Z)}{\sum_{(i,j) \in \Pi} S_{i,j}^2(Z)} S_{i,j}(Z) \right]^2.$$

The experimental study shows that the precision of the resulting estimate of the streak position is very high. Typically the error does not exceed 1 pixel for the SNR$=A/\sigma$ as small as 1.

8.3 Application to Detection of Unauthorized Break-ins in Computer Networks

As mentioned in [164, Ch. 11], cybersecurity has evolved into a critical 21st century problem that affects governments, businesses and individuals. Recently, cyber threats have become more diffuse, more complex, and harder to detect. Current ultra-high-speed networks carry massive aggregate data flows. Malicious events usually produce changes (anomalies) in network traffic profiles, which must be detected and isolated rapidly while keeping a low false alarm rate (FAR). Both requirements are important. However, rapid intrusion detection with minimal FAR and the capability to detect a wide spectrum of attacks is a challenge for modern computer networks.

Consider one of the major computer security risks posed by unauthorized tampering with or break-in to a system. To put it in context, suppose a hacker attempts to break into a computer system. Such a scenario usually involves two phases. In phase 1, the hacker launches a dictionary attack attempting to guess a username and/or password (possibly the root). In phase 2, the hacker performs suspicious activities on the machine, including downloading malware and opening up a backdoor.

We focus only on phase 1 and demonstrate this phase using real-life traces. Schematically a generic scenario of the dictionary attack is shown in Figure 8.5.

To be more specific, with a dictionary attack the hacker attempts to guess a correct user/password combination (ideally one that would gain access to the root shell) to break into a server, typically through SSH. While we illustrate the attack with SSH, this kind of attack applies to any user/password access control method, including web authentication and other similar methods. To achieve this goal the attacker initiates what is essentially a brute force attack: a rapid sequence of SSH authorization requests sent to the server, where each request contains a username/password combination either guessed based on prior partial information about a valid username/password or trying out common usernames and passwords. The word "dictionary" in this context is used figuratively to illustrate that the attacker has a list (dictionary) of "suspected" username/password combinations. In a dictionary attack, the hacker successively tries all of them. This kind of an attack is illustrated in Figure 8.6. This picture corresponds to real-life traffic captured at a regional Internet Service Provider. The traffic involves a busy SSH server, which seems to be frequently subjected to this type of attack.

Figure 8.6(a) shows the intensity of the number of packets passing through the victim server's link. Notice that the server remains idle most of the time, occasionally exhibiting interaction with other computers in the network. Eventually, the server starts to receive suspiciously many SSH requests, so that one can see a distinct traffic pattern that is easy to detect at the network level. The number of SSH packets sent through the link per time unit becomes unusually high as the attacker tries many words in the dictionary, with many unsuccessful authorization requests (except possibly the last one, if the attacker hits a valid user/password combination). Based on this information, one can leverage changepoint detection to look for changes of a positive magnitude in the average number of SSH authorization requests sent per time unit. Changepoint detection is useful in this case to detect the malicious activity rapidly and at the same time to avoid false alarms in very busy SSH servers where many users may be logging in and out. Figure 8.6(b) shows the behavior of the Shewhart detection statistic tuned to the Gaussian model with an unknown mean, i.e., $S(X_n) = \exp\{X_n\}$ (see Chapter 5, Corollary 5.1). As soon as the server is attacked, the statistic exceeds the detection threshold, and an alarm is raised resulting in successful detection of the attack. Recall that,

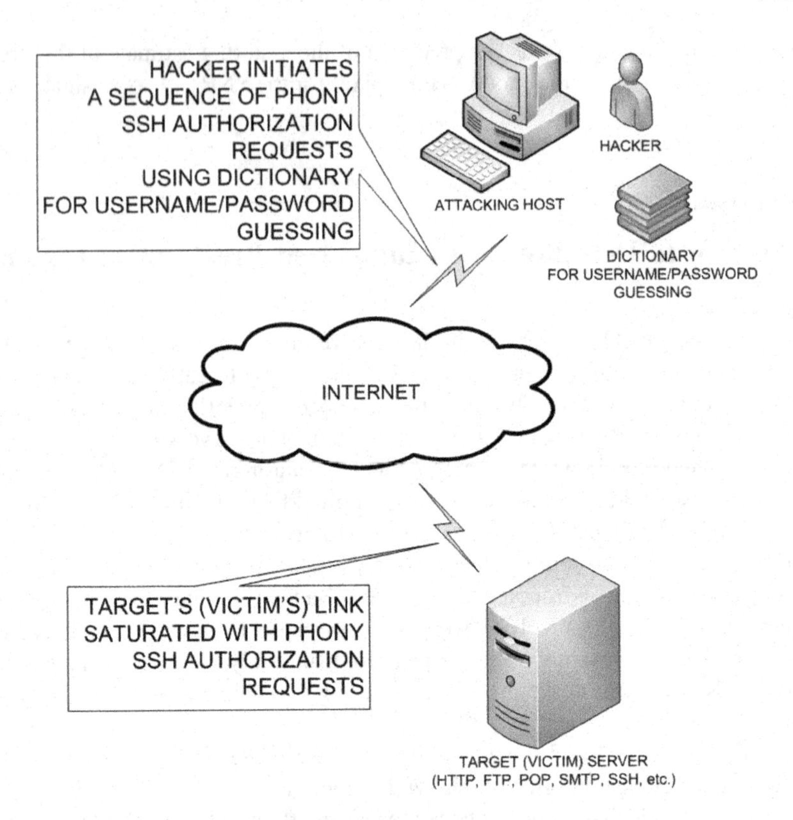

FIGURE 8.5
A generic dictionary attack scenario.

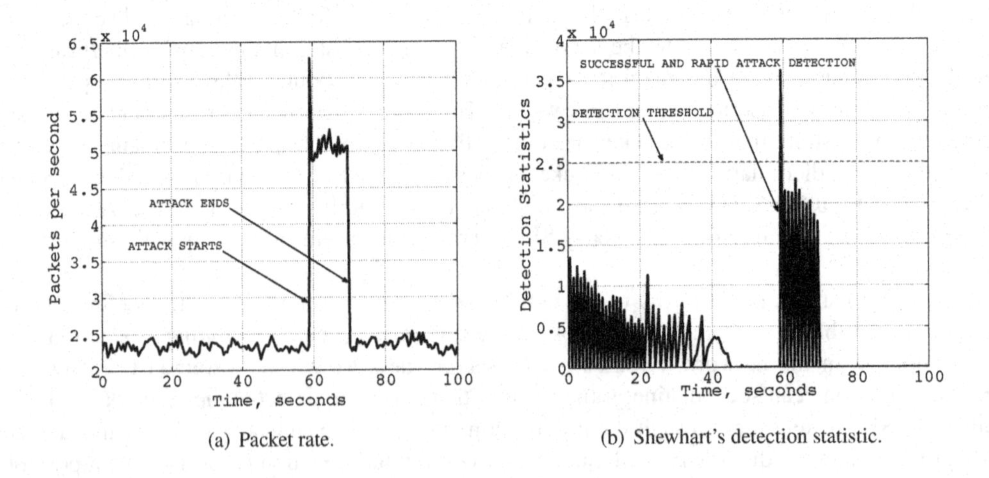

(a) Packet rate. (b) Shewhart's detection statistic.

FIGURE 8.6
SSH dictionary attack traffic pattern and its detection by the Shewhart rule.

as shown in Chapter 5, the Shewhart detection rule has multiple optimality properties (Bayesian and maximin) with respect to the instantaneous probability of detection for the given probability of false alarm and/or mean time to a false alarm. In particular, by Corollary 5.1, this rule maximizes the instantaneous detection probability uniformly for any possible attack intensity. It is also clear that this rule is efficient for detecting contrast changes, in which case the probability of detection will be high. Since for the present trace the attack is indeed quite visible, the Shewhart rule works very well. It detects the attack practically immediately.

To make things more challenging, we intentionally diminished the intensity of the attack. The new dataset is excellent to demonstrate not only the potential of changepoint detection but also that of the hybrid intrusion detection system that exploits spectral techniques described in [164, Sec. 11.3] in detail. The spectral approach is being used in conjunction with change detection in order to filter false positives, which allows us to substantially lower threshold levels. The spectral approach is expected to work because dictionary attacks introduce periodicities in the traffic flow. If one employs a spectral analyzer, the spectral power density will have a high spike. This is exactly the idea behind the hybrid anomaly-spectral intrusion detection system [164, Sec. 11.3].

Figure 8.7(a) shows the data (packet rate) for this modified dataset with a reduced intensity of the attack and Figure 8.7(b) illustrates the detection process with the modified CUSUM rule (5.59)-(5.60), which performs much better in the case of dim attacks, as is obvious from the results of Section 5.3.2. Despite the fact that the intensity of the attack is now far less than before, changepoint detection reveals it almost instantaneously.

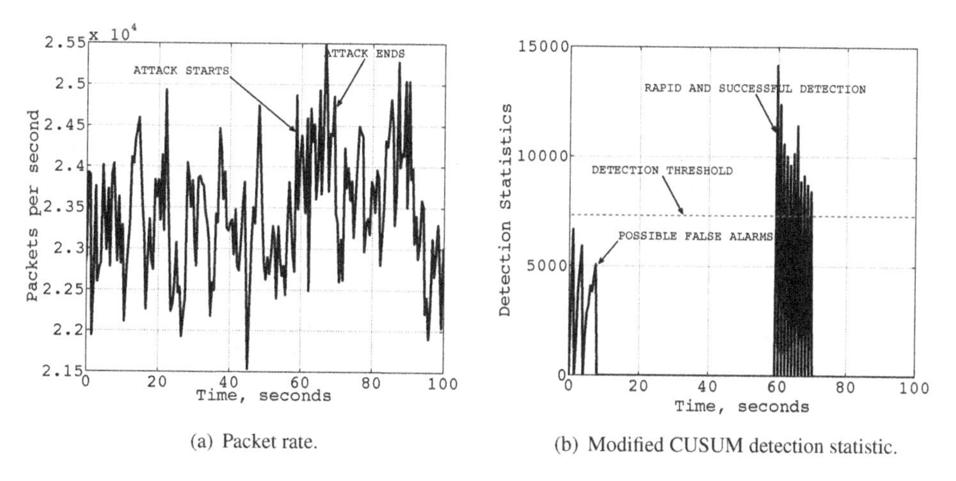

(a) Packet rate.

(b) Modified CUSUM detection statistic.

FIGURE 8.7
Diminished SSH dictionary attack traffic pattern and its detection by the modified CUSUM rule.

The problem occurs when we lower the detection threshold to have an even quicker detection and a higher probability of detection, in which case we inevitably run into numerous false alarms. As we suggested above, to filter these false alarms one can employ a spectral analyzer to uncover hidden periodicities in the traffic flow as traces of the presence of attacks. Figures 8.8(a) and 8.8(b) show the power spectral density for this dataset before and during the attack. Note the spike in the power spectral density under attack. This is exactly because of the aforementioned periodicity. Using the spectral analyzer in conjunction with the changepoint based anomaly detector, i.e., the hybrid anomaly-spectral IDS, one can achieve unprecedented speeds of detection and high probabilities of detection with simultaneously very low false alarm rates.

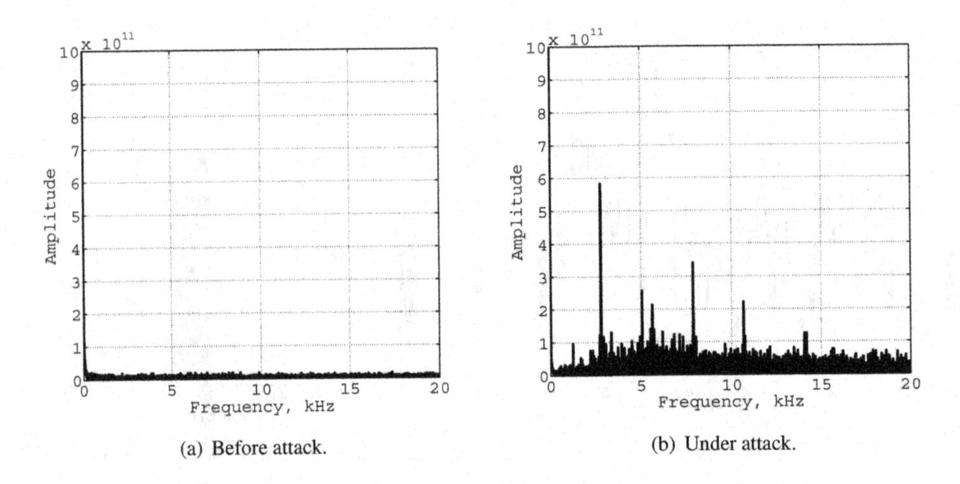

(a) Before attack.

(b) Under attack.

FIGURE 8.8
Power spectral density of traffic before and under dictionary attack.

Appendix A: Useful Auxiliary Results

The following lemma is extensively used for obtaining upper bounds for the moments of the stopping times, which are needed for proving asymptotic optimality properties of sequential tests and change detection rules. In this lemma, P is a generic probability measure and E is a corresponding expectation.

Lemma A.1. *Let τ ($\tau = 0, 1, \dots$) be a non-negative integer-valued random variable and let N ($N \ge 1$) be an integer number. For any $r \ge 1$, the following two inequalities hold*

$$\mathsf{E}[\tau^r] \le N^r + r2^{r-1} \sum_{n=N}^{\infty} n^{r-1} \mathsf{P}(\tau > n); \tag{A.1}$$

$$\mathsf{E}[\tau^r] \le N^r \left(1 + r2^{r-1} \sum_{\ell=1}^{\infty} \ell^{r-1} \mathsf{P}(\tau > \ell N)\right). \tag{A.2}$$

Proof. Inequality (A.1) follows from the following chain of equalities and inequalities

$$\begin{aligned}
\mathsf{E}[\tau^r] &= \int_0^{\infty} r t^{r-1} \mathsf{P}(\tau > t)\, \mathrm{d}t \\
&= \int_0^{N} r t^{r-1} \mathsf{P}(\tau > t)\, \mathrm{d}t + \sum_{n=N}^{\infty} \int_n^{n+1} r t^{r-1} \mathsf{P}(\tau > t)\, \mathrm{d}t \\
&\le N^r + \sum_{n=0}^{\infty} \int_{N+n}^{N+n+1} r t^{r-1} \mathsf{P}(\tau > t)\, \mathrm{d}t \\
&\le N^r + \sum_{n=0}^{\infty} \int_{N+n}^{N+n+1} r t^{r-1} \mathsf{P}(\tau > N+n)\, \mathrm{d}t \\
&= N^r + \sum_{n=0}^{\infty} [(N+n+1)^r - (N+n)^r] \mathsf{P}(\tau > N+n) \\
&= N^r + \sum_{n=N}^{\infty} [(n+1)^r - n^r] \mathsf{P}(\tau > n) \\
&\le N^r + \sum_{n=N}^{\infty} r(n+1)^{r-1} \mathsf{P}(\tau > n) \\
&\le N^r + r2^{r-1} \sum_{n=N}^{\infty} n^{r-1} \mathsf{P}(\tau > n).
\end{aligned}$$

Inequality (A.2) follows from the following chain of equalities and inequalities

$$\begin{aligned}
\mathsf{E}[\tau^r] &= \int_0^{\infty} r t^{r-1} \mathsf{P}(\tau > t)\, \mathrm{d}t \\
&= \int_0^{N} r t^{r-1} \mathsf{P}(\tau > t)\, \mathrm{d}t + \sum_{\ell=1}^{\infty} \int_{\ell N}^{(\ell+1)N} r t^{r-1} \mathsf{P}(\tau > t)\, \mathrm{d}t \\
&\le N^r + \sum_{\ell=1}^{\infty} \int_{\ell N}^{(\ell+1)N} r t^{r-1} \mathsf{P}(\tau > t)\, \mathrm{d}t
\end{aligned}$$

$$
\begin{aligned}
&\leq N^r + \sum_{\ell=1}^{\infty} \int_{\ell N}^{(\ell+1)N} r t^{r-1} \mathsf{P}(\tau > \ell N)\,\mathrm{d}t \\
&= N^r \left(1 + \sum_{\ell=1}^{\infty} [(\ell+1)^r - \ell^r] \mathsf{P}(\tau > \ell N)\right) \\
&\leq N^r \left(1 + \sum_{\ell=1}^{\infty} r(\ell+1)^{r-1} \mathsf{P}(\tau > \ell N)\right) \\
&\leq N^r \left(1 + r 2^{r-1} \sum_{\ell=1}^{\infty} \ell^{r-1} \mathsf{P}(\tau > \ell N)\right).
\end{aligned}
$$

□

Let $(\Omega, \mathscr{F}, \mathscr{F}_n, \mathsf{P}_i)$ $(i = 1,2)$, $n \in \mathbb{Z}_+$, be a filtered probability space with standard assumptions on monotonicity of the sigma-algebras $\mathscr{F}_n = \sigma(X_1, \dots, X_n)$ generated by the stochastic process $\{X_t\}_{t \in \mathbb{Z}_+}$ defined on $(\Omega, \mathscr{F})$. The probability measures P_1 and P_2 are locally mutually continuous, i.e., the restrictions P_1^n and P_2^n of these measures to the sub-σ-algebras $\mathscr{F}_n$ are equivalent for all $n \in \mathbb{Z}_+$. Write $\mathbf{X}^n = (X_0, X_1, \dots, X_n)$ and let $p_i(\mathbf{X}^n)$, $i = 1,2$ denote densities of measures P_i with respect to a non-degenerate sigma-finite measure and $p_i(X_n | \mathbf{X}^{n-1})$ the corresponding conditional densities. Define

$$
\Lambda_n = \frac{\mathrm{d}\mathsf{P}_1^n}{\mathrm{d}\mathsf{P}_2^n} = \prod_{t=1}^{n} \frac{p_1(X_t|\mathbf{X}^{t-1})}{p_2(X_t|\mathbf{X}^{t-1})}, \quad \Lambda_0 = 1 \quad \text{and} \quad R_n = \sum_{k=1}^{n} \prod_{t=k}^{n} \frac{p_2(X_t|\mathbf{X}^{t-1})}{p_1(X_t|\mathbf{X}^{t-1})}, \quad R_0 = 0, \quad n \in \mathbb{Z}_+.
$$

Recall that a stopping time T is called *proper* if it is finite w.p. 1, $\mathsf{P}(T < \infty) = 1$.

Lemma A.2. *Let T be any (possibly randomized) proper stopping time with respect to filtration $\{\mathscr{F}_n\}$. Then*

$$
\mathsf{E}_2 \left[\sum_{n=0}^{T-1} \Lambda_n\right] = \mathsf{E}_1[T], \tag{A.3}
$$

$$
\mathsf{E}_2 \left[\max_{0 \leq n \leq T-1} \Lambda_n\right] \leq \min\left\{\frac{e}{e-1}\left[1 + e^{-1} + I(T-1)\right], \mathsf{E}_1[T]\right\} \tag{A.4}
$$

provided

$$
\lim_{n\to\infty} \mathsf{E}_1 |R_{T\wedge n} - T \wedge n| < \infty, \quad \lim_{n\to\infty} \mathsf{E}_1 \left[(R_n - n) \mathbb{1}_{\{T > n\}}\right] = 0, \tag{A.5}
$$

where E_i is the operator of expectation with respect to the measure P_i and $I(T-1) = \mathsf{E}_1[\log \Lambda_{T-1}]$ is the average Kullback–Leibler information in $T-1$ "observations." In particular, if under P_i the $X_1, X_2, \dots$ are i.i.d. random variables with densities $p_i(x)$, $i = 1,2$, then

$$
\mathsf{E}_2 \left(\max_{0 \leq n \leq T-1} \Lambda_n\right) \leq \min\left\{\frac{e}{e-1}\left[1 + e^{-1} + \mathscr{K}(p_1; p_2) \mathsf{E}_1(T-1)\right], \mathsf{E}_1[T]\right\} \tag{A.6}
$$

where $\mathscr{K}(p_1; p_2) = \int \log[p_1(x)/p_2(x)] p_1(x) \mu(dx)$ is the Kullback–Leibler information number.

Proof. It is easy to see that $R_n = (1/\Lambda_n) \sum_{s=0}^{n-1} \Lambda_s$ and, hence, by the Wald likelihood ratio identity,

$$
\mathsf{E}_2 \left[\sum_{n=0}^{T-1} \Lambda_n; T < \infty\right] \equiv \mathsf{E}_2[\Lambda_T R_T; T < \infty] = \mathsf{E}_1[R_T] \tag{A.7}
$$

where the letter equality holds whenever $\mathsf{P}_1(T < \infty) = 1$. Straightforward calculation shows that the process $\{R_n - n, n \geq 1\}$ is a mean-zero $(\mathsf{P}_1, \mathscr{F}_n)$−martingale. By Proposition C.1 in Appendix C, $\mathsf{E}_1[R_T - T] = 0$ when conditions (A.5) hold, which along with (A.7) proves (A.3).

For any $t > 0$ write $T_t = T \wedge t - 1$. Since the likelihood ratio $\{\Lambda_n\}$ is a $(\mathsf{P}_2, \mathscr{F}_n)$−martingale with unit expectation to prove inequality (A.4) we apply Doob's inequality to the martingale Λ_{T_t} (see Theorem 3.4, Ch. VII in [30]) along with the inequality

$$y \log^+ y \leq e^{-1} + y \log y$$

and obtain

$$\begin{aligned}\mathsf{E}_2\left[\max_{0 \leq n \leq T_t} \Lambda_n\right] &\leq \frac{e}{e-1}\left\{1 + \mathsf{E}_2[\Lambda_{T_t} \log^+ \Lambda_{T_t}]\right\} \\ &\leq \frac{e}{e-1}\left\{1 + e^{-1} + \mathsf{E}_2[\Lambda_{T_t} \log \Lambda_{T_t}]\right\} \\ &= \frac{e}{e-1}\left\{1 + e^{-1} + \mathsf{E}_1[\log \Lambda_{T_t}]\right\}.\end{aligned}$$

Since the sequence $\{\max_{n \leq T_t} \Lambda_{T_t}, t \geq 1\}$ is monotonically non-decreasing, letting $t \to \infty$, by the Beppo Levi theorem we get

$$\mathsf{E}_2\left[\max_{0 \leq n \leq T-1} \Lambda_n\right] \leq \frac{e}{e-1}\left\{1 + e^{-1} + \mathsf{E}_1[\log \Lambda_T]\right\}.$$

On the other hand, by (A.3),

$$\mathsf{E}_2\left[\max_{0 \leq n \leq T-1} \Lambda_n\right] \leq \mathsf{E}_2\left[\sum_{n=0}^{T-1} \Lambda_n\right] = \mathsf{E}_1[T].$$

Combining the last two inequalities, we get (A.4).

In the i.i.d. case, by the Wald identity,

$$\mathsf{E}_1[\log \Lambda_{T_t}] = \mathsf{E}_1\left(\sum_{j=1}^{T_t} \log \frac{p_1(X_j)}{p_2(X_j)}\right) = \mathscr{K}(p_1; p_2)\mathsf{E}_1[T_t],$$

and, consequently,

$$\mathsf{E}_2\left[\max_{0 \leq n \leq T} \Lambda_n\right] \leq \frac{e}{e-1}\left\{1 + e^{-1} + \mathscr{K}(p_1; p_2)\mathsf{E}_1[T]\right\},$$

which along with (A.3) yields (A.6). □

Appendix B: Stochastic Convergence

Since this book deals with asymptotic theories stochastic convergence plays an important role. In this appendix, we start with standard modes of convergence such as convergence in probability and with probability 1, and then we continue with stronger modes – r-complete and r-quick convergence. The latter modes are especially important and extensively used for proving asymptotic results in most of the chapters.

B.1 Standard Modes of Convergence

Let X be a random variable and let $\{X_t\}$ be a continuous-time ($t \in \mathbb{R}_+ = [0,\infty)$) or discrete-time ($t \in \mathbb{Z}_+ = \{0,1,2,\dots\}$) stochastic process, both defined on the probability space $(\Omega, \mathscr{F}, \mathsf{P})$.

Convergence in Distribution (Weak Convergence). Let $F_t(x) = \mathsf{P}(\omega : X_t \le x)$ be the cumulative distribution function (cdf) of X_t and let $F(x) = \mathsf{P}(\omega : X \le x)$ be the cdf of X. We say that the process $\{X_t\}$ converges to X *in distribution* (or *in law* or *weakly*) as $t \to \infty$ and write $X_t \xrightarrow[t\to\infty]{\text{law}} X$ if

$$\lim_{t\to\infty} F_t(x) = F(x)$$

at all continuity points of $F(x)$.

Convergence in Probability. We say that the process $\{X_t\}$ converges to X *in probability* as $t \to \infty$ and write $X_t \xrightarrow[t\to\infty]{\mathsf{P}} X$ if

$$\lim_{t\to\infty} \mathsf{P}(|X_t - X| > \varepsilon) = 0 \quad \text{for all } \varepsilon > 0.$$

Almost Sure Convergence. We say that the process $\{X_t\}$ converges to X *almost surely* (a.s.) or *with probability* 1 (w.p. 1) as $t \to \infty$ and write $X_t \xrightarrow[t\to\infty]{\text{a.s.}} X$ if

$$\mathsf{P}(\omega : \lim_{t\to\infty} X_t = X) = 1.$$

The a.s. convergence implies convergence in probability, and the convergence in probability implies convergence in distribution, while the converse statements are not generally true.

The following implications that establish necessary and sufficient conditions for the a.s. convergence are useful:

$$X_t \xrightarrow[t\to\infty]{\text{a.s.}} X \iff \mathsf{P}\left(\sup_{s\ge t} |X_s - X| > \varepsilon\right) \xrightarrow[t\to\infty]{} 0 \quad \text{for all } \varepsilon > 0. \tag{B.1}$$

The following result is often used in the book.

Lemma B.1. *Let $\{X_t\}$ be either a continuous-time or a discrete-time process. Assume that $t^{-1}X_t$ converges almost surely as $t \to \infty$ to a positive and finite number I. Then, for all $\varepsilon > 0$,*

$$\mathsf{P}\left(\frac{1}{t} \sup_{0\le s\le t} X_s - I > \varepsilon\right) \xrightarrow[t\to\infty]{} 0. \tag{B.2}$$

Proof. For any $\varepsilon > 0$, $t_0 > 0$ and $t > t_0$, we have

$$\begin{aligned}\mathsf{P}\left(\frac{1}{t}\sup_{0\le s\le t} X_s - I > \varepsilon\right) &\le \mathsf{P}\left(\frac{1}{t}\sup_{0\le s\le t_0} X_s - I > \varepsilon\right) + \mathsf{P}\left(\frac{1}{t}\sup_{t_0< s\le t} X_s - I > \varepsilon\right)\\ &\le \mathsf{P}\left(\frac{1}{t}\sup_{0\le s\le t_0} X_s > I+\varepsilon\right) + \mathsf{P}\left(\sup_{s>t_0}\frac{X_s}{s} - I > \varepsilon\right).\end{aligned}$$

Letting $t \to \infty$ and taking into account that $\lim_{t\to\infty} \mathsf{P}(t^{-1}\sup_{0\le s\le t_0} X_s > I+\varepsilon) = 0$, we obtain

$$\limsup_{t\to\infty} \mathsf{P}\left(\frac{1}{t}\sup_{0\le s\le t} X_s - I > \varepsilon\right) \le \mathsf{P}\left(\sup_{s>t_0}\frac{X_s}{s} - I > \varepsilon\right).$$

Since this inequality holds for any $t_0 > 0$, we can let $t_0 \to \infty$. But by assumption $t^{-1}X_t \xrightarrow[t\to\infty]{\text{a.s.}} I$, so by (B.1) the upper bound approaches 0 as $t_0 \to \infty$, which completes the proof of (B.2). □

Remark B.1. The proof of Lemma B.1 shows that the assertion (B.2) is also true under the one-sided condition

$$\mathsf{P}\left(\sup_{s>t}\frac{X_s}{s} - I > \varepsilon\right) \xrightarrow[t\to\infty]{} 0 \quad \text{for all } \varepsilon > 0. \tag{B.3}$$

Often in addition to establishing the a.s. convergence or the convergence in probability one is interested in the convergence of the moments, i.e., $\mathsf{E}|X_t|^m \to \mathsf{E}|X|^m$. The corresponding mode of convergence is the L^p-convergence, $p \ge 1$.

L^p-convergence. We say that the process $\{X_t\}$ converges to X *in L^p* or *in the pth mean* as $t \to \infty$ and write $X_t \xrightarrow[t\to\infty]{L^p} X$ if

$$\lim_{t\to\infty} \mathsf{E}|X_t - X|^p = 0.$$

In general, the a.s. convergence does not guarantee the convergence of the moments. To overcome this difficulty we need an additional uniform integrability condition.

Definition B.1. A process $\{X_t\}$ is said to be *uniformly integrable* if

$$\sup_t \mathsf{E}[|X_t|\mathbb{1}_{\{|X_t|>a\}}] \xrightarrow[a\to\infty]{} 0.$$

Passage to the Limit under the Sign of Expectation. A very useful topic related to convergence in the theory of probability and statistics is establishing conditions under which it is possible to exchange the operations of expectation (i.e., integration) and limit. The following theorem combines three famous results – the *monotone convergence theorem*, *Fatou's lemma*, and *Lebesgue's dominated convergence theorem*. We write $X_t \uparrow X$ if $X_t \to X$ and $\{X_t\}$ is non-decreasing and $X_t \downarrow X$ if $\{X_t\}$ is non-increasing.

Theorem B.1. **(i)** *Monotone convergence theorem. If $X_t \uparrow X$ w.p. 1 as $t \to \infty$ and $\mathsf{E}[X_1^-] < \infty$, then $\mathsf{E}[X_t] \uparrow \mathsf{E}[X]$. If $X_t \downarrow X$ w.p. 1 as $t \to \infty$ and $\mathsf{E}[X_1^+] < \infty$, then $\mathsf{E}[X_t] \downarrow \mathsf{E}[X]$.*

(ii) *Fatou's lemma. Let $\{X_t^+\}$ be uniformly integrable. If $\mathsf{E}[\limsup_t X_t]$ exists, then*

$$\mathsf{E}[\limsup_t X_t] \ge \limsup_t \mathsf{E}[X_t].$$

In particular, this inequality holds if $X_t \le X$, $t \ge 0$, where X is integrable.

(iii) *Lebesgue's dominated convergence theorem. Let* $X_t \xrightarrow[t\to\infty]{\mathsf{P}} X$*. If there exists a random variable* X *such that* $|X_t| \le X$, $t \ge 0$ *and* $\mathsf{E}[X] < \infty$, *then*

$$\lim_{t\to\infty} \mathsf{E}|X_t - X| = 0.$$

Convergence of Moments. The following theorem is useful for establishing the convergence of moments (cf. Loève [81, pp. 165–166]).

Theorem B.2. *Assume that* $X_t \xrightarrow[t\to\infty]{\mathsf{P}} X$ *and* $\mathsf{E}|X_t|^p < \infty$, $p > 0$. *If* $\{|X_t|^p\}$ *is uniformly integrable, then*

$$\mathsf{E}|X_t|^m \xrightarrow[t\to\infty]{} \mathsf{E}|X|^m \quad \text{for all } 0 < m \le p.$$

The converse is also true. If $\mathsf{E}|X_t|^p < \infty$ *and* $X_t \xrightarrow[n\to\infty]{L^p} X$, *then* $\{|X_t|^p\}$ *is uniformly integrable.*

Strong Law of Large Numbers. Let $\{Y_n\}_{n\in\mathbb{N}}$ be a sequence of i.i.d. random variables. Write $S_n = \sum_{i=1}^n Y_i$. The Kolmogorov strong law of large numbers (SLLN) states that if $\mathsf{E}[Y_1]$ exists, then the sample mean S_n/n converges to the mean value $\mathsf{E}[Y_1]$ w.p. 1, i.e.,

$$n^{-1}S_n \xrightarrow[n\to\infty]{\text{a.s.}} \mathsf{E}[Y_1].$$

B.2 Complete and r-Quick Convergence

We now consider modes of convergence that strengthen the almost sure convergence and help to determine the rate of convergence in the SLLN.

r-Complete Convergence. Let $r > 0$. We say that the process $\{X_n\}_{n\in\mathbb{Z}_+}$ converges to X *r-completely* as $n \to \infty$ and write $X_n \xrightarrow[n\to\infty]{r\text{-completely}} X$ if

$$\Sigma(r,\varepsilon) := \sum_{n=1}^{\infty} n^{r-1}\mathsf{P}(|X_n - X| > \varepsilon) < \infty \quad \text{for every } \varepsilon > 0. \tag{B.4}$$

Note that the a.s. convergence of $\{X_n\}$ to X can be equivalently written as

$$\lim_{n\to\infty} \mathsf{P}\left(\sum_{i=n}^{\infty} |X_i - X| > \varepsilon\right) = 0 \quad \text{for every } \varepsilon > 0,$$

so that the r-complete convergence with $r \ge 1$ implies the a.s. convergence, but the converse is not true in general.

For $r = 1$, we simply refer to r-complete convergence as complete convergence, which has been first introduced by Hsu and Robbins [62].

Now, assume that X_n converges a.s. to X. Clearly, if $\Sigma(r,\varepsilon)$ is finite for every $\varepsilon > 0$, then probability $\mathsf{P}(|X_n - X| > \varepsilon)$ decays with the rate faster than $1/n^r$, i.e., $n^r\mathsf{P}(|X_n - X| > \varepsilon) \to 0$ for all $\varepsilon > 0$ as $n \to \infty$. Hence, the r-complete convergence allows one to determine the rate of convergence of X_n to X, i.e., to answer the question on how fast the probability $\mathsf{P}(|X_n - X| > \varepsilon)$ decays to zero.

The following result provides a very useful implication of complete convergence.

Lemma B.2. *Let* $\{X_n\}_{n\in\mathbb{Z}_+}$ *and* $\{Y_n\}_{n\in\mathbb{Z}_+}$ *be two arbitrarily, possibly dependent discrete-time random processes. Assume that there are positive numbers* I_1 *and* I_2 *such that for all* $\varepsilon > 0$

$$\sum_{n=1}^{\infty} \mathsf{P}\left(\left|\frac{1}{n}X_n - I_1\right| > \varepsilon\right) < \infty, \quad \sum_{n=1}^{\infty} \mathsf{P}\left(\left|\frac{1}{n}Y_n - I_2\right| > \varepsilon\right) < \infty, \tag{B.5}$$

i.e., $n^{-1}X_n \xrightarrow[n\to\infty]{\text{completely}} I_1$ *and* $n^{-1}Y_n \xrightarrow[n\to\infty]{\text{completely}} I_2$. *If* $I_1 \geq I_2$, *then for any random time* T *and any* $\delta > 0$

$$\mathsf{P}\left(X_T < b, Y_{T+1} \geq b(1+\delta)\right) \longrightarrow 0 \quad \text{as } b \to \infty. \tag{B.6}$$

Proof. Fix $\delta > 0$, $c \in (0,\delta)$ and let n_b be the smallest integer that is larger than or equal to $(1+c)b/I_2$. Observe that

$$\begin{aligned}\mathsf{P}\left(X_T < b, Y_{T+1} \geq b(1+\delta)\right) &\leq \mathsf{P}\left(X_T \leq b,\, T \geq n_b\right) + \mathsf{P}\left(Y_{T+1} \geq (1+\delta)b,\, T < n_b\right)\\ &\leq \mathsf{P}\left(X_T \leq b,\, T \geq n_b\right) + \mathsf{P}\left(\max_{1\leq n\leq n_b} Y_n \geq (1+\delta)b\right).\end{aligned}$$

Thus, it suffices to show that the two terms on the right-hand side go to 0 as $b\to\infty$. For the first term, we notice that for any $n \geq n_b$,

$$\frac{b}{n} \leq \frac{b}{n_b} \leq \frac{I_2}{1+c} \leq \frac{I_1}{1+c} < I_1,$$

so that

$$\begin{aligned}\mathsf{P}\left(X_T \leq b,\, T \geq n_b\right) &= \sum_{n\geq n_b} \mathsf{P}\left(X_n \leq b,\, T = n\right) \leq \sum_{n\geq n_b} \mathsf{P}\left(\frac{X_n}{n} \leq \frac{b}{n}\right)\\ &\leq \sum_{n\geq n_b} \mathsf{P}\left(\frac{X_n}{n} \leq \frac{I_1}{1+c}\right) = \sum_{n\geq n_b} \mathsf{P}\left(\frac{X_n}{n} - I_1 \leq -\frac{c}{1+c} I_1\right).\end{aligned}$$

Since $n_b \to \infty$ as $b \to \infty$ the upper bound goes to 0 as $b\to\infty$ due to the first condition in (B.5). Moreover, since $c \in (0,\delta)$ there exists $\varepsilon' > 0$ such that

$$\frac{(1+\delta)b}{n_b} = \frac{(1+\delta)b}{\lceil b(1+c)/I_2 \rceil} \geq (1+\varepsilon')I_2.$$

As a result,

$$\mathsf{P}\left(\max_{1\leq n\leq n_b} Y_n \geq (1+\delta)b\right) \leq \mathsf{P}\left(\frac{1}{n_b}\max_{1\leq n\leq n_b} Y_n \geq (1+\varepsilon')I_2\right),$$

where the upper bound goes to 0 as $b\to\infty$ by the second condition in (B.5) (see Lemma B.1). □

Remark B.2. The proof suggests that the assertion (B.6) of Lemma B.2 holds under the following one-sided conditions

$$\mathsf{P}\left(n^{-1}\max_{1\leq s\leq n} Y_s - I_2 > \varepsilon\right) \xrightarrow[n\to\infty]{} 0, \quad \sum_{n=1}^{\infty} \mathsf{P}\left(n^{-1}X_n - I_1 < -\varepsilon\right) < \infty.$$

Complete convergence conditions (B.5) guarantee both these conditions.

Remark B.3. An interesting application of Lemma B.2 is when $X = Y = Z$ and the random time T is the first hitting time of the level b, $T = \inf\{n : Z_n \geq b\}$. Then, Lemma B.2 shows that the *relative overshoot* $(Z_T - b)/b$ converges to 0 in probability as $b \to \infty$ when Z_n/n converges completely to a positive number.

r-Quick Convergence. Let $t \in \mathbb{R}_+$ or $t \in \mathbb{Z}_+$. Let $r > 0$ and for $\varepsilon > 0$ let $L_\varepsilon = \sup\{t : |X_t - X| > \varepsilon\}$ ($\sup\{\varnothing\} = 0$) be the last entry time of the process X_t in the region $(X+\varepsilon, \infty) \cup (-\infty, X-\varepsilon)$. We say that the process $\{X_t\}$ converges to X *r-quickly* as $t\to\infty$ and write $X_t \xrightarrow[t\to\infty]{r\text{-quickly}} X$ if

$$\mathsf{E}[L_\varepsilon^r] < \infty \quad \text{for every } \varepsilon > 0. \tag{B.7}$$

In general, r-quick convergence is stronger than r-complete convergence. Specifically, the following lemma, whose proof can be found in [164, page 37], shows that

$$\max_{1\le i\le n} X_i \xrightarrow[n\to\infty]{r-\text{completely}} I \quad\Longrightarrow\quad X_n \xrightarrow[n\to\infty]{r-\text{quickly}} I \quad\Longrightarrow\quad X_n \xrightarrow[n\to\infty]{r-\text{completely}} I. \tag{B.8}$$

Lemma B.3. *Let X_t, $t\in\mathbb{R}_+$ or $t=n\in\mathbb{Z}_+$, be a random process, $X_0=0$. Define $M_u=\sup_{0\le t\le u}|X_t|$ in the continuous-time case and $M_u=\max_{1\le t\le\lceil u\rceil}|X_t|$ in the discrete-time case, where $\lceil u\rceil$ is an integer part of u. Let $f(t)$ be a nonnegative increasing function, $f(0)=0$, and let for $\varepsilon>0$*

$$L_\varepsilon(f)=\sup\{t:|X_t|>\varepsilon f(t)\},\quad \sup\{\varnothing\}=0$$

be the last time X_t leaves the interval $[-\varepsilon f(t),+\varepsilon f(t)]$.

For any positive number r and any positive ε the following inequalities hold:

$$r\int_0^\infty u^{r-1}\mathsf{P}\{|X_u|\ge\varepsilon f(u)\}\,du\le\mathsf{E}[L_\varepsilon(f)]^r\le r\int_0^\infty t^{r-1}\mathsf{P}\left\{\sup_{t\ge u}\frac{|X_u|}{f(u)}\ge\varepsilon\right\}dt. \tag{B.9}$$

Therefore, finiteness of the integral

$$\int_0^\infty t^{r-1}\mathsf{P}\left\{\sup_{t\ge u}\frac{|X_u|}{f(u)}\ge\varepsilon\right\}dt \quad \text{for all } \varepsilon>0$$

implies r-quick convergence of $X_t/f(t)$ to zero as $t\to\infty$.

When $X_n=S_n/n$, where $\{S_n\}$ is a random walk, then the implications in (B.8) become equivalences. Specifically, the following results hold regarding rates of convergence in SLLN (see, e.g., [164, Theorem 2.4.4] and [12]).

Lemma B.4. *Let $X_n=S_n/n$ and $r>0$. If $\{S_n\}$ is a random walk with $\mathsf{E}[S_1]=I$, then*

$$\begin{aligned}\mathsf{E}[|X_1|^{r+1}]<\infty \Longleftrightarrow X_n \xrightarrow[n\to\infty]{r-quickly} I \quad&\Longleftrightarrow\quad X_n \xrightarrow[n\to\infty]{r-completely} I\\ \Longleftrightarrow \sum_{t=1}^\infty \mathsf{P}\left\{\sup_{k\ge t}\frac{S_k}{k}>I-\varepsilon\right\}<\infty &\quad \text{for all } \varepsilon>0.\end{aligned} \tag{B.10}$$

Appendix C: Identities and Inequalities for Martingales

This appendix gives sample results from the martingale theory that are often used in the book. In particular, we present the optional sampling theorem and Wald's identities for martingales as well as Doob's maximal inequality and moment inequalities for martingales, which are used in the book for deriving asymptotic approximations for operating characteristics of sequential hypothesis tests and change detection rules.

Let $(\Omega, \mathscr{F}, \mathsf{P})$ be a probability space. A family of sigma-algebras $\{\mathscr{F}_t\}$ is called a filtration. A random process $\{X_t\}$ on $(\Omega, \mathscr{F}, \mathsf{P})$ is said to be *adapted* to filtration $\{\mathscr{F}_t\}$ (or simply $\{\mathscr{F}_t\}$*-adapted*) if $\{\omega : X_t(\omega) \in B\} \in \mathscr{F}_t$ for all t. The time index may be either discrete $t = n \in \mathbb{Z}_+$ or continuous $t \in \mathbb{R}_+$.

Let $\{\mathscr{F}_t\}$ be a filtration and let $\{X_t\}$ be a random process on $(\Omega, \mathscr{F}, \mathsf{P})$ adapted to $\{\mathscr{F}_t\}$. A process $\{X_t\}$ is said to be a *martingale* with respect to $\{\mathscr{F}_t\}$ if $\mathsf{E}|X_t| < \infty$ for all t and

$$\mathsf{E}[X_t|\mathscr{F}_u] = X_u \quad \mathsf{P}-\text{a.s. for } u \le t. \tag{C.1}$$

If $\mathsf{E}[X_t|\mathscr{F}_u] \ge X_u$, then $(X_t, \mathscr{F}_t)$ is said to be a *submartingale*; if $\mathsf{E}[X_t|\mathscr{F}_u] \le X_u$, then $(X_t, \mathscr{F}_t)$ is said to be a *supermartingale*. Note that $-X_t$ is a supermartingale whenever X_t is a submartingale. Clearly, the martingale is simultaneously sub- and supermartingale.

In most cases, we deal with martingales with respect to the natural filtration $\mathscr{F}_t = \sigma(X_u, 0 \le u \le t)$ generated by the process X_t.

One typical example of the martingale is the likelihood ratio process

$$\Lambda_t = \frac{\mathsf{dP}_1^t}{\mathsf{dP}_0^t}, \quad t \in \mathbb{R}_+ \text{ or } t \in \mathbb{Z}_+, \tag{C.2}$$

where $\mathsf{P}_i^t = \mathsf{P}_i|_{\mathscr{F}_t}$ $(i = 0, 1)$ are restrictions of probability measures P_i to the sigma-algebra $\mathscr{F}_t$. If P_1^t is absolutely continuous with respect to P_0^t, then Λ_t is the P_0-martingale with unit expectation. Another typical example is the sum $S_n = \sum_{i=1}^n X_i$ of zero-mean i.i.d. random variables X_i, $\mathsf{E}[X_i] = 0$. Also, if θ is a random variable with finite expectation, then $\theta_n = \mathsf{E}[\theta \mid \mathscr{F}_n]$ is a martingale.

If $(X_t, \mathscr{F}_t)$ is a martingale, then $(|X_t|^\gamma, \mathscr{F}_t)$ is a submartingale for $\gamma \ge 1$ and a supermartingale for $0 < \gamma < 1$, which follows from Jensen's inequality.

C.1 The Doob–Wald Identities

The stopping time T is called *proper* if it is finite w.p. 1, $\mathsf{P}(T < \infty) = 1$.

One of the important properties of martingales is that the martingale structure is preserved by optional sampling.[1]

[1]Often optional sampling is called optional stopping.

Theorem C.1 (Optional sampling theorem). *Let $(X_t, \mathscr{F}_t)$ be a* P*-supermartingale and let T and τ be two proper stopping times with respect to $\{\mathscr{F}_t\}$ such that $\tau \le T$. Suppose*

$$\mathsf{E}|X_\tau| < \infty, \quad \mathsf{E}|X_T| < \infty, \tag{C.3}$$
$$\liminf_{t\to\infty} \mathsf{E}[|X_t| \mathbb{1}_{\{T>t\}}] = 0. \tag{C.4}$$

Then

$$\mathsf{E}[X_T \mid \mathscr{F}_\tau] \le X_\tau \quad \mathsf{P}-a.s., \tag{C.5}$$

and if $(X_t, \mathscr{F}_t)$ is a martingale, then

$$\mathsf{E}[X_T \mid \mathscr{F}_\tau] = X_\tau \quad \mathsf{P}-a.s. \tag{C.6}$$

The optional sampling theorem is of fundamental importance in sequential analysis and optimal stopping. In particular, taking $\tau = 0$, we immediately obtain the following identity for martingales that generalizes Wald's identity for sums of i.i.d. random variables. Let $(X_t, \mathscr{F}_t)$ be a P-martingale and let T be a proper stopping time with respect to $\{\mathscr{F}_t\}$. If $\mathsf{E}|X_T| < \infty$ and (C.4) holds, then

$$\mathsf{E}[X_T] = \mathsf{E}[X_0]. \tag{C.7}$$

The following slightly improved version is due to Klass [70, Theorem 4.1].

Proposition C.1. *Let $(X_n, \mathscr{F}_n)_{n\in\mathbb{Z}_+}$ be a mean-zero martingale and let T be any (possibly randomized) proper stopping time with respect to $\{\mathscr{F}_n\}$. Suppose*

$$\lim_{n\to\infty} \mathsf{E}|X_{T\wedge n}| < \infty \quad \textit{and} \quad \lim_{n\to\infty} \mathsf{E}[X_n \mathbb{1}_{\{T>n\}}] = 0.$$

Then

$$\mathsf{E}[X_T] = 0. \tag{C.8}$$

Identities (C.7)–(C.8), which are often referred to as the *Doob–Wald* identities for martingales, imply the following two Wald's identities. If $\{S_n\}_{n\in\mathbb{Z}_+}$ ($S_0 = 0$) is a random walk, i.e., $S_n = \sum_{i=1}^n Y_i$, where $Y_1, Y_2, \dots$ are i.i.d. random variables with finite mean $\mathsf{E}[Y_1] = \mu$, and $\mathsf{E}[T] < \infty$, then

$$\mathsf{E}[S_T] = \mu \mathsf{E}[T], \tag{C.9}$$

and if $\sigma^2 = \mathsf{Var}[Y_1] < \infty$ and $\mathsf{E}[T] < \infty$, then

$$\mathsf{E}[S_T - \mu T]^2 = \sigma^2 \mathsf{E}[T]. \tag{C.10}$$

Another important identity usually referred to as *Wald's Likelihood Ratio Identity* concerns the likelihood ratio process Λ_t defined in (C.2). Denote by E_i the expectations under the probability measures P_i, $i = 0, 1$.

For any stopping time T and nonnegative $\mathscr{F}_T$-measurable random variable Y the following identity holds:

$$\mathsf{E}_1[Y \mathbb{1}_{\{T<\infty\}}] = \mathsf{E}_0[Y \Lambda_T \mathbb{1}_{\{T<\infty\}}]. \tag{C.11}$$

In particular, for $Y = \mathbb{1}_{\{\mathscr{A}\}}$ with $\mathscr{A} \in \mathscr{F}_T$,

$$\mathsf{P}_1(\mathscr{A}, T < \infty) = \mathsf{E}_0[\Lambda_T, \mathscr{A}, T < \infty]. \tag{C.12}$$

Taking $\mathscr{A} = \Omega$, we obtain

$$\mathsf{P}_1(T < \infty) = \mathsf{E}_0(\Lambda_T, T < \infty) = \int_{\{T<\infty\}} \Lambda_T \, d\mathsf{P}_0, \tag{C.13}$$

and hence

$$\mathsf{E}_0(\Lambda_T, T < \infty) = \int_{\{T<\infty\}} \Lambda_T \, d\mathsf{P}_0 = 1$$

whenever $\mathsf{P}_1(T < \infty) = 1$.

C.2 Inequalities for Martingales

Let $(X_t, \mathscr{F}_t)$ be a P-submartingale ($t \in \mathbb{R}_+$ or $t \in \mathbb{Z}_+$). The following Doob's maximal inequality holds:

$$\mathsf{P}\left(\sup_{0\le u\le t} X_u\right) \le \frac{1}{a}\mathsf{E}\left[X_t \mathbb{1}_{\{\sup_{0\le u\le t} X_u \ge a\}}\right] \le \frac{1}{a}\mathsf{E}[X_t^+], \quad a>0. \tag{C.14}$$

If X_t is a uniformly integrable martingale, then this inequality can be extended to stopping times T:

$$\mathsf{P}\left(\sup_{0\le u\le T} |X_u| \ge a\right) \le \frac{1}{a}\mathsf{E}\left[|X_T| \mathbb{1}_{\{\sup_{0\le u\le T} |X_u| \ge a\}}\right] \le \frac{1}{a}\mathsf{E}|X_T|, \quad a>0.$$

Note that if $Y_1, Y_2, \dots$ are i.i.d. zero-mean random variables with $\mathsf{E}|Y_1|^m < \infty$, $m \ge 1$, then taking $X_n = |S_n|^m$, $S_n = \sum_{i=1}^n Y_i$, we obtain from (C.14) that

$$\mathsf{P}\left(\max_{0\le k\le n} |S_k| > a\right) \le \frac{1}{a^m}\mathsf{E}\left[|S_n|^m; \max_{0\le k\le n}|S_k| > a\right] \le \frac{1}{a^m}\mathsf{E}|S_n|^m, \quad a>0,\ n\ge 1. \tag{C.15}$$

For $m = 2$, this gives the Kolmogorov inequality. For this reason, the inequality (C.14) is often referred to as the Doob–Kolmogorov inequality for martingales.

We now provide several useful moment inequalities for martingales.

A martingale $(X_t, \mathscr{F}_t)$ is said to be *square integrable* if $\sup_t \mathsf{E}[X_t^2] < \infty$.

Let $(X_t, \mathscr{F}_t)$ be a nonnegative submartingale and let $M_t = \sup_{0\le u\le t} X_u$. Let $p > 1$. If $\mathsf{E}[X_t^p] < \infty$, then

$$\mathsf{E}\left[\left(\sup_{0\le u\le t} X_u\right)^p\right] \le \left(\frac{p}{p-1}\right)^p \mathsf{E}[X_t^p]. \tag{C.16}$$

Therefore, if $(X_t, \mathscr{F}_t)$ is a square integrable martingale, then

$$\mathsf{E}\left[\sup_{0\le u\le t} X_u^2\right] \le 4\mathsf{E}[X_t^2]. \tag{C.17}$$

Let $(X_t, \mathscr{F}_t)$ be a uniformly integrable martingale and let T be a stopping time. If $\mathsf{E}|X_T|^p < \infty$, then

$$\mathsf{E}\left[\left(\sup_{t\le T}|X_t|\right)^p\right] \le \left(\frac{p}{p-1}\right)^p \mathsf{E}|X_T|^p. \tag{C.18}$$

Let $(X_t, \mathscr{F}_t)$ be a square integrable martingale with independent increments, $X_0 = 0$. For any $0 < p < \infty$ and any stopping time T there are positive constants C_p and c_p, independent of T and X, such that

$$\mathsf{E}\left[\left(\sup_{t\le T}|X_t|\right)^p\right] \le \begin{cases} C_p\,\mathsf{E}[D_T]^{p/2} & \text{if } 0<p\le 2,\\ C_p\,\mathsf{E}[D_T]^{p/2} + c_p\,\mathsf{E}[\sum_{t\le T}|\Delta X_t|^p] & \text{if } p>2, \end{cases} \tag{C.19}$$

where $D_t = \mathsf{Var}[X_t]$ and $\Delta X_t = X_t - \lim_{s\uparrow t} X_s$ is a jump of the process X at time t.

A proof of (C.16)–(C.17) may be found, e.g., in Liptser and Shiryaev [80]. The inequalities (C.19) are particular cases of the Burkholder–Gundy–Novikov inequalities for martingales [20, 21, 102, 103]. A proof of (C.19) may be found in Tartakovsky [154, Lemma 5]. The inequality (C.16) is often referred to as Doob's maximal moment inequality.

Let $S_n = X_1 + \dots + X_n$, $n \ge 1$ be a mean-zero random walk, $\mathsf{E}[X_1] = 0$. If $\mathsf{E}|X_1|^p < \infty$, then (C.19) yields

$$\mathsf{E}|S_T|^p \le \begin{cases} C_p\,\mathsf{E}|X_1|^p\mathsf{E}[T] & \text{if } 0<p\le 2,\\ C_p\,\mathsf{E}[X_1^2]^{p/2}\mathsf{E}[T^{p/2}] + c_p\,\mathsf{E}|X_1|^p\mathsf{E}[T] \le \tilde{C}_p\mathsf{E}|X_1|^p\mathsf{E}[T^{p/2}] & \text{if } p>2, \end{cases} \tag{C.20}$$

where C_p, c_p, and $\tilde{C}_p$ are universal constants depending only on p.

Appendix D: Markov Processes

Markov processes as well as partially observed Markov processes (which are usually referred to as hidden Markov models) are important classes of stochastic processes often used for modeling stochastic systems. In this appendix, we present basic definitions and results that are used in the book.

Let $(\Omega, \mathscr{F}, \mathsf{P})$ be a probability space and let $X = \{X_n\}_{n\in\mathbb{Z}_+}$ be a discrete-time random process on $(\Omega, \mathscr{F}, \mathsf{P})$. Let $\{\mathscr{F}_n\}_{n\in\mathbb{Z}_+}$ be a filtration (i.e., a family of sigma-algebras). Recall that a random process $\{X_n\}_{n\in\mathbb{Z}_+}$ is $\{\mathscr{F}_n\}$-adapted if $\{\omega : X_n(\omega) \in B\} \in \mathscr{F}_n$ for all $n \in \mathbb{Z}_+$.

Let $\{X_n\}_{n\in\mathbb{Z}_+}$ be an $\{\mathscr{F}_n\}$-adapted random process and let P_x be a probability measure on $\mathscr{F}$ given $X_n = x \in \mathscr{X}$.

Definition D.1. A random process $\{X_n\}_{n\in\mathbb{Z}_+}$ is called a *homogeneous Markov process* if

(i) For all $x \in \mathscr{X}$ and all $i, n \in \mathbb{Z}_+$

$$\mathsf{P}_x(X_{n+i} \in B \mid \mathscr{F}_n) = \mathsf{P}_{X_n}(X_{n+i} \in B) \quad \mathsf{P}_x - \text{a.s.}$$

(ii) $\mathsf{P}_x(X_0 = x) = 1, x \in \mathscr{X}$.

The first condition is the Markov property of dependence of the future on the past via the present state.

The probability $P(i,x,B) = \mathsf{P}_{X_n=x}(X_{n+i} \in B) = \mathsf{P}_{X_0=x}(X_i \in B)$ is called the *transition* probability. If it depends on n, the Markov process is *nonhomogeneous*.

If space $\mathscr{X}$ is finite or countable, then the Markov process is called a *Markov chain*.

Example D.1 (Random walk). Let $X_n = \sum_{i=0}^{n} Y_i$, $n \in \mathbb{Z}_+$ be partial sums, where $\{Y_n\}_{n\geq 1}$ is a sequence of i.i.d. random variables with a distribution function $F(y)$ and $X_0 \equiv Y_0 = x$. The process $\{X_n\}_{n\in\mathbb{Z}_+}$ is called a *random walk* and the mean of its increment $\mu = \mathsf{E}[Y_k]$ is called the *drift* of the random walk. It is a homogeneous Markov process with mean $\mathsf{E}[X_n] = x + \mu n$ and transition probability

$$P(x,y) = \mathsf{P}(X_{n+1} \leq y | X_n = x) = \mathsf{P}(X_1 \leq y | X_0 = x) = \mathsf{P}(Y_1 \leq y - x) = F(y - x).$$

If $\mathsf{Var}[Y_k] = \sigma^2 < \infty$, then $\mathsf{Var}[X_n] = \sigma^2 n$. Often it is assumed that the starting point $X_0 = x = 0$.

If $x = 0$ and Y_k takes on two values -1 and $+1$ with equal probabilities $\mathsf{P}(Y_k = -1) = \mathsf{P}(Y_k = +1) = 1/2$, then the random walk is called *simple*. If $x = 0$ and $Y_k \in \{0, 1\}$ with probabilities $\mathsf{P}(Y_k = 1) = p$ and $\mathsf{P}(Y_k = 0) = 1 - p$, then the random walk is called the *Bernoulli random walk*.

Example D.2 (Autoregression – AR(1) process). Define the process X_n recursively

$$X_0 = x, \quad X_n = \rho X_{n-1} + Y_n, \quad n = 1, 2, \dots,$$

where ρ is a finite number and $\{Y_n\}_{n\geq 1}$ is a sequence of i.i.d. random variables. Note that if $\rho = 1$ this process reduces to a random walk. For $\rho \neq 1$ this process is called an *autoregressive process of* 1*st order* (or AR(1) process). This is the homogeneous Markov process with transition probability

$$P(x,y) = \mathsf{P}(X_{n+1} \leq y | X_n = x) = \mathsf{P}(Y_1 \leq y - \rho x | X_0 = x) = F(y - \rho x).$$

If $\mathsf{E}[Y_1] = \mu$ and $\mathsf{Var}[Y_1] = \sigma^2$, then assuming that $|\rho| < 1$,

$$\lim_{n\to\infty} \mathsf{E}[X_n] = \mu/(1-\rho), \quad \lim_{n\to\infty} \mathsf{Var}[X_n] = \sigma^2/(1-\rho^2).$$

D.1 Irreducible, Recurrent and Transient Markov Chains

Let $\{X_n\}_{n\in\mathbb{Z}_+}$ be a homogeneous Markov chain with the state space $\{0,1,2,\dots\}$. It is said that the state j is *reachable* from i if there exists an integer $n \geq 0$ such that $\mathsf{P}_i(X_n = j) = \mathsf{P}(X_n = j|X_0 = i) > 0$, i.e., the probability that the chain will be in state j at time n when it is initialized at i is positive for some $n \geq 0$. If state j is reachable from i and state i is reachable from j, then it is said that states i and j *communicate*. A Markov chain for which *all states* communicate is said to be an *irreducible* Markov chain.

Let $T_{ii} = \inf\{n \geq 1 : X_n = i|X_0 = i\}$ denote the *return time* to state i when the chain starts at i, where $\inf\{\varnothing\} = \infty$, i.e., $T_{ii} = \infty$ if $X_n \neq i$ for all $n \geq 1$. Next, define $\mathsf{P}_i(T_{ii} < \infty)$, the probability of ever returning to state i, given that the chain started in state i. A state i is said to be *recurrent* if $\mathsf{P}_i(T_{ii} < \infty) = 1$ and *transient* if $\mathsf{P}_i(T_{ii} < \infty) < 1$.

By the Markov property, once the chain revisits state i, the future is independent of the past. Hence, after each time state i is visited, it will be revisited with the same probability $p_i = \mathsf{P}_i(T_{ii} < \infty)$ independent of the past. In particular, if $p_i = 1$, the chain will return to state i over and over again (an infinite number of times), and for this reason, it is called recurrent. The transient state will only be visited a finite random number of times.

Formally, let N_i denote the total number of visits to state i, given that $X_0 = i$. The distribution of N_i is geometric, $\mathsf{P}_i(N_i = k) = (1-p_i)p_i^{k-1}, k \geq 1$ (counting the initial visit $X_0 = i$ as the first one). Therefore, the expected number of visits is $\mathsf{E}_i N_i = (1-p_i)^{-1}$, and the state is recurrent if $\mathsf{E}_i N_i = \infty$ and transient if $\mathsf{E}_i N_i < \infty$.

The Markov chain is said to be *recurrent* if it is irreducible and all states are recurrent; otherwise the irreducible Markov chain is said to be *transient*. An irreducible Markov chain with *finite* state space is always recurrent.

A recurrent state i is said to be *positive recurrent* if $\mathsf{E}_i T_{ii} < \infty$ and *null recurrent* if $\mathsf{E}_i T_{ii} = \infty$. The irreducible Markov chain is said to be *positive recurrent* if all states are positive recurrent and *null recurrent* if all states are null recurrent.

D.2 Stationary and Quasi-stationary Distributions

Consider a general discrete-time homogeneous Markov process $\{X_n\}_{n\in\mathbb{Z}_+}$ with state space $\mathscr{X}$ and transition probabilities $P(x,y) = \mathsf{P}(X_{n+1} \leq y|X_n = x)$, $x,y \in \mathscr{X}$. Let P_x denote the probability for the process with initial state $X_0 = x$, i.e., $\mathsf{P}(X_n \in \mathscr{B}|X_0 = x) = \mathsf{P}_x(X_n \in \mathscr{B})$. A *stationary distribution* (or *invariant* or *ergodic*) of the Markov process $\{X_n\}$ is a limit (if it exists)

$$\lim_{n\to\infty} \mathsf{P}_x(X_n \leq y) = \mathsf{Q}_{\mathrm{st}}(y)$$

for every initial state $X_0 = x$ at continuity points of $\mathsf{Q}_{\mathrm{st}}(y)$. This distribution satisfies the following integral equation

$$\mathsf{Q}_{\mathrm{st}}(y) = \int_{\mathscr{X}} P(x,y)\, d\mathsf{Q}_{\mathrm{st}}(x). \tag{D.1}$$

Clearly, if the initial variable X_0 has the probability distribution $\mathsf{Q}_{\mathrm{st}}(x)$, then all the other variables $X_1, X_2, \dots$ have the same distribution, which explains why $\mathsf{Q}_{\mathrm{st}}(x)$ is called the stationary distribution – the Markov process started from the point distributed according the stationary distribution is not only homogeneous but also stationary.

Suppose $\{X_n\}$ is a continuous process. Then the stationary distribution has density $\mathsf{q}_{\mathrm{st}}(y) = \mathrm{d}\mathsf{Q}_{\mathrm{st}}(y)/\mathrm{d}y$ with respect to the Lebesgue measure, and it follows from (D.1) that this stationary density satisfies the equation

$$\mathsf{q}_{\mathrm{st}}(y) = \int_{\mathscr{X}} \mathsf{q}_{\mathrm{st}}(x)\mathscr{K}(x,y)\,\mathrm{d}x, \tag{D.2}$$

where $\mathscr{K}(x,y) = \frac{\partial}{\partial x}P(x,y)$. Thus, the stationary density $\mathsf{q}_{\mathrm{st}}(y)$ is the (left) eigenfunction corresponding to the unit eigenvalue of the linear operator $\mathscr{K}(x,y)$.

A stationary distribution exists for recurrent (more generally Harris-recurrent [13, 56]) Markov processes at least in a generalized sense, i.e., it can be an *improper distribution*, $\int_{\mathscr{X}} \mathrm{d}\mathsf{Q}_{\mathrm{st}}(x) = \infty$.

Example D.3. Let $\{X_n\}$ be given recursively

$$X_0 = x \in [0,\infty), \quad X_{n+1} = (1+X_n)\Lambda_{n+1}, \quad n \in \mathbb{Z}_+, \tag{D.3}$$

where Λ_n, $n \geq 1$, are nonnegative i.i.d. random variables having a Beta-distribution with density

$$p(y) = \frac{y^{\delta-1}(1+y)^{-2\delta-1}}{\mathsf{B}(\delta,\delta+1)}\mathbb{1}_{\{y\geq 0\}}, \quad \mathsf{B}(r,v) = \int_0^1 t^{r-1}(1-t)^{v-1}\mathrm{d}t \ (r,v>0).$$

Using (D.2), we obtain that the stationary pdf is governed by the equation

$$\mathsf{q}_{\mathrm{st}}(y) = \frac{y^{\delta-1}}{\mathsf{B}(\delta+1,\delta)}\int_0^\infty \mathsf{q}_{\mathrm{st}}(x)\frac{(1+x)^{\delta+1}}{(1+x+y)^{1+2\delta}}\,\mathrm{d}x,$$

and the solution is

$$\mathsf{q}_{\mathrm{st}}(y) = \frac{y^{\delta-1}(1+y)^{-1-\delta}}{\mathsf{B}(\delta,1)}\mathbb{1}_{\{y\geq 0\}} = \delta y^{\delta-1}(1+y)^{-1-\delta}\mathbb{1}_{\{y\geq 0\}},$$

which is the density of a Beta-distribution with parameters δ and 1.

Consider now a homogeneous Markov chain $\{X_n\}$ with the state space $\{0,1,2,\dots\}$ and transition probabilities $P_{ij} = \mathsf{P}(X_{n+1}=j|X_n=i)$, $n,i,j \in \mathbb{Z}_+$. A set of the limiting probabilities

$$\lim_{n\to\infty} \mathsf{P}(X_n = j|X_0 = i) = \mathsf{Q}^*_{\mathrm{st}}(j), \quad j = 0,1,2,\dots$$

(if they exist) is said to be the *stationary distribution* of the Markov chain X_n. It satisfies the equation

$$\mathsf{Q}^*_{\mathrm{st}}(i) = \sum_{j=0}^\infty \mathsf{Q}^*_{\mathrm{st}}(j)P_{ji}, \quad i = 0,1,2,\dots; \ \mathsf{Q}^*_{\mathrm{st}}(i) \geq 0; \ \sum_{i=0}^\infty \mathsf{Q}^*_{\mathrm{st}}(i) = 1. \tag{D.4}$$

If $\{X_n\}$ is a positive recurrent Markov chain, then a stationary distribution exists and is given by $\mathsf{Q}^*_{\mathrm{st}}(i) = 1/\mathsf{E}_i T_{ii}$, $i \in \mathbb{Z}_+$. Since every irreducible Markov chain with finite state space is positive recurrent, it follows that a (unique) stationary distribution always exists in the finite state space case. If the Markov chain is null recurrent or transient, then a stationary distribution satisfying (D.4) does not exist. Indeed, the stationary distribution may not exist in the class of probability measures, which is often the case if X_n either goes to ∞ or to 0 (typical for null recurrent and transient chains). However, it may still have a stationary "distribution" in the generalized sense, i.e., an *improper stationary* distribution, satisfying (D.4) with an infinite sum, $\sum_{i=0}^\infty \mathsf{Q}^*_{\mathrm{st}}(i) = \infty$. In this more general sense, a stationary measure always exists for irreducible (indecomposable) recurrent Markov chains and sometimes for transient chains. For further details, see Harris [57, Sec. I.11] and references therein.

As an example, let $\{X_n\}$ be a two-state Markov chain with strictly positive transition probabilities $P_{01} = p$, $P_{10} = q$. Then the unique stationary distribution is $\mathsf{Q}^*_{\rm st}(0) = q/(p+q)$, $\mathsf{Q}^*_{\rm st}(1) = p/(p+q)$.

Another useful class of distributions is the class of *quasi-stationary distributions*, which come up naturally in the context of first-exit times of Markov processes. Let $\{X_n\}$ be a Markov process with state space $\mathscr{X}$ and transition probabilities $P(x,y) = \mathsf{P}(X_{n+1} \le y | X_n = x)$, $x,y \in \mathscr{X}$. If the process is *absorbing*, its *quasi-stationary distribution* is defined to be the limit (if it exists) as $n \to \infty$ of the distribution of X_n, given that absorption has not occurred by the time n,

$$\mathsf{Q}_{\mathscr{A}}(y) = \lim_{n\to\infty} \mathsf{P}_x(X_n \le y | X_1 \notin \mathscr{A}, \dots, X_n \notin \mathscr{A}) \quad \text{for every initial state } X_0 = x,$$

where $\mathscr{A}$ is an absorbing set or state. Equivalently the quasi-stationary distribution can be defined as

$$\mathsf{Q}_{\mathscr{A}}(y) = \lim_{n\to\infty} \mathsf{P}_x(X_n \le y | T_{\mathscr{A}} > n),$$

where $T_{\mathscr{A}} = \inf\{n \ge 1 : X_n \in \mathscr{A}\}$ is the "killing" time. Therefore, the quasi-stationary distribution is nothing but a stationary conditional distribution and also a limiting conditional distribution in that $X_n \to X_\infty$ as $n \to \infty$, and thus it can be used for modeling the long-term behavior of the process (or system).

Of special interest (in particular in certain statistical applications) is the case of a nonnegative Markov process ($\mathscr{X} = [0,\infty)$), where the first time that the process exceeds a fixed positive level A signals that some action is to be taken or a decision should be made. The quasi-stationary distribution is the distribution of the state of the process if a long time has passed and yet no crossover has occurred, i.e.,

$$\mathsf{Q}_A(y) = \lim_{n\to\infty} \mathsf{P}_x(X_n \le y | T_A > n) \quad \text{for every initial state } X_0 = x, \tag{D.5}$$

where $T_A = \inf\{n \ge 1 : X_n \ge A\}$ is the corresponding stopping time which makes the process X_n absorbing.

Various topics pertaining to quasi-stationary distributions are existence, calculation, and simulation. For an extensive bibliography see Pollett [114].

The quasi-stationary distribution defined in (D.5) satisfies the following integral equation

$$\lambda_A \mathsf{Q}_A(y) = \int_0^A P(x,y)\, d\mathsf{Q}_A(x), \tag{D.6}$$

where

$$\lambda_A = \int_0^A P(x,A)\, d\mathsf{Q}_A(x).$$

If the sequence $\{X_n\}_{n\in\mathbb{Z}_+}$ is initialized from the random point $X_0 \sim \mathsf{Q}_A$ distributed according to a quasi-stationary distribution, then all the other variables $X_1, X_2, \dots$ are also distributed according to Q_A,

$$\mathsf{P}_{\mathsf{Q}_A}(X_n \in \mathscr{B} | T_A > n) = \mathsf{Q}_A(\mathscr{B}), \quad n \in \mathbb{Z}_+.$$

In this case, for every $A > 0$ the distribution of the stopping time T_A is strictly geometric with parameter $1 - \lambda_A$,

$$\mathsf{P}_{\mathsf{Q}_A}(T_A = k) = (1-\lambda_A)\lambda_A^{k-1}, \quad k = 1,2,\dots,$$

so $\mathsf{E}_{\mathsf{Q}_A}[T_A] = (1-\lambda_A)^{-1}$.

Let $\{X_n\}$ be a continuous process and let $\mathsf{q}_A(y) = \mathrm{d}\mathsf{Q}_A(y)/\mathrm{d}y$ denote the quasi-stationary density. By (D.6), it satisfies the integral equation

$$\lambda_A \mathsf{q}_A(y) = \int_{\mathscr{X}} \mathsf{q}_A(x) \mathscr{K}(x,y)\, \mathrm{d}x. \tag{D.7}$$

Therefore, the quasi-stationary density $\mathsf{q}_A(y)$ is the (left) eigenfunction corresponding to the eigenvalue λ_A of the linear operator $\mathscr{K}(x,y) = \frac{\partial}{\partial x}P(x,y)$.

The quasi-stationary distribution may not exist. For example, if $\mathsf{P}(X_1 \ge A | X_0 < A) = 1$, there is no quasi-stationary distribution since $T_A = 1$ almost surely, but it must be geometric. If the Markov process $\{X_n\}$ is Harris-recurrent and continuous, then the quasi-stationary distribution always exists [57, Theorem III.10.1].

The above results are also valid for continuous-time Markov processes, in which case the distribution of the stopping time $T_A = \inf\{t \in \mathbb{R}_+ : X_t \ge A\}$ is exponential for all $A > 0$ if the process is started at a quasi-stationary distribution, $X_0 \sim \mathsf{Q}_A$.

Example D.4. Again, let $\{X_n\}$ be given recursively as in (D.3) with nonnegative i.i.d. random variables $\{\Lambda_n\}_{n\ge 1}$ which have the distribution

$$\mathsf{P}(\Lambda_1 \le z) = \begin{cases} 1 & \text{if } z \ge 2 \\ z/2 & \text{if } 0 < z < 2 \\ 0 & \text{if } z \le 0. \end{cases}$$

Let $A < 2$. Then, by (D.7), the quasi-stationary density $\mathsf{q}_A(y)$ satisfies the integral equation

$$\lambda_A\, \mathsf{q}_A(y) = \frac{1}{2}\int_0^A \mathsf{q}_A(x)\, \frac{\mathrm{d}x}{1+x},$$

which yields $\lambda_A = \frac{1}{2}\log(1+A)$ and $\mathsf{q}_A(y) = A^{-1}\mathbb{1}_{\{x\in[0,A)\}}$. Thus, for $A < 2$ the quasi-stationary distribution is uniform, $\mathsf{Q}_A(y) = y/A$. Note that it is attained already for $n = 1$.

An important question is whether the quasi-stationary distribution converges to the stationary distribution as $A \to \infty$, i.e., $\lim_{A\to\infty} \mathsf{Q}_A(y) = \mathsf{Q}_{\mathrm{st}}(y)$ at continuity points of $\mathsf{Q}_{\mathrm{st}}(y)$. Sufficient conditions for stochastically monotone Markov processes, i.e., when the probability $\mathsf{P}_x(X_1 \ge y)$ is non-decreasing and right-continuous in x for all $y \in [0,\infty)$, are given in Theorem 1 of Pollak and Siegmund [111].

D.3 Non-asymptotic Bounds for the Concentration Inequalities

D.3.1 Correlation Inequality

Proposition D.1. *Let $(\Omega, \mathscr{F}, \{\mathscr{F}_n\}, \mathsf{P})$ be a filtered probability space and $(u_j, \mathscr{F}_j)_{1\le j\le n}$ be a sequence of random variables such that $\max_{1\le j\le n} \mathsf{E}|u_j|^r < \infty$ for some $r \ge 2$. Define*

$$\check{b}_{j,n}(r) = \left(\mathsf{E}\left(|u_j| \sum_{k=j}^{n} |\mathsf{E}(u_k|\mathscr{F}_j)|\right)^{r/2}\right)^{2/r}. \tag{D.8}$$

The following correlation inequality holds:

$$\mathsf{E}\left|\sum_{j=1}^{n} u_j\right|^r \le (2r)^{r/2} \left(\sum_{j=1}^{n} \check{b}_{j,n}(r)\right)^{r/2}.$$

The proof may be found in [49].

D.3.2 Uniform Geometric Ergodicity for Homogeneous Markov Processes

We recall some definitions for a homogeneous Markov process $\{X_n\}_{n\in\mathbb{Z}_+}$ defined on a measurable state space $(\mathscr{X},\mathscr{B}(\mathscr{X}))$. Denote by $(P^\theta(\cdot,\cdot))_{\theta\in\Theta}$ the transition probability family of this process, i.e., for any $A\in\mathscr{B}(\mathscr{X}), x\in\mathscr{X}$,

$$P^\theta(x,A) = \mathsf{P}^\theta_x(X_1\in A) = \mathsf{P}^\theta(X_1\in A|X_0=x)\,. \tag{D.9}$$

The n-step transition probability is $P^{n,\theta}(x,A) = \mathsf{P}^\theta_x(X_n\in A)$.

We recall that a measure $\mathsf{Q}^\theta_{\rm st}$ on $\mathscr{B}(\mathscr{X})$ is called *stationary* (or *invariant* or *ergodic*) for this process if it satisfies integral equation (D.1), i.e., for any $A\in\mathscr{B}(\mathscr{X})$,

$$\mathsf{Q}^\theta_{\rm st}(A) = \int_{\mathscr{X}} P^\theta(x,A)\mathsf{Q}^\theta_{\rm st}(\mathrm{d}x)\,. \tag{D.10}$$

If the stationary distribution $\mathsf{Q}^\theta_{\rm st}(x)$ is positive and proper, $\mathsf{Q}^\theta_{\rm st}(\mathscr{X})=1$, then the process is called *positive*.

Assume that the process $\{X_n\}_{n\in\mathbb{Z}_+}$ satisfies the following *minorization* condition:

($\mathbf{D}_1$) *There exist $\delta>0$, a set $C\in\mathscr{B}(\mathscr{X})$ and a probability measure ς on $\mathscr{B}(\mathscr{X})$ with $\varsigma(C)=1$, such that*

$$\inf_{\theta\in\Theta}\left(\inf_{x\in C} P^\theta(x,A) - \delta\,\varsigma(A)\right) > 0$$

for any $A\in\mathscr{B}(\mathscr{X})$, for which $\varsigma(A)>0$.

Obviously, this condition implies that $\eta = \inf_{\theta\in\Theta}\inf_{x\in C} P^\theta(x,C) - \delta > 0$. Now we impose the *uniform drift* condition.

($\mathbf{D}_2$) *There exist some constants $0<\rho<1$ and $D\geq 1$ such that for any $\theta\in\Theta$ there exist a $\mathscr{X}\to[1,\infty)$ function $\mathbf{V}$ and a set C from $\mathscr{B}(\mathscr{X})$ such that*

$$\mathbf{V}^* = \sup_{\theta\in\Theta}\sup_{x\in C}|\mathbf{V}(x)| < \infty$$

and

$$\sup_{\theta\in\Theta}\sup_{x\in\mathscr{X}}\left\{\mathsf{E}^\theta_x[\mathbf{V}(X_1)] - (1-\rho)\mathbf{V}(x) + D\mathbb{1}_{\{C\}}(x)\right\} \leq 0\,. \tag{D.11}$$

In this case, we call $\mathbf{V}$ the *Lyapunov function*.

In this book, we use the following theorem from [50].

Theorem D.1. *Let $\{X_n\}_{n\in\mathbb{Z}_+}$ be a homogeneous Markov process satisfying conditions ($\mathbf{D}_1$) and ($\mathbf{D}_2$) with the same set $C\in\mathscr{B}(\mathscr{X})$. Then $\{X_n,\}_{n\in\mathbb{Z}_+}$ is a positive uniformly geometrically ergodic process, i.e.,*

$$\sup_{n\geq 0} e^{\kappa^* n}\sup_{x\in\mathscr{X}}\sup_{\theta\in\Theta}\sup_{0\leq g\leq \mathbf{V}}\frac{1}{\mathbf{V}(x)}|\mathsf{E}^\theta_x[g(X_n)] - \lambda(\widetilde{g})| \leq R^* \tag{D.12}$$

for some positive constants κ^* and R^* which are given in [50].

D.4 Markov Random Walks

Let $\{X_n,\}_{n\in\mathbb{Z}_+}$ be a Markov process on a general state space $\mathscr{X}$ with σ-algebra $\mathscr{A}$, which is irreducible with respect to a maximal irreducibility measure on $(\mathscr{X},\mathscr{A})$ and is aperiodic. Assume that an additive component $S_n=\sum_{k=1}^n Y_k$ with $S_0=Y_0=0$, taking values on the real line $\mathbb{R}$, is adjoined to X_n and consider the process $\{(X_n, S_n)\}_{n\in\mathbb{Z}_+}$. In most statistical problems that we deal with in this book, $Y_n = g(X_{n-1}, X_n)$, where g is a measurable function on $\mathscr{X} \times \mathscr{X}$ (see, e.g., Sections 3.11 and 3.12).

We need the following definitions.

Definition D.2. We say that the process $\{(X_n, S_n)\}_{n\in\mathbb{Z}_+}$ is an *additive Markov process* and its additive component $\{S_n\}_{n\in\mathbb{Z}_+}$ is a *Markov random walk* (MRW) if $\{(X_n, S_n)\}_{n\in\mathbb{Z}_+}$ is a Markov process on $\mathscr{X} \times \mathbb{R}$ whose transition probability

$$\mathsf{P}\{(X_{n+1}, S_{n+1}) \in A \times (B+s) | (X_n, S_n) = (x, s)\} = \mathsf{P}(x, A \times B) \tag{D.13}$$

does not depend on $n \in \mathbb{Z}_+$ for all $x \in \mathscr{X}$, $A \in \mathscr{A}$ and $B \in \mathscr{B}(\mathbb{R})$ (Borel σ-algebra on $\mathbb{R}$).

Let P_λ and E_λ denote the probability and expectation when the initial distribution of $\{X_n\}$ (i.e., of X_0) is λ. If λ is degenerate at x, we shall simply write P_x and E_x instead of P_λ and E_λ, i.e., for conditional probability and expectation given $X_0 = x$. In what follows, it is assumed that there exists a stationary probability distribution $\pi(A) = \int \mathsf{P}(x, A)\,\mathrm{d}\pi(x)$ for all $A \in \mathscr{A}$ and that $\mu = \mathsf{E}_\pi[Y_1] > 0$. Hereafter, for the sake of brevity, we write $\mathsf{P}(x, A)$ for the transition probability $\mathsf{P}(x, A \times B)$ of (X_n, S_n).

Definition D.3. The Markov process $\{X_n\}_{n\in\mathbb{Z}_+}$ on a state space $\mathscr{X}$ is called *V-uniformly ergodic* if there exists a measurable function $V : \mathscr{X} \to [1, \infty)$ with $\int V(x)\mathrm{d}\pi(x) < \infty$ such that

$$\lim_{n\to\infty} \sup_{x\in\mathscr{X}} \left\{ \frac{\left|\mathsf{E}_x[h(X_n)] - \int h(x)\mathrm{d}\pi(x)\right|}{V(x)} : |h| \le V \right\} = 0. \tag{D.14}$$

By Meyn and Tweedie [90, pp. 382–383], if the process $\{X_n\}$ is irreducible and aperiodic, condition (D.14) implies that there are constants $C > 0$ and $0 < \rho < 1$ such that for all h and $n \ge 1$

$$\sup_{x\in\mathscr{X}} \left\{ \frac{\left|\mathsf{E}_x[h(X_n)] - \int h(x)\mathrm{d}\pi(x)\right|}{V(x)} \right\} \le C\rho^n \sup_{x\in\mathscr{X}} \frac{|h(x)|}{V(x)}.$$

D.4.1 Wald's Identity for Markov Random Walks

Recall that λ stands for an initial distribution of the Markov process $\{X_n\}_{n\in\mathbb{Z}_+}$ and π for its stationary distribution. Define $g = \mathsf{E}(Y_1|X_0, X_1)$ and suppose $\mathsf{E}_\pi|g| < \infty$. Next, define operators $\mathbf{P}$ and $\mathbf{P}_\pi$ by $(\mathbf{P}g)(x) = \mathsf{E}_x[g(x, X_1, Y_1)]$ and $\mathbf{P}_\pi g = \mathsf{E}_\pi[g(X_0, X_1, Y_1)]$ respectively, and set $\overline{g} = \mathbf{P}g$. We shall consider solutions $\Delta(x) = \Delta(x; g)$ of the Poisson equation

$$(\mathbb{I} - \mathbf{P})\Delta = (\mathbb{I} - \mathbf{P}_\pi)\overline{g} \quad \mu^*\text{-a.s.}, \quad \text{with } \mathbf{P}_\pi\Delta = 0, \tag{D.15}$$

where $\mathbb{I}$ is the identity operator.

Assume that

$$\sup_{x\in\mathscr{X}} \{\mathsf{E}_x[V(X_1)]/V(x)\} < \infty. \tag{D.16}$$

By Theorem 17.4.2 of Meyn and Tweedie [91] conditions (D.14) and (D.16) guarantee existence and boundedness of the solution Δ of Poisson equation (D.15) as well as finiteness of $\sup_x[|\Delta(x)|/V(x)]$.

The following theorem, which can be deduced from Theorem 4 of Fuh and Zhang [48], generalizes Wald's identity (C.9) for random walks to Markov random walks.

Theorem D.2 (Wald's identity for MRW). *Let T be a stopping time with finite expectation,* $\mathsf{E}_x[T] < \infty$, $\mathsf{E}_\lambda[T] < \infty$. *If conditions* (D.14) *and* (D.16) *are satisfied and if* $\sup_x \mathsf{E}_x|Y_1| < \infty$, *then*

$$\mathsf{E}_\lambda[S_T] = \mu\,\mathsf{E}_\lambda[T] + \mathsf{E}_\lambda[\Delta(X_T) - \Delta(X_0)] \tag{D.17}$$

and

$$\mathsf{E}_x[S_T] = \mu\,\mathsf{E}_x[T] + \mathsf{E}_x[\Delta(X_T)] - \Delta(x), \tag{D.18}$$

where $\Delta : \mathscr{X} \to \mathbb{R}$ *solves the Poisson equation* (D.15) *for every* $x \in \mathscr{X}$.

Note that the constant $\mathsf{E}_\lambda[\Delta(X_T) - \Delta(X_0)] = 0$ if $\lambda = \pi$, i.e., when the initial distribution λ of the Markov process $\{X_n\}$ is the stationary distribution π of $\{X_n\}$.

D.4.2 Rates of Convergence in the SLLN for Markov Random Walks

We now present the results related to rates of convergence in the SLLN for Markov random walks S_n that generalize the Baum–Katz rates of convergence for random walks given in (B.10). This allows us to deduce sufficient conditions for r-quick and r-complete convergences of $n^{-1}S_n$, which are used in Chapter 3 for proving asymptotic optimality of change detection procedures for hidden Markov models.

Consider a special case of a more general Markov random walk $\{(X_n, S_n)\}_{n\in\mathbb{Z}_+}$ introduced above that has the form

$$S_n = \sum_{t=1}^{n} g(X_{t-1}, X_t), \quad S_0 = 0,$$

where g is a measurable function on $\mathscr{X} \times \mathscr{X}$ and $\{X_n\}_{n\in\mathbb{Z}_+}$ is a Markov process on a general state space $\mathscr{X}$ with σ-algebra $\mathscr{A}$, which is irreducible with respect to a maximal irreducibility measure on $(\mathscr{X}, \mathscr{A})$ and is aperiodic, having a stationary distribution $\pi(A)$, $A \in \mathscr{A}$.

By Lemma B.3,

$$\frac{1}{n} S_n \xrightarrow[n\to\infty]{r\text{-quickly}} I$$

whenever

$$\sum_{n=1}^{\infty} n^{r-1} \mathsf{P}\left\{\sup_{t\ge n} \frac{|S_t - It|}{t} \ge \varepsilon\right\} < \infty \quad \text{for all } \varepsilon > 0, \tag{D.19}$$

which, obviously, implies

$$\sum_{n=1}^{\infty} n^{r-1} \mathsf{P}\{|S_n - In| \ge \varepsilon n\} < \infty \quad \text{for all } \varepsilon > 0,$$

i.e., that

$$\frac{1}{n} S_n \xrightarrow[n\to\infty]{r\text{-completely}} I.$$

Thus, to establish r-complete convergence of $n^{-1}S_n$ to I as $n \to \infty$ it suffices to find sufficient conditions for (D.19).

The following lemma, which can be deduced from Theorems 2 and 6 in [48], shows that certain $(r+1)$th moment conditions are sufficient.

Lemma D.1. *If* $\{X_n\}$ *is initialized from the stationary distribution* π *and*

$$\mathsf{E}_\pi[|g(X_0, X_1)|^{r+1}] < \infty, \tag{D.20}$$

then

$$\sum_{n=1}^{\infty} n^{r-1} \mathsf{P}_\pi\left\{\sup_{t\ge n} \frac{|S_t - \mathscr{K} t|}{t} \ge \varepsilon\right\} < \infty \quad \textit{for all } \varepsilon > 0, \tag{D.21}$$

and therefore,

$$\frac{1}{n} S_n \xrightarrow[n\to\infty]{\mathsf{P}_\pi - r - quickly} \mathscr{K},$$

where $\mathscr{K} = \mathsf{E}_\pi[g(X_0, X_1)]$.

If $\{X_n\}$ *is initialized from a distribution* λ *and, in addition to condition* (D.20), $\sup_{n\ge 0} |\Delta(X_n; g)|^{r+1} < \infty$, *where* $\Delta(X_n; g)$ *is a solution of the Poisson equation* (D.15), *then*

$$\sum_{n=1}^{\infty} n^{r-1} \mathsf{P}_\lambda \left\{ \sup_{t\ge n} \frac{|S_t - \mathscr{K} t|}{t} \ge \varepsilon \right\} < \infty \quad \textit{for all } \varepsilon > 0, \tag{D.22}$$

and therefore,

$$\frac{1}{n} S_n \xrightarrow[n\to\infty]{\mathsf{P}_\lambda - r - quickly} \mathscr{K}.$$

D.5 Hidden Markov Models

We begin with defining a general hidden Markov model (HMM) as a Markov process in a Markovian random environment, in which the underlying environmental Markov process is unobservable, i.e., hidden in observations. Often HMMs are also called partially observable Markov processes. To be more precise, let $\mathbf{Z} = \{Z_n\}_{n\in\mathbb{Z}_+}$ be a homogeneous Markov process on a general state space $\mathscr{Z}$ with transition probability kernel $\mathsf{P}(z, A) = \mathsf{P}(Z_1 \in A | Z_0 = z)$ and stationary probability π. Suppose that a random sequence $\{X_n\}_{n\in\mathbb{Z}_+}$, taking values in $\mathbb{R}^m$, is adjoined to $\mathbf{Z}$ such that $\{(Z_n, X_n)\}_{n\in\mathbb{Z}_+}$ is a Markov process on $\mathscr{Z} \times \mathbb{R}^m$ satisfying $\mathsf{P}\{Z_1 \in A | Z_0 = z, X_0 = x_0\} = \mathsf{P}\{Z_1 \in A | Z_0 = z\}$ for $A \in \mathscr{B}(\mathscr{Z})$, the σ-algebra on $\mathscr{Z}$. Moreover, conditioned on the full $\mathbf{Z}$ sequence, $\{X_n\}$ is a Markov process with probability

$$\mathsf{P}\{X_{n+1} \in B | Z_0, Z_1, \dots; X_0, X_1, \dots, X_n\} = \mathsf{P}\{X_{n+1} \in B | Z_{n+1}, X_n\} \tag{D.23}$$

for each $n \in \mathbb{Z}_+$ and $B \in \mathscr{B}(\mathbb{R}^m)$, the Borel σ-algebra on $\mathbb{R}^m$. Let $f(X_k | Z_k, X_{k-1})$ be the transition probability density of X_k given Z_k and X_{k-1} with respect to a σ-finite measure Q on $\mathbb{R}^m$ and let $p(z, x)$ be a transition probability density for the Markov process $\{X_n\}$ with respect to a sigma-finite measure m on $\mathscr{Z}$ such that

$$\mathsf{P}\{Z_1 \in A, X_1 \in B | Z_0 = z, X_0 = x\} = \int_{z'\in A} \int_{x'\in B} p(z, z') f(x' | z', x) Q(\mathrm{d}x') m(\mathrm{d}z') \tag{D.24}$$

for $B \in \mathscr{B}(\mathbb{R}^m)$. We also assume that the Markov process $\{(Z_n, X_n)\}_{n\in\mathbb{Z}_+}$ has a stationary probability Γ_{st} with probability density function $\pi(z) f(\cdot | z)$ with respect to $Q \times m$.

Definition D.4. The sequence $\{X_n\}_{n\in\mathbb{Z}_+}$ is called a hidden Markov model (or a state-space model) if there is an unobserved, hidden Markov chain $\{Z_n\}_{n\in\mathbb{Z}_+}$ such that the process $\{(Z_n, X_n)\}_{n\in\mathbb{Z}_+}$ satisfies (D.23) and (D.24).

Appendix E: First Exit Times of Random Walks and Markov Random Walks

In this appendix, we present elements of *renewal theory*. Renewal theory is particularly useful for the problem of the excess over a boundary or the "overshoot" problem. Specifically, renewal theory allows us to answer several challenging questions, in particular, to obtain corrections for approximations for the expectation of the first exit times over the boundaries and associated probabilities. These corrections are necessary for higher order approximations for the average sample sizes and probabilities of errors of sequential tests and average detection delays of change detection rules. The proofs are omitted. We give references to the sources where the proofs and further details can be found. An exception is the *General Multidimensional Renewal Theorem* (Theorem E.7) in the non-i.i.d. case in Section E.2 when the statistics which cross the boundaries are not random walks, but rather the sums of dependent and non-identically distributed random variables. This result is novel and we provide detailed proof.

E.1 First Exit Times of Random Walks Over The Boundary

Throughout this section $\Delta S_1, \Delta S_2, \dots$ are i.i.d. random variables with the common distribution function $F(x) = P(\Delta S_i \le x)$. Let $S_n = \Delta S_1 + \dots + \Delta S_n$, $n = 0, 1, 2, \dots$ ($S_0 = 0$) denote partial sums. The discrete-time process $\{S_n\}_{n \in \mathbb{Z}_+}$ is a random walk. However, if the random variables ΔS_i are nonnegative and interpreted as random times between certain events then S_n is the time when the nth renewal occurs, and the process $\{S_n\}_{n \in \mathbb{Z}_+}$ is called a *renewal process*.

E.1.1 The Distribution of the Overshoot

Let $\{S_n\}_{n \in \mathbb{Z}_+}$ be a random walk and, for $a \ge 0$, let

$$T_a = \inf\{n \ge 1 : S_n \ge a\}, \quad \inf\{\varnothing\} = \infty \tag{E.1}$$

be the first time when the random walk exceeds the level (threshold) a and let

$$\kappa_a = S_{T_a} - a \quad \text{on } \{T_a < \infty\} \tag{E.2}$$

denote the excess of the random walk over the level a at the time which it crosses this level. We will refer to κ_a as an *overshoot* (at stopping). When $\{S_n\}$ is a renewal process, κ_n is nothing but the residual waiting time until the next renewal after time n. The evaluation of the distribution of the overshoot $\mathsf{P}(\kappa_a \le y)$, the average overshoot $\varkappa_a = \mathsf{E}[\kappa_a]$ as well as the expectation of certain functions of the overshoot, for instance, $\mathsf{E}[e^{-\lambda \kappa_a}]$, $\lambda > 0$, is of great interest for many statistical applications, including hypothesis testing and changepoint problems. In fact, the key for obtaining good approximations for the probabilities of errors and expected sample sizes of sequential tests is a reasonable approximation for the distribution of the overshoot at least for large values of threshold a.

For example, let P_1 and P_0 be two probability measures and p_1 and p_0 the corresponding densities of the observations X_n, $n \ge 1$. Let $\Delta S_n = \log[p_1(X_n)/p_0(X_n)]$. Applying Wald's identities (C.9) and (C.11), it is easy to show that

$$\mathsf{E}_1[T_a] = (a + \varkappa_a)/\mathsf{E}_1[\Delta S_1], \quad \mathsf{P}_0(T_a < \infty) = e^{-a}\mathsf{E}_1[e^{-\kappa_a}],$$

where E_j stands for the expectation under P_j. Therefore, the primary interest is finding a distribution of the overshoot κ_a or at least an approximation for this distribution for sufficiently large values of a.

The results are somewhat different in the arithmetic and nonarithmetic cases.

Definition E.1. A random variable X is called *arithmetic* if $\mathsf{P}(X \in \{\cdots - 2d, -d, 0, d, 2d, \dots\}) = 1$ for some $d > 0$, i.e., the distribution function $F(x) = \mathsf{P}(X \le x)$ is concentrated on $\{0, \pm d, \pm 2d, \dots\}$. The largest such d is called the *span* of X. In this case, we say that the corresponding random variable is *d-arithmetic*. If there is no such d, then the random variable is called *nonarithmetic* . We shall say that the random walk $\{S_n\}_{n \in \mathbb{Z}_+}$ is nonarithmetic if ΔS_1 is nonarithmetic, and that it is d-arithmetic if ΔS_1 is arithmetic with span d.

It turns out that the overshoot problem can be solved using renewal theory.

Assume that $\mu = \mathsf{E}[\Delta S_1] > 0$, so that by the SLLN $\mathsf{P}(T_a < \infty) = 1$. Define

$$T_+^{(k)} = \inf\left\{n > T_+^{(k-1)} : S_n > S_{T_+^{(k-1)}}\right\}, \quad k = 1, 2 \dots, \tag{E.3}$$

where $T_+^{(0)} = 0$ and

$$T_+^{(1)} = T_+ = \inf\{n \ge 1 : S_n > 0\}. \tag{E.4}$$

The random variable $S_{T_+^{(k)}}$ is called the kth ladder height, and $T_+^{(k)}$ – the kth ladder epoch. Clearly, $(T_+^{(k)} - T_+^{(k-1)}, S_{T_+^{(k)}} - S_{T_+^{(k-1)}})$, $k = 1, 2 \dots$ are i.i.d. Furthermore, $T_a = T_+^{(k)}$ for some k, and hence,

$$T_+^{(\tau)} = T_a \quad \text{and} \quad \kappa_a = S_{T_+^{(\tau)}} - a,$$

where $\tau = \inf\{k \ge 1 : S_{T_+^{(k)}} > a\}$.

Let $F_+(s) = \mathsf{P}(S_{T_+} \le s)$ be the distribution of the positive random variable S_{T_+}. Computing the distribution $F_+(s)$ of S_{T_+} is the subject of renewal theory.

We now present the main results regarding all the necessary distributions and expectations. The proofs can be found in Gut [55], Siegmund [140], and Woodroofe [189].

Let $\mathsf{G}(y) = \lim_{a \to \infty} \mathsf{P}(\kappa_a \le y)$ denote the limiting distribution of the overshoot κ_a and let $\varkappa = \lim_{a \to \infty} \varkappa_a$ denote the limiting average overshoot. In the d-arithmetic case we always assume that $a \to \infty$ through multiples of span d, i.e., $a = dj$, $j \to \infty$. Recall that $\mu = \mathsf{E}[\Delta S_1]$.

The following theorem establishes formulas for the limiting distribution of the overshoot and the limiting average overshoot.

Theorem E.1. *Assume that* $0 < \mu < \infty$.

(i) *If the random walk* $\{S_n\}$ *is nonarithmetic, then*

$$\mathsf{G}(y) = \frac{1}{\mathsf{E}[S_{T_+}]} \int_0^y [1 - F_+(s)]\, ds, \tag{E.5}$$

and if in addition $\mathsf{E}[(\Delta S_1^+)^2] < \infty$, *then*

$$\varkappa = \frac{1}{\mathsf{E}[S_{T_+}]} \int_0^\infty \left\{\int_y^\infty [1 - F_+(s)]\, ds\right\} dy = \frac{\mathsf{E}[S_{T_+}^2]}{2\mathsf{E}[S_{T_+}]}. \tag{E.6}$$

(ii) *If the random walk $\{S_n\}$ is d-arithmetic, then*

$$\lim_{j\to\infty} \mathsf{P}(\varkappa_{a=jd} = id) = \frac{d}{\mathsf{E}[S_{T_+}]}\mathsf{P}\left(S_{T_+} \geq id\right), \quad i \geq 1 \tag{E.7}$$

and if in addition $\mathsf{E}[(\Delta S_1^+)^2] < \infty$, then

$$\varkappa = \frac{\mathsf{E}[S_{T_+}^2]}{2\mathsf{E}[S_{T_+}]} + \frac{d}{2}. \tag{E.8}$$

Note that since

$$\int_0^\infty [1 - F_+(s)]\,\mathrm{d}s = \mathsf{E}[S_{T_+}],$$

(E.5) implies

$$1 - \mathsf{G}(y) = \frac{1}{\mathsf{E}[S_{T_+}]}\int_y^\infty [1 - F_+(s)]\,\mathrm{d}s. \tag{E.9}$$

It is often important to evaluate the value of $\mathsf{E}[e^{-\lambda \varkappa_a}]$ for some $\lambda > 0$, in particular, for $\lambda = 1$. It follows from Theorem E.1(i) that in the nonarithmetic case the limiting value is

$$\lim_{a\to\infty} \mathsf{E}[e^{-\lambda \varkappa_a}] = \int_0^\infty e^{-\lambda y}\,\mathrm{d}\mathsf{G}(y),$$

i.e., it is equal to the Laplace transform.

More generally, if we are interested in higher moments of the overshoot, then the following result holds. See, e.g., Gut [55, Theorem III.10.9].

Theorem E.2. *Let the random walk $\{S_n\}_{n\in\mathbb{Z}_+}$ be nonarithmetic. If $\mathsf{E}[(\Delta S_1^+)^{m+1}] < \infty$ for some $m > 0$, then*

$$\lim_{a\to\infty} \mathsf{E}[\varkappa_a^m] = \frac{\mathsf{E}[S_{T_+}^{m+1}]}{(m+1)\mathsf{E}[S_{T_+}]}.$$

While Theorems E.1 and E.2 are important, to make them indeed useful we need to find a way of computing the limiting distribution and such quantities as moments of ladder variables. We now consider computational aspects related to the overshoot problem. Along with the first ascending ladder variable T_+ in (E.4) defined as the first time the random walk S_n upper-crosses the zero level, we now define the descending ladder variable

$$T_- = \inf\{n \geq 1 : S_n \leq 0\}. \tag{E.10}$$

The following two lemmas allow one to perform computations when explicit forms of the distributions $F_n(s) = \mathsf{P}(S_n \leq s)$, $n = 1, 2, \dots$ can be obtained.

Lemma E.1. *Let $\{S_n\}_{n\in\mathbb{Z}_+}$ be an arbitrarily random walk. If $0 < \mu \leq \infty$, then*

$$\mathsf{E}[T_+] = \frac{1}{\mathsf{P}(T_- = \infty)} = \exp\left\{\sum_{n=1}^\infty \frac{1}{n}\mathsf{P}(S_n \leq 0)\right\},$$

$$\mathsf{E}[T_-] = \frac{1}{\mathsf{P}(T_+ = \infty)} = \exp\left\{\sum_{n=1}^\infty \frac{1}{n}\mathsf{P}(S_n > 0)\right\}.$$

Note that if $\mu = 0$, then both T_+ and T_- are a.s. finite, but $\mathsf{E}[T_+] = \mathsf{E}[T_-] = \infty$.

By Theorem E.1(i), in the nonarithmetic case the asymptotic distribution of the overshoot has density

$$h(y) = \frac{1}{\mathsf{E}[S_{T_+}]}[1 - F_+(y)] = \frac{1}{\mathsf{E}[S_{T_+}]}\mathsf{P}\left(S_{T_+} > y\right), \quad y \geq 0.$$

Introducing the Laplace transform of G,

$$\mathscr{H}(\lambda) = \int_0^\infty e^{-\lambda y}\, \mathrm{dG}(y), \quad \lambda \geq 0,$$

and integrating by parts, we obtain

$$\mathscr{H}(\lambda) = \frac{1 - \mathsf{E}[e^{-\lambda S_{T_+}}]}{\lambda \mathsf{E}[S_{T_+}]}, \quad \lambda > 0.$$

Lemma E.1 allows us to obtain the following useful result. Write $Y^- = -\min(0, Y)$.

Theorem E.3. *Let $\{S_n\}_{n\in\mathbb{Z}_+}$ be a nonarithmetic random walk with drift $\mu > 0$.*

(i) *For every $\lambda > 0$*

$$\lim_{a\to\infty} \mathsf{E}\left[e^{-\lambda \kappa_a}\right] \equiv \mathscr{H}(\lambda) = \frac{1}{\lambda\mu} \exp\left\{-\sum_{n=1}^{\infty} \frac{1}{n} \mathsf{E}\left[e^{-\lambda S_n^+}\right]\right\}.$$

(ii) *Assume in addition that $\mathsf{E}|\Delta S_1|^2 < \infty$. Then*

$$\varkappa = \frac{\mathsf{E}[S_{T_+}^2]}{2\mathsf{E}[S_{T_+}]} = \frac{\mathsf{E}[\Delta S_1^2]}{2\mathsf{E}[\Delta S_1]} - \sum_{n=1}^{\infty} \frac{1}{n} \mathsf{E}[S_n^-].$$

An alternative useful expression for the limiting density $h(y)$ of the overshoot is

$$h(y) = \frac{1}{\mu} \mathsf{P}\left(\min_{n\geq 1} S_n > y\right), \quad y \geq 0, \tag{E.11}$$

which holds for any random walk with positive drift μ. See, e.g., Woodroofe [189, Theorem 2.7] and Gut [55, Theorem III.10.4].

Also, a direct argument shows that for any random walk with positive mean and finite variance

$$\mathsf{E}\left[\min_{0\leq n\leq N} S_n\right] = -\sum_{n=1}^{N} \frac{1}{n} \mathsf{E}[S_n^-],$$

which along with Theorem E.3(ii) yields

$$\varkappa = \frac{\mathsf{E}[S_{T_+}^2]}{2\mathsf{E}[S_{T_+}]} = \frac{\mathsf{E}[\Delta S_1^2]}{2\mathsf{E}[\Delta S_1]} + \mathsf{E}\left[\min_{n\geq 0} S_n\right] \tag{E.12}$$

if $\mu > 0$ and $\mathsf{E}|\Delta S_1|^2 < \infty$.

In the problems of hypothesis testing, the quantity $\mathscr{H}(\lambda = 1)$ is of special interest since it allows one to correct the expressions for error probabilities taking into account an overshoot as well as to optimize sequential tests. More specifically, let P_1 and P_0 be two probability measures with densities p_1 and p_0. Assume that $\{X_n\}_{n\in\mathbb{Z}_+}$ is a sequence of i.i.d. random variables (observations) that come either from P_1 or from P_0. Let $\Delta S_n = \log[p_1(X_n)/p_0(X_n)]$ denote the log-likelihood ratio for the nth observation and $S_n = \Delta S_1 + \cdots + \Delta S_n$ the cumulative log-likelihood ratio. Under P_i ($i = 0, 1$) the log-likelihood ratio process $\{S_n\}$ is a random walk with drifts $\mathscr{K}_1 = \mathsf{E}_1[\Delta S_1]$ and $-\mathscr{K}_0 = \mathsf{E}_0[\Delta S_1]$, where $\mathscr{K}_1$ and $\mathscr{K}_0$ are positive whenever densities $p_1(y)$ and $p_0(y)$ are not the same for almost all y. Introduce the stopping times

$$T_0(a) = \inf\{n \geq 1 : -S_n \geq a\}, \quad T_1(a) = \inf\{n \geq 1 : S_n \geq a\} \tag{E.13}$$

and the associated overshoots $\tilde{\kappa}_0(a) = -S_{T_0} - a$ on $\{T_0 < \infty\}$ and $\tilde{\kappa}_1(a) = S_{T_1} - a$ on $\{T_1 < \infty\}$. If the mean values $\mathscr{K}_0$ and $\mathscr{K}_1$ are positive and finite, then by Theorem E.1 the limiting distributions $\mathsf{G}_0(y) = \lim_{a\to\infty} \mathsf{P}_0(\tilde{\kappa}_0 \le y)$ and $\mathsf{G}_1(y) = \lim_{a\to\infty} \mathsf{P}_1(\tilde{\kappa}_1 \le y)$ exist and are given by (E.5) in the nonarithmetic case and by (E.7) in the d-arithmetic case. Thus, the limiting quantities

$$\gamma_i := \lim_{a\to\infty} \mathsf{E}_i\left[e^{-\tilde{\kappa}_i(a)}\right] = \int_0^\infty e^{-y}\, \mathrm{d}\mathsf{G}_i(y), \quad i = 0, 1 \tag{E.14}$$

are well-defined. Next, define the number

$$\mathscr{L} = \exp\left\{-\sum_{n=1}^\infty \frac{1}{n}[\mathsf{P}_0(S_n > 0) + \mathsf{P}_1(S_n \le 0)]\right\}. \tag{E.15}$$

This $\mathscr{L}$-number plays an important role in optimization of hypothesis tests and changepoint problems.

Note that for any $n \ge 1$

$$\mathsf{P}_1(S_n = 0) = \mathsf{E}_0\left[e^{S_n} \mathbb{1}_{\{S_n=0\}}\right] = \mathsf{P}_0(S_n = 0),$$

so that

$$\sum_{n=1}^\infty \frac{1}{n}[\mathsf{P}_0(S_n > 0) + \mathsf{P}_1(S_n \le 0)] \equiv \sum_{n=1}^\infty \frac{1}{n}[\mathsf{P}_0(S_n \ge 0) + \mathsf{P}_1(S_n < 0)]. \tag{E.16}$$

Therefore, the $\mathscr{L}$-number is symmetric in the sense it can be also defined as

$$\mathscr{L} = \exp\left\{-\sum_{n=1}^\infty \frac{1}{n}[\mathsf{P}_0(S_n \ge 0) + \mathsf{P}_1(S_n < 0)]\right\}. \tag{E.17}$$

Using this property, in the following theorem, we establish a useful relationship between the constants γ_0 and γ_1 via the $\mathscr{L}$-number.

Theorem E.4. *If $0 < \mathscr{K}_0 < \infty$ and $0 < \mathscr{K}_1 < \infty$, then*

$$\gamma_0 = \mathscr{L}/\mathscr{K}_0 \quad \text{and} \quad \gamma_1 = \mathscr{L}/\mathscr{K}_1, \tag{E.18}$$

and hence, the following identity holds:

$$\mathscr{L} = \gamma_0 \mathscr{K}_0 = \gamma_1 \mathscr{K}_1. \tag{E.19}$$

Equality (E.19) is indeed useful since if one of the constants γ_0 or γ_1 is computed, another one can be computed immediately using this identity. However, we stress that this result holds only for LLR-based random walks but not for arbitrarily random walks.

Example E.1. Let $\{S_n\}$ be the Gaussian random walk with mean $\mathsf{E}[S_n] = \mu n$ and variance $\mathsf{Var}[S_n] = \sigma^2 n$, where $\mu > 0, \sigma^2 > 0$. Write $\varphi(x) = (2\pi)^{-1/2} e^{-x^2/2}$ and $\Phi(x)$ for the standard normal density and distribution function, respectively. Let $q = \mu^2/\sigma^2$ and denote $\beta_n = \mathsf{E}[e^{-\lambda S_n^+}]$. Direct computations show that for $\lambda \ge 0$ and $n \ge 1$

$$\beta_n = \Phi(-\sqrt{qn}) + \Phi\left[-(\lambda\sigma - \sqrt{q})\sqrt{n}\right] \exp\left\{\left[\lambda\sigma\left(\frac{\lambda\sigma}{2} - \sqrt{q}\right)\right] n\right\}, \tag{E.20}$$

so

$$\mathscr{H}(\lambda) = \frac{1}{\lambda\mu} \exp\left\{-\sum_{n=1}^\infty \frac{1}{n}\beta_n\right\}$$

is easily computed numerically. Computations become especially simple when $\mu/\sigma^2 = 1/2$, which is the case when ΔS_1 is the log-likelihood ratio in the problem of testing two hypotheses related to the mean of the Gaussian i.i.d. sequence.

The limiting average overshoot is also easily computable

$$\varkappa = \mu \frac{1+q}{2q} - \frac{\mu}{\sqrt{q}} \sum_{n=1}^{\infty} \frac{1}{\sqrt{n}} \left[\varphi(\sqrt{qn}) - \Phi(-\sqrt{qn})\sqrt{qn}\right]. \tag{E.21}$$

Example E.2. We now consider another example where all computations can be performed precisely. Assume that the distribution function $F(x)$ of ΔS_1 has an exponential right tail, i.e., $F(x) = 1 - C_0\, e^{-C_1 x}$, $x \ge 0$ for some positive constants C_0 and C_1. In this case, the distribution of the overshoot κ_a is exactly exponential with the parameter C_1 for all $a \ge 0$ assuming that $\mu > 0$. Indeed, for $y \ge 0$ and $a \ge 0$,

$$\begin{aligned} \mathsf{P}(\kappa_a > y, T_a < \infty) &= \sum_{n=1}^{\infty} \mathsf{P}(S_n > a+y, T_a = n) = \sum_{n=1}^{\infty} \mathsf{P}(\Delta S_n > a+y-S_{n-1}, T_a = n) \\ &= \sum_{n=1}^{\infty} \mathsf{E}\left[C_0 e^{-C_1(a+y-S_{n-1})} \mathbb{1}_{\{T_a=n\}}\right] = e^{-C_1 y} \sum_{n=1}^{\infty} \mathsf{E}\left[C_0 e^{-C_1(a-S_{n-1})} \mathbb{1}_{\{T_a=n\}}\right]. \end{aligned}$$

Setting $y = 0$, yields

$$\sum_{n=1}^{\infty} \mathsf{E}\left[C_0 e^{-C_1(a-S_{n-1})} \mathbb{1}_{\{T_a=n\}}\right] = \mathsf{P}(T_a < \infty).$$

Hence

$$\mathsf{P}(\kappa_a > y, T_a < \infty) = \mathsf{P}(T_a < \infty) e^{-C_1 y}, \quad y, a \ge 0.$$

If $\mu > 0$, then $\mathsf{P}(T_a < \infty) = 1$, and therefore,

$$\mathsf{P}(\kappa_a > y) = e^{-C_1 y} \quad \text{for all } y \ge 0,\ a \ge 0. \tag{E.22}$$

In particular, for all $a \ge 0$

$$\mathsf{E}[\kappa_a] = 1/C_1, \quad \mathsf{E}\left[e^{-\lambda \kappa_a}\right] = C_1/(\lambda + C_1). \tag{E.23}$$

E.1.2 Approximations for the Expectation of the Stopping Time

The overshoot problem is directly related to the evaluation of moments of the stopping time T_a defined in (E.1). Indeed, since

$$S_{T_a} = a + \kappa_a \quad \text{on } \{T_a < \infty\},$$

assuming that $0 < \mu < \infty$ and using Wald's identity $\mathsf{E}[S_{T_a}] = \mu \mathsf{E}[T_a]$, we obtain

$$\mathsf{E}[T_a] = \frac{1}{\mu}(a + \varkappa_a). \tag{E.24}$$

This equality is true if $\mathsf{E}[T_a] < \infty$, so we need to assume that $\mu > 0$.

Since in general we cannot compute the average overshoot $\varkappa_a = \mathsf{E}[\kappa_a]$ for every $a \ge 0$, the natural question is "how accurate an approximation $\mathsf{E}[T_a] \approx (a+\varkappa)/\mu$ with $\varkappa_a$ replaced by its limiting value $\varkappa = \lim_{a\to\infty} \varkappa_a$ is?" From Theorem E.3 we may expect that the second moment condition (or at least finiteness of the second moment of the positive part) is required.

Since the one-sided stopping time T_a is extremely important in hypothesis testing and change-point detection problems, we now present some precise answers to the above questions.

Lemma E.2. *Let $\mu > 0$ and $\mathsf{E}[(\Delta S_1^-)^m] < \infty$, $m \geq 1$. Then $\mathsf{E}[T_a^m] < \infty$ for all $a \geq 0$. Moreover, the family $\{(T_a/a)^m, a \geq 1\}$ is uniformly integrable.*

The following theorem establishes the SLLN for $\{T_a, a > 0\}$ and the first order expansion for the moments of the stopping time T_a.

Theorem E.5. **(i)** *Let $0 < \mu < \infty$. Then*

$$\frac{T_a}{a} \xrightarrow[a\to\infty]{\mathsf{P}-a.s.} \frac{1}{\mu} \tag{E.25}$$

and

$$\frac{\mathsf{E}[T_a]}{a} \xrightarrow[a\to\infty]{} \frac{1}{\mu}. \tag{E.26}$$

(ii) *Let $r > 1$. If in addition $\mathsf{E}[(\Delta S_1^-)^r] < \infty$, then*

$$\left(\frac{\mathsf{E}[T_a]}{a}\right)^m \xrightarrow[a\to\infty]{} \frac{1}{\mu^m} \quad \text{for all } 1 < m \leq r. \tag{E.27}$$

Theorem E.5 provides a first-order expansion for the expected value of the stopping time T_a:

$$\mathsf{E}[T_a] = \frac{a}{\mu}(1 + o(1)) \quad \text{as } a \to \infty$$

(see (E.26)), which cannot be improved as long as only the first moment condition is assumed. However, such an improvement is possible under the second moment condition. The following theorem makes the approximation (E.24) precise up to the vanishing term $o(1)$.

Theorem E.6. *Assume that $\mu > 0$ and $\mathsf{E}[(\Delta S_1^+)^2] < \infty$.*

(i) *If the random walk $\{S_n\}$ is nonarithmetic, then*

$$\mathsf{E}[T_a] = \frac{1}{\mu}(a + \varkappa) + o(1) \quad \text{as } a \to \infty, \tag{E.28}$$

where $\varkappa = \mathsf{E}[S_{T_+}^2]/2\mathsf{E}[S_{T_+}]$.

(ii) *If the random walk $\{S_n\}$ is d-arithmetic, then*

$$\mathsf{E}[T_{a=jd}] = \frac{1}{\mu}(jd + \varkappa) + o(1) \quad \text{as } j \to \infty, \tag{E.29}$$

where $\varkappa = \mathsf{E}[S_{T_+}^2]/2\mathsf{E}[S_{T_+}] + d/2$.

The condition of finiteness of $\mathsf{E}[(\Delta S_1^+)^2]$ cannot be relaxed in the asymptotic approximations (E.28) and (E.29), since, otherwise, the ladder heights will have an infinite second moment. However, finiteness of the variance $\mathsf{Var}[\Delta S_1]$ is not required.

The rate of the small term $o(1)$ seems to be difficult to find in general. In the exponential case considered in Example E.2 this formula is exact with $\varkappa = 1/C_1$, i.e.,

$$\mathsf{E}[T_a] = (a + 1/C_1)/\mu \quad \text{for every } a \geq 0,$$

assuming that $\mu > 0$.

Finally, we note that if $\mu > 0$ and $\mathsf{Var}[\Delta S_1] = \sigma^2 < \infty$, then

$$\mathsf{E}[T_a] \sim a/\mu, \quad \mathsf{Var}[T_a] \sim a\sigma^2/\mu^3 \quad \text{as } a \to \infty$$

and the asymptotic distribution (as $a \to \infty$) of the properly normalized stopping time T_a is standard normal:

$$\lim_{a\to\infty} \mathsf{P}\left(\frac{T_a - a/\mu}{\sqrt{a\sigma^2/\mu^3}} \leq x\right) = \Phi(x) \quad \text{for all } -\infty < x < \infty \tag{E.30}$$

See, e.g., Gut [55, Theorem III.5.1].

E.2 A General Multidimensional Renewal Theorem

In the multidimensional case, classical renewal theory provides first-order asymptotic approximations for the moments of the first time such that multiple random walks are simultaneously above a large threshold [36, 37, 54, 64]. In this section, we present a generalization of a multidimensional renewal theorem extending classical results to more general processes that are not necessarily random walks. This generalization is used in Chapter 1 for developing an asymptotic sequential hypothesis testing theory (Theorem 1.1) for general stochastic processes, but is also of independent interest.

All random variables are defined on a probability space $(\Omega, \mathscr{F}, \mathsf{P})$, and we denote by E the expectation that corresponds to P.

Let $\{S_i(n)\}_{n \in \mathbb{N}}$, $i \in \mathscr{N} = \{1, \dots, N\}$, be (possibly dependent) sequences of random variables. Write $\Delta S_i(n) = S_i(n) - S_i(n-1)$ and $M_i(n) = \max_{1 \le s \le n} M_i(s)$. For $i \in \mathscr{N}$, consider the first-hitting times

$$T_i(b) = \inf\{n \ge 1 : S_i(n) \ge b\}, \quad b > 0.$$

Our goal is to obtain first-order asymptotic approximations as $b \to \infty$ for the moments of the following stopping times

$$T_{\min}(b) = \min_{1 \le i \le N} T_i(b), \quad T_{\max}(b) = \max_{1 \le i \le N} T_i(b), \quad T(b) = \inf\left\{n \ge 1 : \min_{1 \le i \le N} S_i(n) \ge b\right\}.$$

To this end, we assume that the SLLN holds for each of these processes: there are positive numbers $\mu_1, \dots, \mu_N$ such that

$$\frac{S_i(n)}{n} \xrightarrow{\text{a.s.}} \mu_i \quad \text{as} \quad n \to \infty \quad \text{for all } i \in \mathscr{N}. \tag{E.31}$$

Given this assumption, our goal is to provide sufficient conditions for the following asymptotic approximations for hold as $b \to \infty$:

$$\mathsf{E}[T_{\min}^m(b)] \sim \left(\frac{b}{\mu_{\max}}\right)^m, \tag{E.32}$$

$$\mathsf{E}[T_{\max}^m(b)] \sim \left(\frac{b}{\mu_{\min}}\right)^m \sim \mathsf{E}[T^m(b)], \tag{E.33}$$

where $\mu_{\max} = \max_{1 \le i \le N} \mu_i$ and $\mu_{\min} = \min_{1 \le i \le N} \mu_i$.

The following lemma establishes asymptotic lower bounds on these moments under SLLN (E.31).

Lemma E.3. *Suppose* (E.31) *holds. Then, for every $i \in \mathscr{N}$ as $b \to \infty$*

$$\frac{T_i(b)}{b} \xrightarrow{a.s.} \frac{1}{\mu_i}, \quad \frac{T(b)}{b} \xrightarrow{a.s.} \frac{1}{\mu_{\min}}, \tag{E.34}$$

and consequently, for every $m > 0$

$$\begin{aligned} \mathsf{E}[T_{\min}^m(b)] &\ge \left(\frac{b}{\mu_{\max}}\right)^m (1 + o(1)), \\ \mathsf{E}[T^m(b)] \ge \mathsf{E}[T_{\max}^m(b)] &\ge \left(\frac{b}{\mu_{\min}}\right)^m (1 + o(1)). \end{aligned} \tag{E.35}$$

Proof. The asymptotic lower bounds in (E.35) follow directly from (E.34) and Fatou's lemma. Therefore, it suffices to establish (E.34). In fact, it suffices to establish the second convergence in (E.34), which is more general. Obviously, $T(b) \ge T_i(b)$ for every $i \in \mathscr{N}$. From (E.31) it follows that

$T_i(b)$ is a.s. finite for any $b > 0$ and that $T_i(b) \to \infty$ almost surely as $b \to \infty$. Then, $S_i(T_i(b)) \geq b$ w.p. 1 and

$$\frac{T(b)}{b} \geq \frac{T_i(b)}{b} \geq \frac{T_i(b)}{S_i(T_i(b))} \xrightarrow[b\to\infty]{} \frac{1}{\mu_i},$$

where the convergence follows from (E.31). Since this is true for every $i \in \mathscr{N}$, we conclude that

$$\liminf_{b\to\infty} \frac{T(b)}{b} \geq \frac{1}{\mu_{\min}} \quad \text{a.s.} \tag{E.36}$$

In order to prove the reverse inequality, we observe that

$$S_i(T_i(b)) \leq b + \Delta S_i(T_i(b)) \quad \text{for every} \quad i \in \mathscr{N}$$

so that

$$\min_{i\in\mathscr{N}} \left[S_i(T(b))\, \mathbb{1}_{\{T(b)=T_i(b)\}}\right] \leq b + \max_{i\in\mathscr{N}} \left[\Delta S_i(T(b))\, \mathbb{1}_{\{T(b)=T_i(b)\}}\right],$$

and consequently,

$$\frac{\min_{i\in\mathscr{N}} S_i(T(b))}{T(b)} \leq \frac{b}{T(b)} + \frac{\max_{i\in\mathscr{N}} \Delta S_i(T(b))}{T(b)}.$$

But from (E.31) it follows that $\Delta S_i(n)/n \to 0$ almost surely as $n \to \infty$ for every $i \in \mathscr{N}$, which implies that

$$\mu_{\min} \leq \liminf_{b\to\infty} \frac{b}{T(b)} \quad \text{a.s.} \tag{E.37}$$

From (E.36) and (E.37) we obtain the second convergence in (E.34). This completes the proof. □

In order to show that the asymptotic lower bounds in (E.35) are sharp, we need either to impose some conditions on the rate of convergence in (E.31) or assume a special structure (independence) of the increments of the processes. In what follows, we use the standard notation $x^- = -\min(0,x)$.

Theorem E.7. *Suppose that the SLLN* (E.31) *holds.*

(i) *Let the processes $S_1,\dots,S_N$ have arbitrarily, possibly dependent increments. Let $r \geq 1$. Then, the asymptotic approximations* (E.32)–(E.33) *hold for all $1 \leq m \leq r$, if for every $i \in \mathscr{N}$*

$$\sum_{n=1}^{\infty} n^{r-1} \mathsf{P}\left(\frac{1}{n} S_i(n) < \mu_i - \varepsilon\right) < \infty \quad \text{for all } 0 \leq \varepsilon < \mu_{\min}. \tag{E.38}$$

(ii) *Let the processes $S_1,\dots,S_N$ have independent increments. Then, the asymptotic approximations* (E.32)–(E.33) *hold for all $m > 0$, if there is a $\lambda \in (0,1)$ such that*

$$\max_{i\in\mathscr{N}} \sup_{n\in\mathbb{N}} \mathsf{E}\left[\exp\{-\lambda(\Delta S_i(n))^-\}\right] < 1. \tag{E.39}$$

Proof. Lemma E.3 implies that in order to establish the asymptotic approximations (E.32)–(E.33) it suffices to prove that under conditions (E.38) and (E.39) the following asymptotic upper bounds hold as $b \to \infty$:

$$\mathsf{E}[T_i^r(b)] \leq \left(\frac{b}{\mu_i}\right)^r (1+o(1)), \quad \mathsf{E}[T^r(b)] \leq \left(\frac{b}{\mu_{\min}}\right)^r (1+o(1)). \tag{E.40}$$

Moreover, it suffices to prove the second inequality, which is more general.

Proof of (i). For any $\varepsilon \in (0,\mu_{\min})$, let $M_b(\varepsilon) = M_b$ be the smallest integer strictly greater than $b/(\mu_{\min} - \varepsilon)$. Setting $\tau = T(b)$ and $N = M_b$ in Lemma A.1, we obtain the following inequality

$$\mathsf{E}[T^r(b)] \leq M_b^r + r2^{r-1} \sum_{n=M_b}^{\infty} n^{r-1} \mathsf{P}(T(b) > n). \tag{E.41}$$

By the definition of $T(b)$, for every $n \in \mathbb{N}$ we have

$$\{T(b) > n\} \subset \bigcup_{i=1}^{N} \{S_i(n) < b\} = \bigcup_{i=1}^{N} \left\{ \frac{S_i(n)}{n} - \mu_i < \frac{b}{n} - \mu_i \right\},$$

and by the definition of M_b, for every $n \geq M_b$

$$\bigcup_{i=1}^{N} \left\{ \frac{S_i(n)}{n} - \mu_i < \frac{b}{n} - \mu_i \right\} \subset \bigcup_{i=1}^{N} \left\{ \frac{S_i(n)}{n} - \mu_i < -\varepsilon \right\},$$

so that

$$\mathsf{P}(T(b) > n) \leq \sum_{i=1}^{N} \mathsf{P}\left(\frac{S_i(n)}{n} - \mu_i < -\varepsilon \right) \quad \text{for } n \geq M_b.$$

Hence, we obtain

$$\sum_{n=M_b}^{\infty} n^{r-1} \mathsf{P}(T(b) > n) \leq \sum_{i=1}^{N} \sum_{n=M_b}^{\infty} n^{r-1} \mathsf{P}\left(\frac{S_i(n)}{n} < \mu_i - \varepsilon \right) \xrightarrow[b \to \infty]{} 0, \tag{E.42}$$

where the convergence to 0 follows from condition (E.38). Therefore, from (E.41) and (E.42) we obtain that as $b \to \infty$

$$\mathsf{E}[T^r(b)] \leq M_b^r + o(1) = \left(\frac{b}{\mu_{\min} - \varepsilon} \right)^r (1 + o(1)). \tag{E.43}$$

Since (E.43) holds for an arbitrarily $\varepsilon \in (0, \mu_{\min})$, letting $\varepsilon \to 0$ proves (E.40).

Proof of (ii). Due to the almost sure convergence in (E.34), it suffices to show that, for every $r > 0$, the family $\{(T(b)/b)^r, b > 0\}$ is uniformly integrable when (E.39) holds. Without loss of generality, we restrict ourselves to $b \in \mathbb{N}$. Observe that for any $b, c \in \mathbb{N}$ we have $T(b+c) \leq T(b) + T(c;b)$, where

$$T(c;b) = \inf\{n > T(b) : S_i(n) - S_i(T(b)) > c \quad \forall i \in \mathcal{N}\}.$$

Obviously, for every $b \in \mathbb{N}$, $T(b) \leq \sum_{n=0}^{b-1} T(1;n)$, and consequently,

$$||T(b)||_r \leq \sum_{n=0}^{b-1} ||T(1;n)||_r \leq b \sup_{n \in \mathbb{N}} ||T(1;n)||_r,$$

so

$$||T(b)/b||_r \leq \sup_{n \in \mathbb{N}} ||T(1;n)||_r.$$

It remains to show that $\sup_{n \in \mathbb{N}} ||T(1;n)||_r$ is finite when (E.39) holds. Note that, for any $\ell \in \mathbb{N}$, we have

$$\begin{aligned}
\{T(1;n) > \ell\} &= \left\{ \max_{T(n) < n \leq T(n) + \ell} \min_{i \in \mathcal{N}} (S_i(n) - S_i(T(n)) \leq 1 \right\} \\
&\subset \left\{ \min_{i \in \mathcal{N}} (S_i(T(n) + \ell) - S_i(T(n)) \leq 1 \right\} \\
&= \bigcup_{i=1}^{N} \{S_i(T(n) + \ell) - S_i(T(n)) \leq 1\},
\end{aligned}$$

and therefore,

$$\mathsf{P}\{T(1;n) > \ell\} \le \sum_{i=1}^{N} \mathsf{P}\{S_i(T(n)+\ell) - S_i(T(n)) \le 1\}. \tag{E.44}$$

By Markov's inequality, for any $\lambda \in (0,1)$,

$$\begin{aligned}
\mathsf{P}\{S_i(T(n)+\ell)) - S_i(T(n)) \le 1\} &= \mathsf{P}\left(\exp\{-\lambda(S_i(T(n)+\ell) - S_i(T(n))\} \ge e^{-\lambda}\right) \\
&\le e^{\lambda}\,\mathsf{E}\left[\exp\{-\lambda(S_i(T(n)+\ell) - S_i(T(n))\}\right] \\
&\le e^{\lambda}\,\mathsf{E}\left[\prod_{u=T(n)+1}^{T(n)+\ell} \exp\{-\lambda\Delta S_i(u)\}\right].
\end{aligned}$$

Let $\beta_i(\lambda) := \sup_{n\in\mathbb{N}} \mathsf{E}[\exp\{-\lambda(\Delta S_i(n))^-\}]$. We have

$$\begin{aligned}
\mathsf{E}\left[\prod_{u=T(n)+1}^{T(n)+\ell} \exp\{-\lambda\Delta S_i(u)\}\,\Big|\, T(n)\right] &= \prod_{u=T(n)+1}^{T(n)+\ell} \mathsf{E}\left[\exp\{-\lambda\Delta S_i(u)\}\right] \\
&\le \prod_{u=T(n)+1}^{T(n)+\ell} \mathsf{E}\left[\exp\{-\lambda(\Delta S_i(u))^-\}\right] \\
&\le (\beta_i(\lambda))^{\ell},
\end{aligned} \tag{E.45}$$

where the first equality holds due to the fact that the event $\{T(b) = n\}$ does not depend on $\{\Delta S_i(u), n+1 \le u \le \ell\}$ since $T(b)$ is the stopping time generated by the sequences ΔS_i, $i \in \mathscr{N}$. From (E.44) and (E.45) we conclude that

$$\mathsf{P}(T(1;n) > \ell) \le e^{\lambda} \sum_{i=1}^{N} (\beta_i(\lambda))^{\ell} \le (Ne^{\lambda}) \left(\max_{i\in\mathscr{N}} \beta_i(\lambda)\right)^{\ell}.$$

This inequality together with condition (E.39) implies $\sup_{n\in\mathbb{N}} ||T(1;n)||_r < \infty$ for every $r > 0$, which completes the proof. □

Remark E.1. Clearly, condition (E.38) is satisfied when each $S_i(n)/n$ converges r-completely to μ_i, i.e., when for every $1 \le i \le N$

$$\sum_{n=1}^{\infty} n^{r-1}\mathsf{P}\left(\left|\frac{1}{n}S_i(n) - \mu_i\right| > \varepsilon\right) < \infty \quad \text{for all } 0 \le \varepsilon < \mu_i. \tag{E.46}$$

E.3 Expectation of First Exit Times for Markov Random Walks

We use notation and definitions from Section D.4. Specifically, consider an additive Markov process $\{(X_n, S_n)\}_{n\in\mathbb{Z}_+}$ with an additive component $S_n = \sum_{t=1}^{n} Y_t$, $S_0 = 0$, which is a Markov random walk, where X_0 is distributed according to a measure λ. Let P_λ and E_λ denote the probability and expectation under the initial distribution of X_0 being λ. When λ is degenerate at x we write P_x and E_x. Assume that there exists a stationary probability distribution $\pi(A) = \int \mathsf{P}(x,A)\,\mathrm{d}\pi(x)$ for all $A \in \mathscr{A}$ and $\mu = \mathsf{E}_\pi[Y_1] > 0$.

For $a > 0$, define the first exit time

$$T_a = \inf\{n \ge 1 : S_n \ge a\}$$

and the corresponding ladder epoch

$$T_+ = \inf\{n \geq 1 : S_n > 0\}.$$

Definition E.2. A Markov random walk is called *lattice* with span $d > 0$ if d is the maximal number for which there exists a measurable function $\gamma: \mathscr{X} \to [0, \infty)$ called the shift function, such that $\mathsf{P}\{Y_1 - \gamma(x) + \gamma(y) \in \{\cdots, -2d, -d, 0, d, 2d, \cdots\} | X_0 = x, X_1 = y\} = 1$ for almost all $x, y \in \mathscr{X}$. If no such d exists, the Markov random walk is called *non-lattice*. A lattice random walk whose shift function γ is identically 0 is called *arithmetic*. Otherwise, it is *nonarithmetic*.

By $\mathsf{P}_+(x, A \times B) = \mathsf{P}_x(X_{T_+} \in A, S_{T_+} \in B)$ denote the transition probability associated with the Markov random walk generated by the ascending ladder variable S_{T_+} and by π_+ the corresponding invariant measure.

Define the overshoot $\kappa_a = S_{T_a} - a$ on $T_a < \infty$ and the limiting distribution of the overshoot $\mathsf{G}(y) = \lim_{a\to\infty} \mathsf{P}(\kappa_a \leq y)$. In the non-arithmetic case, the limiting average overshoot

$$\varkappa = \lim_{a\to\infty} \mathsf{E}_x[\kappa_a] = \int y \, d\mathsf{G}(y)$$

can be written as

$$\varkappa = \frac{\mathsf{E}_{\pi_+}[S_{T_+}^2]}{2\mathsf{E}_{\pi_+}[S_{T_+}]}.$$

Define $\lambda^*(B) = \sum_{n=0}^{\infty} \mathsf{P}_\lambda(X_n \in B)$ on $\mathscr{A}$. Let $g = \mathsf{E}(Y_1 | X_0, X_1)$ and $\mathsf{E}_\pi|g| < \infty$. Define operators $\mathbf{P}$ and $\mathbf{P}_\pi$ by $(\mathbf{P}g)(x) = \mathsf{E}_x[g(x, X_1, Y_1)]$ and $\mathbf{P}_\pi g = \mathsf{E}_\pi[g(X_0, X_1, Y_1)]$ respectively, and set $\overline{g} = \mathbf{P}g$. Consider the Poisson equation

$$(\mathbb{I} - \mathbf{P})\Delta = (\mathbb{I} - \mathbf{P}_\pi)\overline{g} \quad \lambda^*\text{-a.s.}, \quad \mathbf{P}_\pi \Delta = 0, \tag{E.47}$$

where $\mathbb{I}$ is the identity operator.

The following theorem provides an asymptotic expansion for the expectation $\mathsf{E}_\lambda[T_a]$, generalizing Theorem E.6 to the case of Markov random walks. A proof can be found in Fuh and Lai [46, Theorem 3.4].

Theorem E.8. *Assume that $\{X_n\}$ is V-uniformly ergodic, i.e., condition* (D.14) *holds. Also, suppose that $\mu = \mathsf{E}_\pi[Y_1] > 0$, $\sup_x \mathsf{E}_x[|Y_1|^2] < \infty$ and $\sup_x\{\mathsf{E}[|Y_1|^2 V(X_1)]/V(x)\} < \infty$.*

(i) *If the Markov random walk is nonarithmetic, then*

$$\mathsf{E}_\lambda[T_a] = \frac{1}{\mu}\left(a + \varkappa - \int \Delta(x)[d\pi_+(x) - d\lambda(x)]\right) + o(1) \quad \text{as } a \to \infty, \tag{E.48}$$

and therefore,

$$\mathsf{E}_x[T_a] = \frac{1}{\mu}\left(a + \varkappa - \int \Delta(x) d\pi_+(x) - \Delta(x)\right) + o(1) \quad \text{as } a \to \infty, \tag{E.49}$$

where $\varkappa = \mathsf{E}_{\pi_+}[S_{T_+}^2]/2\mathsf{E}_{\pi_+}[S_{T_+}]$ and $\Delta(x)$ is a solution of the Poisson equation (E.47).

(ii) *If the Markov random walk is d-arithmetic, then asymptotic expansions* (E.48) *and* (E.49) *hold as $a \to \infty$ through integral multiples of d with $\varkappa = \mathsf{E}[S_{T_+}^2]/2\mathsf{E}[S_{T_+}] + d/2$.*

Appendix F: Nonlinear Renewal Theory

In this appendix, we present results from the Nonlinear Renewal Theory for random walks perturbed by a slowly changing term and for more general cases with curved boundaries as well as from the Markov Nonlinear Renewal Theory for perturbed Markov random walks. These results are used in Chapter 1 and Chapter 3 for deriving higher order approximations for operating characteristics of sequential hypothesis tests and changepoint detection rules. We provide nonlinear renewal theorems without proofs. The proofs and further details can be found in [45, 77, 78, 140, 164, 188, 189, 194].

F.1 Nonlinear Renewal Theory for Perturbed Random Walks

Nonlinear renewal theory extends renewal theory, which deals with a first passage of a random walk $\{S_n = \sum_{i=1}^n \Delta S_i\}$ to a constant threshold, to the perturbed random walks of the form

$$Z_n = S_n + \xi_n, \quad n \in \mathbb{Z}_+, \tag{F.1}$$

assuming certain smoothness conditions on the sequence of "perturbations" $\{\xi_n\}$. Moreover, in the general case, a threshold a_n can be time-varying with certain restrictions on the rate of increase. We begin by considering a particular case of a constant threshold $a_n = a$. A more general case will be considered later on.

Specifically, for $a > 0$, define the stopping times

$$T_a = \inf\{n \geq 1 : Z_n \geq a\} \tag{F.2}$$

and

$$\tau_a = \inf\{n \geq 1 : S_n \geq a\}. \tag{F.3}$$

Let

$$\kappa_a = S_{\tau_a} - a \text{ on } \{\tau_a < \infty\} \quad \text{and} \quad \tilde{\kappa}_a = Z_{T_a} - a \text{ on } \{T_a < \infty\} \tag{F.4}$$

denote the overshoots of the statistics S and Z over threshold a at stopping. Also, let $\tau_+ = \inf\{n \geq 1 : S_n > 0\}$ denote the first ascending ladder variable.

One of the main results of nonlinear renewal theory for perturbed random walks is that the limiting distribution of the overshoot as $a \to \infty$ does not change when the random walk is perturbed with an additive slowly changing nonlinear term, i.e., $\lim_{a\to\infty} \mathsf{P}(\tilde{\kappa}_a < y) = \lim_{a\to\infty} \mathsf{P}(\kappa_a < y)$. Another important result is that the asymptotic approximation for the expectation of the stopping time remains exactly the same if the original threshold value a is replaced with $a - \lim_{n\to\infty} \mathsf{E}[\xi_n]$.

To proceed we have to define the notion of "slowly changing" random variables. The sequence of random variables $\{\xi_n\}_{n\in\mathbb{Z}_+}$ is called *slowly changing* if

$$\frac{1}{n} \max_{1\leq k\leq n} |\xi_k| \xrightarrow[n\to\infty]{\mathsf{P}} 0, \tag{F.5}$$

and for every $\varepsilon > 0$ there are $n_0 \geq 1$ and $\delta > 0$ such that

$$\mathsf{P}\left(\max_{1 \leq k \leq n\delta} |\xi_{n+k} - \xi_{n_0}| > \varepsilon\right) < \varepsilon \quad \text{for all } n \geq n_0. \tag{F.6}$$

The following theorem, which we refer to as the *First Nonlinear Renewal Theorem* (NRT) *for perturbed random walks*, shows that asymptotically (as $a \to \infty$) the overshoot $\tilde{\kappa}_a$ has the same limiting distribution as κ_a, so that adding a slowly changing term to the random walk does not change the distribution for large a.

Theorem F.1 (First NRT). *Let $\{S_n\}_{n \in \mathbb{Z}_+}$ be a nonarithmetic random walk with a positive drift $\mu = \mathsf{E}[\Delta S_1] > 0$. Assume that $\{\xi_n\}_{n \in \mathbb{Z}_+}$ is a slowly changing sequence. Then $\lim_{a \to \infty} \mathsf{P}(\tilde{\kappa}_a \leq y) = \lim_{a \to \infty} \mathsf{P}(\kappa_a \leq y) = \mathsf{G}(y)$, where $\mathsf{G}(y)$ is defined as*

$$\mathsf{G}(y) = \frac{1}{\mathsf{E}[S_{\tau_+}]} \int_0^y \mathsf{P}(S_{\tau_+} > t)\, dt, \quad y \geq 0, \tag{F.7}$$

and for every $\lambda > 0$

$$\lim_{a \to \infty} \mathsf{E}\left[e^{-\lambda \tilde{\kappa}_a}\right] = \frac{1}{\lambda \mu} \exp\left\{-\sum_{n=1}^{\infty} \frac{1}{n} \mathsf{E}\left[e^{-\lambda S_n^+}\right]\right\}. \tag{F.8}$$

If in addition $\mathsf{E}[(\Delta S_1^+)^2] < \infty$, then the limiting average overshoot $\varkappa = \lim_{a \to \infty} \mathsf{E}[\tilde{\kappa}_a] = \lim_{a \to \infty} \mathsf{E}[\kappa_a]$ is given by

$$\varkappa = \frac{\mathsf{E}[S_{\tau_+}^2]}{2\mathsf{E}[S_{\tau_+}]} = \frac{\mathsf{E}[\Delta S_1^2]}{2\mathsf{E}[\Delta S_1]} - \sum_{n=1}^{\infty} \frac{1}{n} \mathsf{E}[S_n^-]. \tag{F.9}$$

Under some additional conditions Theorem F.1 is valid in the arithmetic case too [79].

We now proceed with the approximations for the expected sample size $\mathsf{E}[T_a]$. Note that $S_{T_a} + \xi_{T_a} = a + \tilde{\kappa}_a$. Taking expectations on both sides and using Wald's identity yields

$$\mu \mathsf{E}[T_a] = a + \mathsf{E}[\tilde{\kappa}_a] - \mathsf{E}[\xi_{T_a}].$$

By Theorem F.1, $\mathsf{E}[\tilde{\kappa}_a] = \varkappa + o(1)$ for large a. If $\mathsf{E}[\xi_n] \to \bar{\xi}$ as $n \to \infty$, then we expect that

$$\mathsf{E}[T_a] = \frac{1}{\mu}\left(a + \varkappa - \bar{\xi}\right) + o(1) \quad \text{as } a \to \infty. \tag{F.10}$$

Although this heuristic argument is simple, a rigorous treatment is quite tedious.

We now present the *Second Nonlinear Renewal Theorem for perturbed random walks* that deals with a detailed higher-order asymptotic approximation for the expected sample size $\mathsf{E}[T_a]$ that was conjectured in (F.10). To this end, we need the following additional conditions. Assume there exist events $\mathscr{A}_n \in \mathscr{F}_n^X$, $n \geq 1$, constants ℓ_n, $n \geq 1$, $\mathscr{F}_n^X$-measurable random variables η_n, $n \geq 1$, and an integrable random variable η such that

$$\sum_{n=1}^{\infty} \mathsf{P}\left(\bigcup_{k=n}^{\infty} \mathscr{A}_k^c\right) < \infty, \tag{F.11}$$

$$\xi_n = \ell_n + \eta_n \quad \text{on } \mathscr{A}_n, \; n \geq 1, \tag{F.12}$$

$$\limsup_{\varepsilon \to 0} \sup_{n \geq 1} \max_{0 \leq k \leq n\varepsilon} |\ell_{n+k} - \ell_n| = 0, \tag{F.13}$$

$$\max_{0 \leq k \leq n} |\eta_{n+k}|, \; n \geq 1 \quad \text{are uniformly integrable}, \tag{F.14}$$

$$\sum_{n=1}^{\infty} \mathsf{P}(\eta_n \leq -\varepsilon n) < \infty \quad \text{for some } 0 < \varepsilon < \mu, \tag{F.15}$$

$$\eta_n \xrightarrow[n\to\infty]{\text{law}} \eta, \tag{F.16}$$

$$\lim_{a\to\infty} a\,\mathsf{P}(T_a \le \varepsilon a/\mu) = 0 \quad \text{for some } 0<\varepsilon<1. \tag{F.17}$$

Note that if $\{\eta_n\}_{n\ge 1}$ is a slowly changing sequence, then $\{\xi_n\}_{n\ge 1}$ is also slowly changing when conditions (F.11)–(F.13) hold.

Theorem F.2 (Second NRT). *Assume that* $\mu = \mathsf{E}[\Delta S_1] > 0$, *that* $\mathsf{E}[\Delta S_1^2] < \infty$, *that conditions* (F.11)–(F.17) *hold, and that the sequence* $\{\eta_n\}_{n\in\mathbb{Z}_+}$ *is slowly changing.*

(i) *If the random walk* $\{S_n\}_{n\in\mathbb{Z}_+}$ *is nonarithmetic, then*

$$\mathsf{E}[T_a] = \frac{1}{\mu}\left(a + \varkappa - \mathsf{E}[\eta] - \ell_{N_a}\right) + o(1) \quad \text{as } a\to\infty, \tag{F.18}$$

where $N_a = \lfloor a/\mu \rfloor$ *and* $\varkappa$ *is the limiting average overshoot given by* (F.9).

(ii) *If the random walk* $\{S_n\}_{n\in\mathbb{Z}_+}$ *is d-arithmetic and if in addition* $\ell_n = 0$ *and the random variable* η *is continuous, then*

$$\mathsf{E}[T_{a=jd}] = \frac{1}{\mu}\left(jd + \varkappa - \mathsf{E}[\eta]\right) + o(1) \quad \text{as } j\to\infty, \tag{F.19}$$

where $\varkappa = \mathsf{E}[S_{T_+}^2]/2\mathsf{E}[S_{T_+}] + d/2$.

While at first glance conditions (F.6), (F.13)–(F.16) look complicated, in many statistical problems conditions (F.14)–(F.16) and (F.5)–(F.6) may be reduced to a single moment condition on ΔS_1, and therefore, easily checked. Usually, the verification of condition (F.17) causes the main difficulty. The following example illustrates this point.

Example F.1. Consider the stopping time T_a defined in (F.2) where the statistic Z_n is of the CUSUM-type form

$$Z_n = S_n - \min_{0\le k\le n} S_k, \quad S_0 = 0,$$

so that $\ell_n = 0$ and $\xi_n = \eta_n = -\min_{0\le k\le n} S_k$. Conditions (F.11)–(F.13) hold trivially. Condition (F.16) is satisfied with $\eta = -\min_{k\ge 0} S_k$. Recall that Theorem F.2 assumes the second moment condition $\mathsf{E}[\Delta S_1^2] < \infty$, so that the uniform integrability in (F.14) follows. It remains to establish (F.17). Even in this simple example verification of this condition is not a trivial task (see [164, pages 55–56]).

The following result can be used for verification of condition (F.17).

Lemma F.1. *Let* $\{S_n\}_{n\in\mathbb{Z}_+}$ *be a zero-mean random walk with finite moment of order* $2+\alpha$ *for some* $\alpha \ge 0$, *i.e.,* $\mathsf{E}[\Delta S_1] = 0$, $\mathsf{E}[|\Delta S_1|^{2+\alpha}] = \mu_\alpha < \infty$. *Then for all* $\varepsilon > 0$

$$\lim_{n\to\infty} n^{1+\alpha/2}\,\mathsf{P}\left(\max_{1\le t\le n} |S_t| > \varepsilon n\right) = 0.$$

Proof. Applying Doob's maximal submartingale inequality to the submartingale $|S_n|^{2+\alpha}$, we obtain

$$\begin{aligned}\mathsf{P}\left(\max_{1\le t\le n} |S_t| \ge \varepsilon n\right) &\le \frac{1}{(\varepsilon n)^{2+\alpha}}\,\mathsf{E}\left[|S_n|^{2+\alpha}\mathbb{1}_{\{\max_{1\le t\le n} S_t \ge \varepsilon n\}}\right]\\ &= \frac{1}{\varepsilon^{2+\alpha} n^{1+\alpha/2}}\,\mathsf{E}\left[\left(\frac{|S_n|^{2+\alpha}}{n^{1+\alpha/2}}\right)\mathbb{1}_{\{\max_{1\le t\le n} S_t \ge \varepsilon n\}}\right].\end{aligned}$$

Since by inequality (C.20), for some universal constant C_α,

$$\mathsf{E}[|S_n|^{2+\alpha}] \le C_\alpha \mu_\alpha n^{1+\alpha/2}$$

it follows that

$$\mathsf{P}\left(\max_{1\le t\le n}|S_t|\ge \varepsilon n\right)\le \frac{C_\alpha \mu_\alpha}{\varepsilon^{2+\alpha} n^{1+\alpha/2}} \xrightarrow[n\to\infty]{} 0.$$

Hence, in order to prove the assertion of the lemma it suffices to show that

$$\mathsf{E}\left[\left(\frac{|S_n|}{\sqrt{n}}\right)^{2+\alpha} \mathbb{1}_{\{\max_{1\le t\le n} S_t \ge \varepsilon n\}}\right] \xrightarrow[n\to\infty]{} 0, \tag{F.20}$$

which obviously implies that $\mathsf{P}(\max_{1\le t\le n}|S_t|>\varepsilon n)=o(1/n^{1+\alpha/2})$ as $n\to\infty$.

Note that

$$\left(\frac{|S_n|}{\sqrt{n}}\right)^{2+\alpha} \le \left(\frac{\sum_{i=1}^n |\Delta S_i|}{\sqrt{n}}\right)^{2+\alpha} \le \frac{\sum_{i=1}^n |\Delta S_i|^{2+\alpha}}{n},$$

so that

$$\mathsf{E}\left[\left(\frac{|S_n|}{\sqrt{n}}\right)^{2+\alpha}\right] \le \mathsf{E}\left[|\Delta S_1|^{2+\alpha}\right] = \mu_\alpha < \infty.$$

Consequently, the family $\{(S_n/\sqrt{n})^{2+\alpha}, n\ge 1\}$ is uniformly integrable. This immediately yields (F.20) and the proof is complete. □

Note that Lemma F.1 allows us to immediately conclude that condition (F.17) is satisfied if $\xi_n=0$, i.e., for the stopping time $T_a=\inf\{n: S_n\ge a\}$ whenever $\mu>0$ and $\mathsf{E}[S_1^2]<\infty$. Indeed, writing $N_a=(1-\varepsilon)a/\mu$ we obtain

$$\begin{aligned} a\mathsf{P}(T_a\le N_a) &= a\mathsf{P}\left(\max_{1\le n\le N_a} S_n\ge a\right) = a\mathsf{P}\left(\max_{1\le n\le N_a} S_n-\mu N_a\ge a-\mu N_a\right)\\ &= a\mathsf{P}\left(\max_{1\le n\le N_a}(S_n-\mu N_a)\ge \varepsilon a\right) = a\mathsf{P}\left(\max_{1\le n\le N_a}(S_n-\mu N_a)\ge \varepsilon\mu N_a\right)\\ &\le a\mathsf{P}\left(\max_{1\le n\le N_a}(S_n-\mu n)\ge \varepsilon\mu N_a\right) \xrightarrow[a\to\infty]{} 0 \quad \text{for all } 0<\varepsilon<1. \end{aligned} \tag{F.21}$$

In the following lemma, we provide sufficient conditions for (F.17).

Lemma F.2. *Let $\{S_n\}_{n\in\mathbb{Z}_+}$ be a random walk with $\mathsf{E}[\Delta S_1]=\mu>0$ and the finite m-th absolute moment, $\mathsf{E}[|\Delta S_1|^m]<\infty$ for some $m\ge 2$. Let*

$$T_a=\inf\{n\ge 1: S_n+\xi_n\ge a\},\quad a>0,$$

where ξ_n satisfies condition (F.5). If in addition the probability $\mathsf{P}(\max_{1\le t\le n}\xi_t>\varepsilon n)$ approaches zero with the rate faster than $1/n^{m/2}$, i.e.,

$$\lim_{n\to\infty} n^{m/2}\mathsf{P}\left(\max_{1\le t\le n}\xi_t>\varepsilon n\right)=0 \quad \textit{for all } \varepsilon>0, \tag{F.22}$$

then

$$\lim_{a\to\infty} a^{m/2}\mathsf{P}\{T_a\le(1-\varepsilon)a/\mu\}=0 \quad \textit{for all } 0<\varepsilon<1.$$

Proof. Again, let $N_a=(1-\varepsilon)a/\mu$ and let $\widetilde{S}_n=S_n-\mu n$. Obviously,

$$\mathsf{P}(T_a\le N_a)=\mathsf{P}\left(\max_{1\le n\le N_a}(S_n+\xi_n)\ge a\right)$$

and in just the same way as in (F.21), we obtain

$$\mathsf{P}\left(\max_{1\le n\le N_a}(S_n+\xi_n)\ge a\right)\le \mathsf{P}\left(\max_{1\le n\le N_a}(\widetilde{S}_n+\xi_n)\ge \varepsilon\mu N_a\right),$$

where the latter probability can be evaluated from above as follows:

$$\mathsf{P}\left(\max_{1\le n\le N_a}(\widetilde{S}_n+\xi_n)\ge \varepsilon\mu N_a\right)\le \mathsf{P}\left(\max_{1\le n\le N_a}\widetilde{S}_n\ge \varepsilon\mu N_a/2\right)+\mathsf{P}\left(\max_{1\le n\le N_a}\xi_n\ge \varepsilon\mu N_a/2\right).$$

Therefore,

$$N_a^{m/2}\,\mathsf{P}(T_a\le N_a)\le N_a^{m/2}\,\mathsf{P}\left(\max_{1\le n\le N_a}\widetilde{S}_n\ge \varepsilon\mu N_a/2\right)+N_a^{m/2}\,\mathsf{P}\left(\max_{1\le n\le N_a}\xi_n\ge \varepsilon\mu N_a/2\right).$$

However, by Lemma F.1, for any $0<\varepsilon<1$,

$$N_a^{m/2}\,\mathsf{P}\left(\max_{1\le n\le N_a}\widetilde{S}_n\ge \varepsilon\mu N_a/2\right)\to 0 \quad \text{as } a\to\infty$$

and by the assumption (F.22),

$$N_a^{m/2}\,\mathsf{P}\left(\max_{1\le n\le N_a}\xi_n\ge \varepsilon\mu N_a/2\right)\to 0 \quad \text{as } a\to\infty,$$

also for any $0<\varepsilon<1$, which proves the assertion of the lemma. □

F.2 General Nonlinear Renewal Theory

The nonlinear renewal methods presented above for the stopping time (F.2) with a constant threshold are based on the expansion of a nonlinear function of the random walk and then applying the classical renewal argument to the leading term. An alternative approach is to expand a boundary around an appropriate point. We now consider this more general approach that allows us to handle the stopping time

$$T_a=\inf\{n\ge 1: Z_n\ge b_n(a)\},\quad a\in\mathscr{A} \tag{F.23}$$

with a time-varying boundary $b_n(a)$. As before in (F.1), the statistic $Z_n=S_n+\xi_n$ is the random walk $S_n=\sum_{i=1}^n \Delta S_i$ with positive and finite drift $\mu=\mathsf{E}[\Delta S_i]$ perturbed by random variables ξ_n, $n\ge 1$, which do not depend on the increments of the random walk $\Delta S_{n+1},\Delta S_{n+2},\dots$. In most statistical applications, we will be interested in cases where a is a constant part of the threshold and $\mathscr{A}=[0,\infty)$, i.e., $b_n(a)=a+g_n$. We provide exact statements with no proofs. The proofs may be found in Zhang [194].

For $c\ge 0$ and $\nu<\mu$, define the stopping time

$$\tau_c(\nu)=\inf\left\{n\ge 1:\widetilde{S}_n(\nu)\ge c\right\}, \tag{F.24}$$

where $\widetilde{S}_n(\nu)=S_n-\nu n$. For $\nu=0$ and $c=a$ this is the stopping time τ_a defined in (F.3), which is a subject of classical renewal theory, discussed in Chapter E. Let $c=c_a$ and $\nu=\nu_a$ be suitable values of c and ν that depend on a and write $\tau_a=\tau_{c_a}(\nu_a)$. The basic idea is to consider powers of absolute differences between the stopping times, $|T_a-\tau_a|^p$, $p\ge 1$, and to establish uniform integrability of the family $|T_a - \tau_a|^p, a \in \mathscr{A}$. Then nonlinear renewal theorems may be established using classical renewal theorems for τ_a defined by the random walk $\widetilde{S}_n(\nu) = S_n - \nu_a n$ with the drift $\mu - \nu_a$.

Let N_a denote the point of intersection of the boundary $b_t(a)$ with the line μt, i.e.,

$$N_a = \sup\{n \ge 1 : b_n(a) \ge \mu n\}, \quad \sup\{\varnothing\} = 1.$$

In the following, we assume that $b_t(a)$ is twice differentiable in t. Define

$$\nu_a = \frac{\partial b_t(a)}{\partial t}\Big|_{t=N_a}, \quad \nu_{\max} = \sup_{t \ge N_a, a \in \mathscr{A}} \frac{\partial b_t(a)}{\partial t},$$

$$\kappa_c(\nu) = \tilde{S}_{\tau_c(\nu)}(\nu) - c \quad \text{on } \{\tau_c(\nu) < \infty\}, \quad \tilde{\kappa}_a = Z_{T_a} - b_{T_a}(a) \quad \text{on } \{T_a < \infty\},$$

$$\tau_+(\nu) = \inf\{n \ge 1 : \tilde{S}_n(\nu) > 0\}, \tag{F.25}$$

$$\varkappa(\nu) = \frac{\mathsf{E}[\tilde{S}_{\tau_+(\nu)}(\nu)]^2}{2\mathsf{E}[\tilde{S}_{\tau_+(\nu)}(\nu)]}, \tag{F.26}$$

$$\mathsf{G}(y,\nu) = \frac{1}{\mathsf{E}[\tilde{S}_{\tau_+(\nu)}(\nu)]} \int_0^y \mathsf{P}\{\tilde{S}_{\tau_+(\nu)}(\nu) > s\} \, ds, \quad y \ge 0. \tag{F.27}$$

Note that $\kappa_c(\nu)$ and $\tilde{\kappa}_a$ are the overshoots in the linear and nonlinear schemes (F.24) and (F.23), respectively. By renewal theory, $\varkappa(\nu) = \lim_{c\to\infty} \mathsf{E}[\kappa_c(\nu)]$ is the limiting average overshoot and $\mathsf{G}(y,\nu) = \lim_{c\to\infty} \mathsf{P}\{\kappa_c(\nu) \le y\}$ is the limiting distribution of the overshoot in the linear case.

Suppose that

$$\lim_{N_a\to\infty} \nu_a = \nu^* < \mu \tag{F.28}$$

and that there are functions $\rho(\varepsilon) > 0$ and $\sqrt{y} \le \gamma(y) \le y$, $\gamma(y) = o(y)$ as $y \to \infty$ such that

$$\frac{T_a - N_a}{\gamma(N_a)} = O(1) \quad \text{as } N_a \to \infty, \tag{F.29}$$

$$\lim_{n\to\infty} \mathsf{P}\left\{\max_{1\le i\le \rho(\varepsilon)\gamma(n)} |\xi_{n+i} - \xi_n| \ge \varepsilon\right\} = 0 \quad \text{for every } \varepsilon > 0, \tag{F.30}$$

$$\sup_{\{|t-N_a|\le K\gamma(N_a), a\in\mathscr{A}\}} \left|\gamma(N_a)\frac{\partial^2 b_t(a)}{\partial t^2}\right| < \infty \quad \text{for all } K < \infty. \tag{F.31}$$

The following theorem extends Theorem F.1, the first NRT for perturbed random walks, to the general case.

Theorem F.3 (First general NRT). *Assume that conditions* (F.28)–(F.31) *are satisfied. Let* $\mathsf{G}(y,\nu)$ *be as in* (F.27) *and* $\mathsf{G}^*(y) = \mathsf{G}(y,\nu^*)$. *If the random walk* $\{S_n - \nu^* n\}_{n\in\mathbb{Z}_+}$ *is nonarithmetic with a positive and finite drift* $\mu - \nu^*$, *then the limiting distribution of the overshoot* $\tilde{\kappa}_a$ *is*

$$\lim_{N_a\to\infty} \mathsf{P}(\tilde{\kappa}_a \le y) = \mathsf{G}^*(y) \quad \textit{for every } y \ge 0. \tag{F.32}$$

Let $\sigma^2 = \mathsf{Var}[\Delta S_1] < \infty$. Under certain additional conditions on the behavior of threshold $b_n(a)$ and the sequence $\{\xi_n\}$, in particular when $\xi_n/\sqrt{n} \to 0$ w.p. 1 as $n \to \infty$,

$$\widetilde{T}_a = \frac{T_a - N_a}{\sqrt{N_a \sigma^2/(\mu - \nu^*)^2}} \xrightarrow[N_a\to\infty]{\text{law}} \Phi(x) \quad \text{for all } -\infty < x < \infty,$$

and $\lim_{N_a\to\infty} \mathsf{P}(\widetilde{T}_a \le x, \tilde{\kappa}_a \le y) = \Phi(x) \cdot \mathsf{G}^*(y)$ for all $-\infty < x < \infty$, $y \ge 0$, where $\Phi(x)$ is the standard normal distribution function. See Theorem 1 and Proposition 1 in Zhang [194].

We now turn to the higher-order asymptotic approximations for the expected sample size.

Definition F.1. The sequence of random variables $\{\xi_n\}_{n\in\mathbb{Z}_+}$ is called *regular* if there exist a nonnegative random variable L with finite expectation $\mathsf{E}[L] < \infty$, a deterministic sequence $\{\ell_n\}_{n\in\mathbb{Z}_+}$, and a random sequence $\{\eta_n\}_{n\in\mathbb{Z}_+}$ such that the following conditions hold:

$$\xi_n = \ell_n + \eta_n \quad \text{for } n \geq L, \tag{F.33}$$

$$\max_{1\leq i\leq n^{1/2}} |\ell_{n+i} - \ell_n| \leq K \quad \text{for some } 0 < K < \infty, \tag{F.34}$$

$$\left\{\max_{1\leq i\leq n} |\eta_{n+i}|, n \geq 1\right\} \quad \text{is uniformly integrable,} \tag{F.35}$$

$$\lim_{n\to\infty} n\,\mathsf{P}\left(\max_{0\leq i\leq n} \eta_{n+i} \geq \varepsilon n\right) = 0 \quad \text{for all } \varepsilon > 0, \tag{F.36}$$

$$\sum_{n=1}^{\infty} \mathsf{P}(\eta_n \leq -\varepsilon n) < \infty \quad \text{for some } 0 < \varepsilon < \mu - \nu_{\max}. \tag{F.37}$$

Note that in many statistical applications conditions (F.35)–(F.37) may be reduced to a single moment condition.

In order to obtain a higher-order approximation for $\mathsf{E}[T_a]$ the following set of conditions is needed:

$$\frac{\partial b_t(a)}{\partial t} \leq \nu^* \quad \text{for } t \geq \delta N_a,\ a \in \mathscr{A}; \tag{F.38}$$

$$\sup_{\{1-K\leq t/N_a\leq 1+K, a\in\mathscr{A}\}} \left| N_a \frac{\partial^2 b_t(a)}{\partial t^2} \right| < \infty \quad \text{for any } K > 0; \tag{F.39}$$

$$\lim_{N_a\to\infty} \sup_{\{(t-N_a)^2\leq KN_a\}} \left| \frac{N_a}{2} \frac{\partial^2 b_t(a)}{\partial t^2} - d \right| = 0 \quad \text{for any } K > 0 \text{ and some } d; \tag{F.40}$$

$$\lim_{n\to\infty} \mathsf{P}\left(\max_{1\leq i\leq \sqrt{n}} |\eta_{n+i} - \eta_n| \geq \varepsilon\right) = 0 \quad \text{for every } \varepsilon > 0 \tag{F.41}$$

(i.e., condition (F.30) holds for η_n with $\gamma(n) = \sqrt{n}$ and $\rho(\varepsilon) = 1$);

$$\max_{1\leq i\leq \sqrt{n}} |\ell_{n+i} - \ell_n| \xrightarrow[n\to\infty]{} 0 \tag{F.42}$$

(cf. (F.34));

$$\eta_n \xrightarrow[n\to\infty]{\text{law}} \eta, \tag{F.43}$$

where η is an integrable random variable with expectation $\mathsf{E}[\eta]$; and

$$\lim_{N_a\to\infty} N_a \mathsf{P}(T_a \leq \varepsilon N_a) = 0 \quad \text{for some } 0 < \varepsilon < 1. \tag{F.44}$$

The following theorem is the *Second General Nonlinear Renewal Theorem.*

Theorem F.4 (Second General NRT). *Suppose that the sequence $\{\xi_n\}_{n\in\mathbb{Z}_+}$ is regular, i.e., conditions* (F.33)–(F.37) *are satisfied, and the random walk $\{S_n - \nu^* n\}_{n\in\mathbb{Z}_+}$ is nonarithmetic. If $\mathsf{Var}[\Delta S_1] = \sigma^2 < \infty$ and conditions* (F.38)–(F.44) *hold, then as $N_a \to \infty$,*

$$\mathsf{E}[T_a] = N_a - \frac{\ell_{N_a}}{\mu - \nu_a} + \frac{1}{\mu - \nu^*}\left\{\varkappa(\nu^*) + \frac{d\sigma^2}{(\mu - \nu^*)^2} - \mathsf{E}[\eta]\right\} + o(1). \tag{F.45}$$

It is not difficult to verify that Theorem F.4 implies Theorem F.2(i) for perturbed random walks. Indeed, since threshold $b_t(a) = a$ is constant, $\partial b_t(a)/\partial t = 0$, and we need to set $\nu_a = \nu^* = 0$, $d = 0$ and $N_a = a/\mu$. Therefore, the following corollary for perturbed random walks holds true.

Corollary F.1. *Let $b_n(a) = a > 0$, let the stopping time T_a be defined as in* (F.2) *and let $N_a = a/\mu$. Suppose that the sequence $\{\xi_n\}_{n\in\mathbb{Z}_+}$ is regular and the random walk $\{S_n\}_{n\in\mathbb{Z}_+}$ is nonarithmetic. If* $\mathsf{Var}[\Delta S_1] = \sigma^2 < \infty$ *and conditions* (F.41)–(F.44) *hold, then*

$$\mathsf{E}[T_a] = \frac{1}{\mu}\{a - \ell_{N_a} - \mathsf{E}[\eta] + \varkappa\} + o(1) \quad \text{as } a \to \infty, \tag{F.46}$$

where $\varkappa = \varkappa(0) = \mathsf{E}[S_{T_+}^2]/2\mathsf{E}[S_{T_+}]$.

It is seen that condition (F.33) is almost identical and conditions (F.35)–(F.37), (F.43) are identical to the corresponding conditions required in Theorem F.2. However, conditions (F.34), (F.36) (F.41) and (F.42) required for Corollary F.1 are weaker than those of Theorem F.2. This is particularly true for the deterministic sequence $\{\ell_n\}$, which in Theorem F.2 cannot increase faster than $O(\log n)$, while in Theorem F.4 it can grow at the rate $O(\sqrt{n})$.

In certain interesting cases, not all sufficient conditions postulated in Theorem F.4 hold. In particular, in Section 1.5.3.2 of Chapter 1 uniform integrability condition (F.35) is not satisfied. For this reason, we now present a modified version of the Second General NRT, which is useful in such cases. This theorem can be deduced from the paper by Zhang [194] noting that certain conditions in Theorem F.4 are needed for establishing uniform integrability of the overshoot.

Theorem F.5 (Modified General NRT). *Suppose that conditions* (F.33), (F.34), (F.36) *are satisfied, and the random walk $\{S_n - \nu^* n\}_{n\in\mathbb{Z}_+}$ is nonarithmetic. Assume further that* $\mathsf{Var}[\Delta S_1] = \sigma^2 < \infty$ *and conditions* (F.38)–(F.40), (F.42), (F.44) *hold as well as the following conditions hold:*

$$\mathsf{E}[\chi_a] = \mathsf{E}[\kappa_a] + o(1) \quad \text{as } N_a \to \infty \quad \text{and} \quad \lim_{n\to\infty} \mathsf{E}[\eta_n] = \mathsf{E}[\eta].$$

Then as $N_a \to \infty$, the asymptotic expansion (F.45) *holds.*

Remark F.1. The intermediate case between the first-order approximation

$$\mathsf{E}[T_a] = N_a(1 + o(1)) \quad \text{as } N_a \to \infty,$$

which holds if $\mathsf{E}|\Delta S_1|^2 < \infty$ under condition (F.37), and the third-order approximation (F.45) is the following second-order approximation

$$\mathsf{E}[T_a] = N_a - \ell_{N_a}/(\mu - \nu_a) + O(1) \quad \text{as } N_a \to \infty.$$

This second-order approximation holds when $\{\xi_n\}$ is regular and conditions (F.39) and (F.44) are satisfied.

F.3 Markov Nonlinear Renewal Theory

We give a brief summary of the Markov nonlinear renewal theory developed in Fuh [45]. We provide a simpler version which is more transparent and useful for this book. It covers the case of stopping times when a perturbed Markov random walk exceeds a constant threshold, not a time-varying boundary.

Let the process $\{(X_n, S_n)\}_{n\in\mathbb{Z}_+}$ be an additive Markov process and $\{S_n\}_{n\in\mathbb{Z}_+}$ a Markov random walk (see Definition D.2). To be more specific, the Markov process $\{X_n\}_{n\in\mathbb{Z}_+}$ is defined on a general state space $\mathscr{X}$ with σ-algebra $\mathscr{A}$, which is irreducible with respect to a maximal irreducibility measure on $(\mathscr{X}, \mathscr{A})$ and is aperiodic; $S_n = \sum_{k=1}^n Y_k$ is the additive component, taking values on the real line $\mathbb{R}$, such that$\{(X_n,S_n)\}_{n\in\mathbb{Z}_+}$ is a Markov process on $\mathscr{X}\times\mathbb{R}$ with transition probability given in (D.13).

Let P_λ (E_λ) denote the probability (expectation) under the initial distribution of X_0 being λ. If λ is degenerate at x, we shall simply write P_x (E_x) instead of P_λ (E_λ). We assume that there exists a stationary probability distribution π, $\pi(A)=\int \mathsf{P}(x,A)\,\mathrm{d}\pi(x)$ for all $A\in\mathscr{A}$ and $\mathscr{K}=\mathsf{E}_\pi[Y_1]>0$.

Let $\{Z_n=S_n+\xi_n,\}_{n\in\mathbb{Z}_+}$ be a perturbed Markov random walk in the following sense: $\{S_n\}$ is a Markov random walk, ξ_n is $\mathscr{F}_n$-measurable, where $\mathscr{F}_n$ is the σ-algebra generated by $\{(X_k,S_k),0\le k\le n\}$, and $\{\xi_n\}$ is *slowly changing*, that is, $\max_{1\le t\le n}|\xi_t|/n\to 0$ in probability as $n\to 0$. For $a\ge 0$ and $\mu\ge 0$, define the stopping times

$$T_a=\inf\{n\ge 1: Z_n>a-\mu n\},\quad \tau_a(\mu)=\inf\left\{n\ge 1:\widetilde{S}_n(\mu)\ge a\right\},\quad \inf\{\varnothing\}=\infty,$$

where $\widetilde{S}_n(\mu)=S_n+\mu n$. Since $\mathscr{K}>0$ and $\mu\ge 0$ it follows that $T_a<\infty$ and $\tau_a<\infty$ w.p. 1 for all $a>0$.

Assume that the Markov process $\{X_n\}_{n\in\mathbb{Z}_+}$ on a state space $\mathscr{X}$ is V-uniformly ergodic (see Definition D.3), i.e., there exists a measurable function $V:\mathscr{X}\to[1,\infty)$ with $\int V(x)\pi(dx)<\infty$ such that (D.14) holds.

The following assumptions for Markov chains are used. Recall that λ stands for an initial distribution of the Markov chain $\{X_n\}_{n\in\mathbb{Z}_+}$.

A1. $\sup_x \mathsf{E}_x|Y_1|^2<\infty$ and $\sup_x\{\mathsf{E}[|Y_1|^r V(X_1)]/V(x)\}<\infty$ for some $r\ge 1$.

A2. For some $r\ge 1$,

$$\sup_{\|h\|_V\le 1}\left|\int_{x\in\mathscr{X}} h(x)\mathsf{E}_x|Y_1|^r\,\mathrm{d}\lambda(x)\right|<\infty. \tag{F.47}$$

Note that A1 implies that $\sup_x\{\mathsf{E}[V(X_1)]/V(x)\}<\infty$.

To establish the Markov nonlinear renewal theorem, we shall use (D.13) in conjunction with the following extension of Cramer's (strongly non-lattice) condition: There exists $\delta>0$ such that for all $m,n=1,2,\ldots,\ \delta^{-1}<m<n$, and all $\theta\in R$ with $|\theta|\ge\delta$

$$\mathsf{E}_\pi|\mathsf{E}\{\exp(i\theta(Y_{n-m}+\cdots+Y_{n+m}))|X_{n-m},\cdots,X_{n-1},X_{n+1},\cdots,X_{n+m},X_{n+m+1}\}|\le e^{-\delta}.$$

Let $\tau_+=\inf\{n\ge 1:\widetilde{S}_n>0\}$ and let $\mathsf{P}_+(x,B\times R)=\mathsf{P}_x\{X_{\tau_+}\in B\}$ denote the transition probability associated with the Markov random walk generated by the ascending ladder variable $\widetilde{S}_{\tau_+}$. Under the V-uniform ergodicity condition and $\mathscr{K}>0$, a similar argument as on page 255 of Fuh and Lai [46] yields that the transition probability $\mathsf{P}_+(x,B\times R)$ has an invariant measure π_+. Let E_+ denote expectation when X_0 has the initial distribution π_+. Define $\kappa_a=Z_{T_a}-a$ (on $\{T_a<\infty\}$) and

$$\mathsf{G}(y)=\frac{1}{\mathsf{E}_+[\widetilde{S}_{\tau_+}]}\int_0^y \mathsf{P}_+\{\widetilde{S}_{\tau_+}>t\}\,\mathrm{d}t,\quad y\ge 0.$$

Recall the definition of arithmetic and nonarithmetic Markov random walks given in Definition E.2.

The following theorem is the *First Markov Nonlinear Renewal Theorem* (MNRT).

Theorem F.6 (First MNRT). *Suppose* A1 *and* A2 *hold with* $r=1$ *and* $\mathscr{K}=\mathsf{E}_\pi[Y_1]\in(0,\infty)$, $\mu\in[0,\infty)$. *Let λ be an initial distribution of X_0. Suppose for every $\varepsilon>0$ there is $\delta>0$ such that*

$$\lim_{n\to\infty}\mathsf{P}_\lambda\left(\max_{1\le j\le n\delta}|\xi_{n+j}-\xi_n|\ge\varepsilon\right)=0.$$

If $Y_1+\mu$ is non-arithmetic under P_λ, then $\lim_{a\to\infty}\mathsf{P}_\lambda\{\kappa_a\le y\}=\mathsf{G}(y)$ *for any* $y\ge 0$.

If, in addition, $(T_a - a)/\sqrt{a}$ converges in distribution to a random variable W as $a \to \infty$, then for every real number t with $P_+\{W = t\} = 0$

$$\lim_{a\to\infty} \mathsf{P}_\lambda\{\kappa_a \le y,\, T_a \le a + t\sqrt{a}\} = \mathsf{G}(y)\mathsf{P}_+\{W \le t\}.$$

To study the expected value of the stopping time T_a, we first give the regularity conditions on the perturbation $\xi = \{\xi_n, n \ge 1\}$, which for the Markov case generalize conditions given in Definition F.1.

Definition F.2. The sequence of random variables $\{\xi_n\}_{n\in\mathbb{Z}_+}$ is said to be *regular* if there exist a non-negative random variable L, a deterministic sequence $\{\ell_n\}_{n\in\mathbb{Z}_+}$ and a random sequence $\{\eta_n\}_{n\in\mathbb{Z}_+}$ such that the following conditions hold:

$$\xi_n = \ell_n + \eta_n \quad \text{for } n \ge L \text{ and } \sup_{x\in\mathscr{X}} \mathsf{E}_x[L] < \infty, \tag{F.48}$$

$$\max_{1\le j\le\sqrt{n}} |\ell_{n+j} - \ell_n| \le K, \quad K < \infty, \tag{F.49}$$

$$\{\max_{1\le j\le n} |\eta_{n+j}|,\, n \ge 1\} \text{ is uniformly integrable}, \tag{F.50}$$

$$n \sup_{x\in\mathscr{X}} \mathsf{P}_x\{\max_{0\le j\le n} \eta_{n+j} \ge \varepsilon n\} \to 0 \text{ as } n \to \infty \text{ for all } \varepsilon > 0, \tag{F.51}$$

$$\sum_{n=1}^{\infty} \sup_{x\in\mathscr{X}} \mathsf{P}_x\{-\eta_n \ge \varepsilon n\} < \infty \quad \text{for some } 0 < \varepsilon < \mu. \tag{F.52}$$

Write $N_a = a/(\mu + \mathscr{K})$, where $\mathscr{K} = \mathsf{E}_\pi[Y_1] \in (0,\infty)$ and $\mu \in [0,\infty)$. The following theorem is the *Second Markov Nonlinear Renewal Theorem.*

Theorem F.7 (Second MNRT). *Assume* A1 *and* A2 *hold with $r = 2$. Let λ be an initial distribution such that $\mathsf{E}_\lambda[V(X_0)] < \infty$. Suppose that the perturbation ξ is regular, i.e., conditions* (F.48)–(F.52) *hold and the following conditions hold as well:*

$$\eta_n \xrightarrow[n\to\infty]{\mathsf{P}_\pi-law} \eta, \tag{F.53}$$

where η is an integrable random variable with expectation $\mathsf{E}_\pi[\eta]$; there exists $0 < \varepsilon < 1$ such that

$$\sup_{x\in\mathscr{X}} \mathsf{P}_x(T_a \le \varepsilon N_a) = o(1/a) \quad \text{as } a \to \infty. \tag{F.54}$$

Then, as $a \to \infty$,

$$\mathsf{E}_\lambda[T_a] = \frac{1}{\mu + \mathscr{K}}\left(a + \varkappa - \ell_{N_a} - \mathsf{E}_\pi[\eta] - \int \Delta(y)\, \mathrm{d}[\pi_+(y) - \lambda(y)]\right) + o(1),$$

and therefore,

$$\mathsf{E}_x[T_a] = \frac{1}{\mu + \mathscr{K}}\left(a + \varkappa - \ell_{N_a} - \mathsf{E}_\pi[\eta] - \int \Delta(y)\, \mathrm{d}\pi_+(y) + \Delta(x)\right) + o(1),$$

where $\Delta(x)$ is a solution of the Poisson equation (E.47) *and*

$$\varkappa = \frac{\mathsf{E}_{\pi_+}[\widetilde{S}^2_{\tau_+}]}{2\mathsf{E}_{\pi_+}[\widetilde{S}_{\tau_+}]}.$$

Bibliography

[1] P. Armitage. Sequential analysis with more than two alternative hypotheses, and its relation to discriminant function analysis. *Journal of the Royal Statistical Society - Series B Methodology*, 12(1):137–144, 1950.

[2] P. Armitage, C. K. McPherson, and B. C. Rowe. Repeated significance tests on accumulating data. *Journal of the Royal Statistical Society - Series A General*, 132(2):235–244, 1969.

[3] S. Asmussen and P. W. Glynn. *Stochastic Simulation: Algorithms and Analysis*. Stochastic Modelling and Applied Probability. Springer, New York, 2007.

[4] P. A. Bakut, I. A. Bolshakov, B. M. Gerasimov, A. A. Kuriksha, V. G. Repin, G. P. Tartakovsky, and V. V. Shirokov. *Statistical Radar Theory*, volume 1 (G. P. Tartakovsky, Editor). Sovetskoe Radio, Moscow, USSR, 1963. In Russian.

[5] A. V. Balakrishnan. *Kalman Filtering Theory (Enlarged 2nd ed.)*. Series in Communications and Control Systems. Optimization Software, Inc., Publications Division, 1987.

[6] R. K. Bansal and P. Papantoni-Kazakos. An algorithm for detecting a change in a stochastic process. *IEEE Transactions on Information Theory*, 32(2):227–235, Mar. 1986.

[7] M. Baron and A. G. Tartakovsky. Asymptotic optimality of change-point detection schemes in general continuous-time models. *Sequential Analysis*, 25(3):257–296, Oct. 2006. Invited Paper in Memory of Milton Sobel.

[8] W. Bartky. Multiple sampling with constant probability. *Annals of Mathematical Statistics*, 14:363–377, 1943.

[9] M. Basseville. Detecting changes in signals and systems - A survey. *Automatica*, 24(3):309–326, May 1988.

[10] M. Basseville. On-board component fault detection and isolation using the statistical local approach. *Automatica*, 34(11):1391–1416, Nov. 1998.

[11] M. Basseville and I. V. Nikiforov. *Detection of Abrupt Changes – Theory and Application*. Information and System Sciences Series. Prentice-Hall, Inc, Englewood Cliffs, NJ, USA, 1993. Online.

[12] L. E. Baum and M. Katz. Convergence rates in the law of large numbers. *Transactions of the American Mathematical Society*, 120(1):108–123, Oct. 1965.

[13] P. Baxendale. T. E. Harris's contributions to recurrent Markov processes and stochastic flows. *Annals of Probability*, 39(2):417–428, Mar. 2011.

[14] R. N. Bhattacharya and R. R. Rao. *Normal Approximations and Asymptotic Expansions*, volume 64 of *Classics in Applied Mathematics*. SIAM, Philadelphia, PA, USA, 2010.

[15] A. Bissell. CUSUM techniques for quality control. *Journal of the Royal Statistical Society - Series C Applied Statistics*, 18(1):1–30, 1969.

[16] W. R. Blanding, P. K. Willett, Y. Bar-Shalom, and S. Coraluppi. Target detection and tracking for video surveillance. *IEEE Transactions on Aerospace and Electronic Systems*, 45(4):1275–1292, Oct. 2009.

[17] P. Bougerol. *Products of Random Matrices with Applications to Schrödinger Operators*. Birkhäuser, Boston, USA, 1995.

[18] P. Bougerol. Théorèmes limite pour les systèmes linéaires à coefficients markoviens. *Probability Theory and Related Fields*, 78:193–221, 1998.

[19] G. E. Box, A. Luceno, and M. del Carmen Paniagua-Quinones. *Statistical Control by Monitoring and Adjustment (2nd ed.)*. John Wiley & Sons, Inc, New York, USA, 2009.

[20] D. L. Burkholder. Distribution function inequalities for martingales. *Annals of Probability*, 1(1):19–42, Feb. 1973.

[21] D. L. Burkholder, B. J. Davis, and R. F. Gundy. Integral inequalities for convex functions of operators on martingales. In L. M. Le Cam, J. Neyman, and E. L. Scott, editors, *Proceedings of the Sixth Berkeley Symposium on Mathematical Statistics and Probability, June 21–July 18, 1970*, volume 2: Probability Theory, pages 223–240. University of California Press, Berkeley, CA, USA, 1972.

[22] H. P. Chan. Optimal sequential detection in multi-stream data. *Annals of Statistics*, 45(6):2736–2763, Dec. 2017.

[23] F.-K. Chang. Structural health monitoring: Promises and challenges. In *Proceedings of the 30th Annual Review of Progress in Quantitative NDE (QNDE), Green Bay, WI, USA*. American Institute of Physics, July 2003.

[24] A. Chen, T. Wittman, A. G. Tartakovsky, and A. L. Bertozzi. Efficient boundary tracking through sampling. *Applied Mathematics Research Express*, 2(2):182–214, 2011.

[25] H. Chernoff. Sequential design of experiments. *Annals of Mathematical Statistics*, 30(3):755–770, Sept. 1959.

[26] G. A. Churchill. Stochastic models for heterogeneous DNA sequences. *Bulletin of Mathematical Biology*, 51:79–94, 1989.

[27] S. V. Crowder, D. M. Hawkins, M. R. Reynolds Jr., and E. Yashchin. Process control and statistical inference. *Journal of Quality Technology*, 29(2):134–139, Apr. 1997.

[28] S. Dayanik, W. B. Powell, and K. Yamazaki. Asymptotically optimal Bayesian sequential change detection and identification rules. *Annals of Operations Research*, 208(1):337–370, Jan. 2013.

[29] H. F. Dodge and H. G. Roming. A method of sampling inspection. *The Bell System Technical Journal*, 8:613–631, 1929.

[30] J. L. Doob. *Stochastic Processes*. Wiley Series in Probability and Statistics. John Wiley & Sons, Inc, New York, 1953.

[31] V. P. Dragalin. Asymptotic solution of a problem of detecting a signal from k channels. *Russian Mathematical Surveys*, 42(3):213–214, 1987.

[32] V. P. Dragalin. Asymptotic solutions in detecting a change in distribution under an unknown parameter. *Statistical Problems of Control*, 83:45–52, 1988. In Russian.

[33] V. P. Dragalin and A. A. Novikov. Adaptive sequential tests for composite hypotheses. In *Statistics and Control of Random Processes: Proceedings of the Steklov Institute of Mathematics*, volume 4, pages 12–23. TVP Science Publ., Moscow, 1995.

[34] V. P. Dragalin, A. G. Tartakovsky, and V. V. Veeravalli. Multihypothesis sequential probability ratio tests–Part I: Asymptotic optimality. *IEEE Transactions on Information Theory*, 45(11):2448–2461, Nov. 1999.

[35] V. P. Dragalin, A. G. Tartakovsky, and V. V. Veeravalli. Multihypothesis sequential probability ratio tests–Part II: Accurate asymptotic expansions for the expected sample size. *IEEE Transactions on Information Theory*, 46(4):1366–1383, Apr. 2000.

[36] R. H. Farrell. Limit theorems for stopped random walks. *Ann. Math. Statist.*, 35(3):1332–1343, Sep 1964.

[37] R. H. Farrell. Limit theorems for stopped random walks II. *Ann. Math. Statist.*, 37(4):860–865, Aug 1966.

[38] P. Feigin and R. Tweedie. Random coefficient autoregressive processes: A Markov chain analysis of stationarity and finiteness of moments. *Journal of Time Series Analysis*, 6(1):1–14, Jan. 1985.

[39] G. Fellouris and G. Sokolov. Second-order asymptotic optimality in multichannel sequential detection. *IEEE Transactions on Information Theory*, 62(6):3662–3675, June 2016.

[40] G. Fellouris and A. G. Tartakovsky. Almost optimal sequential tests of discrete composite hypotheses. *Statistica Sinica*, 23(4):1717–1741, 2013.

[41] T. S. Ferguson. Who solved the secretary problem? *Statistical Science*, 4(3):282–289, Aug. 1989.

[42] S. E. Fienberg and G. Shmueli. Statistical issues and challenges associated with rapid detection of bio-terrorist attacks. *Statistics in Medicine*, 24(4):513–529, July 2005.

[43] M. Frisén. Optimal sequential surveillance for finance, public health, and other areas (with discussion). *Sequential Analysis*, 28(3):310–393, July 2009.

[44] C.-D. Fuh. SPRT and CUSUM in hidden Markov models. *Annals of Statistics*, 31(3):942–977, June 2003.

[45] C.-D. Fuh. Asymptotic operating characteristics of an optimal change point detection in hidden Markov models. *Annals of Statistics*, 32(5):2305–2339, Oct. 2004.

[46] C. D. Fuh and T. L. Lai. Asymptotic expansions in multidimensional Markov renewal theory and first passage times for Markov random walks. *Advances in Applied Probability*, 33(3):652–673, 2001.

[47] C. D. Fuh and A. G. Tartakovsky. Asymptotic Bayesian theory of quickest change detection for hidden Markov models. *IEEE Transactions on Information Theory*, 65(1):511–529, Jan. 2018.

[48] C. D. Fuh and C. H. Zhang. Poisson equation, maximal inequalities and r-quick convergence for Markov random walks. *Stochastic Processes and their Applications*, 87(1):53–67, Jan. 2000.

[49] L. Galthouk and S. Pergamenshchikov. Uniform concentration inequality for ergodic diffusion processes observed at discrete times. *Stochastic Processes and their Applications*, 123(1):91–109, Jan. 2013.

[50] L. Galthouk and S. Pergamenshchikov. Geometric ergodicity for classes of homogeneous Markov chains. *Stochastic Processes and their Applications*, 124(9):3362–3391, Sept. 2014.

[51] G. K. Golubev and R. Z. Khas'minskii. Sequential testing for several signals in Gaussian white noise. *Theory of Probability and its Applications*, 28(3):573–584, 1984.

[52] B. K. Guéppié, L. Fillatre, and I. V. Nikiforov. Sequential detection of transient changes. *Sequential Analysis*, 31(4):528–547, Dec. 2012.

[53] B. K. Guéppié, L. Fillatre, and I. V. Nikiforov. Detecting a suddenly arriving dynamic profile of finite duration. *IEEE Transactions on Information Theory*, 63(5):3039–3052, May 2017.

[54] A. Gut. Complete convergence and convergence rates for randomly indexed partial sums with an application to some first passage times. *Acta Mathematica Hungarica*, 42(3-4):225–232, 1983.

[55] A. Gut. *Stopped Random Walks: Limit Theorems and Applications*, volume 5 of *Series in Applied Probability*. Springer-Verlag, New York, USA, 1988.

[56] T. E. Harris. The existence of stationary measures for certain Markov processes. In *Proceedings of the Third Berkeley Symposium on Mathematical Statistics and Probability, December 1954 and July–August 1955*, volume 2: Contributions to Probability Theory, pages 113–124. University of California Press, Berkeley, CA, USA, 1956.

[57] T. E. Harris. *The Theory of Branching Processes*. Springer-Verlag, Berlin, DE, 1963.

[58] D. M. Hawkins and D. H. Olwell. *Cumulative Sum Charts and Charting for Quality Improvement*. Series in Statistics for Engineering and Physical Sciences. Springer-Verlag, USA, 1998.

[59] D. M. Hawkins, P. Qiu, and C. W. Kang. The changepoint model for statistical process control. *Journal of Quality Technology*, 35(4):355–366, Oct. 2003.

[60] W. Hoeffding. A lower bound for the average sample number of a sequential test. *Annals of Mathematical Statistics*, 24(1):127–130, Mar. 1953.

[61] W. Hoeffding. Lower bounds for the expected sample size and the average risk of a sequential procedure. *Annals of Mathematical Statistics*, 31(2):352–368, June 1960.

[62] P. L. Hsu and H. Robbins. Complete convergence and the law of large numbers. *Proceedings of the National Academy of Sciences of the United States of America*, 33(2):25–31, Feb. 1947.

[63] J. Hu, M. Brown, and W. Turin. HMM based on-line handwriting recognition. *IEEE Transactions on Pattern Analysis and Machine Intelligence*, 18:1039–1045, 1996.

[64] J. J. Hunter. Renewal theory in two dimensions: asymptotic results. *Advances in Applied Probability*, 6(3):546–562, 009 1974.

[65] D. R. Jeske, N. T. Steven, A. G. Tartakovsky, and J. D. Wilson. Statistical methods for network surveillance. *Applied Stochastic Models in Business and Industry*, 34(4):425–445, Oct. 2018. Discussion Paper.

[66] D. R. Jeske, N. T. Steven, J. D. Wilson, and A. G. Tartakovsky. Statistical network surveillance. *Wiley StatsRef: Statistics Reference Online*, pages 1–12, Aug. 2018.

[67] B. H. Juang and L. Rabiner. *Fundamentals of Speech Recognition*. Prentice Hall, New York, New York, USA, 1993.

[68] S. Kent. On the trail of intrusions into information systems. *IEEE Spectrum*, 37(12):52–56, Dec. 2000.

[69] J. Kiefer and J. Sacks. Asymptotically optimal sequential inference and design. *Annals of Mathematical Statistics*, 34(3):705–750, Sept. 1963.

[70] M. J. Klass. A best possible improvement of Wald's equation. *Annals of Probability*, 16(2):840–853, Mar. 1988.

[71] C. Klüppelberg and S. Pergamenshchikov. The tail of the stationary distribution of a random coefficient AR(q) process with applications to an ARCH(q) process. *Annals of Applied Probability*, 14(2):971–1005, June 2004.

[72] A. Kunda, Y. He, and P. Bahl. Recognition of handwritten word: First and second order hidden Markov model based approach. *Pattern Recognition*, 18:283–297, 1989.

[73] T. L. Lai. Asymptotic optimality of invariant sequential probability ratio tests. *Annals of Statistics*, 9(2):318–333, Mar. 1981.

[74] T. L. Lai. Sequential changepoint detection in quality control and dynamical systems (with discussion). *Journal of the Royal Statistical Society - Series B Methodology*, 57(4):613–658, 1995.

[75] T. L. Lai. Information bounds and quick detection of parameter changes in stochastic systems. *IEEE Transactions on Information Theory*, 44(7):2917–2929, Nov. 1998.

[76] T. L. Lai. Sequential multiple hypothesis testing and efficient fault detection-isolation in stochastic systems. *IEEE Transactions on Information Theory*, 46(2):595–608, Mar. 2000.

[77] T. L. Lai and D. Siegmund. A nonlinear renewal theory with applications to sequential analysis I. *Annals of Statistics*, 5(5):946–954, Sept. 1977.

[78] T. L. Lai and D. Siegmund. A nonlinear renewal theory with applications to sequential analysis II. *Annals of Statistics*, 7(1):60–76, Jan. 1979.

[79] S. Lalley. Non-linear renewal theory for lattice random walks. *Sequential Analysis*, 1(3):193–205, July 1982.

[80] R. S. Liptser and A. N. Shiryaev. *Theory of Martingales*. Kluwer Academic Publishers, Dordrecht, NL, 1989.

[81] M. Loève. *Probability Theory (4th ed.)*. Springer-Verlag, New York, USA, 1977.

[82] G. Lorden. Integrated risk of asymptotically Bayes sequential tests. *Annals of Mathematical Statistics*, 38(5):1399–1422, Oct. 1967.

[83] G. Lorden. On excess over the boundary. *Annals of Mathematical Statistics*, 41(2):520–527, Apr. 1970.

[84] G. Lorden. Procedures for reacting to a change in distribution. *Annals of Mathematical Statistics*, 42(6):1897–1908, Dec. 1971.

[85] G. Lorden. 2-SPRT's and the modified Kiefer-Weiss problem of minimizing an expected sample size. *Annals of Statistics*, 4(2):281–291, Mar. 1976.

[86] G. Lorden. Nearly-optimal sequential tests for finitely many parameter values. *Annals of Statistics*, 5(1):1–21, Jan. 1977.

[87] J. Marage and Y. Mori. *Sonar and Underwater Acoustics*. STE Ltd and John Wiley & Sons, London, Hoboken, 2013.

[88] R. L. Mason and J. C. Young. *Multivariate Statistical Process Control with Industrial Application*. SIAM, Philadelphia, PA, USA, 2001.

[89] Y. Mei. Efficient scalable schemes for monitoring a large number of data streams. *Biometrika*, 97(2):419–433, Apr. 2010.

[90] S. Meyn and R. Tweedie. *Markov Chains and Stochastic Stability*. Springer Verlag, Berlin, New York, 1993.

[91] S. P. Meyn and R. L. Tweedie. *Markov Chains and Stochastic Stability*. Springer-Verlag, New York, USA, 2009. (Second Edition).

[92] D. C. Montgomery. *Introduction to Statistical Quality Control (6th ed.)*. John Wiley & Sons, Inc, 2008.

[93] G. V. Moustakides. Optimal stopping times for detecting changes in distributions. *Annals of Statistics*, 14(4):1379–1387, Dec. 1986.

[94] G. V. Moustakides. Sequential change detection revisited. *Annals of Statistics*, 36(2):787–807, Mar. 2008.

[95] G. V. Moustakides. Multiple optimality properties of the Shewhart test. *Sequential Analysis*, 33(3):318–344, July 2014.

[96] G. V. Moustakides. Optimum Shewhart tests for Markovian data. In *Proceedings of the Fifty-third Annual Allerton Conference, UIUC, Illinois, USA*, pages 822–826, Sept. 2015.

[97] G. V. Moustakides, A. S. Polunchenko, and A. G. Tartakovsky. A numerical approach to performance analysis of quickest change-point detection procedures. *Statistica Sinica*, 21(2):571–596, Apr. 2011.

[98] N. Mukhopadhyay, S. Datta, and S. Chattopadhyay. *Applied Sequential Methodologies: Real-World Examples with Data Analysis*, volume 173 of *Statistics Textbooks and Monographs*. Marcel Dekker, Inc, New York, USA, 2004.

[99] I. V. Nikiforov. A generalized change detection problem. *IEEE Transactions on Information Theory*, 41(1):171–187, Jan. 1995.

[100] I. V. Nikiforov. A simple recursive algorithm for diagnosis of abrupt changes in random signals. *IEEE Transactions on Information Theory*, 46(7):2740–2746, July 2000.

[101] I. V. Nikiforov. A lower bound for the detection/isolation delay in a class of sequential tests. *IEEE Transactions on Information Theory*, 49(11):3037–3046, Nov. 2003.

[102] A. A. Novikov. On discontinuous martingales. *Theory of Probability and its Applications*, 20(1):11–26, 1975.

[103] A. A. Novikov. Martingale identities and inequalities and their applications in nonlinear boundary-value problems for random processes. *Mathematical Notes*, 35(3):241–249, 1984.

[104] E. S. Page. Continuous inspection schemes. *Biometrika*, 41(1–2):100–114, June 1954.

[105] E. Paulson. A sequential decision procedure for choosing one of k hypotheses concerning the unknown mean of a normal distribution. *Annals of Mathematical Statistics*, 34(2):549–554, June 1963.

[106] I. V. Pavlov. A sequential decision rule for the case of many composite hypotheses. *Engineering Cybernetics*, 22:19–23, 1984.

[107] I. V. Pavlov. Sequential procedure of testing composite hypotheses with applications to the Kiefer-Weiss problem. *Theory of Probability and its Applications*, 35(2):280–292, 1990.

[108] S. Pergamenchtchikov and A. G. Tartakovsky. Asymptotically optimal pointwise and minimax quickest change-point detection for dependent data. *Statistical Inference for Stochastic Processes*, 21(1):217–259, Jan. 2018.

[109] M. Pollak. Optimal detection of a change in distribution. *Annals of Statistics*, 13(1):206–227, Mar. 1985.

[110] M. Pollak and A. M. Krieger. Shewhart revisited. *Sequential Analysis*, 32(2):230–242, June 2013.

[111] M. Pollak and D. Siegmund. Convergence of quasi-stationary to stationary distributions for stochastically monotone Markov processes. *Journal of Applied Probability*, 23(1):215–220, Mar. 1986.

[112] M. Pollak and A. G. Tartakovsky. Asymptotic exponentiality of the distribution of first exit times for a class of Markov processes with applications to quickest change detection. *Theory of Probability and its Applications*, 53(3):430–442, 2009.

[113] M. Pollak and A. G. Tartakovsky. Optimality properties of the Shiryaev–Roberts procedure. *Statistica Sinica*, 19(4):1729–1739, Oct. 2009.

[114] P. K. Pollet. Quasi-stationary distributions: a bibliography. Technical report, School of Mathematics and Physics, The University of Queensland, Australia, July 2012. Online.

[115] A. S. Polunchenko, G. Sokolov, and A. G. Tartakovsky. Optimal design and analysis of the exponentially weighted moving average chart for exponential data. *Sri Lankan Journal of Applied Statistics, Special Issue: Modern Statistical Methodologies in the Cutting Edge of Science*, 15(4):57–80, Dec. 2014.

[116] A. S. Polunchenko and A. G. Tartakovsky. On optimality of the Shiryaev–Roberts procedure for detecting a change in distribution. *Annals of Statistics*, 38(6):3445–3457, Dec. 2010.

[117] H. V. Poor. Quickest detection with exponential penalty for delay. *Annals of Statistics*, 26(6):2179–2205, Dec. 1998.

[118] L. Rabiner. A tutorial on hidden Markov models and selected applications in speech recognition. *Proceedings of IEEE*, 77:257–286, 1989.

[119] V. Raghavan, A. Galstyan, and A. G. Tartakovsky. Hidden Markov models for the activity profile of terrorist groups. *Annals of Applied Statistics*, 7(6):2402–2430, Dec. 2013.

[120] V. Raghavan, G. V. Steeg, A. Galstyan, and A. G. Tartakovsky. Modeling temporal activity patterns in dynamic social networks. *IEEE Transactions on Computational Social Systems*, 1(1):89–107, Jan. 2013.

[121] V. G. Repin. Detection of a signal with unknown moments of appearance and disappearance. *Problems of Information Transmission*, 27(1):61–72, Jan. 1991.

[122] M. A. Richards. *Fundamentals of Radar Signal Processing.* 2nd edition. McGraw-Hill Education Europe, USA, 2014.

[123] Y. Ritov. Decision theoretic optimality of the CUSUM procedure. *Annals of Statistics*, 18(3):1464–1469, Sept. 1990.

[124] S. W. Roberts. A comparison of some control chart procedures. *Technometrics*, 8(3):411–430, Aug. 1966.

[125] H. Rolka, H. Burkom, G. F. Cooper, M. Kulldorff, D. Madigan, and W. K. Wong. Issues in applied statistics for public health bioterrorism surveillance using multiple data streams: research needs. *Statistics in Medicine*, 26(8):1834–1856, 2007.

[126] J. S. Sadowsky and J. A. Bucklew. On large deviations theory and asymptotically efficient monte carlo estimation. *IEEE transactions on Information Theory*, 36(3):579–588, 1990.

[127] W. A. Shewhart. *Economic Control of Quality of Manufactured Products.* D. Van Nostrand Co, New York, USA, 1931.

[128] W. A. Shewhart. *Statistical Method from the Viewpoint of Quality Control.* Washington D.C.: Graduate School of the Department of Agriculture, Washington D.C., USA, 1939.

[129] A. N. Shiryaev. The detection of spontaneous effects. *Soviet Mathematics – Doklady*, 2:740–743, 1961. Translation from Doklady Akademii Nauk SSSR, **138**:799–801, 1961.

[130] A. N. Shiryaev. The problem of the most rapid detection of a disturbance in a stationary process. *Soviet Mathematics – Doklady*, 2:795–799, 1961. Translation from Doklady Akademii Nauk SSSR, **138**:1039–1042, 1961.

[131] A. N. Shiryaev. On optimum methods in quickest detection problems. *Theory of Probability and its Applications*, 8(1):22–46, Jan. 1963.

[132] A. N. Shiryaev. On the detection of disorder in a manufacturing process - I. *Theory of Probability and its Applications*, 8(3):247–265, 1963.

[133] A. N. Shiryaev. On the detection of disorder in a manufacturing process - II. *Theory of Probability and its Applications*, 8(4):402–413, 1963.

[134] A. N. Shiryaev. Some exact formulas in a "disorder" problem. *Theory of Probability and its Applications*, 10(2):348–354, 1965.

[135] A. N. Shiryaev. *Statistical Sequential Analysis: Optimal Stopping Rules.* Nauka, Moscow, RU, 1969. In Russian.

[136] A. N. Shiryaev. *Optimal Stopping Rules*, volume 8 of *Series on Stochastic Modelling and Applied Probability.* Springer-Verlag, New York, USA, 1978.

[137] A. N. Shiryaev. From disorder to nonlinear filtering and martingale theory. In A. A. Bolibruch, Y. S. Osipov, and Y. G. Sinai, editors, *Mathematical Events of the Twentieth Century*, pages 371–397. Springer-Verlag, Berlin Heidelberg DE, 2006.

[138] A. N. Shiryaev. Quickest detection problems: Fifty years later. *Sequential Analysis*, 29(4):345–385, Oct. 2010.

[139] D. Siegmund. Importance sampling in the Monte Carlo study of sequential tests. *Ann. Statist.*, 4(4):673–684, July 1976.

[140] D. Siegmund. *Sequential Analysis: Tests and Confidence Intervals.* Series in Statistics. Springer-Verlag, New York, USA, 1985.

[141] D. Siegmund. Change-points: From sequential detection to biology and back. *Sequential Analysis*, 32(1):2–14, Jan. 2013.

[142] D. O. Siegmund and B. Yakir. Minimax optimality of the Shiryayev–Roberts change-point detection rule. *Journal of Statistical Planning and Inference*, 138(9):2815–2825, Sept. 2008.

[143] G. Simons. Lower bounds for average sample number of sequential multihypothesis tests. *Annals of Mathematical Statistics*, 38(5):1343–1364, Oct. 1967.

[144] M. Sobel and A. Wald. A sequential decision procedure for choosing one of three hypotheses concerning the unknown mean of a normal distribution. *Annals of Mathematical Statistics*, 20(4):502–522, Dec. 1949.

[145] C. Sonesson and D. Bock. A review and discussion of prospective statistical surveillance in public health. *Journal of the Royal Statistical Society A*, 166:5–21, 2003.

[146] C. Stein. A two sample test for a linear hypothesis whose power is independent of the variance. *Annals of Mathematical Statistics*, 16(1):243–258, 1945.

[147] C. Stein. Some problems in sequential estimation. *Econometrica*, 17(1):77–78, 1949.

[148] P. Szor. *The Art of Computer Virus Research and Defense.* Addison-Wesley Professional, Upper Saddle River, NJ, USA, 2005.

[149] A. G. Tartakovskii. Sequential composite hypothesis testing with dependent non-stationary observations. *Problems of Information Transmission*, 17(1):29–42, 1981.

[150] A. G. Tartakovskii. Optimal detection of random-length signals. *Problems of Information Transmission*, 23(3):203–210, July 1987.

[151] A. G. Tartakovskii. Detection of signals with random moments of appearance and disappearance. *Problems of Information Transmission*, 24(2):39–50, Apr. 1988.

[152] A. G. Tartakovsky. *Sequential Methods in the Theory of Information Systems.* Radio i Svyaz', Moscow, RU, 1991. In Russian.

[153] A. G. Tartakovsky. Asymptotic optimality of certain multihypothesis sequential tests: Non-i.i.d. case. *Statistical Inference for Stochastic Processes*, 1(3):265–295, Oct. 1998.

[154] A. G. Tartakovsky. Asymptotically optimal sequential tests for nonhomogeneous processes. *Sequential Analysis*, 17(1):33–62, Jan. 1998.

[155] A. G. Tartakovsky. Asymptotic performance of a multichart CUSUM test under false alarm probability constraint. In *Proceedings of the 44th IEEE Conference Decision and Control and European Control Conference (CDC-ECC'05), Seville, SP*, pages 320–325. IEEE, Omnipress CD-ROM, 2005.

[156] A. G. Tartakovsky. Multidecision quickest change-point detection: Previous achievements and open problems. *Sequential Analysis*, 27(2):201–231, Apr. 2008.

[157] A. G. Tartakovsky. Asymptotic optimality in Bayesian changepoint detection problems under global false alarm probability constraint. *Theory of Probability and its Applications*, 53(3):443–466, 2009.

[158] A. G. Tartakovsky. Discussion on "Change-points: From sequential detection to biology and back" by David Siegmund. *Sequential Analysis*, 32(1):36–42, Jan. 2013.

[159] A. G. Tartakovsky. Rapid detection of attacks in computer networks by quickest changepoint detection methods. In N. Adams and N. Heard, editors, *Data Analysis for Network Cyber-Security*, pages 33–70. Imperial College Press, London, UK, 2014.

[160] A. G. Tartakovsky. On asymptotic optimality in sequential changepoint detection: Non-iid case. *IEEE Transactions on Information Theory*, 63(6):3433–3450, June 2017.

[161] A. G. Tartakovsky. Asymptotically optimal quickest change detection in multistream data—part 1: General stochastic models. *Methodology and Computing in Applied Probability*, online first, 11 July 2019.

[162] A. G. Tartakovsky and J. Brown. Adaptive spatial-temporal filtering methods for clutter removal and target tracking. *IEEE Transactions on Aerospace and Electronic Systems*, 44(4):1522–1537, Oct. 2008.

[163] A. G. Tartakovsky and G. V. Moustakides. State-of-the-art in Bayesian changepoint detection. *Sequential Analysis*, 29(2):125–145, Apr. 2010.

[164] A. G. Tartakovsky, I. V. Nikiforov, and M. Basseville. *Sequential Analysis: Hypothesis Testing and Changepoint Detection*. Monographs on Statistics and Applied Probability. Chapman & Hall/CRC Press, Boca Raton, London, New York, 2014.

[165] A. G. Tartakovsky, M. Pollak, and A. S. Polunchenko. Third-order asymptotic optimality of the generalized Shiryaev–Roberts changepoint detection procedures. *Theory of Probability and its Applications*, 56(3):457–484, Sept. 2012.

[166] A. G. Tartakovsky and A. S. Polunchenko. Minimax optimality of the Shiryaev–Roberts procedure. In *Proceedings of the 5th International Workshop on Applied Probability (IWAP'10), Madrid, SP*, Universidad Carlos III de Madrid, Colmenarejo Campus, July 2010.

[167] A. G. Tartakovsky, A. S. Polunchenko, and G. Sokolov. Efficient computer network anomaly detection by changepoint detection methods. *IEEE Journal of Selected Topics in Signal Processing*, 7(1):4–11, Feb. 2013.

[168] A. G. Tartakovsky, B. L. Rozovskii, R. B. Blaźek, and H. Kim. Detection of intrusions in information systems by sequential change-point methods. *Statistical Methodology*, 3(3):252–293, July 2006.

[169] A. G. Tartakovsky, B. L. Rozovskii, R. B. Blaźek, and H. Kim. A novel approach to detection of intrusions in computer networks via adaptive sequential and batch-sequential change-point detection methods. *IEEE Transactions on Signal Processing*, 54(9):3372–3382, Sept. 2006.

[170] A. G. Tartakovsky and V. V. Veeravalli. Change-point detection in multichannel and distributed systems. In N. Mukhopadhyay, S. Datta, and S. Chattopadhyay, editors, *Applied Sequential Methodologies: Real-World Examples with Data Analysis*, volume 173 of *Statistics: a Series of Textbooks and Monographs*, pages 339–370. Marcel Dekker, Inc, New York, USA, 2004.

[171] A. G. Tartakovsky and V. V. Veeravalli. General asymptotic Bayesian theory of quickest change detection. *Theory of Probability and its Applications*, 49(3):458–497, July 2005.

[172] K. Tsui, W. Chiu, P. Gierlich, D. Goldsman, X. Liu, and T. Maschek. A review of healthcare, public health, and syndromic surveillance. *Quality Engineering*, 20(4):435–450, 2008.

[173] S. Tugac and M. Elfe. Hidden Markov model based target detection. In *Proceedings of the 13th International Conference on Information Fusion (FUSION 2010), Edinburgh, Scotland*, pages 1–7, 26-29 July 2010.

[174] C. S. Van Dobben de Bruyn. *Cumulative Sum Tests: Theory and Practice*, volume 24 of *Statistics Monograph*. Charles Griffin and Co. Ltd, London, UK, 1968.

[175] S. Vasuhi and Vaidehi. Target detection and tracking for video surveillance. *WSEAS Transactions on Signal Processing*, 10:2168–177, 2014.

[176] N. V. Verdenskaya and A. G. Tartakovskii. Asymptotically optimal sequential testing of multiple hypotheses for nonhomogeneous Gaussian processes in an asymmetric situation. *Theory of Probability and its Applications*, 36(3):536–547, 1991.

[177] A. Wald. Sequential tests of statistical hypotheses. *Annals of Mathematical Statistics*, 16(2):117–186, June 1945.

[178] A. Wald. *Sequential Analysis*. John Wiley & Sons, Inc, New York, USA, 1947.

[179] A. Wald and J. Wolfowitz. Optimum character of the sequential probability ratio test. *Annals of Mathematical Statistics*, 19(3):326–339, Sept. 1948.

[180] A. Wald and J. Wolfowitz. Bayes solutions of sequential decision problems. *Annals of Mathematical Statistics*, 21(1):82–99, Mar. 1950.

[181] G. B. Wetherill and D. W. Brown. *Statistical Process Control: Theory and Practice (3rd ed.)*. Texts in Statistical Science. Chapman and Hall, London, UK, 1991.

[182] D. J. Wheeler and D. S. Chambers. *Understanding Statistical Process Control (2nd ed.)*. SPC Press, Inc, 1992.

[183] J. Whitehead. *The Design and Analysis of Sequential Clinical Trials*. Statistics in Practice. John Wiley & Sons, Inc, Chichester, UK, 1997.

[184] A. S. Willsky and H. L. Jones. A generalized likelihood ratio approach to the detection and estimation of jumps in linear systems. *IEEE Transactions on Automatic Control*, 21(1):108–112, Feb. 1976.

[185] W. H. Woodall. Control charts based on attribute data: Bibliography and review. *Journal of Quality Technology*, 29(2):172–183, Apr. 1997.

[186] W. H. Woodall. Controversies and contradictions in statistical process control. *Journal of Quality Technology*, 32(4):341–350, Oct. 2000.

[187] W. H. Woodall and D. C. Montgomery. Research issues and ideas in statistical process control. *Journal of Quality Technology*, 31(4):376–386, Oct. 1999.

[188] M. Woodroofe. A renewal theorem for curved boundaries and moments of first passage times. *Annals of Probability*, 4(1):67–80, Feb. 1976.

[189] M. Woodroofe. *Nonlinear Renewal Theory in Sequential Analysis*, volume 39 of *CBMS-NSF Regional Conference Series in Applied Mathematics*. SIAM, Philadelphia, PA, USA, 1982.

[190] M. Woodroofe and H. Takahashi. Asymptotic expansions for the error probabilities of some repeated significance tests. *Annals of Statistics*, 10(3):895–908, Sept. 1982.

[191] R. H. Woodward and P. L. Goldsmith. *Cumulative Sum Techniques*, volume 3 of *Mathematical and Statistical Techniques for Industry*. Oliver and Boyd for Imperial Chemical Industries, Ltd., Edinburgh, UK, 1964.

[192] Y. Xie and D. Siegmund. Sequential multi-sensor change-point detection. *Annals of Statistics*, 41(2):670–692, Mar. 2013.

[193] J. Yamato, J. Ohya, and K. Ishii. Recognizing human action in time-sequential images using hidden Markov model. In *Proceedings of the IEEE Computer Society Conference on Computer Vision and Pattern Recognition*, pages 379–385, 1992.

[194] C. H. Zhang. A nonlinear renewal theory. *Annals of Probability*, 16(2):793–825, Apr. 1988.

Index

Printed in the United States
by Baker & Taylor Publisher Services